Lian Tongshu

THEORY OF CONJUGATION FOR REFLECTING PRISMS

Adjustment and Image Stabilization of Optical Instruments

With 237 Figures and 54 Tables

International Academic Publishers
A Pergamon-CNPIEC Joint Venture

Pergamon Press
Oxford · New York · Beijing · Frankfurt
São Paulo · Sydney · Tokyo · Toronto

DISTRIBUTORS:

U.K.	Pergamon Press plc , Headington Hill Hall , Oxford OX3 0BW, England
U.S.A.	Pergamon Press , Inc., Maxwell House, Fairview Park , Elmsford, New York 10523 , U.S.A.
FEDERAL REPUBLIC OF GERMANY	Pergamon Press GmbH , Hammerweg 6, D-6242 Kronberg , Federal Republic of Germany
BRAZIL	Pergamon Editora Ltda , Rua Eca de Queiros , 346 , CEP 04011 , Sao Paulo , Brazil
AUSTRALIA	Pergamon Press (Australia) Pty Ltd , PO Box 544, Potts Point , NSW 2011 , Australia
JAPAN	Pergamon Press , 5th Floor , Matsuoka Central Building, 1-7-1 Nishishinjuku, Shinjuku-ku , Tokyo 160 , Japan
CANADA	Pergamon Press Canada Ltd , Suite No 271 , 253 College Street , Toronto , Ontario , Canada M5T 1R5
PEOPLE'S REPUBLIC OF CHINA	International Academic Publishers , Xizhimenwai Dajie, Beijing Exhibition Center , Beijing , 100044, People's Republic of China

First edition 1991

Library of Congress Cataloging-in-Publication Data

Lien, T'ung-shu

Theory of conjugation for reflecting prisms : adjustment and image stabilization of optical instruments/Lian Tongshu

p.cm.

Includes bibliographical references.

1. Prisms, 2. Reflection (Optics) , 3. Optical phase conjugation.

I. Title.

QC425.4 L53   1989

681', 42-dc20   89-16181

CIP

British Library Cataloguing in Publication Data

Tongshu , Lian

Theory of conjugation for reflecting prisms : adjustment and image stabilization of optical instruments.

1. Optical systems. Geometric aspects

I. Title

535', 32

ISBN 7-80003-102-0/O.36   International Academic Publishers
ISBN 0-08-037935-4   Pergamon Press

Printed by the Printing House of China Building Industry Press

To My Mother
and
To My Wife

A Few Words on the Book
"Theory of Conjugation for Reflecting Prisms"

The present volume deals with optical characteristics of reflecting prisms as usually incorporated in various types of optical instruments and should be referred to as a branch of geometrical optics. It is worthy of mention that the author, with thirty odd years of experience in teaching and designing optical instruments, puts forth in this book a unified theory mathematically on all kinds of prism systems not only from the point of view of design and usage but also taking into account effects due to manufacturing errors. The theory may thus facilitate the designers as well as users to master the basic principles involved in various types of prism systems. As a new achievement of the author I should like to regard this work as a useful and valuable addition to the area of geometrical optics.

为《反射棱镜的共轭理论》- 序述字

本书内容，探讨在光学仪器中所用各种型式的反射棱镜的性能，是属于几何光学范畴的。作者以其在光学仪器设计及教学上卅余年的实践经验，建立了有关各种反射棱镜性能的统一理论，将有助于光学仪器设计者和使用者对棱镜若干理论的了解。这是作者的一项新成就，就其实际价值，应予肯定。这为几何光学增添了新的篇章。

Wang Daheng 王大珩

Wang Daheng
Professor of Applied Optics
President of the Chinese Optical Society
Member of the Chinese Academy of Sciences

July 1, 1990

Foreword

Reflecting prisms and lenses are two most important types of optical components in many optical systems, especially in conventional optical instruments. However, as compared with lenses, there are only a few monographs on mirror and prism systems, and material in this field is most likely to be only one chapter in a book entitled *Applied Optics* or *Engineering Optics* or something similar.

It is felt that people engaged in the design and adjustment of optical instruments very often have difficulty in determining the ray path and image orientation through a reflecting prism, particularly when the latter is moving, and the problem becomes even more complicated if the reflecting system is located in convergent light. The difficulty is caused simply by a three-dimensional motion of the reflecting prism and the spatial structure of its reflecting surfaces.

It is apparent therefore that many researchers, engineers, teachers and students who are interested in optical instrumentation and applied optics, will need to obtain an in-depth understanding of the basic theory for reflecting prisms as well as of its practical aspects.

Fortunately *Theory of Conjugation for Reflecting Prisms (Adjustment and Image Stabilization of Optical Instruments)* is a book which goes far towards answering this need.

The book evolved mainly from the author's teaching experience and research achievements over nearly forty years. The work presents a sound theoretical development, which contains a series of new concepts, theorems, formulas and characteristic parameters, as well as a new principle for classification and a new system of graphical tabulation of reflecting prisms.

The author effectively ties mechanics to geometrical optics with an insight that is unique. This new method of treatment by creatively developing the so-called *rigid body's kinematics model* to simulate the real physical phenomena for reflecting prisms has proved to be very successful in establishing the present theory in a systematic and comprehensive way.

Professor Lian graduated in mechanical engineering from Tsinghua University in 1952, and has been working in optics ever since. Now I can say that Lian Tongshu does not have to apologize to his professors who taught him the theory of mechanics in the past, since he has had a wonderful record in extending the potential application of mechanics to other disciplinary areas. Incidentally, the significance in this regard is far-reaching pedagogically.

Without doubt, the presentation is a new, important and practical contribution to geometrical optics as well as to optical engineering.

Liang Jinwen (J.W. Liang)
Professor at Tsinghua University
Vice President of the Chinese Society for Measurement

Preface

The *Theory of Conjugation for Reflecting Prisms* contains mainly the material which is intended to provide a thorough, mathematical description of the relationships between the object and image spaces for mirror and prism systems as well as a few basic theorems which underlie them.

Unlike lenses, mirror and prism systems are often used as moving parts, most likely rotating, in optical systems, and only in this way can optical systems be designed to perform a great diversity of functions.

Even though a reflecting prism is rigidly fixed in an optical system, and really appears to be stationary while the instrument embodying the system is in use, it might also be subject to a slight tilt or a small shift while the whole optical system as well as the prism itself is being adjusted at the final stage of assembly. Then, one still has to consider the effect resulting from a small displacement, either linear or angular, of a reflecting prism.

Therefore, it is reasonable that this theory should also cover that part which deals with the object and image relationship for the mirror and prism systems in movement.

Thus, the material included in this text has a close bearing on the principle, design and manufacture of optical instruments, and will prove to be useful especially for the adjustment of prisms, and in the design of scanners, trackers, image stabilizers, prism mountings and so on.

The present volume is in three parts. In Part I, we present mainly a review of various relevant topics of mathematics and kinematics, such as vector formulas of reflection and rotation, matrix of transformation of coordinates and its relation with a matrix of rotation, kinematics of rigid bodies and the concept of gradient. In Part II, we provide a complete theoretical treatment of the problem of the relation between two conjugate spaces for reflecting prisms. It is again divided into two sections: one, referred to as the theoretical foundation, which covers the material of image formation for reflecting prisms, image motion due to a finite rotation of a mirror system and image motion caused by a small displacement, either linear or angular, of a reflecting prism; and the other, regarded as the applied foundation, in which the theory of adjustment for reflecting prisms and the theory of image stabilization by using reflecting prisms as compensators are developed. The importance attached to these theoretical topics becomes apparent with the observation that they comprise nearly 75% of the material in this text. Following the development of these strong theoretical and applied foundations, several previously and currently manufactured typical optical instruments are examined and discussed in Part III in full detail with regard to the problems of adjustment and image stabilization, thus showing how to apply the basic theory developed in Part II as well as the mathematical and kinematical tools mentioned in Part I to solve practical engineering tasks. Fifty-four commonly used reflecting prisms are tabulated in the Appendix, where for each prism a table is given with one adjustment diagram, one matrix, six formulas and twenty characteristic parameters. A carefully designed set of problems is provided at the end of each chapter to help guide the reader through the learning process in an orderly manner. Some of these problems might hopefully lead to the reader's thinking being in harmony with that of the author.

This book is the final result of about thirty-eight years of continuous devotion to teaching and research work in the fields of optical instrumentation and applied optics, plus some contributions by my colleagues.

An imaginary physical model called *rigid body's kinematics model* has been designed to simulate the real physical phenomena of both image formation and image motion for reflecting prisms. Thanks to this approach to research, the theory presented here includes a series of new theorems, concepts, formulas and parameters as well as new types of classification and tabulation of reflecting prisms.

The theory of conjugation for mirror and reflecting prism systems is referred to as a branch of the discipline of geometrical optics. However, it can also be very useful in some related disciplinary areas. In fact, it has already been employed in interferometry in regard to the adjustment of optical path difference (OPD), and it can also be of great value in designing a rotating reflecting prism for changing the Q factor of the resonator cavity of lasers.

The material is planned for a graduate level course, and also presented in a form which should make it useful for reference, for individual study, and for short courses.

I should like to thank all my former esteemed teachers. In particular, I am indebted to Professor Zhou Peiyuan, Professor Chien Weizang, Professor Wang Daheng, Professor Liang Jinwen, Mr. Wu Zhuyun, and Mr. Jiang Mengqi, who educated me in fundamental and specialized knowledge.

I would like to acknowledge my special debt to Professor Hou Qian, one of my treasured friends, for his consistent help, encouragement and numerous invaluable suggestions regarding organization, terminology and various specialized subjects. I also owe a debt of gratitude to Professor Jian Xiuwen, who carefully examined the initial portions of the manuscript and kept me from being stranded upon the shoals of bad grammar. My gratitude is due to Professor Cai Jiahua, whose support at the very beginning was decidedly important. I also owe thanks to Associate Professor Xu Yaohui, Associate Professor You Dinghua, Senior Engineer Tang Zhiyi, and my wife, Senior Engineer Xu Huiying, who designed an excellent program for computing all the data necessary for the tabulation and prepared the illustrations beautifully. My thanks for contributions to the development of this book are extended to Professor Ding Hanzhang, Professor He Shoyu, Professor Francis T. S. Yu, Doctor Philip Lam, Senior Engineer Tang Jiafen, Professor Wo Xinneng, Senior Engineer Qi Kangnan, Associate Professor Peng Liming, Senior Engineer Jie Deer, Associate Professor Hao Shuying, and Associate Professor Wang Qi. I am grateful to Senior Engineer Dai Chuanheng, Senior Engineer Zhang Zhongyuan, Senior Engineer Guo Yuxiang, and Senior Engineer Zhu Changchen, who arranged for me to practise on a variety of optical products in their factories during the preparation of the first draft. Acknowledgment should be made to the National Defence Industry Press, Beijing Institute of Technology Press, Editorial Board of Acta Optica Sinica of the Chinese Optical Society, Editorial Board of Yunguang Jishu, and Editorial Board of Instrumentation-Analysis-Monitoring which have published most of my books and papers in this field.

As for my undergraduate and graduate students, I appreciate their constant interest, enthusiasm, motivation and feedback, especially, the active participation of my Ph. D. and M. S. candidates: Mr. Zhao Yuejin, Mr. Yao Wu, Mr. Yang Yanfeng, Mr. Lin Ning, Ms. Wan Fang, and Mr. Zhang Jianmin, in making the last correction of proofs and checking some of the complicated formulas.

Special mention must be made of that incomparable and beloved couple, Cheng Musheng (Xenia), Professor of English, Tsinghua University and Zhu Enhuai, Visiting Professor of English, Tsinghua Graduate School. They spent much of their valuable time carefully examining and improving the written language of the body of the manuscript all ten chapters. It is impressive that they embody all the qualities one hopes to find in the academic world —— earnestness, responsibility, profound scholarship and dedication.

My thanks, again, to Professor Wang Daheng, Professor Liang Jinwen and Professor Hou Qian for their cordial attention and many words of encouragement.

I would also like to take this opportunity to refer to Ms. Zheng Taoying, my affectionate mother, deeply remembering her unbounded love, uniform kindness, utmost patience, and earnest teachings which exercised a lasting influence on my life and work and gave me the devotion needed to write a series of books including the present one.

And, finally, my thanks go to all the workers of International Academic Publishers for their efforts in bringing the book to fruition on time; to the late Editor-in-Chief, Professor Wu Zhenxin, who was good enough to sign the agreement with me as to publication; and particularly to the Senior Editor, Professor Liu Qiyuan, the evidence of whose wisdom and thoroughness appears on every page. I have learned much from him about how to be "ruthless" with verbiage.

Lian Tongshu

Beijing Institute of Technology

Beijing, China

June 1990

Contents

List of Symbols

| Symbol | Quantity Represented |

N — Unit vector of mirror normal

A — Direction of entering parallel beam; object vector; input vector (usually being unit vector, but not necessarily)

A' — Direction of emerging parallel beam; image vector; output vector

P — Unit vector of rotation axis

θ — Angle of rotation about axis P; signs of θ and P according to right-handed rule

R — Reflection matrix of single plane mirror, mirror system or reflecting prism

(A) — Column matrix of vector A (or called column vector)

$(A)_1$ — Column matrix of vector A expressed relative to coordinate system $x_1 y_1 z_1$ (or, column vector A expressed with respect to coordinate system $x_1 y_1 z_1$)

$(A)_{1'}$ — Column matrix of vector A expressed relative to coordinate system $x_1' y_1' z_1'$

$(A)_0$ — Column matrix of vector A expressed relative to coordinate system $x y z$*

$(A)_{0'}$ — Column matrix of vector A expressed relative to coordinate system $x' y' z'$*

For instance,

$$(A)_0 = \begin{pmatrix} A_x \\ A_y \\ A_z \end{pmatrix}, \quad (A)_{0'} = \begin{pmatrix} A_{x'} \\ A_{y'} \\ A_{z'} \end{pmatrix}, \quad (A_1)_2 = \begin{pmatrix} A_{1,x_2'} \\ A_{1,y_2'} \\ A_{1,z_2'} \end{pmatrix}$$

$[R]_2$ — Reflection matrix R expressed relative to coordinate system $x_2 y_2 z_2$

S — Rotation matrix

$S_{P,\theta}$ — Matrix representation of rotation about axis P through angle θ

G — Matrix of transformation of coordinates

$(G)_{01}$ — Matrix of transformation of coordinates from coordinate system xyz to coordinate system $x_1 y_1 z_1$

$(G)_{10}$ — Matrix of transformation of coordinates from coordinate system $x_1 y_1 z_1$ to coordinate system $x y z$

D — Aperture of reflecting prism

n — Prism index

---

*A subscript '0' is imaginarily attached to the most frequently used designation of coordinate systems xyz and $x'y'z'$, although they actually do not have any subscript at all.

L Thickness of plane-parallel plate or tunnel length of unfolded reflecting prism

t Number of reflections or reflecting surfaces

M_0, M' A pair of basic conjugate points of reflecting prism

T Characteristic vector

2φ Characteristic angle

$S_{T,2\varphi}$ Matrix representation of rotation about axis T through angle 2φ

$\vec{\lambda}$ Screw axis of equivalent helical motion of image formation

r_0 Vector dropped from M' perpendicularly onto screw axis $\vec{\lambda}$ for locating it in space

I Center of inversion

r_i Position vector of center of inversion measured from M'

$2d$ Axial shift of equivalent helical motion of image formation

g Position vector of object point

g' Position vector of image point

g'_θ Position vector of image point formed by mirror system after having rotated through angle θ about fixed axis

A'_θ Image vector formed by mirror system after having rotated through angle θ about fixed axis

P Unit vector representing the rotation axis of reflecting prism or mirror system

P' Unit vector representing the image of P formed by reflecting prism or mirror system prior to its rotation

q, q' A pair of conjugate points lying on P and P', respectively

F' Image point lying on exit optical axis of reflecting prism

r_q Position vector of q

b' Distance from M' to F'

b' Vector directed from M' to F'

$\vec{\rho}_{F'}, \vec{\rho}'_{F'}$ Vectors directed from q and q' to F', respectively

$\Delta\theta$ Magnitude of small angular displacement ($\Delta\vec{\theta}$ or $\Delta\theta P$) of prism

$\Delta\vec{\mu}'$ Image rotation (due to $\Delta\theta P$)

$\Delta S'_{F'}$ Image point displacement (due to $\Delta\theta P$)

D Unit vector representing the direction of small linear displacement of prism

D' Unit vector representing the image of D formed by prism

Δg Magnitude of small linear displacement (Δg or $\Delta g\, D$) of prism

$\Delta S'$ Image displacement (due to $\Delta g\, D$)

$J_{\mu P}$ Matrix of image rotation

J_{SD} Matrix of image displacement

J_{SP} Matrix of image point displacement

C_q Cross-product-matrix of vector r_q

C_m Cross-product-matrix of vector $\overrightarrow{M_0 M}'$

$C_{b'}$ Cross-product-matrix of vector b'

E Unit matrix of order 3

r' Arbitrary direction in image space of prism

r' Unit vector along arbitrary direction r'

$\Delta\mu_r'$ r' image rotation, i.e., component of image rotation $\Delta\vec{\mu}'$ along direction r'

$\Delta\mu_x'$ x' image rotation

$\Delta\mu_y'$ y' image rotation

$\Delta\mu_z'$ z' image rotation

$\vec{\delta}_h$ Extreme-valued characteristic vector of r' image rotation

$\vec{\delta}_u$ Extreme-valued characteristic vector of x' image rotation

$\vec{\delta}_v$ Extreme-valued characteristic vector of y' image rotation

$\vec{\delta}_w$ Extreme-valued characteristic vector of z' image rotation

$\vec{\eta}_h$ Gradient axis direction of r' image rotation

$\vec{\eta}_u$ Gradient axis direction of x' image rotation

$\vec{\eta}_v$ Gradient axis direction of y' image rotation

$\vec{\eta}_w$ Gradient axis direction of z' image rotation

h Extreme-valued axis direction of r' image rotation

u Extreme-valued axis direction of x' image rotation

v Extreme-valued axis direction of y' image rotation

w Extreme-valued axis direction of z' image rotation

δ_h Extreme value of r' image rotation

δ_u Extreme value of x' image rotation

δ_v Extreme value of y' image rotation

δ_w Extreme value of z' image rotation

$\Delta S_r'$ r' image displacement, i.e., component of image displacement $\Delta S'$ along direction r'

$\Delta S_x'$ x' image displacement

$\Delta S_y'$ y' image displacement

$\Delta S_z'$ z' image displacement

$\vec{\delta}_e$ Extreme-valued characteristic vector of r' image displacement

$\vec{\delta}_a$ Extreme-valued characteristic vector of x' image displacement

$\vec{\delta}_b$ Extreme-valued characteristic vector of y' image displacement

$\vec{\delta}_c$ Extreme-valued characteristic vector of z' image displacement

$\vec{\eta}_e$ Gradient shift direction of r' image displacement

$\vec{\eta}_a$ Gradient shift direction of x' image displacement

$\vec{\eta}_b$ Gradient shift direction of y' image displacement

$\vec{\eta}_c$ Gradient shift direction of z' image displacement

e Extreme-valued shift direction of r' image displacement

a Extreme-valued shift direction of x' image displacement

b Extreme-valued shift direction of y' image displacement

c Extreme-valued shift direction of z' image displacement

δ_e Extreme value of r' image displacement

δ_a Extreme value of x' image displacement

δ_b Extreme value of y' image displacement

Symbol	Description
δ_c	Extreme value of z' image displacement
$\vec{\eta}_{F'hq}$	Gradient axis of r' image point displacement
$\vec{\eta}_{F'uq}$	Gradient axis of x' image point displacement
$\vec{\eta}_{F'vq}$	Gradient axis of y' image point displacement
$\vec{\eta}_{F'wq}$	Gradient axis of z' image point displacement
$\vec{\eta}_{F'eP}$	Gradient (translation) direction of r' image point displacement
$\vec{\eta}_{F'aP}$	Gradient (translation) direction of x' image point displacement
$\vec{\eta}_{F'bP}$	Gradient (translation) direction of y' image point displacement
$\vec{\eta}_{F'cP}$	Gradient (translation) direction of z' image point displacement
$\vec{\zeta}_{dq}$	One-dimensional zero-valued axis of r' image point displacement
$\vec{\zeta}_{lq}$	One-dimensional zero-valued axis of x' image point displacement
$\vec{\zeta}_{mq}$	One-dimensional zero-valued axis of y' image point displacement
$\vec{\zeta}_{nq}$	One-dimensional zero-valued axis of z' image point displacement
$\vec{\zeta}_{lmq}$	Two-dimensional zero-valued axis of $x'\text{-}y'$ image point displacement
$\vec{\zeta}_{mnq}$	Two-dimensional zero-valued axis of $y'\text{-}z'$ image point displacement
$\vec{\zeta}_{nlq}$	Two-dimensional zero-valued axis of $z'\text{-}x'$ image point displacement
$\vec{\zeta}_q$	Three-dimensional zero-valued axis passing through point q
Π_l	Plane of one-dimensional zero-valued axis of x' image point displacement
Π_m	Plane of one-dimensional zero-valued axis of y' image point displacement
Π_n	Plane of one-dimensional zero-valued axis of z' image point displacement
Π_{lm}	Plane of two-dimensional zero-valued axis of $x'\text{-}y'$ image point displacement
Π_{mn}	Plane of two-dimensional zero-valued axis of $y'\text{-}z'$ image point displacement
Π_{nl}	Plane of two-dimensional zero-valued axis of $z'\text{-}x'$ image point displacement
Π_{lmn}	Plane of three-dimensional zero-valued axis
C_Π	Planar three-dimensional zero-valued pole C_Π with associated plane Π
δ	Image lean
$\vec{\psi}$	Tilt of image plane
$\vec{\xi}$	Optical axis deviation
$\vec{\gamma}$	Image shift
$\Delta\varepsilon\,\boldsymbol{P}_\varepsilon$	Angular perturbation
$\Delta\theta_i\,\boldsymbol{P}_i$	Angular compensation of i compensator

Notes on the term *image orientation*:

The term image orientation used in this book represents the direction of a image body and thus is a quantity with three degrees of freedom, while the same term used in other books conventionally represents the direction of a planar image in the plane of its own and thus is a quantity with only one degree of freedom.

Part I

Fundamental Knowledge

Chapter 1

Mathematical and Kinematical Foundations

Compared with lenses, mirror and reflecting prism systems have two main features: one, their asymmetric form, and the other their nature of movability as indicated in the Preface. These factors lead to the result that drawings to illustrate the ray paths and the relationships between two conjugate spaces appear, in many cases, to have extremely complex spatial structure.

Such problems can no longer be solved simply by graphical means, and it is advisable instead to use mathematical methods.

It is well known that vector treatment and matrix method are two closely related powerful tools for this purpose.

Since the so-called *rigid body's kinematics model* has been used to simulate the real physical phenomena of both image formation and image motion for reflecting prisms, the method of kinematics treatment will be employed throughout this book.

The above touches upon at least three major branches of mathematics and kinematics. It is our aim in this chapter to review some relevant topics of these fields. However, because of the orientated goal of this text and its limited space, we shall restrict our attention to only the material essential for the study of the later chapters, foregoing a complete consideration of the fields.

1.1 Formula of Reflection in Vector Form

A reflecting prism may contain one or more reflecting surfaces, and each of them acts like a plane mirror. Therefore, the formula of reflection in vector form for a single plane mirror is considered as one of the most fundamental formulas. For short, we will call it the *reflection vector formula*.

Referring to Fig. 1.1, unit vectors A, A', N are used to represent the incident ray direction, the reflected ray direction, and the normal to the mirror, respectively.

The geometrical relation among the three unit vectors is self-explanatory from the diagram in Fig. 1.1, so the resulting equation can be written directly as follows:

$$A' = A - 2(A \cdot N)N. \quad \bullet \tag{1.1}$$

This is the so-called *reflection vector formula*.

We can easily see from Eq. (1.1) that it makes no difference whether the normal vector N is toward or away from the mirror.

It is also clear from Eq. (1.1) that vectors A and A' are not necessarily unit vectors, and

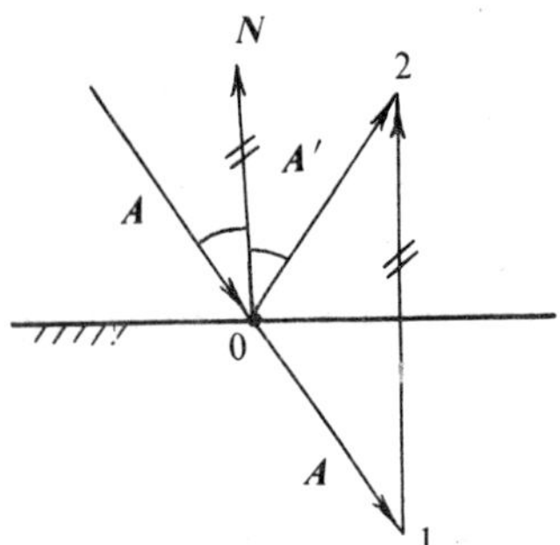

Fig . 1 . 1 . A diagram illustrating the reflection of a ray of light

the reflection vector equation (1.1) will still hold true even if A is assumed to be different from a unit vector. In this case, the unknown vector A' determined by solving Eq. (1.1) will always have the same magnitude as the given vector A. The invariance in this regard may be referred to as a feature associated with the property of the reflection vector formula. However, we usually assume vectors A and A' to be unit vectors, just for the sake of convenience.

Another problem is whether the vectors A and A' in Eq . (1.1) can also be employed to specify an object vector and its conjugate image vector. The answer is absolutely positive for at least one reason, i.e., the relation of the image vector to its conjugate object vector is the same as that of the reflected ray direction to the incident ray direction. In other words, that is because the relationship between the object vector A and its conjugate image vector A' also follows the law of reflection. ·

A better understanding at this point can be acquired once the kinematics model of a rigid body has been adopted to describe the relationship between two conjugate spaces or two conjugate bodies for reflecting prisms.

Furthermore, the reflection vector formula (1.1) can be expanded using $A = A_x i + A_y j + A_z k$, $N = N_x i + N_y j + N_z k$, and $A' = A_x' i + A_y' j + A_z' k$, where i , j , k are unit vectors along the x , y , z axes of an orthogonal coordinate system and the rest are the direction cosines of the three unit vectors. Thus, we have

$$\left.\begin{aligned}
A_x' &= (1 - 2N_x^2)A_x - 2N_xN_yA_y - 2N_xN_zA_z \\
A_y' &= -2N_xN_yA_x + (1 - 2N_y^2)A_y - 2N_yN_zA_z \\
A_z' &= -2N_xN_zA_x - 2N_yN_zA_y + (1 - 2N_z^2)A_z
\end{aligned}\right\}, \qquad (1.2)$$

which shows that a reflection performs mathematically a linear transformation, and hence the equation (1.2) can be written in matrix form

$$(A')_0 = R (A)_0 , \qquad (1.3)$$

where R denotes the *reflection matrix* for a single plane mirror

$$R = \begin{pmatrix}
(1 - 2N_x^2) & -2N_xN_y & -2N_xN_z \\
-2N_xN_y & (1 - 2N_y^2) & -2N_yN_z \\
-2N_xN_z & -2N_yN_z & (1 - 2N_z^2)
\end{pmatrix} , \qquad (1.4)$$

4

and $(A)_0$ and $(A')_0$ denote *column matrices* of vectors A and A' relative to a coordinate system designated by xyz, respectively

$$(A)_0 = \begin{pmatrix} A_x \\ A_y \\ A_z \end{pmatrix}, \quad (A')_0 = \begin{pmatrix} A_x' \\ A_y' \\ A_z' \end{pmatrix}. \tag{1.5}$$

For short, $(A)_0$ and $(A')_0$ are called *column vectors*, whilst their transposed matrices $(A)_0^t$ and $(A')_0^t$ are called *row vectors* and can be written in a more compact form

$$(A)_0^t = (A_x, A_y, A_z), \quad (A')_0^t = (A_x', A_y', A_z'). \tag{1.6}$$

It should be noted that we have here adopted a notation $(A)_i$ to specify a column vector by using parentheses to distinguish it from an ordinary vector, and a subscript, for example i, if necessary plus a corner mark, say a prime, to relate three components of the column vector to the chosen coordinate system $x_i' y_i' z_i'$, for instance:

$$(A)_{1'}^t = (A_{x_1'}, A_{y_1'}, A_{z_1'}), \quad (A')_2 = \begin{pmatrix} A_{x_2}' \\ A_{y_2}' \\ A_{z_2}' \end{pmatrix}.$$

In addition, we have to define a zero number " 0 ", " $0'$ ", etc. as the subscripts of the notation for the column vectors relative to those frequently used coordinate systems xyz, $x'y'z'$, etc. which do not have a subscript at all, for example :

$$(A)_{0''}^t = (A_{x''}, A_{y''}, A_{z''}).$$

It is also stipulated that an ordinary vector A is to be recovered from its corresponding column vector (A) by underlining the latter,

$$\underline{(A)} = A. \tag{1.7}$$

Finally, note that Eqs. (1.1) and (1.3) are valid only when the mirror normal is assumed to be a unit vector.

1.2 Formula of Rotation in Vector Form

Let a unit vector P represent an axis of rotation, as shown in Fig 1.2, and a vector A be rotated through a finite angle θ about the axis P to a new orientation denoted by a vector A'.

Now let us try to derive the formula for determining the vector A'.

In this case, we only care about the orientation of the vectors, instead of their location, so it is allowed to translate the vector P until it intersects vector A at its tail O.

The head B of the vector A moves on a circular path centered at O' as A rotates around the axis of rotation P.

We begin by drawing the solid polygon $OO'DCO$ to resolve the unknown vector A' into three components; thus,

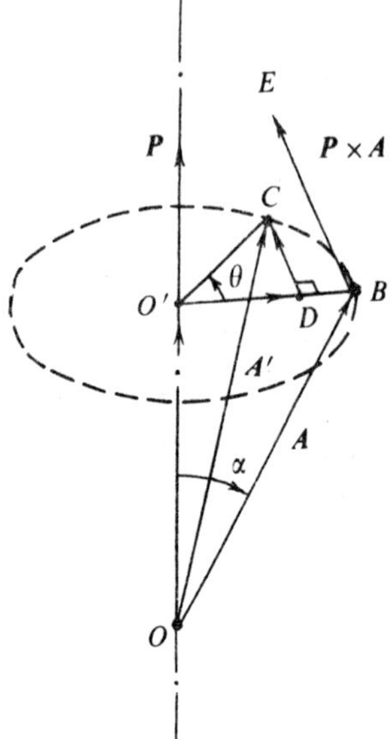

Fig . 1 . 2 . A diagram illustrating a vector A rotating about the axis P through a finite angle θ

$$A' = \overrightarrow{OO'} + \overrightarrow{O'D} + \overrightarrow{DC} \ . \tag{1.8}$$

A simple principle to be observed here is to project the vector A' onto certain directions which are easily determined. In this case, the directions of three components on the right side of Eq. (1.8) are parallel to the following three vectors : P, $[A - (A \cdot P)P]$ and $P \times A$, respectively. It is apparent that the above three directions are, correspondingly, axial, radial and tangential.

By adjusting these three vectors to suitable length, we then obtain

$$\overrightarrow{OO'} = (A \cdot P)P \ ,$$

$$\overrightarrow{O'D} = [\,A - (A \cdot P)P]\cos\theta \quad ,$$

$$\overrightarrow{DC} = -\,(A \times P)\sin\theta \ .$$

Finally, substituting all the above equations into Eq . (1.8) and arranging it in an orderly form yields the required formula

$$A' = A\cos\theta + (1 - \cos\theta)\,(A \cdot P)P - \sin\theta\,(A \times P) \ . \tag{1.9}$$

This is the formula of rotation in vector form, and for brevity, we call it the *rotation vector formula.*

We can also expand Eq . (1.9) by using $A = A_x i + A_y j + A_z k$, $P = P_x i + P_y j + P_z k$, $A' = A_x' i + A_y' j + A_z' k$, and consequently arrive at the following set of linear equations

$$A_x' = \left(\cos\theta + 2P_x^2\sin^2\frac{\theta}{2}\right)A_x + \left(-P_z\sin\theta + 2P_x P_y\sin^2\frac{\theta}{2}\right)A_y + \left(P_y\sin\theta + 2P_x P_z\sin^2\frac{\theta}{2}\right)A_z$$

$$A_y' = \left(P_z \sin\theta + 2P_x P_y \sin^2 \frac{\theta}{2} \right) A_x + \left(\cos\theta + 2P_y^2 \sin^2 \frac{\theta}{2} \right) A_y + \left(-P_x \sin\theta + 2P_y P_z \sin^2 \frac{\theta}{2} \right) A_z$$

$$A_z' = \left(-P_y \sin\theta + 2P_x P_z \sin^2 \frac{\theta}{2} \right) A_x + \left(P_x \sin\theta + 2P_y P_z \sin^2 \frac{\theta}{2} \right) A_y + \left(\cos\theta + 2P_z^2 \sin^2 \frac{\theta}{2} \right) A_z$$

$$(1.10)$$

Moreover, Eq. (1.10) can be written in matrix form

$$(A')_0 = S_{P,\theta} (A)_0 \ , \tag{1.11}$$

where

$$(A)_0 = \begin{pmatrix} A_x \\ A_y \\ A_z \end{pmatrix} , \qquad (A')_0 = \begin{pmatrix} A_x' \\ A_y' \\ A_z' \end{pmatrix},$$

and

$$S_{P,\theta} = \begin{pmatrix} \cos\theta + 2P_x^2 \sin^2 \dfrac{\theta}{2} & -P_z \sin\theta + 2P_x P_y \sin^2 \dfrac{\theta}{2} & P_y \sin\theta + 2P_x P_z \sin^2 \dfrac{\theta}{2} \\[2mm] P_z \sin\theta + 2P_x P_y \sin^2 \dfrac{\theta}{2} & \cos\theta + 2P_y^2 \sin^2 \dfrac{\theta}{2} & -P_x \sin\theta + 2P_y P_z \sin^2 \dfrac{\theta}{2} \\[2mm] -P_y \sin\theta + 2P_x P_z \sin^2 \dfrac{\theta}{2} & P_x \sin\theta + 2P_y P_z \sin^2 \dfrac{\theta}{2} & \cos\theta + 2P_z^2 \sin^2 \dfrac{\theta}{2} \end{pmatrix}$$

$$(1.12)$$

The symbol $S_{P,\theta}$ denotes such a matrix, which represents mathematically the transformation of a rotation about the axis P through an angle θ, and for short it is called the *rotation matrix*.

We have treated the cross product $A \times P$ and the sign of θ in terms of P in Eqs. (1.10) and (1.12) according to the right-handed rule, and correspondingly we have designated a right-handed coordinate system xyz. These generally accepted conventions will consistently be kept throughout this book.

In particular, if the axis of rotation P is parallel to any one of the coordinate axes x, y, z, then we have in turn three matrices of rotation $S_{i,\theta}$, $S_{j,\theta}$, and $S_{k,\theta}$

$$S_{i,\theta} = \begin{pmatrix} 1 & 0 & 0 \\ 0 & \cos\theta & -\sin\theta \\ 0 & \sin\theta & \cos\theta \end{pmatrix}, \quad S_{j,\theta} = \begin{pmatrix} \cos\theta & 0 & \sin\theta \\ 0 & 1 & 0 \\ -\sin\theta & 0 & \cos\theta \end{pmatrix}, \quad S_{k,\theta} = \begin{pmatrix} \cos\theta & -\sin\theta & 0 \\ \sin\theta & \cos\theta & 0 \\ 0 & 0 & 1 \end{pmatrix}.$$

$$(1.13)$$

Very often, the rotation considered above is quite small, for instance a slight tilt, therefore substituting $\Delta\theta$ for θ, and neglecting small quantities of second and higher powers by putting $\sin\Delta\theta \approx \Delta\theta$ and $\cos\Delta\theta \approx 1$, Eq. (1.10) will become greatly simplified

$$A' = A - \Delta\theta \, (A \times P) . \tag{1.14}$$

Furthermore, by letting $\Delta A = A' - A$, we have thus

$$\Delta A = -\Delta\theta \, (A \times P) = \Delta\theta \, P \times A . \tag{1.15}$$

Again, it should be noticed that Eqs. (1.10), (1.12), and (1.15) are valid only when P is assumed to be a unit vector.

1.3 Relation Between Rotation and Transformation of Coordinates

First of all let us introduce a matrix symbol G_{21} defined by the following matrix formula

$$(A)_1 = G_{21} \, (A)_2 \quad , \tag{1.16}$$

where G_{21} is obviously the matrix of transformation of coordinates from the rectangular coordinate system $x_2 y_2 z_2$ to the rectangular coordinate system $x_1 y_1 z_1$, or in other words, using this matrix G_{21} the coordinates of a vector A relative to $x_1 y_1 z_1$ can be given in terms of the coordinates of the same vector A relative to $x_2 y_2 z_2$.

It is well known that this matrix G_{21} can be found in the following table

Table 1.1   Direction cosines of one set of coordinate axes relative to another set of coordinate axes

	$i_2 \, (x_2)$	$j_2 \, (y_2)$	$k_2 \, (z_2)$
$i_1 \, (x_1)$	$g_{11} = \cos(\widehat{i_1 , i_2})$	$g_{12} = \cos(\widehat{i_1 , j_2})$	$g_{13} = \cos(\widehat{i_1 , k_2})$
$j_1 \, (y_1)$	$g_{21} = \cos(\widehat{j_1 , i_2})$	$g_{22} = \cos(\widehat{j_1 , j_2})$	$g_{23} = \cos(\widehat{j_1 , k_2})$
$k_1 \, (z_1)$	$g_{31} = \cos(\widehat{k_1 , i_2})$	$g_{32} = \cos(\widehat{k_1 , j_2})$	$g_{33} = \cos(\widehat{k_1 , k_2})$

and then (Ref. 1)

$$G_{21} = \begin{pmatrix} g_{11} & g_{12} & g_{13} \\ g_{21} & g_{22} & g_{23} \\ g_{31} & g_{32} & g_{33} \end{pmatrix} . \tag{1.17}$$

Obviously, we have also

$$(A)_2 = G_{12} \, (A)_1 \quad , \tag{1.18}$$

and

$$G_{12} = (G_{21})^{-1} = (G_{21})^{t} . \tag{1.19}$$

The equation (1.19) is correct because of the orthogonality of the matrix of transformation of rectangular coordinates G_{21}. In fact, the reflection matrix R as well as the rotation matrix $S_{P,\theta}$ is also orthogonal.

Problems of moving bodies, for example a rotating reflecting prism used in a beam- steering system, will be dealt with from time to time in the chapters to follow.

In solving the problems of this type, as a rule it is advisable to assume two rectangular coordinate systems: one, which is connected to the fixed space and consequently is called the *fixed coordinate system*, say $x_f y_f z_f$, and the other, which is rigidly connected to the rotating body and is therefore named the *moving coordinate system*, say $x_m y_m z_m$.

Now suppose that the reflecting prism in question has rotated about a given fixed axis P (unit vector) through a certain angle θ. Meanwhile, the moving coordinate system $x_m y_m z_m$ joined to the prism as a single body has also rotated around P through θ.

For convenience, as is always possible, the moving coordinate system is so orientated relative to the body of the prism that the initial directions of its coordinate axes x_m, y_m, z_m would be parallel to the directions of the fixed coordinate axes x_f, y_f, z_f, respectively.

Thus, the moving coordinate system $x_m y_m z_m$ at any instant can be considered as a result obtained by turning the fixed coordinate system $x_f y_f z_f$ around the axis P through an angle θ.

Only referring to the conditions prescribed above, one could ask what the relation will be between the rotation and the related transformation of coordinates from the moving (or fixed) coordinate system to the fixed (or moving) coordinate system. This is exactly the question which the present section is intended to answer.

For this purpose, let us assume a vector A to represent the initial orientation of an arbitrary fixed line in the body of the prism prior to its rotating, and another vector A' the final (not necessarily the final) orientation of the same line after its having turned through an angle θ about the axis P.

The above relationship indicates that the vector A' may also be taken as a result obtained by rotating the vector A around the axis P through an angle θ. Hence, by using Eq. (1.10) the vector A' can be expressed as

$$(A')_f = (S_{P,\theta})_f (A)_f \;, \tag{1.20}$$

where the subscript 'f' shows that the two column matrices and the rotation matrix are all expressed relative to the fixed coordinate system $x_f y_f z_f$.

Obviously, the relative orientation of the rotating vector A' with respect to the moving coordinate system $x_m y_m z_m$ remains unchanged as it used to be at the initial moment when the vector A was expressed relative to the fixed coordinate system $x_f y_f z_f$, therefore

$$(A)_f = (A')_m \;. \tag{1.21}$$

Substituting Eq. (1.21) into Eq. (1.20) yields

$$(A')_f = (S_{P,\theta})_f (A')_m \;. \tag{1.22}$$

By definition, we also have

$$(A')_f = G_{mf} (A')_m \;. \tag{1.23}$$

Comparing the last two equations, we then arrive at the following expression

$$G_{mf} = (S_{P,\theta})_f \;. \tag{1.24}$$

Similarly, if we replace Eq. (1.20) by

$$(A')_m = (S_{P,\theta})_m (A)_m ,$$ (1.25)

the expression (1.24) will be derived in another form

$$G_{mf} = (S_{P,\theta})_m .$$ (1.26)

Finally, Eqs. (1.24) and (1.26) can be written as a single:

$$G_{mf} = (S_{P,\theta})_{f \text{ or } m}$$ (1.27)

or just

$$G_{mf} = S_{p,\theta} ,$$ (1.28)

and this result can readily verify the equality of their inverses

$$G_{fm} = S_{P,-\theta} .$$ (1.29)

Stated in words, then "the matrix G_{mf} of transformation of coordinates from the moving coodinate system $x_m y_m z_m$ to the fixed coordinate system $x_f y_f z_f$ is equal to the matrix $S_{P,\theta}$ of rotation around P through θ", or "the matrix G_{fm} of transformation of coordinates from the fixed coordinate system $x_f y_f z_f$ to the moving coordinate system $x_m y_m z_m$ is equal to the matrix $S_{P,-\theta}$ of rotation around P through $(-\theta)$, i.e., the matrix of rotation in reverse direction."

Of course, the validity of these statements is based on the fact that the moving coordinate system $x_m y_m z_m$ is exactly the coordinate system resulting from turning the fixed coordinate system $x_f y_f z_f$ around the axis P through an angle θ.

The important remarks made above will later be applied so frequently that it is instructive to keep them in mind in a certain logical way, and hence for a better understanding it seems necessary to give some additional explanations as follows:

When going further into the physical meaning of a matrix of the transformation of coordinates from one coordinate system to another coordinate system, for instance G_{12}, we find that according to the rule of convention adopted in this text, it implies that the coordinate system related to the first subscript $x_1 y_1 z_1$ may be considered to be connected to an observed space, with respect to which the primary coordinates of A to be transformed, namely $(A)_1$, are measured, while the coordinate system related to the second subscript $x_2 y_2 z_2$ may be assumed to be connected to an observing space, relative to which the resulting coordinates of A, namely $(A)_2$, are expressed.

Let G_{12} be replaced by G_{fm}. Then, the coordinate systems $x_f y_f z_f$ and $x_m y_m z_m$ will have not only the property just described for the coordinate systems $x_1 y_1 z_1$ and $x_2 y_2 z_2$, but also the rotation-relationship, which is merely associated with such a pair of fixed and moving coordinate systems.

Again, to aid in visualizing the physical aspects of Eq. (1.29) an observer is assumed for the time being to be located in the moving coordinate system $x_m y_m z_m$ rigidly attached to the observing space.

Now, we have an arbitrary column vector $(A)_f$, whose coordinates are measured with respect to the fixed coordinate system $x_f y_f z_f$, and it is required to convert these coordinates A_{x_f}, A_{y_f}, and A_{z_f} from an old basis of $x_f y_f z_f$ to a new basis of $x_m y_m z_m$, the latter being the moving coordinate system. Obviously, this can be done in two different ways. First, it can be generally treated as a problem of transformation of coordinates, and hence the matrix of transformation of coordinates G_{fm} is needed for this calculation. Secondly, the above case may also be handled particularly as a problem of kinematics, where the column vector $(A)_f$ should be turned in the reverse direction, that is, rotated through an angle $(-\theta)$ about the axis P to a new orientation relative to the observer as well as relative to the moving coordinate system $x_m y_m z_m$ which is assumed to be fixed temporarily, and thus the rotation matrix $S_{P,-\theta}$ has to be applied to this computation.

Therefore, we have accomplished the logic deduction leading to Eq. (1.29) and consequently to Eq. (1.28), i.e., $G_{fm} = S_{P,-\theta}$ or $G_{mf} = S_{P,\theta}$, so as to make it easy to memorize them.

This result enables us to apply the regular equation (1.12) as well as its simple forms, namely Eq. (1.13), to most cases in performing the transformation of coordinates, and these orderly computations in conjunction with the memorized rule will help us avoid some careless mistakes, which may sometimes occur due to the intricate geometrics involved.

It has been seen that combining Eqs. (1.24) and (1.26) leads to the following equation

$$(S_{P,\theta})_f = (S_{P,\theta})_m \ . \tag{1.30}$$

and this is simply because the axis of rotation P has the same respective direction cosines relative to the two coordinate systems $x_f y_f z_f$ and $x_m y_m z_m$.

1.4　General Motion of a Rigid Body

General motion of a rigid body may be considered as a combination of translation and rotation in a certain way, and is more exactly called *space motion* or *three-dimensional motion*.

For our purpose, we are concerned more with the displacement of a rigid body than those instantaneous quantities such as velocity, acceleration, and so forth.

Moreover, a displacement of a rigid body resulting from a three-dimensional motion of the body can also be regarded as a *linear displacement* combined with another *angular displacement* in some forms, as will be discussed in more detail below.

1.4.1　Equivalent of a Displacement of a Rigid Body in General Motion

In Fig. 1.3, let two rectangular parallelepipeds $oabcdefg$ and $o'a'b'c'd'e'f'g'$, respectively, represent the initial and final locations and orientations of a rigid body in general motion. The actual process of how the rigid body reaches the final location and orientation has been omitted in the illustration since it is irrelevant to the problem here.

To find the equivalent displacement, let us try to bring the initial figure $oabcdefg$ to coincide

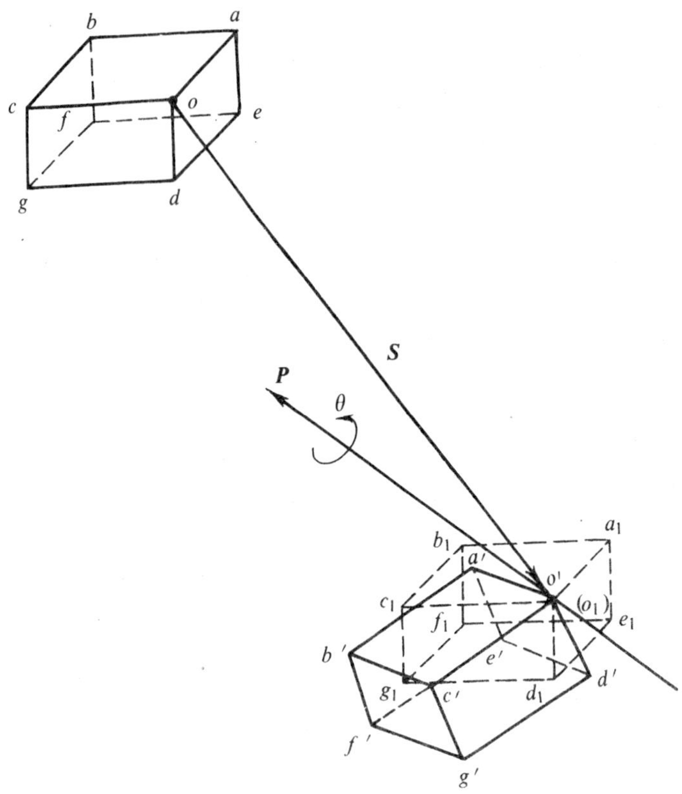

with the final figure $o'a'b'c'd'e'f'g'$! First of all, choose an arbitrary particle of the rigid body, belonging to the initial figure, say o, as a reference point, which is customarily called the *base point*, then translate the figure $oabcdefg$ rectilinearly by a linear displacement $\overrightarrow{oo'}$ to an intermediate place $o_1a_1b_1c_1d_1e_1f_1g_1$ with one of its vertexes o_1 coincident with point o', and furthermore rotate this intermediate figure $o_1a_1b_1c_1d_1e_1f_1g_1$ about the fixed point o' (o_1) properly until the rest of the intermediate figure coincides completely with the final figure $o'a'b'c'd'e'f'g'$.

Now it seems that we have to deal with a rigid body in rotation around a fixed point.

It is well known that the possibilities of motion of a rigid body can be expressed in terms of degrees of freedom. There are six possible degrees of freedom in all: translation in the three rectangular coordinate directions, and rotation about the three coordinate axes. Thus, the position of a rigid body, including its location and orientation can be determined by the locations of three of its component particles not lying in a straight line.

The conditions of constraints imposed upon a rigid body by fixing one particle of it will prevent it from translating in three coordinate directions (reduced by 3 degrees of freedom), then with another particle to be fixed the body is prevented from rotating about two coordinate axes perpendicular to the line joining the two chosen particles (reduced by 2 more), and finally the only possibility left to rotate the body around the fixed axis will also be removed by holding a third particle, which should not be coincident with the axis in question.

Suppose that a rigid body designated with a symbol B, as shown in Fig. 1.4, is attached at a fixed point o, and thus has three degrees of freedom of rotation about three rectangular axes intersecting at o.

Let S be a sphere drawn about o with an arbitrary radius. Then, we have both the sphere S and the center o as a part of the rigid body B.

Although the rotation about a fixed point is a special case of three-dimensional motion, it is also subject to the general law mentioned above in regard to the degrees of either freedom or constraint. Therefore, the position of the whole rigid body, in this special case only the orientation of the rigid body B, is also to be determined by locating its three particles not lying in a straight line. The only difference is that the center of rotation o must be one of the three

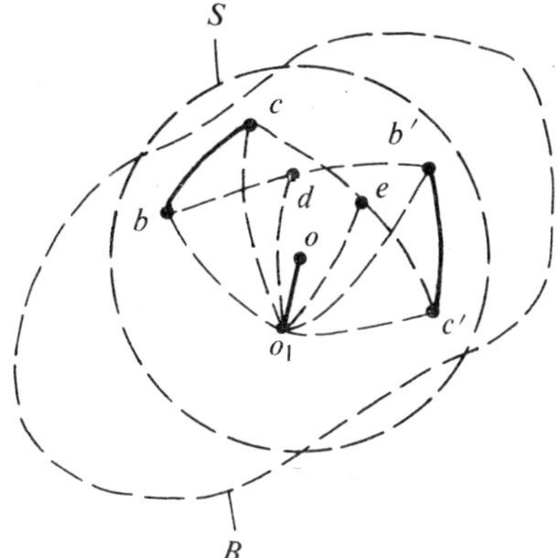

Fig.1.4.   Rotation of a rigid body B about a fixed point o

particles. The others may be chosen anywhere on the sphere S except on a pair of diametrically opposite points. Then, let the other two particles be b and c in Fig. 1.4.

Join the points b and c by a circular arc $\overset{\frown}{bc}$ of the great circle of the sphere (a circle cut out on the sphere by the plane through the center) passing through these two points; thus the orientation of the rigid body B, or what comes to the same thing, of the sphere S will be uniquely determined by the position of the arc $\overset{\frown}{bc}$.

Now, suppose that the arc $\overset{\frown}{bc}$ represents the initial place of the sphere S, and after the body is moved in a certain manner around o, the same arc comes to its final place $\overset{\frown}{b'c'}$.

Describe arcs $\overset{\frown}{bb'}$ and $\overset{\frown}{cc'}$ of great circles, and through the mid-points of these two arcs d and e, respectively, erect perpendicular bisectors, intersecting at o_1 and its diametrically opposite o_2 (o_2 is omitted from the diagram). Then, construct spherical triangles bo_1c and $b'o_1c'$ by drawing arcs from points b, c, b' and c' to the vertex o_1, which is joined to the center o by a straight line oo_1.

Obviously, the point o_1 is equidistant from b and b', and also equidistant from c and c', so we have

$$\overset{\frown}{o_1 b} = \overset{\frown}{o_1 b'} \ , \quad \overset{\frown}{o_1 c} = \overset{\frown}{o_1 c'} .$$

In addition,

$$\overset{\frown}{bc} = \overset{\frown}{b'c'} .$$

Therefore, $bo_1 c$ and $b'o_1 c'$ are two congruent triangles, and consequently $\angle bo_1 c = \angle b'o_1 c'$.

Moreover, due to the following equalities

$$\angle bo_1 b' = \angle bo_1 c + \angle co_1 b' \ ,$$

$$\angle co_1 c' = \angle b'o_1 c' + \angle co_1 b' \ ,$$

we obtain

$$\angle bo_1 b' = \angle co_1 c' .$$

Thus, it will be seen that if the rigid body B rotates about the axis oo_1 through an angle $\angle bo_1 b'$ (or $\angle co_1 c'$), then the arc $\overset{\frown}{bc}$ will be brought to coincide with the arc $\overset{\frown}{b'c'}$ completely, that is to say, the equivalent of a displacement of a rigid body in motion around a fixed point can be considered as an angular displacement about some axis passing through this fixed point. For example, in the case under consideration the angle of rotation is $\angle bo_1 b'$ (or $\angle co_1 c'$) while the axis of rotation is oo_1.

It is obvious from the deducton made above that there always exists such an axis of rotation, say oo_1 (or $o_2 o_1$); also, the solutions to this axis as well as to the angle of rotation around it are unique, and their direction and magnitude can be determined solely in terms of the relative orientation of the initial and final places of the rigid body regardless of whatever the intermediate process of the motion might be.

Now, employing the conclusion just deduced and going back to the beginning of this section, we can make a general statement as follows:

Any displacement resulting from space motion of a rigid body (see Fig.1.3) may be reduced to a linear displacement $\overrightarrow{oo'}$ of the body to bring an arbitrary particle o of it, namely a base point, into its final location o', followed by an angular displacement θ of the body about a certain axis P passing through the base point o' (final location), to move the remainder of the rigid body to its final position.

As the displacement of the rigid body approaches zero as a limit, we can make the statement still simpler, as follows:

A general motion of a rigid body is equivalent to a translation in regard to an arbitrary particle o of it, called a base point, plus a rotation about some axis through the selected base point o.

Here we have replaced the word 'followed' by 'plus'.

One can also verify the equivalent displacement of the general motion of a rigid body from the viewpoint of degrees of freedom. It has been apparent that three parameters are required to specify such an equivalent displacement: the linear displacement of rectilinear translation $S = \overrightarrow{oo'}$, the direction of the axis of rotation P (unit vector), and the angle of rotation θ, through which the rigid body is rotated about this axis.

14

And these three parameters, in turn, can be expressed in terms of six independent generalized coordinates. Thus, the equivalent displacement concerned has the same number of degrees of freedom as that of the motion of a free rigid body.

It should be pointed out that even though both the initial and final positions of the rigid body have been given, the linear displacement S (or $\overrightarrow{oo}\,'$), and consequently the point o' for locating the axis of rotation all depend on the arbitrarily selected point o, whereas the orientation of the rotation axis P and the angle of rotation θ are independent of the choice of the base point o at all.

Therefore, a given change in position including location and orientation of a rigid body can be accomplished by an infinite number of combinations of linear displacement and angular displacement of the body. The number of versions of combination will still be doubled if we reverse the order of the linear displacement and the angular displacement by shifting the point of location of the rotation axis from o' to o. Of course, the order of constituent displacements will no longer be important as they approach zero as the limit.

Finally, we will again emphasize the fact that the change in orientation of a rigid body is only related to the orientation of the axis of rotation P and the angle of rotation θ about this axis, while it has nothing to do with the location of the axis P and the translation.

1.4.2 Equivalent Helical Motion

It has been readily foreseen from the discussion in the last paragraph that the equivalent displacement or the equivalent motion of a rigid body can still be simplified by changing the choice of a base point until the direction of the linear displacement S ($\overrightarrow{oo}'$), which varies with the choice, becomes parallel to the orientation of the rotation axis P, which remains unchanged once the finite space motion or the change in orientation of a rigid body has been given.

Thus, a helical motion is obtained corresponding to a finite space motion of a rigid body and for clarity it may be called the *equivalent helical motion*. In this situation, the axis of rotation is named the axis of screw, or just the *screw axis*. To distinguish this axis from the ordinary one, we particularly use a vector symbol $\overrightarrow{\lambda}$ to designate it.

It is evident that the screw axis $\overrightarrow{\lambda}$ remains parallel to the ordinary rotation axis P, so all what we have to do is to find the location of the screw axis, and also the linear displacement along this axis, which for short is called the *axial shift*.

As we did before, in Fig. 1.5 two rectangular parallelepipeds $oabcdefg$ and $o'a'b'c'd'e'f'g'$ are used to denote the initial and final locations and orientations of a moving rigid body. Also, it is possible to suppose that an equivalent displacement, as a combination of a linear displacement and an angular displacement, expressed in terms of a set of parameters S, P, θ as well as a corresponding base point o, or equally a point o', has been found and assigned in some way to represent the given finite space motion of the rigid body.

The base point o was arbitrarily selected in the body, so in nine cases out of ten, the corresponding linear displacement S we have here will not be parallel to the axis of rotation P.

Now, we need to derive the formula for determining a new set of parameters ($S_\lambda = \overrightarrow{hh}'$, $\overrightarrow{\lambda}$, θ, h or h') relying on the original set of parameters ($S = \overrightarrow{oo}'$, P, θ, o or o'), where h is

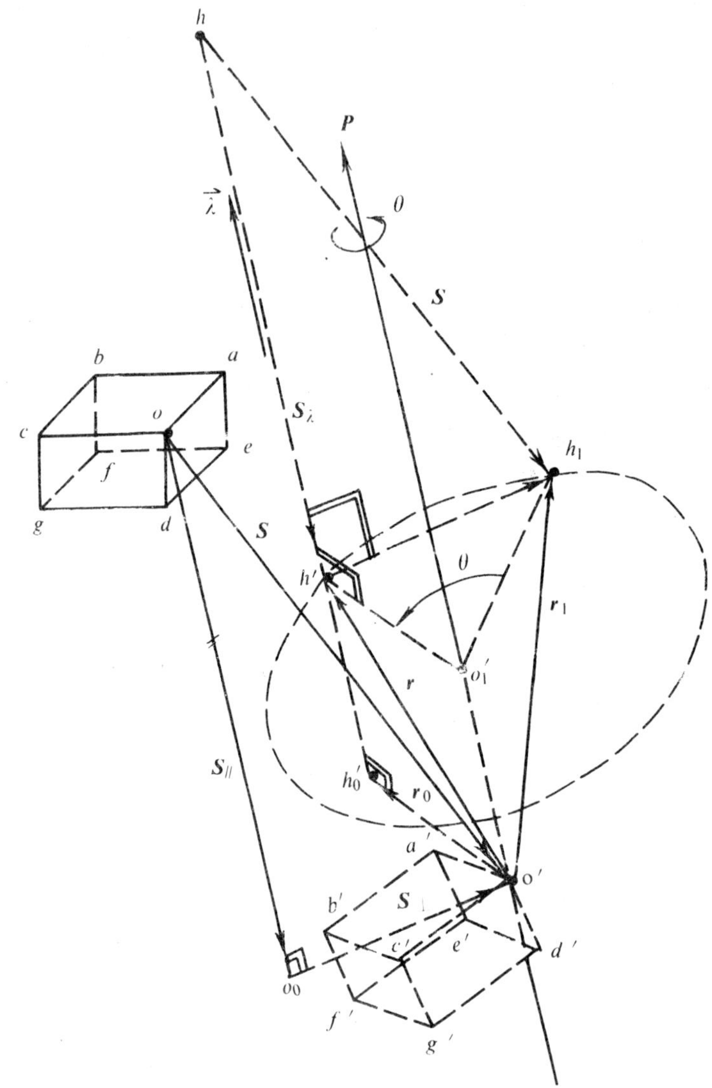

another point of the rigid body at the initial place, assigned to be a new base point of the body in such a way that the corresponding linear displacement S_λ or $\overrightarrow{hh'}$ becomes parallel to the axis of rotation $\vec{\lambda}$ (the latter two are actually coincident in this case), and consequently the new version of the equivalent displacement turns out to be an equivalent helical motion.

   To solve this problem, we have to locate the new base points h and h' of the rigid body at its initial and final places, respectively, by using the known parameters. Referring to Fig.1.5,

16

locate the point h at a certain position, then translate this point rectilinearly to its intermediate location h_1 by a linear displacement S equal to $\overrightarrow{oo'}$, and next revolve the intermediate point through an angle θ about the axis P, thus point h_1 moves along the circular path centered at o_1' to the final position h'.

Now, the vector drawn from h to h', denoted by $S_\lambda = \overrightarrow{hh'}$, will represent the linear displacement of the new version of equivalent motion corresponding to the base point h. According to the requirement prescribed above, the vector S_λ (or $\overrightarrow{hh'}$) must be parallel to P, and no doubt the unit vector $\vec{\lambda}$, directed from h' toward h and thus parallel to P, will be exactly the screw axis of the equivalent helical motion to be determined.

Draw vectors from o' toward h_1 and h', denoted by r_1 and r, respectively, and join their extremities by a vector $\overrightarrow{h'h_1}$. Here, the vector r can be considered as a vector resulting from turning the vector r_1 through an angle θ about the axis of rotation P. Drop a perpendicular from o' onto the screw axis $\vec{\lambda}$, cutting its extension at h_0', and designate the vector $\overrightarrow{o'h_0'}$ as r_0. Resolve S into components along and perpendicular to P, getting $S_{//}$ and $S_\perp$, as shown in the diagram.

Now, we obtain two congruent right triangles oo_0o' and $hh'h_1$, therefore we have

$$\overrightarrow{h'h_1} = S \quad . \tag{1.31}$$

It is easy to see that the above equation expresses the basic requirement, a starting point, from which one will be able to derive the formula to locate the base point h', whose position may be represented by the position vector $r = \overrightarrow{o'h'}$, where o' has been reasonably taken as an origin.

According to the rotation vector formula (1.9),

$$r_1 = r\cos(-\theta) + [1 - \cos(-\theta)](r \cdot P)P - \sin(-\theta)(r \times p) . \tag{1.32}$$

The left side of Eq. (1.31) can be obtained, subtracting r from r_1

$$\overrightarrow{h'h_1} = r_1 - r . \tag{1.33}$$

Putting r_1 from Eq. (1.32) into Eq. (1.33) gives

$$\overrightarrow{h'h_1} = (\cos\theta - 1)r + (1 - \cos\theta)(r \cdot p) + \sin\theta\,(r \times p) . \tag{1.34}$$

On the other hand, considering the geometrical meaning of the vector product, from the right triangle oo_0o', the right side of Eq. (1.31) can be expressed as a vector triple product

$$S_\perp = P \times (S \times P) - S - (S \cdot P)P . \tag{1.35}$$

Then, equating the right sides of both Eq. (1.34) and Eq. (1.35) yields

$$(\cos\theta - 1)r + (1 - \cos\theta)(r \cdot P)P + \sin\theta\,(r \times P)$$
$$= S - (S \cdot P)P . \tag{1.36}$$

Performing the vector product of both sides of Eq. (1.36) on the right by P, we obtain

$$[(\cos\theta - 1)r + (1 - \cos\theta)(r \cdot P)P + \sin\theta\,(r \times P)] \times P$$
$$= [S - (S \cdot P)P] \times P.$$

Still further by using the well-known formula of vector algebra,

$$(\cos\theta - 1)\,r \times P + \sin\theta\,(r \cdot P)\,P - r\sin\theta = S \times P \; . \tag{1.37}$$

Combining Eqs. (1.36) and (1.37) to solve for r yields

$$r = (r \cdot P)P + \frac{1}{2}[\,(S \cdot P)P - S] + \frac{1}{2}\left(\frac{\sin\theta}{\cos\theta - 1}\right)S \times P \; . \tag{1.38}$$

Now, introduce two parameters and define them below

$$p = (r \cdot P)\; , \tag{1.39}$$

$$r_0 = \frac{1}{2}[\,(S \cdot P)P - S] + \frac{1}{2}\left(\frac{\sin\theta}{\cos\theta - 1}\right)S \times P \; , \tag{1.40}$$

where p is a varying scalar parameter and r_0 a constant vector parameter, since S, P, and θ are given.

Then, substituting Eqs. (1.39) and (1.40) in Eq. (1.38), we arrive at

$$r = r_0 + p\,P \; . \tag{1.41}$$

It is clear from analytic geometry that the above formula (1.41) represents the

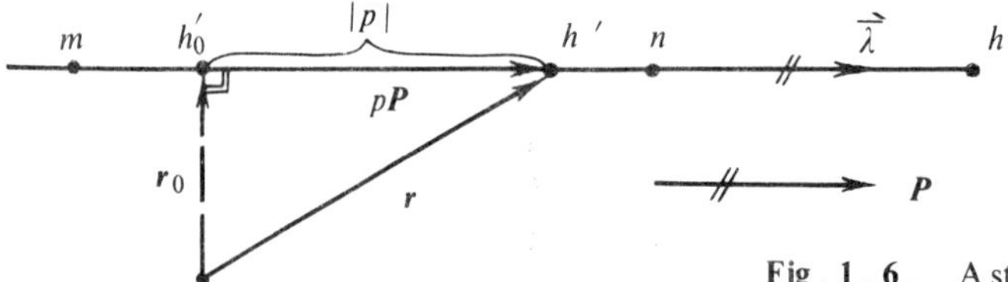

Fig. 1.6. A straight line parallel to a given vector P and passing through a given point h_0'

equation in vector form of a straight line parallel to a given vector P and passing through a given point h_0' determined by a position vector r_0, as shown in Fig. 1.6. An arbitrary point h', determined by this vector equation r, is moving on the straight line mn while changing the value of the parameter $p = r \cdot P$, and the vector r will be converted into r_0, as $p = 0$. Since the absolute value $|p|$ is a measure of the projection of r on P, r_0 will certainly be the vector-perpendicular dropped from o' to the line mn.

The illustration also indicates that there exist an infinite number of base points which correspond to an equivalent helical motion for a given finite space motion of a rigid body, however all these base points form a single straight line which is none other than the line of action of the required screw axis $\vec{\lambda}$.

To locate $\vec{\lambda}$, we prefer to utilize r_0 instead of the varying r. Since r_0 is uniquely determined by Eq. (1.40) in terms of a given set of parameters including S, P, θ, and a base point

o', there in general exists a unique solution of equivalent helical motion for a given finite space motion of a rigid body.

It then remains to give the formula of determining the axial shift of the equivalent helical motion. Let l denote this shift, then from Fig 1.5,

$$l = S \cdot P. \tag{1.42}$$

Now, we will investigate some special cases.

In the case of pure translation of a rigid body, putting $\theta = 0$ in Eq. (1.40) results in $r_0 = \infty$, i.e., the rotation axis is located at infinity.

In the case of pure rotation of a rigid body, we can always substitute $S = 0$ into Eq. (1.40) and then obtain $r_0 = 0$. This result implies that an arbitrary particle on the rotation axis has been chosen in the body as a base point, and thus the rotation axis itself is the screw axis. However, it is also clear from Eq. (1.42) that all equivalent motions of a rigid body in pure rotation need only to meet the following requirement: $S \cdot P = 0$ where S may not necessarily be zero but perpendicular to P.

It is instructive to point out that the same result, namely $r_0 = 0$, also happens when $S \times P = 0$ where S may not necessarily be zero but parallel to P, that is to say, the given set of kinematic parameters is already in accordance with an equivalent helical motion.

1.4.3 Addition of Angular Velocities

The method developed for the addition of angular velocities will play an important role in what follows, although we are not concerned about the instantaneous quantities themselves of a moving rigid body.

Suppose that a rigid body has simultaneously a number of angular velocities $\vec{\omega}_1$, $\vec{\omega}_2$, $\cdots$, $\vec{\omega}_n$, as shown in Fig. 1.7. In general, the action lines of these angular velocities are spatial and nonconcurrent.

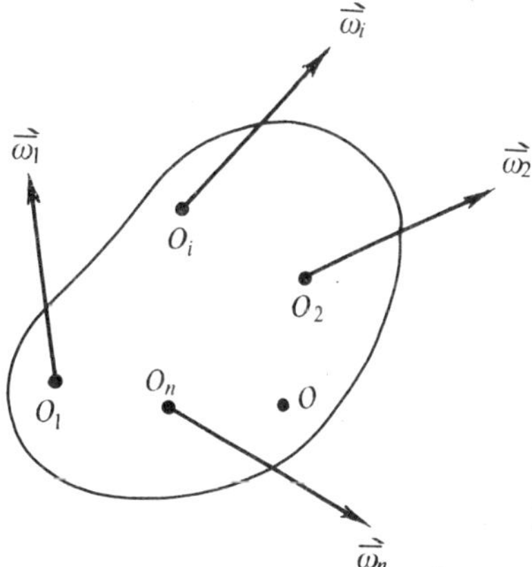

Fig.1.7. Addition of a number of angular velocities

To add up these angular velocities, as was done in the previous paragraph, here again we have to choose an arbitrary point O of the body as a base point.

Now , let us try to treat any one of the given angular velocities, for example $\vec{\omega}_i$. with respect to the selected base point O, illustrated in Fig. 1.8. Through O we add a pair of angular

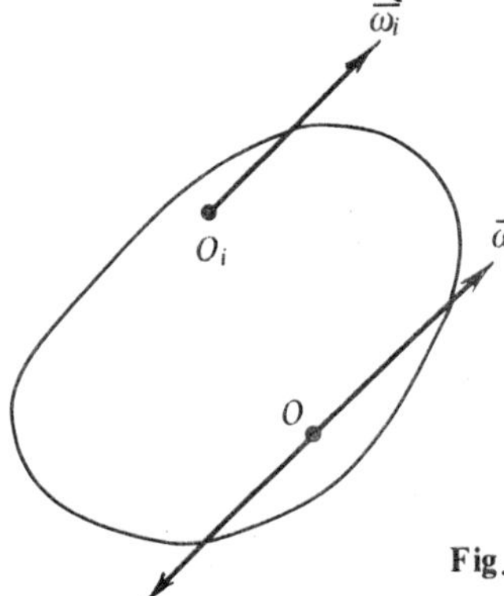

Fig.1.8 . Resolving an arbitrary angular velocity

with respect to a chosen point

velocities, opposite in sense , equal and parallel to $\vec{\omega}_i$. As this pair balances itself, no change takes place in the motion of the body due to the addition. Then, we have three angular velocity vectors. By rearranging them, the original angular velocity $\vec{\omega}_i$ passing through O_i has been resolved into two components, a couple of angular velocities constituted by $\vec{\omega}_i$ passing through O_i and $-\vec{\omega}_i$ passing through O, and another single angular velocity $\vec{\omega}_i$ just like the original but passing through O.

Next, we proceed to explore the physical meaning of a couple of angular velocities.

In Fig. 1.9, lettered to correspond with Fig. 1.8, let E be an arbitrary particle of the body. According to the well-known formula in kinematics, the velocity $v_{E,i}$ of particle E due to the

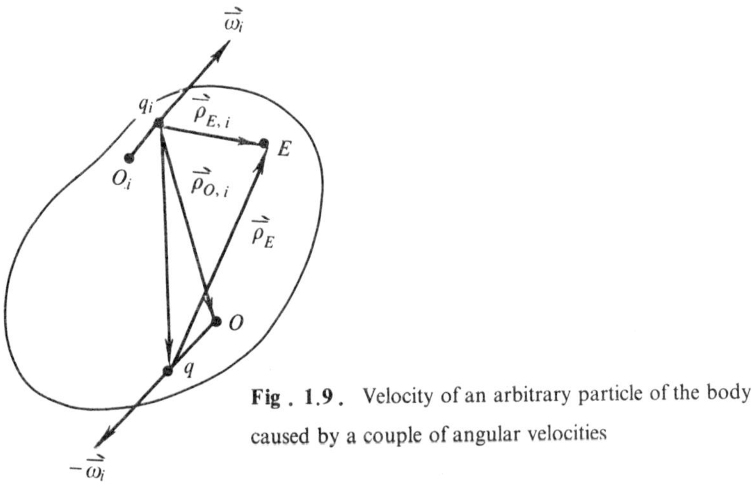

Fig . 1.9 . Velocity of an arbitrary particle of the body

caused by a couple of angular velocities

effect of the couple of angular velocities can be found as

$$v_{E,i} = \vec{\omega}_i \times \vec{\rho}_{E,i} + (-\vec{\omega}_i) \times \vec{\rho}_E = \vec{\omega}_i \times (\vec{\rho}_{E,i} - \vec{\rho}_E)$$

$$= \vec{\omega}_i \times \overrightarrow{q_i q} \, , \tag{1.43}$$

20

where q_i and q are arbitrary points on the action lines of $\vec{\omega}_i$ and $-\vec{\omega}_i$, respectively.

The resulting equation shows that the expression of $v_{E,i}$ is independent of the location of particle E; it holds for any point in the body. Therefore, a couple of angular velocities substantially represents a translation of the body that is subjected to the couple.

If q is replaced by the base point O, Eq. (1.43) can be rewritten

$$v_{E,i} = \vec{\omega}_i \times \overrightarrow{q_i O} = \vec{\omega}_i \times \vec{\rho}_{o,i} \; , \tag{1.44}$$

where $\vec{\rho}_{o,i}$ denotes the vector drawn from an arbitrary point q_i on the action line of the angular velocity in question to the selected base point O. Thus, the couple caused by $\vec{\omega}_i$ with respect to O is equal to the moment of $\vec{\omega}_i$ about O.

The procedures done above will be applied in turn to each of the other angular velocities. Eventually, we have two groups of components: one group, which consists of n concurrent angular velocities intersecting at the common base point O, and the other, which includes n couples of angular velocities.

The components in both groups can be composed in the usual way. Thus,

$$\vec{\Omega}_0 = \sum_{i=1}^{n} \vec{\omega}_i \; , \tag{1.45}$$

$$v_0 = \sum_{i=1}^{n} \vec{\omega}_i \times \vec{\rho}_{o,i} \; , \tag{1.46}$$

where $\vec{\Omega}_0$ represents the resultant angular velocity passing through O, and v_0 the resultant

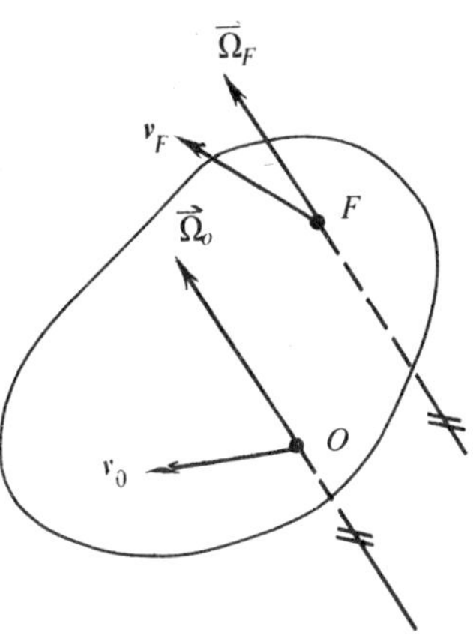

Fig. 1.10. Resultant linear and angular velocities with respect to two different base points O and F

linear velocity of the base point O. It should be noted that these two resultant velocities completely specify the motion of any point in the rigid body as they in fact represent a superposition of translation and rotation of the whole body.

Obviously, Eqs. (1.45) and (1.46) are valid for the chosen base point O only. If considering another base point F, as shown in Fig. 1.10, we have accordingly

$$\vec{\Omega}_F = \sum_{i=1}^{n} \vec{\omega}_i \; , \tag{1.47}$$

$$v_F = \sum_{i=1}^{n} \vec{\omega}_i \times \vec{\rho}_{F,i} \ , \tag{1.48}$$

where we can easily see that $\vec{\Omega}_F$ is exactly equal to $\vec{\Omega}_0$ but passing through different points, while v_F in general differs from v_0.

1.4.4 Addition of Small Rotations

Suppose that a rigid body has accomplished in succession a number of small rotations or slight tilts about the axes of rotation P_1, P_2, $\cdots$, P_n through small angles $\Delta\theta_1$, $\Delta\theta_2$, $\cdots$, $\Delta\theta_n$, respectively. It is then required to find the resultant motion of the body.

A hint possibly given us from Eq. (1.15) by comparing it with Eqs. (1.9) and (1.44) is that we can employ ordinary vectors $\Delta\theta_1 P_1$, $\Delta\theta_2 P_2$, $\cdots$, $\Delta\theta_n P_n$ to represent these small rotations, or small angular displacements, and add them up according to the conventional rules of vector algebra as they are considered to be vectors of angular velocities. This method of treatment is said to be reasonable if errors of second and higher powers introduced thereof are practically negligible.

Similarly, taking a convenient particle O of the body as a base point, we obtain

$$\Delta\vec{\theta}_o = \sum_{i=1}^{n} \Delta\theta_i P_i \ , \tag{1.49}$$

$$\Delta S_o = \sum_{i=1}^{n} \Delta\theta_i P_i \times \vec{\rho}_{o,i} \ , \tag{1.50}$$

where $\Delta\vec{\theta}_o$ represents the vector of small resultant angular displacement passing through the base point O, and ΔS_o the vector of small resultant linear displacement of O .

To end this section, it is instructive to point out that the student is warned against using such a symbol as 'θP' to denote a finite rotation of a rigid body, because substantially a finite angular displacement is by no means a vector; especially, this conceptually unsuccessful notation might lead to critical mistakes if he really takes it as a vector in calculation.

1.5 Gradient

Let a certain physical quantity u be a scalar function of position in space. Thus, $u = u\ (x,y,z)$ or the alternatives $u = u\ (r)$ and $u = u\ (q)$ are called a *scalar field*, where x,y,z are the rectangular coordinates of an arbitrary point q in space and r the position vector of q.

In order to learn the rate of variation of the quantity u in all directions at any point in space, we start with the term *directional derivative* defined as the rate of variation of u in an arbitrary direction l and denoted by $\dfrac{\partial u}{\partial l}$.

It is easily understood that such a derivative can be formulated as follows:

$$\frac{\partial u}{\partial l} = \frac{\partial u}{\partial x}\cos\alpha + \frac{\partial u}{\partial y}\cos\beta + \frac{\partial u}{\partial z}\cos\gamma \ , \tag{1.51}$$

where $\cos\alpha$, $\cos\beta$ and $\cos\gamma$ are the direction cosines of the vector l.

Now, introduce a new term, the gradient of u, denoted by 'grad u' or G_u and defined by the following expression:

$$G_u = \mathrm{grad}\ u = \frac{\partial u}{\partial x}\,i + \frac{\partial u}{\partial y}\,j + \frac{\partial u}{\partial z}\,k\ . \qquad (1.52)$$

Comparing this with Eq. (1.51), we find that the derivative of u in the direction of l is equal to the scalar product as

$$\frac{\partial u}{\partial l} = G_u \cdot l^0 = |G_u|\cos(\widehat{G_u,l})\ , \qquad (1.53)$$

where l^0 represents the unit vector along the direction of l.

From Eq. (1.53), we can see that once l becomes parallel to grad u, the directional derivative has a maximum

$$\left.\frac{\partial u}{\partial l}\right|_{l\parallel G_u} = |G_u|\cos 0^0 = |G_u| = \left(\frac{\partial u}{\partial l}\right)_{max}\ . \qquad (1.54)$$

The resulting formula tells clearly that the gradient grad u of a scalar field $u(q)$ is a vector function that indicates a direction in which the rate of variation of the scalar function $u(q)$ reaches the extreme values (the maximum as well as the minimum) at a point q; and meanwhile these extreme values of rate of variation are exactly equal to the magnitude (or modulus) of the gradient $|\mathrm{grad}\ u|$.

Problems

1.1   Prove that the scalar product of two vectors A and B can be expressed in matrix form as

$$(A \cdot B) = (A)'(B) = (B)'(A)\ . \qquad (P1.1)$$

Note: The right part of Eq. (P1.1) is correct if and only if both (A) and (B) are column matrixes, and in the general case,

$$C'D = (D'C)'\ .$$

1.2   Transform the component formula from the vector form $A_p = [A \cdot P]\,P$ into the matrix form $(A_p) = M_p(A)$ so as to determine the projection matrix M_p corresponding to the unit vector P.

Hint: Use Eq. (P1.1).

1.3   Derive the matrix equation to represent the vector product of two vectors $D = A \times B$ in the following form

$$(D) = C_A(B) \qquad (P1.2)$$

giving the cross-product-matrix C_A of the vector A in detail.

1.4   Verify the expression of the rotation matrix $S_{P,0}$ in Eq. (1.12) by applying the results, obtained from the previous problems, to Eq. (1.9).

1.5   Suppose that a universal joint (Hooke's joint) is used to unite two shafts making an angle of α. If the driving shaft turns through a known angle θ, determine the angle φ through which the driven shaft is caused to turn, by using the formula of scalar product instead of graphical treatment.

Hint: two pivot axes of the coupling member are cross each other at right angle.

Part Ⅱ

Theory of Conjugation for Reflecting Prisms

Chapter 2
Image Formation for Reflecting Prisms

As mentioned in the Preface, the theory of conjugation for reflecting prisms contains methods and laws governing the relationship between the object and image spaces for a prism regardless of whether the prism is stationary or moving. However, to distinguish them from each other, the former case is customarily called the image formation while the latter is called the image motion, although the image formation can be referred to as a special case of the problem of image motion and the image motion may also be regarded as a problem of image formation in a broad sense.

We also realize that the theory of image formation for a prism is the more basic knowledge and is part of the theory of image motion caused by a moving prism.

Therefore, we may say that the material included in Chapters 1 and 2 is devoted to providing a theoretical foundation for this text in the following three aspects: mathematical, kinematical, and optical.

2.1 Terminology

For the convenience of the later discussion, we will first define some terms.

Entrance face It denotes the first refracting surface at which the entering ray is refracted into the prism, as designated by *abcd* in Fig. 2.1.

Exit face It denotes the last refracting surface at which the emerging ray is refracted away from the prism, as indicated by *defg* in the figure.

Optical axis It represents the path of the axial ray of a cylindrical beam entering perpendicular to the entrance face of the prism, then passing through it, and then emerging perpendicular to the exit face, as signified by 1 2 3 4 5 6 7 in Fig. 2.1.

Entrance optical axis It is a part of the optical axis before reflection, i.e., the incident part 12.

Exit optical axis It is a part of the optical axis after reflection, i.e., the emergent part 67.

The last three definitions regarding optical axes are restricted to those reflecting prisms which can be used in either collimated light or convergent (divergent) light. For some other reflecting prisms used in a collimated beam only, their optical axes should be treated individually. For instance, the optical axis of a Dove prism is illustrated in Fig. 2.2.

As a matter of fact, any suchlike individual prism does have a definite optical axis as long

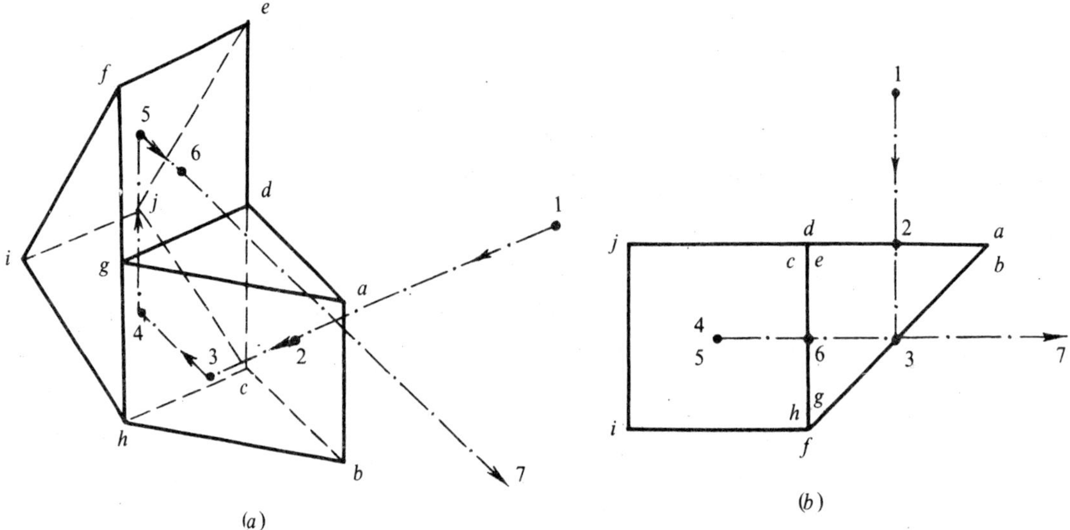

(a)

(b)

Fig .2.1. An arbitrary reflecting prism illustrating
some specialized terms of a prism

Fig.2.2. Optical axis of a Dove prism

as it has been applied in combination with a certain lens system, and this is supposed to be
most cases of the usage of reflecting prisms.

Optical-axis plane   A plane is said to be an optical-axis plane if it is simultaneously parallel to the entrance and exit optical axes. It is clear from the definition that the optical-axis
plane never appears singly, therefore any reflecting prism has at least a family of optical-axis
planes parallel to one another, and simultaneously parallel to both the entrance and exit optical
axes. Consequently, for those prisms where the exit optical axis happens to be parallel to or
coincident with the entrance optical axis, there exist an infinite number of such families of optical-axis planes. Since the entrance optical axis and the exit optical axis of the Dove prism are
coaxial, it can also be taken as an example showing an infinite number of families of optical-axis planes; each family oriented differently but all parallel to either the entrance optical axis or the exit optical axis of it .

Conjugate optical-axis plane Let a certain optical-axis plane be an object plane. If
the corresponding image plane formed by the reflecting prism under discussion is found to be
parallel to that optical-axis plane, then the latter is said to be a conjugate optical-axis plane.
This definition can readily be extended to the whole family of planes parallel to that optical-axis

plane, therefore we have also a family of conjugate optical- axis planes.

Planar reflecting prism and spatial reflecting prism A reflecting prism is said to be a planar reflecting prism if it contains at least one family of conjugate optical-axis planes, while all the others are spatial reflecting prisms. In accordance with this definition, a spatial reflecting prism does not have any conjugate optical- axis plane. Some planar reflecting prisms have only one family of conjugate optical- axis planes as their entrance optical axis is neither parallel to nor coincident with their exit optical axis; as an example, we can also take the prism shown in Fig. 2.1 . Some planar reflecting prisms have more than one family of conjugate optical-axis planes when the exit optical axis is either parallel to or coincident with the entrance optical axis. For instance, the Dove prism possesses two families of conjugate optical- axis planes, one including all the planes parallel to the paper of that figure (2.2), and the other including all the planes perpendicular to the paper and parallel to either the entrance optical axis or the exit optical axis of the prism. Another example, the corner prism, as shown in Fig. 2.3, is possessed of an infinite number of families of conjugate optical- axis planes which are parallel to either the entrance optical axis or the exit optical axis of the prism.

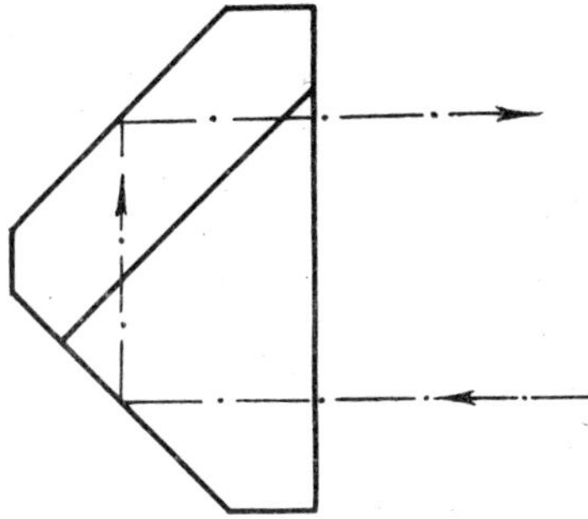

Fig . 2.3. A corner prism indicating an infinite number of families of conjugate optical- axis planes

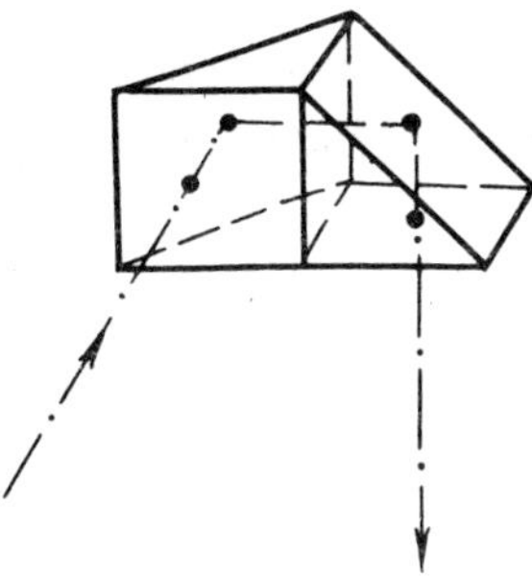

Fig.2.4. A combination of two rectangular prisms as an example of a spatial reflecting prism

A compound prism indicated in Fig.2.4 can be taken as an example of a spatial prism.

Optical- axis section If the entrance optical axis and the exit optical axis of a reflecting prism are either intersecting or parallel to each other, then the plane formed by these two axes is called the optical- axis section of the prism. Hence, obviously some reflecting prisms have an infinite number of optical- axis sections as their exit optical axis is coincident with the entrance optical axis. For example, a Pechan prism as shown in Fig. 2.5 belongs to the case just mentioned.

Conjugate optical- axis section If the image plane of the optical- axis sectional plane formed by the reflecting prism under discussion appears to be the optical- axis section itself, then the latter is said to be the conjugate optical- axis section.

The last two definitions regarding optical- axis sections are restricted to those reflecting prisms which can be used in either collimated light or convergent (divergent) light, because it will become clear from the later chapters that for other reflecting prisms used in a collimated

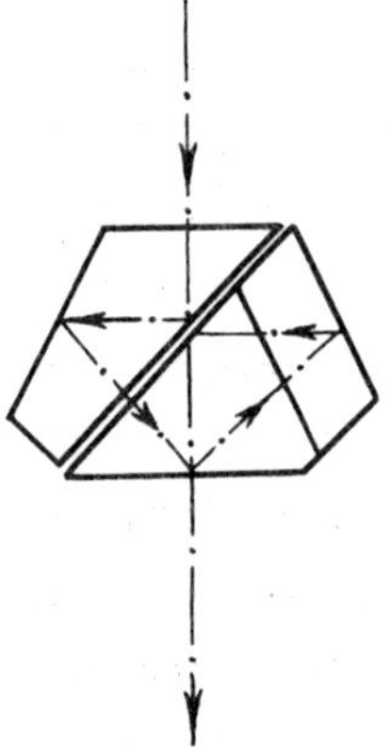

Fig. 2.5. A Pechan prism indicating an infinite number of optical-axis sections passing through either the entrance or the exit optical axis

beam only, the conjugate optical-axis section does not mean anything more than the conjugate optical-axis plane does.

Coplanar reflecting prism A reflecting prism is said to be a coplanar reflecting prism if it contains at least one conjugate optical-axis section. In other words, a coplanar reflecting prism is a planar reflecting prism whose entrance optical axis and exit optical axis are lying in the same plane. The rest of the planar reflecting prisms where the entrance optical axis and the exit optical axis are neither intersecting nor parallel to one another may thus be called ordinary planar reflecting prisms.

Even number (e.n.) reflecting prism For short, a reflecting prism with an even number of reflecting surfaces may be named the even number reflecting prism.

Odd number (o.n.) reflecting prism Similarly, the odd number reflecting prism represents a reflecting prism with an odd number of reflecting surfaces.

Object body and image body When a reflecting prism is used alone as an optical system, the object and the corresponding image formed by the reflecting prism may be considered three dimensional, therefore we adopt the terms *object body and image body* to stress their spatial nature.

The readers are not asked to understand all the definitions listed above very well until they have read all the chapters in Part Ⅱ.

2.2 Structural Analysis of a Prism in Regard to Its Image Formation

2.2.1 The Optical Tunnel

A prism consists essentially of two refracting surfaces and one or more reflecting

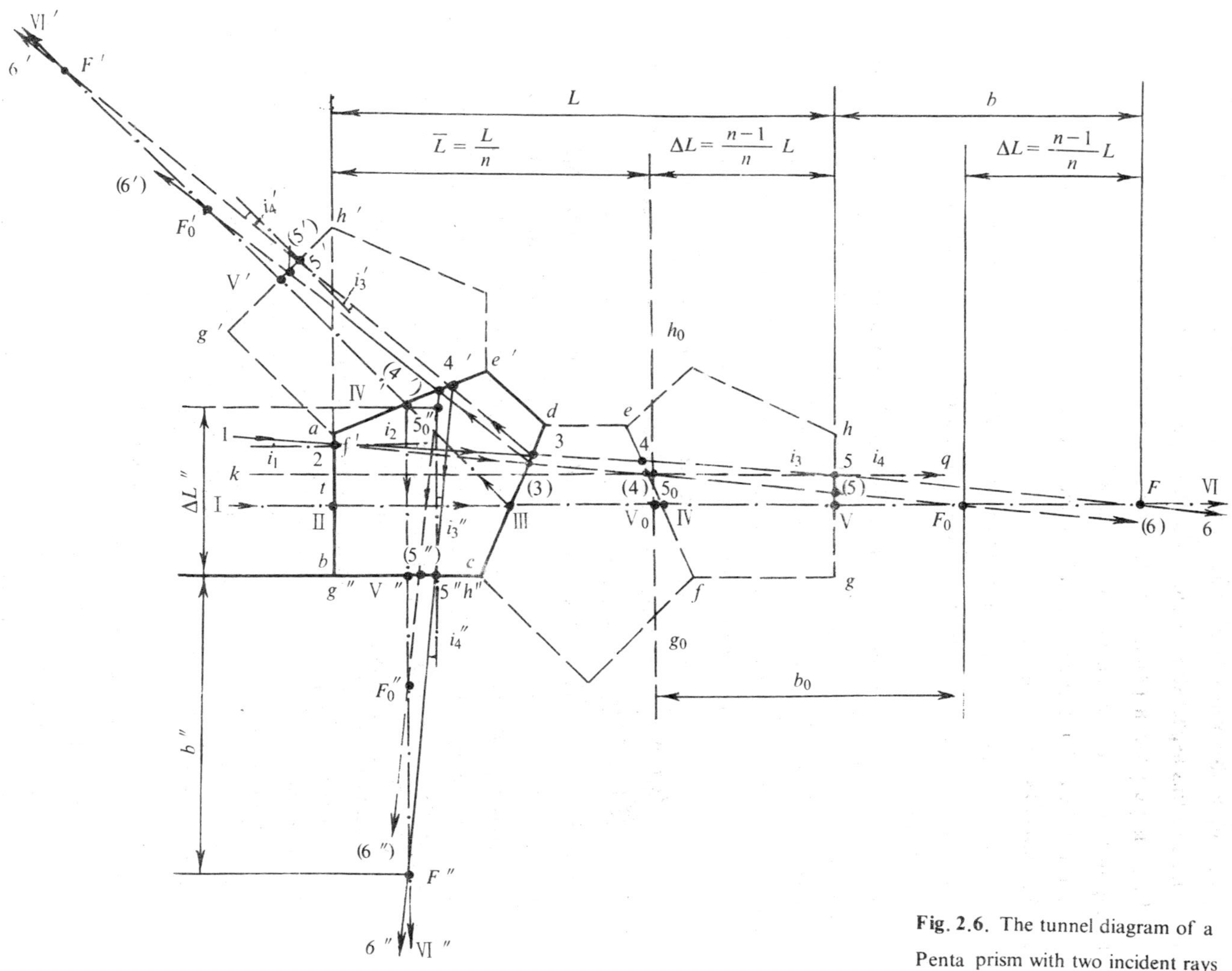

Fig. 2.6. The tunnel diagram of a Penta prism with two incident rays

surfaces. For the convenience of later discussions,we may distinguish two parts of a prism, the refracting part and the reflecting part, the latter acting merely as a mirror system.

To examine the principle of image formation for a prism, we will begin with the contributions made to this aspect due to each of the two parts of the prism.

For this purpose, a prism can be unfolded by reflecting it successively in each of the reflecting surfaces, and it will generate a so-called *tunnel diagram*.

Let us take a Penta prism as an example, without any loss of generality, to see how to obtain the *optical tunnel* of this prism.

In Fig. 2.6 , suppose that the Penta prism is denoted by $abcde'f'g''h''$, where ab represents the entrance face, cd the first reflecting surface, $e'f'$ the second reflecting surface, and $g''h''$ the exit face.

To aid in acquiring a better understanding of the unfolding process, we introduce first an axial incident ray I II , which enters perpendicular to the entrance face ab and passes through the prism along the path III IV $'$V $''$VI $''$. This one is none other than tracing the optical axis of the prism. Moreover, we assume an oblique incident ray 12, which travels through the prism along the path $34'5''6''$. The emergent rays $5''6''$ and $V''VI''$ meet at the point F''. Also the same incident ray 12 would travel along the other path $(3)(4')(5'')(6'')$ if only the reflecting part of the prism were taken into consideration. In the latter case, no refractions happen at all. Similarly, the emergent rays $(5'')(6'')$ and $V''VI''$ meet at the point F_0''.

It should be noted that all the symbols with double prime '$''$' at the right upper corner, in this case, represent the configuration including points, lines, and planes, which is located in the image space of the reflecting part of the prism as well as of the prism itself,symbols with a single prime "$'$" represent the configuration which is located in the image space of the reflecting surface cd, and plain symbols (without a prime) represent the configuration which is located in the object space of the reflecting part of the prism as well as of the prism itself.

Now, we may proceed to unfold the prism in the following way.

Firstly, consider the configuration, including the succeeding rays and points after the last reflecting surface $e'f'$, namely, IV $'$V $''$VI $''$, $4'5''6''$, $(4')(5'')(6'')$, F_0'', F'', and the exit face itself $g''h''$, as an image located in the image space of this mirror surface, and then find in it the corresponding object (probably virtual object), getting consequently IV $'$V $'$VI $'$, $4'5'6'$, $(4')(5')(6')$, F_0',F' and $g'h'$.

The same treatment will be done at each successive reflecting surface in a reverse order. In this case, we will right deal with the first reflecting surface cd.

Now, the configuration, including III IV $'$V $'$VI $'$, $34'5'6'$, $(3)(4')(5')(6')$, F_0',F' and $g'h'$, is considered as an image located in the image space of the mirror surface cd, and hence the corresponding object (most likely virtual) in the mirror can easily be found to be III IV V VI , 3456, $(3)(4)(5)(6)$,F_0 , F and gh.

Let us stress the results which have been obtained from unfolding the prism.

The reflecting part of the prism is transformed into a portion of 'air', that is to say, it becomes nothing but the medium surrounding the prism. Hence, the ray path 12 $(3)(4')(5'')(6'')$

is converted to a straight line $12(3)(4)(5)(6)$, as if the incident ray 12 passes through the 'air' without any deviation.

As a result the prism itself turns out to be a plane-parallel plate $abgh$, or in other words, all the reflecting surfaces disappear and the exit face $g''h''$ of the prism is converted to the exit face gh of the plane-parallel plate. The path of the axial incident ray $\text{I}\ \text{II}\ \text{III}\ \text{IV}'\text{V}''\text{VI}''$ becomes a straight line $\text{I}\ \text{II}\ \text{III}\ \text{IV}\ \text{V}\ \text{VI}$ as if a perpendicularly incident ray will penetrate straight through a parallel plate, neither laterally displaced nor deviated. The path of the oblique incident ray $1\ 2\ 3\ 4\ '5''6''$ becomes a broken line 123456, where the emergent ray 56 keeps parallel to the incident 12, although laterally displaced, as if it happens while an oblique incident ray passes through a parallel sided glass block.

The reasoning for being able to unfold a reflecting prism into a plane-parallel plate is simply that the insertion of a prism in imaging systems should not cause any color preferences.

It is well known that the plane-parallel plate can still be reduced to an equivalent air plate (or customarily called equivalent air thickness). As indicated in Fig. 2.6, this can be done by drawing a line ktq passing through the exit point 5 of the emergent ray 56 and parallel to the straightened optical axis $\text{I}\ \text{VI}$, cutting the extension of the incident ray 12 at 5_0, and then from the point 5_0 draw a plane g_0h_0 perpendicular to $\text{I}\ \text{VI}$. Thus, we obtain the equivalent air plate abg_0h_0.

No doubt, rays of light pass through this equivalent air plate in straight lines. In this case, two refractions at the entrance and exit faces vanish as well.

Let L and $\overline{L}$ denote the lengths of the plane-parallel plate and the equivalent air plate, respectively, and ΔL the axial image shift caused by the plane-parallel plate. For the paraxial rays, we have

$$\Delta L = \frac{n-1}{n}\, L \tag{2.1}$$

or

$$\overline{L} = \frac{L}{n}\,, \tag{2.2}$$

where n denotes the refractive index of the glass.

It is worthwhile to examine the above two processes, unfolding and reducing, in more detail.

Substantially, unfolding is a process to cancel all the reflecting effect of a prism; also it is considered as a reverse process of image formation of the reflecting part of a prism, so that any figure in the image space of the relfecting part will be converted into its correspondent in the object space of that part. In this way, the exit face $g''h''$ of the prism is brought back to the object space of the reflecting part and is transformed into the face gh, which combines then with the entrance face ab of the prism to form a plane-parallel plate.

To get also an insight into the process of reducing the plane-parallel plate again to an equivalent air plate, a relevant part of the drawing in Fig. 2.6 has been singled out and reproduced in Fig. 2.7.

Assume here the plane-parallel plate *abgh* to be an optical system. Two incident rays 12 and *kt* meet at point 5_0, which is taken as an object point. After passing through the glass plate, two corresponding emergent rays 56 and 5*q* meet at point 5, which will thus be the

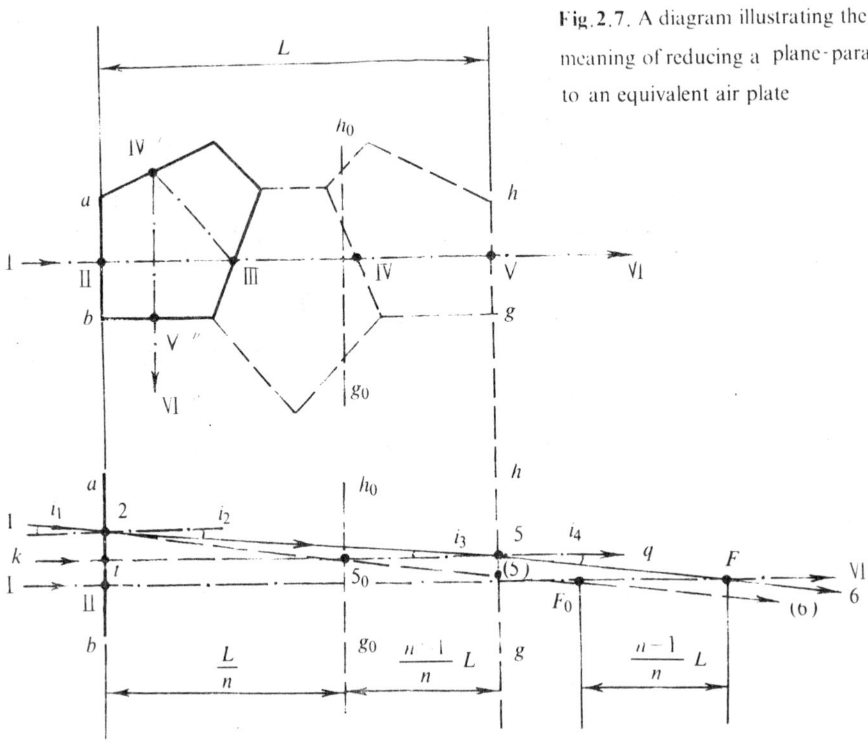

Fig.2.7. A diagram illustrating the physical meaning of reducing a plane-parallel plate to an equivalent air plate

conjugate image point of the object point 5_0 formed by the plane-parallel plate. Consequently, reducing is also a reverse process to convert any figure in the image space of the plane-parallel plate into its correspondent in the object space of that glass plate by simply a longitudinal shift along the entrance optical axis but in opposite direction. In this way, the exit face *gh* of the plane-parallel plate is still further brought to its object space and is thus transformed into the exit face $g_0 h_0$ of the equivalent air plate.

To sum up, by the unfolding and the subsequent reducing processes, rays starting from an image in the image space of a prism will be traced through the prism in the reverse direction in order to find the corresponding object in the object space. On the other hand, a quite effective measure, which has been adopted in unfolding, is to ignore the refraction at the exit face of the prism until the face has been brought to the object space of the reflecting part, so that the sandwich structure[*] of the prism has been transformed into a much more easily handled form.

According to the point developed above, we may resolve a prism structurally into two constituent systems, and as far as the effect of image formation is concerned we may relate them to

---

[*] A reflecting system is sandwiched between two refracting surfaces.

34

each other in the following logical equation

$$\text{Prism} = \text{Plane-parallel plate (in the object space of the reflecting part)}$$
$$\longrightarrow \text{Reflecting part},\tag{2.3}$$

where the special operation symbol '$\longrightarrow$' represents a combination of two optical systems in series arranged in such an order that the system appearing at the position before the operation symbol in Eq. (2.3) will be the preceding one in the series, as indicated by the numerical order in Fig. 2.8.

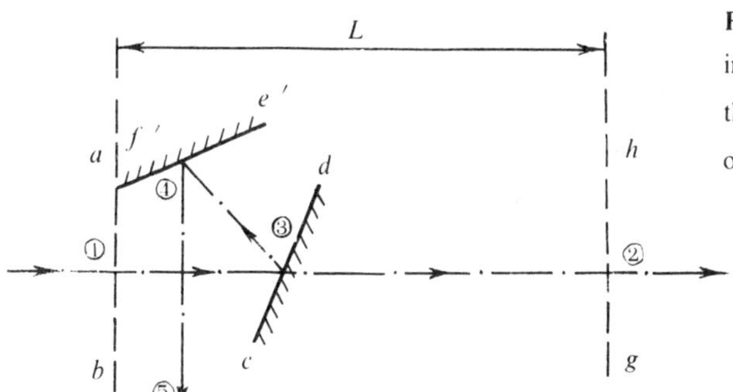

Fig. 2.8. Resolving a Penta prism into two constituent systems with the plane-parallel plate in the object space of the reflecting part

It should be noted that the combination of two optical systems on the right side of Eq. (2.3) is said to be equal to the prism only in the sense of that the combination will be able to maintain the original relationship between two conjugate spaces with respect to the prism regardless of the fact that the same combination can no longer truly describe the ray path within the prism.

Obviously, if unfolding the prism to the image space of its reflecting part, we will obtain a similar logical equation as

$$\text{Prism} = \text{Reflecting part} \longrightarrow \text{Plane-parallel plate}$$
$$\text{(in the image space of the reflecting part)}.\tag{2.4}$$

The structural resolution of a Penta prism according to this idea is illustrated in Fig. 2.9, where the plane-parallel plate is denoted by $a''b''g''h''$.

It is evident from both Eqs. (2.3) and (2.4) that the image-forming effect of the refracting part of a prism is equivalent to a plane-parallel plate; besides, it is necessary to specify the space in which the plate is formed. However, the refracting part of a prism will cause an image shift only, and any change of image orientation is due to the reflecting part. Hence, if a prism is used in a collimated beam, we need not consider two refractions while discussing the image formation.

The unfolding and reducing processes as well as their corresponding concepts, the optical tunnel and the equivalent air thickness, were originally introduced for the purpose of determining prism sizes and to aid in visualizing or sketching ray paths through the prism. The idea of an equivalent air plate has proved to be of great usefulness from the fact that ray paths can be accurately constructed through the system by merely drawing straight lines without

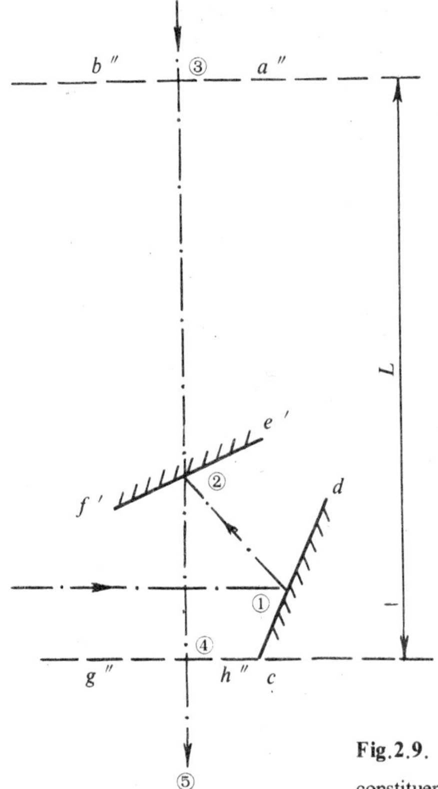

Fig.2.9. Resolving a Penta prism into two constituent systems with the plane-parallel plate in the image space of the reflecting part

considering neither reflections nor refractions.

Now, the method and concepts developed for determining prism sizes will be extented to the analysis of image formation for prisms, which ought to be considered as a more basic problem so long as they are used as components in imaging systems. Also, the more you have been familiar with the image formation of prisms, the better you will know how to determine their sizes.

Either of Eqs. (2.3) and (2.4) can be used to find the image formed by a prism with the object given.

Referring to Eq. (2.3), we can first find the image of the given object formed through the plane-parallel plate by simply shifting the object along the entrance optical axis of the prism by

$$\Delta L = \frac{n-1}{n} L \;,$$

and then find again the image of this shifted object formed by the reflecting part, thus obtaining the final image of the original object with respect to the prism as a whole.

While referring to Eq. (2.4), we can first find the image of the given object formed by the reflecting part, and then the obtained image will still be shifted along the exit optical axis by

$\Delta L'' = \dfrac{n-1}{n} L$ caused by the plane-parallel plate, thus obtaining the final image required.

Since a shift ΔL along the entrance optical axis is imaged by the reflecting part as a shift $\Delta L''$ along the exit optical axis, the results from both two Eqs. (2.3) and (2.4) will be the same.

The above conclusion governing the general relationships of image formation for prisms can also be verified alternatively as follows:

In Fig. 2.6, prior to unfolding we have two optical systems, one being a prism and the other its reflecting part; after unfolding, these two optical systems have become a plane-parallel plate and 'air', respectively. Hence, there are altogether four optical systems.

It was clearly seen from the drawing that we intentionally introduced a figure consisting of two incident rays 12 and I II with their intersection point F_0 which can be referred to as an arbitrary object common for the four optical systems. Now, each of the four optical systems will form a corresponding image of its own. Obviously, that must be the emerging portions of two prescribed incident rays, and their intersection. For the 'air' we have the image as two emerging rays (5) (6) and (V) (VI) meeting at the point F_0; for the plane-parallel plate two emerging rays 5 6 and V VI meeting at the point F; for the reflecting part two emerging rays (5'')(6'') and (V'')(VI'') with their intersection point F_0''; finally for the prism itself two emerging rays 5'' 6'' and V'' VI'' with the point of intersection F''. By what has essentially happened during the unfolding process, we realize that the image formed by the 'air' and the image formed by the reflecting part appear to be a pair of conjugates for the reflecting part when the same object is involved, the former being the object and the latter the image; the image formed by the plane-parallel plate and the image formed by the prism constitute also a pair of conjugates for the reflecting part relative to the same object, the former being the object and the latter the image. Therefore, the same relationship will hold for the difference in position $\overrightarrow{F_0 F}$ between two images in the object space and the difference in position $\overrightarrow{F_0'' F''}$ between two corresponding images in the image space of the reflecting part.

One of the purposes of this section is to find the image of a given object for a prism. We presume that the image of the same object for the reflecting part of the prism has already been determined by some means or other. Then all we need to know in addition is the difference in position between two images formed relative to the same object by the prism and its reflecting part, respectively. This can be done either directly by calculating the image shift $\overrightarrow{F_0'' F''}$ in the image space of the reflecting part along the exit optical axis of the prism or indirectly by computing the image shift $\overrightarrow{F_0 F}$ in the object space of the reflecting part along the entrance optical axis and then transforming it again into the image space of the reflecting part, so that we will go back to Eqs. (2.4) and (2.3).

These equations can also be used in the reverse direction. For instance, by Eq. (2.3) we should find the corresponding object of the given image for the reflecting part first, and then shift the obtained object along the entrance optical axis but in the opposite direction, thus getting the original object.

The tunnel diagram is also very useful in detecting the presence of unwanted reflections.

The optical tunnel of a Penta prism is shown again in Fig. 2. 10 , where the non-working surface *de ′* and the second reflecting surface *e′f ′* are imaged in the first reflecting surface, as a mirror, at *de* and *ef* (actually virtual object).In the diagram, a ray 12 strikes the entrance face

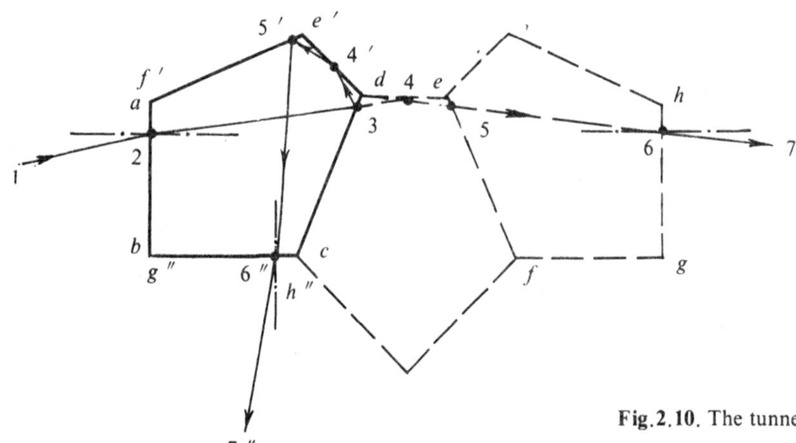

Fig.2.10. The tunnel diagram for a
Penta prism illustrating a ghost ray

at greater obliquity. Its refracted ray 23 is reflected off the plane *de* and cut *ef* again in the optical tunnel, therefore the actual ray passing through the prism will be reflected at the non-working surface *de ′* and consequently subjected to three reflections. Since the prism is intended to be used with two reflections, rays like 12 with the extra reflection are called ghost rays. The ghost reflections may be eliminated in some cases by cutting a notch in prisms (Ref. 14).

It seems to be instructive to make a few remarks upon the wording. To be consistent with those published books on optical engineering, we have also used the word 'unfold' in this text while discussing the process to form an optical tunnel of a prism. However, a real fear is in our minds that there might be some misunderstanding when such wording is introduced. If people explained the word 'unfold' here as a procedure to straighten the prism 'axis' at each reflecting surface simply by rotating the reflected axis and the rest of the prism about the folding axis formed by the intersection of the plane of incidence (usually the paper of the diagram) with the reflecting surface, it would not be very strict. We all know that a mere rotation can not change the handedness of an image while imaging through a single mirror will do so .

2.2.2 A Pair of Basic Conjugate Points — Theorem of Image Location for Reflecting Prisms

It has been mentioned in the previous paragragh that we need to find the image of a given object formed by the reflecting part of a prism in order to get the final image formed by the prism. However, such a treatment does not make the problem any the less difficult since how to solve the question of image formation for the reflecting part of a prism completely, i. e. , the problem of image formation for a pure mirror system, still remains a complicated one.

To simplify the method and also to meet the need for a sound theoretical development,

38

we are now trying to define a pair of conjugate points for an arbitrary prism.

For this purpose, let us apply the logical Eq. (2.3) to one of the structural elements of a prism in the reverse direction. Suppose the exit face, say $g''h''$, Fig. 2.6 or 2.7, is considered as an image plane in relation to the Penta prism. To determine the corresponding object plane of the image plane $g''h''$ for the prism, we find first the object plane gh corresponding to the image plane $g''h''$ for the reflecting part of the prism, and then shift the plane gh further by $\Delta L = \dfrac{n-1}{n} L$ along the entrance optical axis but in the opposite direction, thus getting the original object plane $g_0 h_0$ for the prism. Therefore, we come to a general conclusion stated as follows:

"The exit face $g_0 h_0$ of the equivalent air plate of a prism and the exit face $g''h''$ of the prism itself form a pair of conjugate planes for the prism; the former being the object plane, the latter the image plane."

To prove this by logically formulated deduction, introduce a special symbol 'object $\xrightarrow{\text{optical system}}$ image' to relate a pair of conjugate object and image to the optical system concerned, and let the abbreviations PR, RP, PPP, EAP, O.S., and I.S. denote the terms 'Prism, Reflecting part, Plane-parallel plate, Equivalent air plate, Object space, and Image space', respectively. Then, the results obtained in the previous paragraph can be written as

$$g_0 h_0 \xrightarrow{\text{PPP in O.S. of RP}} gh , \tag{2.5}$$

$$gh \xrightarrow{\text{RP}} g''h''. \tag{2.6}$$

Combining the above in series

$$g_0 h_0 \xrightarrow{\text{PPP in O.S. of RP}} gh \xrightarrow{\text{RP}} g''h'' \tag{2.7}$$

or

$$g_0 h_0 \xrightarrow{\text{PPP in O.S. of RP} \dashrightarrow \text{RP}} g''h''. \tag{2.8}$$

The logical equation (2.3) can also be rewritten as

$$PR = PPP \text{ in O.S. of RP} \dashrightarrow RP . \tag{2.3}$$

Putting Eq. (2.3) into Eq. (2.8) yields

$$g_0 h_0 \xrightarrow{\text{PR}} g''h'' \tag{2.9}$$

or

$$g_0 h_0 \xrightarrow{\text{Prism}} g''h''. \tag{2.10}$$

Therefore, the logical deduction leads to the same conclusion.

On the other hand, the entrance optical axis and the exit optical axis are, no doubt, a pair of conjugate lines for the prism. Hence, we will readily be able to get for the Penta prism a pair of conjugate points M_0 and M', at which the entrance and exit optical axes intersect with

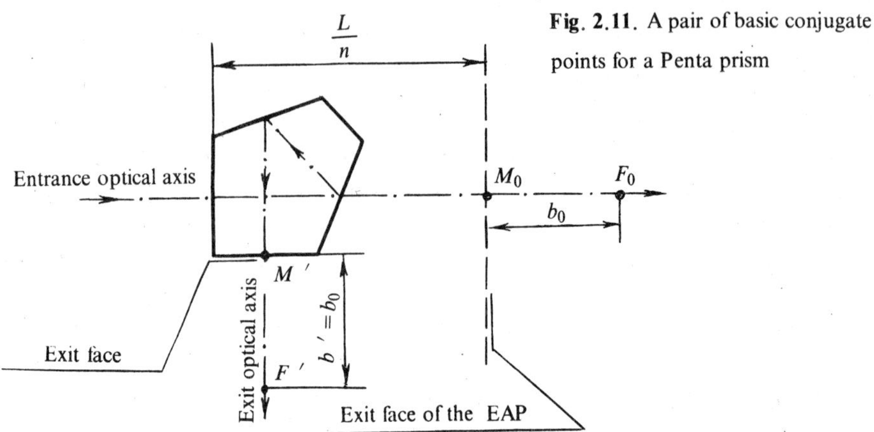

Fig. 2.11. A pair of basic conjugate points for a Penta prism

the exit face of the equivalent air plate and the exit face of the prism, respectively, as shown in Fig.2.11.

Obviously, the relation of the point M_0 with respect to the point M' for the Penta prism holds also for any other reflecting prisms.

To emphasize this fact, we shall formulate it as a theorem.

Theorem of image location for reflecting prisms:

"The intersection point M_0 of the entrance optical axis of a reflecting prism with the exit face of its equivalent air plate and the intersection point M' of the exit optical axis of the same reflecting prism with its exit face form a pair of basic conjugate points for the prism, the former being the object point, the latter the image point."

There are an infinite number of pairs of conjugate points for a prism, nevertheless the specific pair among them, namely, points M_0 and M' are in general most easily to be determined due to the well-prepared values of thickness L for all commonly used prisms in many handbooks of optics.

Locations of these two distinct points relative to the prism are dependent only on the structure of the latter, and they will to a certain extent characterize the prism in terms of its image formation. So, we call them in the theorem *a pair of basic conjugate points*, starting from which one can also locate other pairs of conjugate points, in particular, points lying on the entrance and exit optical axes at the same distance from and on the same side of the points M_0 and M', respectively, for example, points F_0 and F' in Fig.2.11.

As compared with lenses, these two basic conjugate points may be referred to as two cardinal points of a prism.

2.2.3 A Pair of Completely Conjugate Coordinate Systems
— Graphical Means of Determining the Image

Making use of two basic conjugate points of a prism, we can determine the image of a given object formed by the prism by graphical means. We are able to do so at least for all the planar reflecting prisms.

40

This can be illustrated through an example as in Fig. 2.12. Take a roof prism with its two

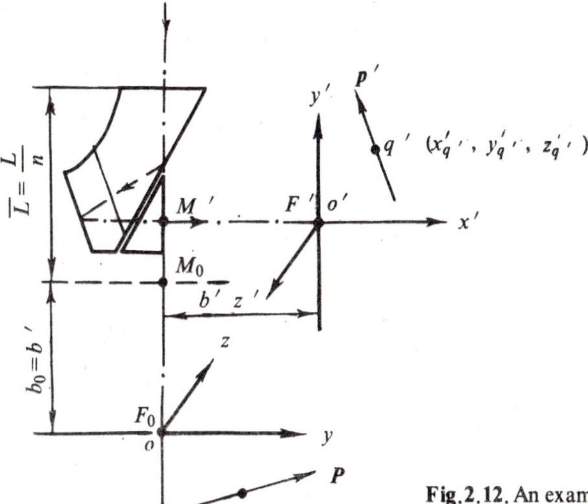

Fig.2.12. An example illustrating graphical means of determining the image

basic conjugate points M_0 and M' having been located. Suppose that the prism is in a convergent beam, most likely it is positioned after an objective in a telescope, and let F' be a known image point for the prism on its exit optical axis at a distance b' from M'. Thus, we can readily locate the object point F_0 conjugate to F' by laying off on the entrance optical axis $b_0 = \overline{M_0 F_0} = \overline{M'F'} = b'$.

Now, assume an image coordinate system $o'x'y'z'$, usually right-handed, with its origin o' located at F', the x' axis coincident with the exit optical axis, the y' axis lying in the conjugate optical-axis plane, i.e., the plane of the diagram, and the z' axis perpendicular to the paper and directed toward the reader. Then, construct by graphical means the conjugate object coordinate system $oxyz$ with its origin o located at F_0. Since an odd number reflecting prism, like a single mirror, will change the handedness of coordinate systems from the object space to the image space and *vice versa*, the object coordinate system $oxyz$ obtained in this case must be left-handed.

Thus, we have drawn a pair of conjugate coordinate systems $oxyz$ and $o'x'y'z'$ for this roof prism. These two are said to be a pair of completely conjugate coordinate systems because they are conjugate to each other not only in orientation but also in location. While discussing problems in the collimated beam we need only a pair of orientationally conjugate coordinate systems xyz and $x'y'z'$ where the origins o and o' will no longer be significant and have consequently been dropped.

Now, we have had anything prepared. By utilizing such a pair of completely conjugate coordinate systems, we will be able to find the image of an arbitrary object.

Given an arbitrary unit vector p (not necessarily a unit vector) by its rectangular components in the object coordinate system $oxyz$

$$P = P_x\,i + P_y\,j + P_z\,k \ . \tag{2.11}$$

Suppose that the corresponding image vector $P\,'$ will be defined in terms of its rectangular components in the image coordinate system $o\,'x\,'y\,'z\,'$

$$P\,' = P_x'\,i\,' + P_y'\,j\,' + P_z'\,k\,' \ . \tag{2.12}$$

Owing to the self evident relation between the two conjugate spaces for the prism, we have

$$P_x' = P_x \quad , \quad P_y' = P_y \quad , \quad P_z' = P_z \ . \tag{2.13}$$

Substituting these into Eq. (2.12), we find finally

$$P\,' = P_x\,i\,' + P_y\,j\,' + P_z\,k\,' \ . \tag{2.14}$$

Now, if we should know also the location of the image vector $P\,'$, then we may choose an arbitrary known point q on the action line of the given object vector P, and according to the object and image relationship, the corresponding image point $q\,'$ will certainly lie on the action line of the image vector $P\,'$ so that we can locate the latter.

Let the position vectors r_q and r_q' of these two points be expressed in terms of their rectangular coordinates in the $oxyz$ and $o\,'x\,'y\,'z\,'$ systems, respectively,

$$r_q = x_q\,i + y_q\,j + z_q\,k \quad , \tag{2.15}$$

$$r_q' = x_q'\,i\,' + y_q'\,j\,' + z_q'\,k\,' \ . \tag{2.16}$$

Similarly, we have

$$x_q' = x_q \quad , \quad y_q' = y_q \quad , \quad z_q' = z_q \ . \tag{2.17}$$

Putting Eq. (2.17) into Eq. (2.16) gives finally

$$r_q' = x_q\,i\,' + y_q\,j\,' + z_q\,k\,' \ . \tag{2.18}$$

The above mentioned method is said to be a graphical means because the two completely conjugate coordinate systems on which the method is based have been constructed mainly by drawing. However, it enables us to determine the image point of an arbitrary object point for all planar and some spatial reflecting prisms, and will thus prove to be of practical use.

Of course, we also realize that the present graphical means has its shortcoming too for the reason that it can not be used as a general method suitable for all cases and in particular the theory relying on such a graphical method will hardly be developed to deal with problems concerning a moving prism. Therefore more rigorous analytic methods will be discussed later on.

Finally, I hope that you, the reader, would enjoy answering Problems. 2.1 through 2.7 on your own. The solutions are given in the back of the book.

2.3 Kinematics Approach to the Problem of Image Formation for a Prism

2.3.1 An Analogy Between the Finite Space Motion of a Rigid Body and the Image Formation for a Prism or for a Mirror System

We now observe the image formation of two prisms as indicated in Fig. 2.13 and Fig. 2.14.

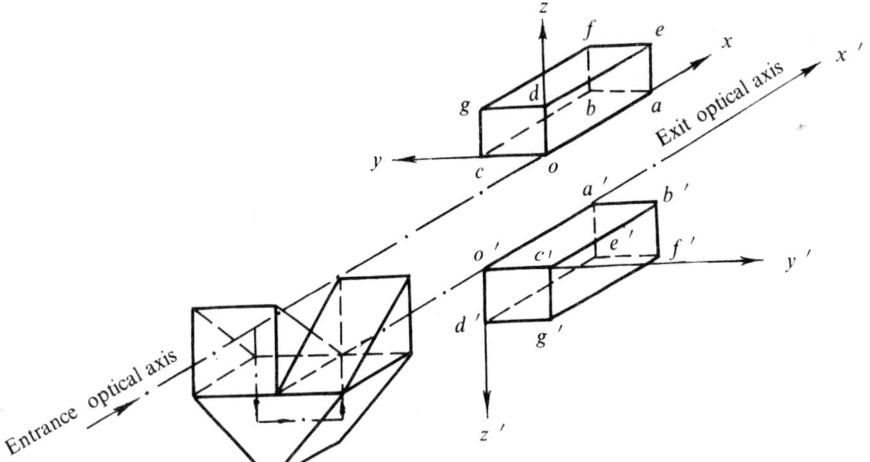

Fig.2.13. Image formation of a reflecting prism with an even number of reflecting surfaces

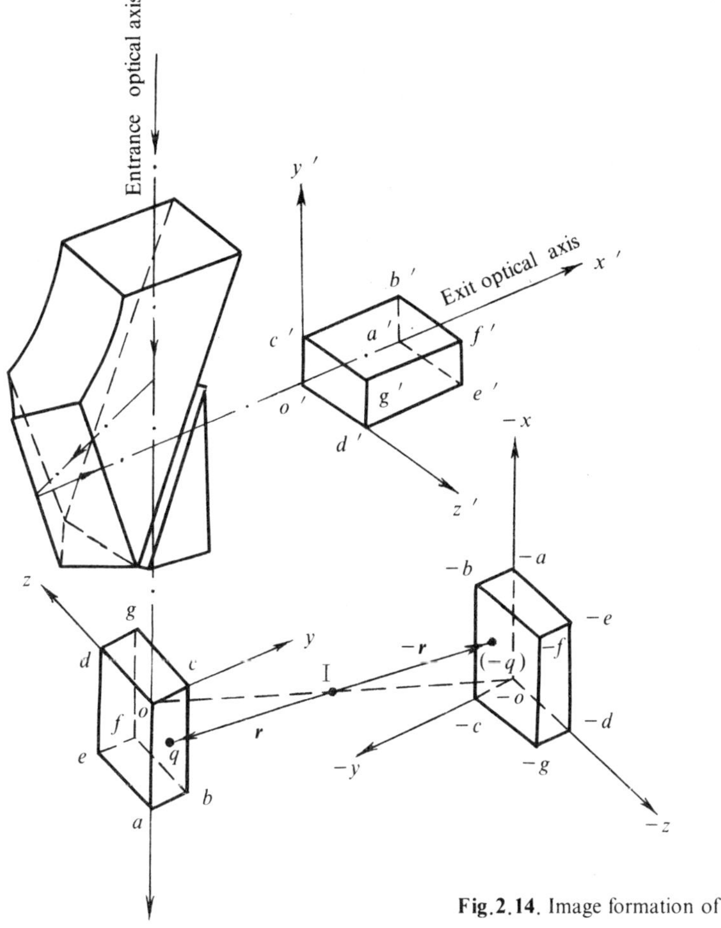

Fig.2.14. Image formation of a reflecting prism with an odd number of reflecting surfaces

The former named the Porro prism system 2, Abbe modification, has an even number of reflecting surfaces ($t=4$) while the latter called the roof prism of the Carl Zeiss system has an odd number of reflecting surfaces ($t=3$), where t denotes the number of reflections.

As mentioned in the previous section, for both cases $oxyz$ and $o'x'y'z'$ are used to represent a pair of completely conjugate coordinate systems, where the coordinate axes xyz and $x'y'z'$ indicate the orientation of the object and the image, respectively, while the origins o and o' express respectively the location of the object and the image. As a matter of fact, o and o' can locate only one pair of object and image points instead of the whole.

Moreover, based on these two conjugate coordinate systems we can construct two parallelepipeds $oabcdefg$ and $o'a'b'c'd'e'f'g'$ which represent a pair of conjugate object and image bodies, giving a vivid and bodily sensation.

It should be noted that all the imaging rays will be restricted to the paraxial region if the prism under consideration is located in convergent light, so that the aberrations become negligibly small or may be compensated by the centered lens system. Therefore, the image formed by the prism can be considered practically idealized.

Let us investigate the first case of an e.n. prism (the abbreviation of a reflecting prism with an even number of reflecting surfaces) in Fig. 2.13, where a right-handed object coordinate system $oxyz$ will generate a right-handed image coordinate system $o'x'y'z'$; the image is thus said to be congruent to the original object. Making a comparison between the two diagrams in Fig. 2.13 and in Fig. 1.3, we can easily come to the conclusion that a so-called *rigid body's kinematics model*, or for short just a *kinematics model* can be established to simulate the real physical problem of image formation for an e.n. prism by simply assuming the initial space position of a moving rigid body to represent the location and the orientation of the object and the final space position of the same rigid body to represent the location and the orientation of the corresponding image.

Then, examine the second case of an o.n. prism (the abbreviation of a reflecting prism with an odd number of reflecting surfaces) in Fig. 2.14, where a left-handed object coordinate system $oxyz$ will generate a right-handed image coordinate system $o'x'y'z'$. Since the process that converts a right-handed (left-handed) coordinate system in the object space into a left-handed (right-handed) coordinate system in the image space is defined as *inversion*, the image in this case is said to be inverted to the original object, or in other words, an o.n. prism forms an inverted image.

Now, let q be an arbitrary point of the object; again take any point I in the object space as the point of symmetry, and then convert the point q to its counterpart ($-q$) by assuming $\overrightarrow{I(-q)} = -\overrightarrow{Iq} = -r$. Thus, an inverted object $(-o)(-a)(-b)(-c)(-d)(-e)(-f)(-g)$, consisting of all the points $(-q)$, as well as an inverted object coordinate system $(-o)(-x)(-y)(-z)$ has been obtained, where the axes $(-x)(-y)(-z)$ become opposite to the directions of the axes xyz. Consequently, the inverted object coordinate system $(-o)(-x)(-y)(-z)$ and the image coordinate system $o'x'y'z'$ for an o.n. prism are also the same in handedness. Obviously, when a reflecting prism with an odd number of reflecting surfaces is encountered, only an inversion operation will be necessary in addition to the kinematics treatment. The point I, here, may also be called the *center of inversion*.

Therefore, no matter what the number of reflections t will be, the problem of image formation for a prism can be investigated by using the same methods developed for the analysis of the finite space motion of a rigid body.

The advantages of doing so will become apparent in what follows.

2.3.2 Characteristic Parameters of Image Formation for a Prism — Theorem of Image Orientation for Reflecting Prisms

From Sections 2.3.1 and 1.4.1, it is reasonable to expect certain counterparts in the image formation of a prism, that resemble their similarities in kinematics.

Since a finite space motion of a free rigid body can be considered as a combination of a translation and a rotation in series, you may divide the nature of conjugation for a reflecting prism into two categories: locational conjugation and orientational conjugation.

Locational conjugation is a problem corresponding to the translation part in the equivalent motion of a rigid body, or in other words, it is a matter of choosing a base point of the rigid body, as the point o, or equally, point o ' in Fig. 1.3, therefore the question of locational conjugation may be settled by selecting a certain pair of conjugate points for the prism under consideration. We will stress once again that the locational conjugation means to locate only one pair of object and image points instead of the whole.

No doubt, we will adopt those two basic conjugate points M_0 and M '. Quantitatively, the locational conjugation is thus represented by the vector $\overrightarrow{M_0 M}$ ', which corresponds to the rectilinearly translational displacement of a rigid body ' s equivalent motion.

For specifying the relative position of the vector $\overrightarrow{M_0 M}$ ' with respect to the prism, a rectangular coordinate system o 'x'y 'z ' of that prism is defined as follows :   ,

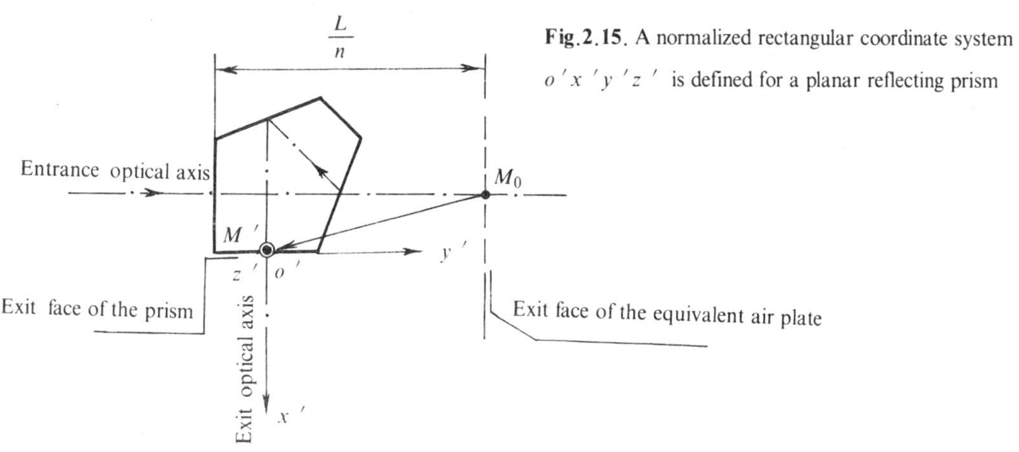

Fig.2.15. A normalized rectangular coordinate system o 'x 'y 'z ' is defined for a planar reflecting prism

In Fig. 2.15, the assumed coordinate system $o'x$ 'y 'z ' should first be right-handed since it will very often act as a working coordinate system in which many calculations including cross product are performed; its origin o' is located at point M'; the x' axis is coincident with the exit optical axis; the y' axis lies in the conjugate optical-axis plane so orientated as to make the z' axis facing the reader.

Such a reference frame is called the *normalized coordinate system* of a planar reflecting prism and it will consistently be employed for all planar reflecting prisms throughout this book.

Hence, for the Penta prism

$$\overrightarrow{M_0 M'} = 0.5\, D\, i' + \left(-0.5D + 3.414\, \frac{D}{n} \right) j',$$

where D is the aperture of the prism, n the refractive index of the glass, and i', j', k' the unit vectors along coordinate axes x', y', z', respectively.

In general cases, we have

$$\overrightarrow{M_0 M'} = \left(A_1 \cdot D + B_1 \cdot \frac{D}{n} \right) i' + \left(A_2 \cdot D + B_2 \cdot \frac{D}{n} \right) j'$$
$$+ \left(A_3 \cdot D + B_3 \cdot \frac{D}{n} \right) k', \tag{2.19}$$

in which the coefficients A_1, B_1, A_2, B_2, A_3, and B_3 are constant for each specific prism. We have listed these data for about fifty commonly used prisms in the Appendix.

Orientational conjugation is a problem corresponding to the rotation part in the equivalent motion of a rigid body, or in other words, it is a matter of rotational transform.

Relying on the principles stated in Sections 1.4.1 and 2.3.1, we can give, straightaway, the following theorem governing the object and image relationship in orientation.

Theorem of image orientation for reflecting prisms:

"There exist such a unique axis of rotation T (unit vector) and such an angle of rotation 2φ for each reflecting prism that in the case of an even number of reflections an arbitrary image vector A' formed by this prism can be defined as the vector obtained by turning the conjugate object vector A through an angle 2φ about the axis T while in the case of an odd number of reflections an arbitrary image vector A' formed by this prism can be regarded as the vector resulting from turning the inverted object vector $-A$ through an angle 2φ about the axis T, where the inverted object vector $-A$ is in the direction opposite to the conjugate object vector A."

Since the orientation of the axis T and the magnitude of the angle 2φ are closely related to

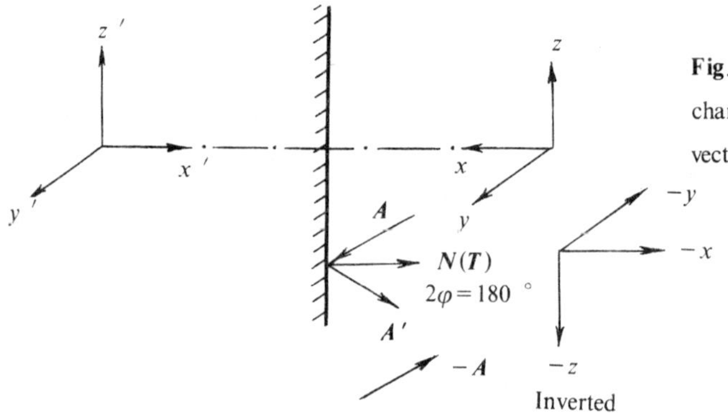

Fig.2.16. A plane mirror having its characteristic vector T and normal vector N in the same direction.

46

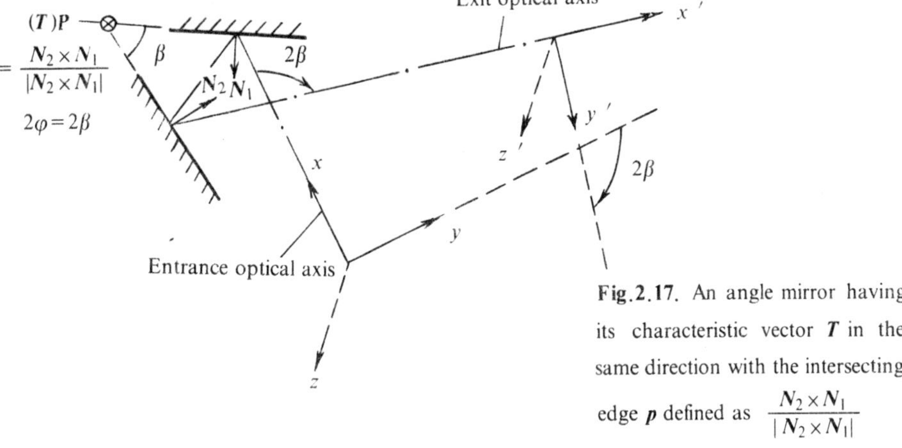

Fig.2.17. An angle mirror having its characteristic vector T in the same direction with the intersecting edge p defined as $\dfrac{N_2 \times N_1}{|N_2 \times N_1|}$

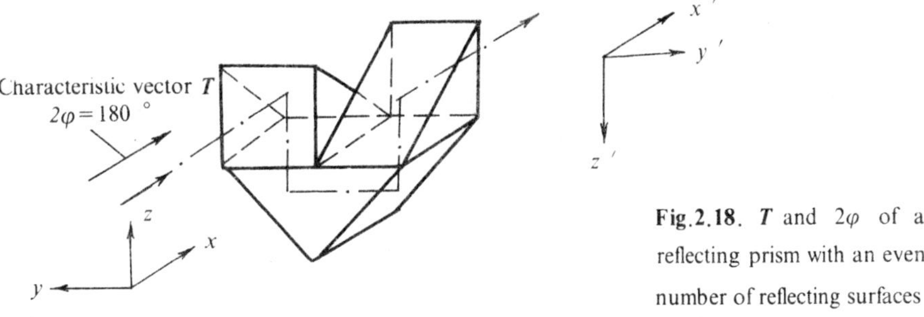

Fig.2.18. T and 2φ of a reflecting prism with an even number of reflecting surfaces

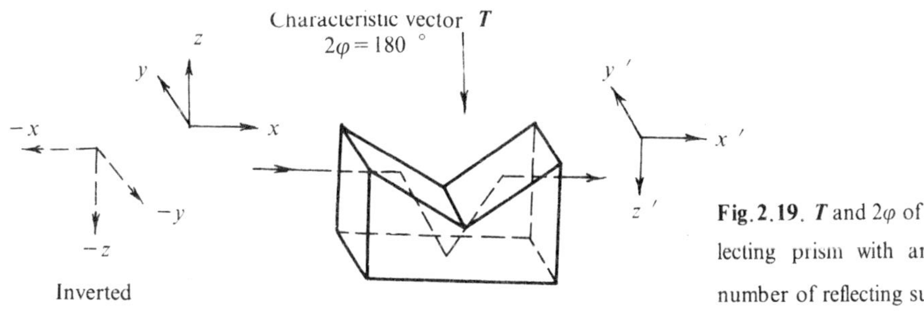

Fig.2.19. T and 2φ of a reflecting prism with an odd number of reflecting surfaces

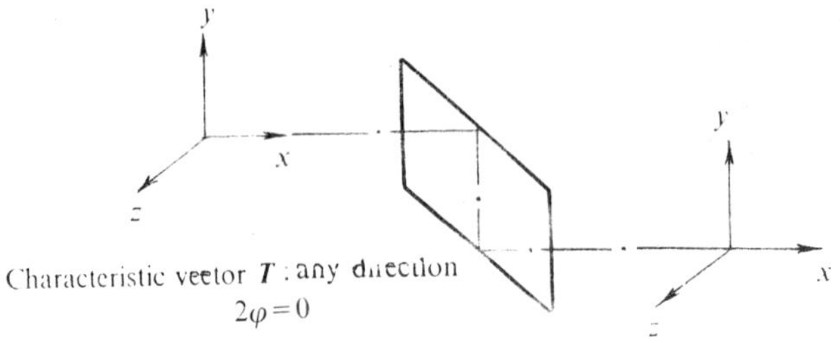

Fig.2.20. Rhomb prism

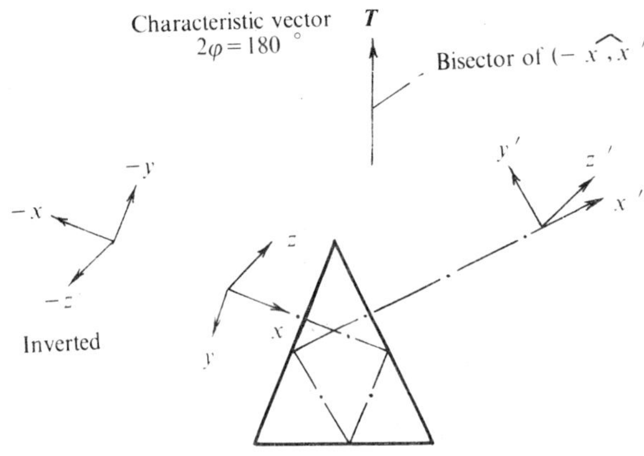

Fig.2.21. A planar reflecting prism having its characteristic vector *T* parallel to the conjugate optical-axis plane and 2φ = 180°

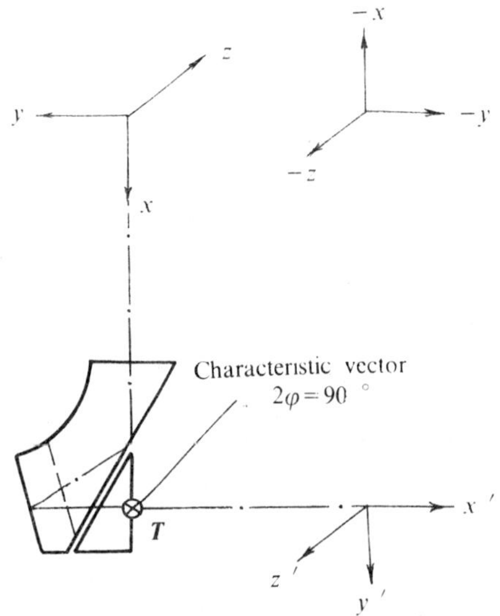

Fig.2.22. A planar reflecting prism having its characteristic vector *T* perpendicular to the conjugate optical- axis plane

the structure of the prism, they will thus be termed the *characteristic vector* and the *characteristic angle*, respectively.

The theorem seems easily to be accepted, so only a few examples for a better understanding are given below. For most of them , the illustrations are self-evident , therefore description will be omitted or simplified.

For a single plane mirror indicated in Fig. 2. 16, one can easily construct a pair of object and image coordinate systems xyz and $x'y'z'$. Then, it is easily seen that the inverted object coordinate system $(-x)(-y)(-z)$ will be rotated through $180°$ about the normal vector N to new orientations parallel to the image coordinate system $x'y'z'$, and also an arbitrary image vector A' can be obtained as a vector resulting from turning the inverted object vector $-A$ through $180°$ about N. Therefore, the normal vector N of a plane mirror plays the part of its characteristic vector T, while the characteristic angle $2\varphi = 180°$.

Inspection of Fig. 2. 17 verifies in the same manner that the intersecting edge P of an angle mirror will act as its characteristic vector T while the characteristic angle 2φ will be equal to twice the angle between the mirrors, i. e., $2\varphi = 2\beta$. In this case, the sense of T, namely P, should be so stipulated that $T(P)$ and $N_2 \times N_1$ will be of the same sign, where N_1 and N_2 may be both toward or both away from the mirrors.

The characteristic parameters T and 2φ for those prisms shown in Figs. (2.18) through (2.22) are left to be proved by the readers.

Some important properties concerning the characteristic vector T are stated as follows :

"With the object fixed, the conjugate image formed by a prism will not change in orientation no matter through what angle the prism has rotated around its characteristic vector T."

The reason for this phenomenon is simply because the orientation of T and the magnitude of 2φ for the prism remain unchanged while the prism is rotating around T.

"For a prism with an even number of reflecting surfaces, if the object vector A is orientated along the vector T, namely $A = T$, then the conjugate image vector A' will also be orientated along T , namely $A' = T$."

"For a prism with an odd number of reflecting surfaces, if the object vector A is orientated along the vector T, namely $A = T$, then the conjugate image vector A' will be orientated against T, namely $A' = -T$."

These phenomena can readily be explained directly by the theorem of image orientation for reflecting prisms.

The characteristic vector T will later be discussed in more detail.

According to the above discussion, we can introduce two characteristic parameters of image formation for a prism.

One is the well known vector $\overrightarrow{M_0M'}$ which will reasonably be assigned to represent the parameter of locational conjugation for a prism.

The other is a matrix R which relying on the theorem of image orientation for reflecting prisms will be defined as

$$R = (-1)^t S_{T,2\varphi} \ ,$$

$$(2.20)$$

so that it can be used to represent the parameter of orientational conjugation for a prism.

It is clear from Eq. (2.20) that the expression of R contains a rotation matrix $S_{T,2\varphi}$ which represents mathematically the transformation of a rotation about the axis T through an angle 2φ, and a multiplying factor $(-1)^t$ which takes account of an inversion operation while $t=$odd number.

No doubt, the matrix R is precisely the reflection matrix for a prism or the reflection matrix for a series of reflections.

The introduction of two characteristic parameters $\overrightarrow{M_0M}\,'$ and $R=(-1)^t S_{T,2\varphi}$ of image formation for a prism will prove to be of great utility for the further development of the theoretical work.

As mentioned before, the problem regarding the parameter $\overrightarrow{M_0M}\,'$ has already been settled, therefore it remains to deal with the reflection matrix R as well as the characteristic vector T and the characteristic angle 2φ for reflecting prisms.

2.4 The Reflection Matrix for a Prism

2.4.1 The Analytical Method of Determining the Reflection Matrix R for a Prism

Using the reflection matrix for a single plane mirror, Eq. (1.4), one can immediately obtain the reflection matrix for a prism as a product of reflection matrices for each reflecting surfaces

$$R = R_t\, R_{t-1} \cdots R_i \cdots R_2 R_1 \,, \tag{2.21}$$

where R_i denotes the reflection matrix for the ith reflecting surface

$$R_i = \begin{pmatrix} (1 - 2N_{ix}^2) & -2N_{ix}N_{iy} & -2N_{ix}N_{iz} \\ -2N_{ix}N_{iy} & (1 - 2N_{iy}^2) & -2N_{iy}N_{iz} \\ -2N_{ix}N_{iz} & -2N_{iy}N_{iz} & (1 - 2N_{iz}^2) \end{pmatrix}. \tag{2.22}$$

2.4.2 The Treatment of a Roof at 90° as a Single Plane Mirror Combined with an Inversion Operation

Very often, we encounter a roof prism where a certain reflecting surface has been replaced with a roof at 90°. In this case, we prefer to transform the roof mathematically in such a way that its reflection matrix will easily be determined.

A roof can be considered as a special case of the two-mirror systems where two mirrors are at 90° to each other. It is known from Sect. 2.3.2 that the characteristic vector T of the roof will be parallel to the intersecting edge P, and the characteristic angle 2φ will be equal to $2\beta = 180°$, as shown in Fig. 2.23.

50

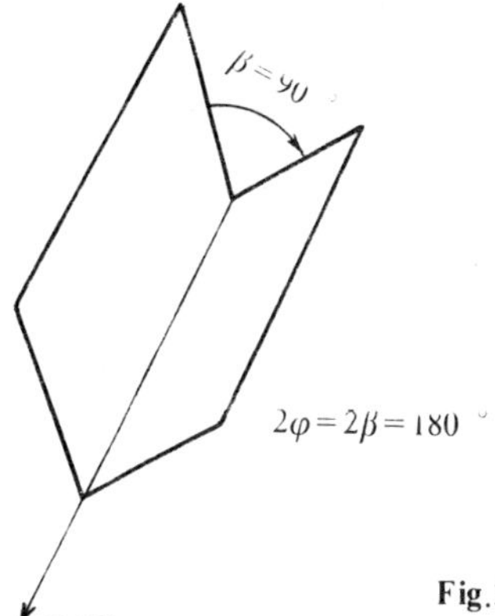

Fig.2.23. A pair of mirrors mounted at a fixed angle $\beta = 90$ °— a roof

Let A and A' be the object vector and its conjugate image vector for the roof, respectively.

By employing the rotation vector formula (1.9), the theorem of image orientation for reflecting prisms can be expressed mathematically as

$$A' = (-1)^t [A\cos2\varphi + (1 - \cos2\varphi)(A \cdot T)T - \sin2\varphi (A \times T)] . \qquad (2.23)$$

Now putting $T = P$, $2\varphi = 180$ °, and $t = 2$ into Eq. (2.23), we obtain

$$A' = -A - 2[(-A) \cdot P]P$$

or

$$A' = (-1)[A - 2 (A \cdot P)P] . \qquad (2.24)$$

Rewrite the reflection vector formula (1.1)

$$A' = A - 2 (A \cdot N)N. \qquad (1.1)$$

Then, comparing Eqs. (2.24) and (1.1) will lead to the conclusion that a roof at 90 ° can be replaced with a mirror by taking the roof edge P as the normal vector N, i.e., $N = P$, and also changing the input vector A to its inverted $-A$ prior to entering the mirror.

Therefore, an alternative form of Eq. (2.21) is given below.

$$R = (-1)^r R_{t-r} R_{t-r-1} \cdots R_2 R_1 , \qquad (2.25)$$

where r denotes the number of roofs in a prism; in most cases, $r = 1$ or 0.

2.4.3 The Graphical Method of Determining the Reflection Matrix R for a Prism

For all the planar reflecting prisms, also some spatial reflecting prisms of simple form, we can easily construct a pair of conjugate coordinate systems by graphical methods.

Since the function of each reflection matrix R_i of Eq. (2.21) has been implied in the relationship of a pair of conjugate coordinate systems, we will be able to find the overall reflection matrix R for the prism directly from the known coordinate systems, instead of doing a big job of calculating t individual matrices and their successive products.

Let $x_1 y_1 z_1$ and $x_1{}' y_1{}' z_1{}'$ be a pair of conjugate object and image coordinate systems of a prism; A and A' be a pair of conjugate object and image vectors; and according to the convention accepted in this text $G_{1'1}$ represents the matrix of transformation of coordinates from the image coordinate system $x_1{}' y_1{}' z_1{}'$ to the object coordinate system $x_1 y_1 z_1$.

Using the definition of transformation of coordinates gives

$$(A)_1 = G_{1'1} (A)_{1'} . \tag{2.26}$$

From the object and image relationship,

$$(A)_1 = (A')_{1'} . \tag{2.27}$$

Then, substituting Eq. (2.27) into Eq. (2.26) yields

$$(A')_{1'} = G_{1'1} (A)_1 \tag{2.28}$$

We have also

$$(A')_{1'} = [R]_{1'} (A)_{1'} , \tag{2.29}$$

where the subscript '1'' shows up that the reflection matrix R for the prism is expressed relative to the image coordinate system $x_1{}' y_1{}' z_1{}'$.

Comparing the last two equations, we then arrive at the following expression

$$[R]_{1'} = G_{1'1} . \tag{2.30}$$

This means that the matrix of transformation of coordinates $G_{1'1}$ from the image coordinate system to its conjugate object coordinate system is equal exactly to the reflection matrix R expressed relative to the image coordinate system.

Now let us start the same problem in another way by replacing A with A' in Eq. (2.26); then the latter becomes

$$(A')_1 = G_{1'1} (A')_{1'} . \tag{2.31}$$

Similarly, substituting Eq. (2.27) into Eq. (2.31) gives

$$(A')_1 = G_{1'1} (A)_1 . \tag{2.32}$$

Again, we have

$$(A')_1 = [R]_1 (A)_1 , \tag{2.33}$$

where the subscript '1' shows up that the reflection matrix R for the same prism is expressed relative to the object coordinate system $x_1 y_1 z_1$.

Comparing Eqs. (2.32) and (2.33) provides a similar expression

$$[R]_1 = G_{1'1} . \tag{2.34}$$

The above equation (2.34) indicates that the matrix of transformation of coordinates $G_{1'1}$ from the image coordinate system to the object coordinate system is also equal to the reflection matrix R expressed relative to the object coordinate system.

Obviously, Eqs. (2.34) and (2.30) may be written together

$$[R]_1 = G_{1'1} = [R]_{1'} , \tag{2.35}$$

which can readily be extended for a series of object and image coordinate systems with respect to the same prism

$$[R]_1 = G_{1'1} = [R]_{1'} = G_{1''1'} = [R]_{1''} = G_{1'''1''} = [R]_{1'''} = \cdots , \qquad (2.36)$$

where the $x_1{'}y_1{'}z_1{'}$ acts once as the image coordinate system relative to $x_1y_1z_1$, and again as the object coordinate system relative to $x_1{''}y_1{''}z_1{''}$, and so on.

Example

Determine the reflection matrix R for the prism indicated in Fig. (2.24) by graphical methods.

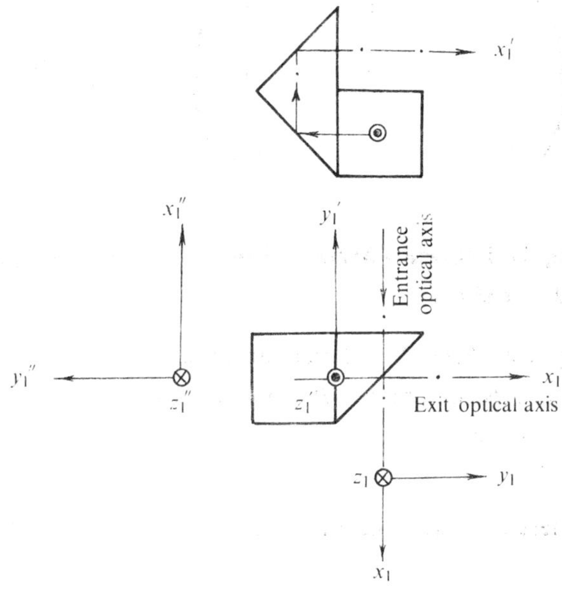

Fig. 2.24. A diagram of determining the reflection matrix R for a prism by graphical methods

Solution

Assume a coordinate system (normalized) $x_1{'}\ y_1{'}\ z_1{'}$; then take it as an image coordinate system and find the corresponding object coordinate system $x_1y_1z_1$ by drawing. During this process, we have actually considered functions of each reflecting surfaces.

Now, from Eq. (2.35) the reflection matrix R for the prism can be determined as the matrix of transformation of coordinates $G_{1'1}$ from $x_1{'}y_1{'}z_1{'}$ to $x_1y_1z_1$ by an arrangement of direction cosines according to Table 1.1 (see Table 2.1).

To verify Eq. (2.36), we may also consider the $x_1'y_1'z_1'$ as an object coordinate system relative to the prism and then construct the corresponding image coordinate system $x_1''\ y_1''\ z_1''$. Thus, the same reflection matrix R can be obtained (see Table 2.2).

Table 2.1

	x_1'	v_1'	z_1'	
$-[R]_1 = [R]_{1'} = G_{1'1} =$	0	-1	0	x_1
	1	0	0	y_1
	0	0	-1	z_1

Table 2.2

	x_1''	y_1''	z_1''	
$[R]_{1'} = [R]_{1''} = G_{1''1'} =$	0	-1	0	x_1'
	1	0	0	y_1'
	0	0	-1	z_1'

2.4.4 The Method of Determining the Reflection Matrix R for a Prism by Using Its Characteristic Parameters T, 2φ, and t

It has already been known from Sect. 2.3.2 that according to the theorem of image orientation for reflecting prisms, the reflection matrix R of a prism can be related to its characteristic parameters by the following equation:

$$R = (-1)^t S_{T,2\varphi} \, . \tag{2.20}$$

Therefore, whenever these parameters T, 2φ, and t are known, one may determine the reflection matrix R from Eq. (2.20).

Example

Determine the reflection matrix R for the prism as in Fig. 2.19 by using Eq. (2.20).

Solution

As has been seen in Fig. 2.19 the characteristic vector T was found to be in the z' direction, i.e., $T = k'$, $2\varphi = 180°$, and $t = 3$.

Then, from the third matrix of Eq. (1.13), and Eq. (2.20)

$$[R]_0 = (-1)^t S_{T,2\varphi} = (-1)^3 S_{k',180°}$$

$$= -\begin{pmatrix} \cos 180° & -\sin 180° & 0 \\ \sin 180° & \cos 180° & 0 \\ 0 & 0 & 1 \end{pmatrix} = \begin{pmatrix} 1 & 0 & 0 \\ 0 & 1 & 0 \\ 0 & 0 & -1 \end{pmatrix} \, .$$

2.5 The Characteristic Vector T and Characteristic Angle 2φ for a Prism

Since the characteristic parameters T and 2φ for a prism will continue to prove to be highly significant in many aspects, we try to investigate various methods of determining these two parameters.

2.5.1. The Graphical Method of Determining T and 2φ

This method can be used to determine the characteristic vector T and the characteristic angle 2φ for all planar reflecting prisms.

Suppose we have a planar reflecting prism as in Fig.2.25, and the paper of the diagram represents the conjugate optical-axis plane of the prism, i.e., both the entrance and exit optical axes are parallel to the plane of the diagram.

Now, if we draw an object plane containing (not necessarily containing)the entrance optical axis and parallel to the conjugate optical-axis plane, then according to the definition of a planar reflecting prism we will find a corresponding image plane containing (not necessarily)the exit optical axis and parallel to the same conjugate optical-axis plane, that is to say,these two

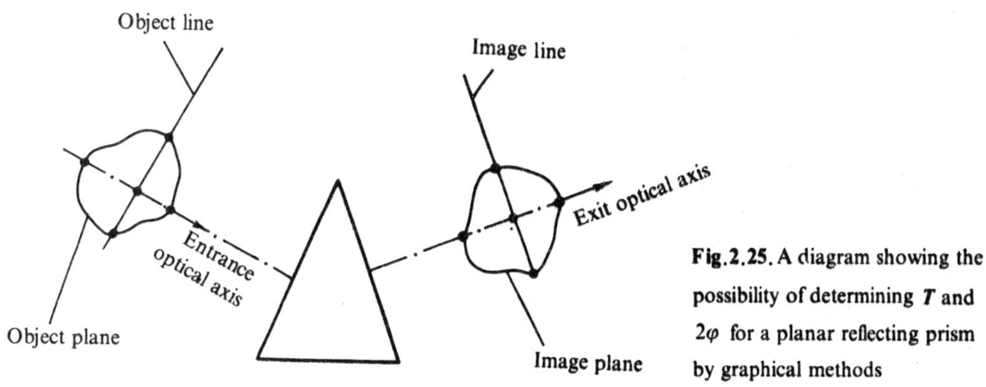

Fig.2.25. A diagram showing the possibility of determining T and 2φ for a planar reflecting prism by graphical methods

conjugate planes are parallel to each other, and both are all parallel again to the conjugate optical-axis plane.

Under this condition, the orientation of the characteristic vector T can be found to be only one of the following two posibilities: parallel to or perpendicular to the conjugate optical-axis plane, and no other choice can be made.

Thus, the simple geometrical relation enable us to find the characteristic vector T and characteristic angle 2φ for any planar reflecting prism graphically.

This method has already been used many times for those prisms shown in Figs. 2. 16 through 2.22, so we do not have to discuss it any more here.

2.5.2 The Partly Graphical Method of Determining T and 2φ

This method is suitable for some spatial reflecting prisms of such a simple type that one can easily obtain a pair of related coordinate systems by graphical methods while the rest of the work for solving T and 2φ will be completed by calculation.

We will illustrate the method by means of an example.

The spatial reflecting prism as shown in Fig. 2.26 is a combination of two rectangular prisms. Assume $x'y'z'$ to be an image coordinate system. Since the prism has an even number

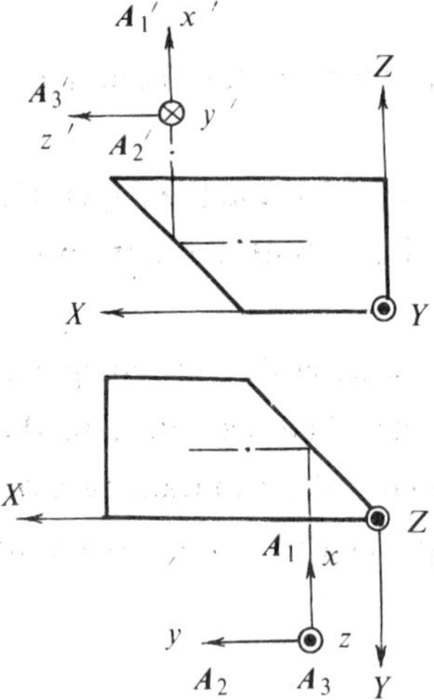

Fig.2.26. A pair of related coordinate systems xyz and $x'y'z'$, and the working coordinate system XYZ

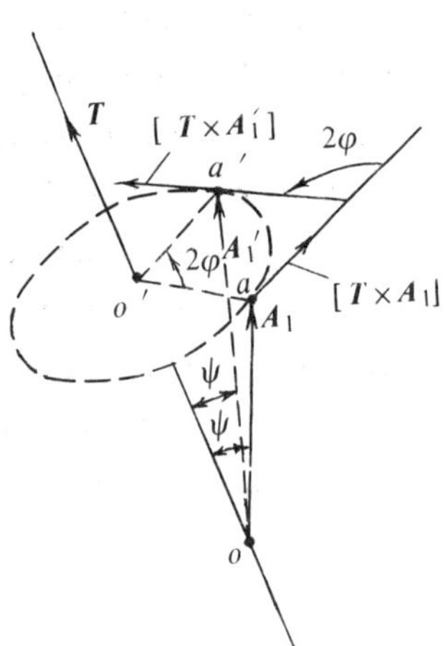

Fig.2.27. The diagram illustrating the rotation relationship of A_1 and A_1'

of reflections, we draw just the conjugate object coordinate system xyz. To avoid confusion, we construct additionally a working coordinate system XYZ, i.e., the reference frame for calculation. Let (A_1, A_2, A_3), (A_1', A_2', A_3'), and (i, j, k) be three sets of unit vectors along the coordinate axes xyz, $x'y'z'$ and XYZ, respectively. Again, let $\cos\alpha$, $\cos\beta$ and $\cos\gamma$ be the direction cosines of the characteristic vector T relative to the frame XYZ.

In Fig.2.27, the vectors A_1 and A_1' have been translated to meet the action line of T at a common point o.

According to the physical meaning of the vector T, we can readily see that both the vectors A_1 and A_1' intersect the vector T at same angles ψ, and in like manner the vectors A_2 and A_2' will both meet the vector T at other equal angles. Thus, we have

$$(A_1 \cdot T) = (A_1' \cdot T), \tag{2.37}$$

$$(A_2 \cdot T) = (A_2' \cdot T). \tag{2.38}$$

In addition,

$$\cos^2\alpha + \cos^2\beta + \cos^2\gamma = 1. \tag{2.39}$$

These three equations can be used to solve for the direction cosines of T.

The characteristic angle 2φ may be found in the following way. From Fig. 2.27, we can also see that 2φ is the angle formed by the lines $o'a$ and $o'a'$, which in turn will be the projections of the vectors A_1 and A_1' onto the plane perpendicular to the vector T, respectively, that is to say, 2φ equals the angle between two cross products $(T \times A_1)$ and $(T \times A_1')$. Therefore,

$$\cos 2\varphi = \frac{(T \times A_1) \cdot (T \times A_1')}{|(T \times A_1)||(T \times A_1')|} . \tag{2.40}$$

Finally, substituting $|(T \times A_1)| = |(T \times A_1')|$ into the above equation gives

$$\cos 2\varphi = \frac{(T \times A_1) \cdot (T \times A_1')}{|(T \times A_1)|^2} . \tag{2.41}$$

We now proceed to find the numerical solutions for T and 2φ. From Fig. 2.26, we have

$$\left.\begin{array}{l} A_1 = 0 \cdot i - j + 0 \cdot k = -j \\[4pt] A_1' = 0 \cdot i + 0 \cdot j + k = k \\[4pt] A_2 = i + 0 \cdot j + 0 \cdot k = i \\[4pt] A_2' = 0 \cdot i - j + 0 \cdot k = -j \end{array}\right\} , \tag{2.42}$$

and

$$T = \cos\alpha \cdot i + \cos\beta \cdot j + \cos\gamma \cdot k . \tag{2.43}$$

Putting Eqs. (2.42) and (2.43) into Eqs. (2.37) and (2.38), and rewriting Eq. (2.39) yields

$$-\cos\beta = \cos\gamma ,$$

$$\cos\alpha = -\cos\beta ,$$

$$\cos^2\alpha + \cos^2\beta + \cos^2\gamma = 1 .$$

Solving the last three equations gives

$$\cos\beta = \pm \frac{1}{\sqrt{3}} .$$

Then, we have the first solution for T by taking $\cos\beta = + \dfrac{1}{\sqrt{3}}$

$$(\cos 125°\,16', \cos 54°\,44', \cos 125°\,16') ,$$

and the second solution for T by taking $\cos\beta = - \dfrac{1}{\sqrt{3}}$

$$(\cos 54°\,44', \cos 125°\,16', \cos 54°\,44') .$$

Again, putting Eqs. (2.42) and (2.43) into Eq. (2.41) leads to

$$(T \times A_1) = \cos\gamma \cdot i - \cos\alpha \cdot k ,$$

$$(T \times A_1') = \cos\beta \cdot i - \cos\alpha \cdot j ,$$

and thus

$$(\boldsymbol{T} \times \boldsymbol{A}_1) \cdot (\boldsymbol{T} \times \boldsymbol{A}_1') = \cos \beta \cos \gamma = -\cos^2 \alpha ,$$

$$|(\boldsymbol{T} \times \boldsymbol{A}_1)|^2 = 2 \cos^2 \alpha .$$

Consequently,

$$\cos 2\varphi = \frac{-\cos^2\alpha}{2\cos^2\alpha} = -\frac{1}{2} .$$

Also, we have two solutions for 2φ : $2\varphi = 120°$ (corresponding to the first solution of $\boldsymbol{T}$), and $2\varphi = 240°$ (corresponding to the second solution of $\boldsymbol{T}$).

2.5.3 The Analytical Method of Determining $\boldsymbol{T}$ and 2φ

Whenever the previous two methods fail to work, we may adopt the analytical method. Most likely, the approach will be suitable for some spatial reflecting prisms in irregular form as well as for some complicated mirror systems.

In this case, one should first find the reflection matrix R for a prism by using the mathematical method, which was discussed in Secs. 2.4.1 and 2.4.2.

Therefore, we will here discuss only how to determine the characteristic parameters $\boldsymbol{T}$ and 2φ for a prism by means of its known reflection matrix R.

Ler us now begin with the principle of the method.

The behaviour of the characteristic vector $\boldsymbol{T}$, as mentioned at the end of Sect. 2.3.2, will suggest that the vector $\boldsymbol{T}$ for a prism can be found as one of the eigenvectors of the reflection matrix R of the prism if and only if the corresponding eigenvalue should be assigned to be $\lambda = (-1)'$.

The above statement will be formulated as follows:

Rewrite the basic formula

$$(A') = R(A). \tag{2.44}$$

Suppose

$$(A) = (\boldsymbol{T}) , \quad (A') = \lambda(\boldsymbol{T}) . \tag{2.45}$$

Substituting Eq. (2.45) into Eq. (2.44) yields

$$\lambda(\boldsymbol{T}) = R(\boldsymbol{T}) .$$

Thus,

$$(R - \lambda E)(\boldsymbol{T}) = 0 , \tag{2.46}$$

where λ is the eigenvalue associated with the eigenvector $\boldsymbol{T}$; E is a unit matrix of order 3.

The final matrix equation (2.46) will be used to solve for $\boldsymbol{T}$, and λ will then be replaced with $(-1)'$.

We now proceed to derive the equation for determining 2φ.

Using Eq. (1.12), we may rewrite Eq. (2.20) in its expanded form.

$$R = (-1)' \begin{pmatrix} \cos2\varphi + 2T_x^2 \sin^2\varphi & -T_z\sin2\varphi + 2T_xT_y\sin^2\varphi & T_y\sin2\varphi + 2T_xT_z\sin^2\varphi \\ T_z\sin2\varphi + 2T_xT_y\sin^2\varphi & \cos2\varphi + 2T_y^2\sin^2\varphi & -T_x\sin2\varphi + 2T_yT_z\sin^2\varphi \\ -T_y\sin2\varphi + 2T_xT_z\sin^2\varphi & T_x\sin2\varphi + 2T_yT_z\sin^2\varphi & \cos2\varphi + 2T_z^2\sin^2\varphi \end{pmatrix} . \tag{2.47}$$

In the generalized form, R may also be written as

$$R = \begin{pmatrix} r_{11} & r_{12} & r_{13} \\ r_{21} & r_{22} & r_{23} \\ r_{31} & r_{32} & r_{33} \end{pmatrix} . \tag{2.48}$$

Summing up all the elements of the main diagonal of R in both Eqs. (2.47) and (2.48), respectively, and then equating these two sums, we obtain

$$\sum_{i=1}^{3} r_{ii} = (-1)^{l} [3\cos 2\varphi + 2 (T_x^2 + T_y^2 + T_z^2)\sin^2\varphi] .$$

Noting $T_x^2 + T_y^2 + T_z^2 = 1$, we thus arrive at the final expression

$$\cos 2\varphi = \frac{1}{2} \left[(-1)^{l} \sum_{i=1}^{3} r_{ii} - 1 \right] . \tag{2.49}$$

Example

Determine T and 2φ for the prism indicated in Fig. 2.28 by the analytical method.

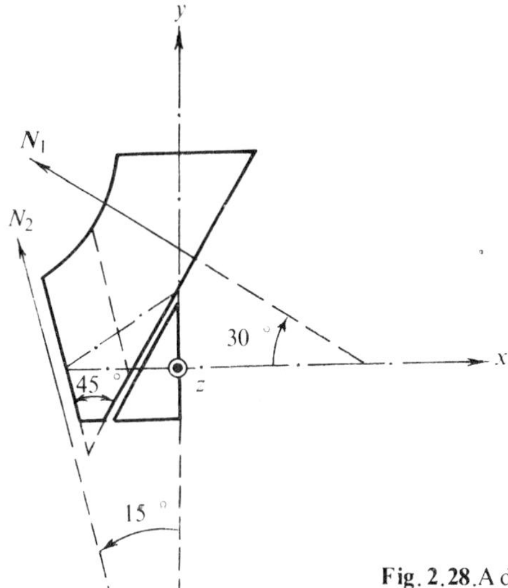

Fig. 2.28. A diagram of determining T and 2φ for a prism by analytical methods

Solution

Let N_1 and N_2 denote the normal vector and the roof-edge vector, respectively, and assume a coordinate system xyz. Then, the unit vectors N_1 and N_2 are expressed relative to the coordinate system xyz

$$N_{1x} = -\cos 30\,° = -0.8660 \ , \ N_{1y} = \cos 60\,° = 0.50 \ , \ N_{1z} = 0 \ ,$$

$$N_{2x} = -\cos 75\,° = -0.2588 \ , \ N_{2y} = \cos 15\,° = 0.9659 \ , \ N_{2z} = 0 \ .$$

Putting these data into Eq. (2.22), we determine the reflection matrices R_1 and R_2 for the reflecting surface and the equivalent mirror of the roof. Then from Eq. (2.25), assuming $r=1$ we have

$$R = (-1)^r R_2 R_1 = -R_2 R_1$$

$$= - \begin{pmatrix} 0.866 & 0.5 & 0 \\ 0.5 & -0.866 & 0 \\ 0 & 0 & 1 \end{pmatrix} \begin{pmatrix} -0.5 & 0.866 & 0 \\ 0.866 & 0.5 & 0 \\ 0 & 0 & 1 \end{pmatrix}$$

$$= \begin{pmatrix} 0 & -1 & 0 \\ 1 & 0 & 0 \\ 0 & 0 & -1 \end{pmatrix} .$$

Now, substituting the obtained reflection matrix R into Eq. (2.46) and considering $\lambda = (-1)^3 = -1$ will yield

$$T_x - T_y = 0 \ ,$$

$$T_x + T_y = 0 \ ,$$

$$0 \cdot T_z = 0 \ .$$

Also,

$$T_x^2 + T_y^2 + T_z^2 = 1 \ .$$

Solving the above set of equations gives

$$T_x = 0 \ , \ T_y = 0 \ , \ T_z = \pm 1 \ .$$

On the other hand, from Eq. (2.49) we have

$$\cos 2\varphi = \frac{1}{2} \, [\,(-1)^3 \, (-1) - 1] = 0 \ ,$$

and thus

$$2\varphi = 90\,° \quad \text{or} \quad 270\,° .$$

Combining the above results gives the solutions as : either $T = k$, $2\varphi = 270\,°$ or $T = -k$, $2\varphi = 90\,°$.

It has already been pointed out that a prism may have more than one eigenvectors. Then probably the reader would ask what the other eigenvetors of the prism will be. Let us investigate them by taking a single plane mirror as an example. In this case, all the vectors parallel to the plane of the mirror belong to its eigenvectors, where the corresponding eigenvalues are all equal to 1, i.e,

$$(J) = R\,(J) \ .$$

The situation mentioned above has given the reason for the fact that we adopted a special term *characteristic vector* to distinguish *T* from the other eigenvectors of a prism.

In general cases, the solutions for the eigenvalue λ in both real and complex fields of a reflection matrix R will be found to be

$$\lambda_1 = (-1)^l , \qquad\qquad\qquad (2.50)$$

$$\lambda_{2,3} = e^{\pm i \left\{ (\pi - 2\varphi) - \frac{[(-1)^l + 1]}{2} \pi \right\}} . \qquad\qquad\qquad (2.51)$$

Those wishing to know more about this may refer to Ref. 19.

2.6 The Types of Problems of Image Formation for a Prism

Traditionally, the problems of image formation for a prism are divided into two types: one, the problems in a collimated beam, and the other the problems in a convergent beam.

In the former case, the prism is placed in parallel light and both the object and image points are located infinitely distant, therefore quantitatively the positions of these points can only be expressed in the form of angular quantities, i.e., the directions of the incident parallel rays as well as of the emergent parallel rays. Thus under this condition the problem of location has tranformed itself into a problem of orientation. Consequently, for problems of this type our discussions will be restricted to the calculation of image orientation only.

In the latter case, the prism is placed in non-parallel light and then both the object and image points are located at finite distances. Hence, for problems of this type our discussions will contain the calculation of image orientation as well as of image location.

It is instructive and convenient to formulate the above assignments in some different way:
(1) for problems in a collimated beam, it is required to determine the image vector of an arbitrary object vector formed by a prism.
(2) For problems in a convergent beam, it is required to determine the image point of an arbitrary object point formed by a prism.

For the same reason as mentioned in Sect. 2.3.1, the phenomena in the above two cases both agree exactly with what happens in kinematics of a rigid body, and their counterparts in kinematics will be the problem of rotation of a rigid body attached at a fixed point and the problem of general finite space motion of a free rigid body, respectively.

It is clear that the basic problem of general finite space motion of a rigid body is to determine the position of any particle of the rigid body at its final place, and this is nothing substantially different from the fact that the main problem of image formation for a prism in a convergent beam is to find the image point of an arbitrary object point formed by the prism.

Actually, an arbitrary object vector is none other than a line joining two arbitrary object points, therefore as viewed from kinematics the problem in a collimated beam can also be referred to as a special case of the problem in a convergent beam. However, from the point of view of optics, there do exist some distinct points in problems of each type, so that we will keep the traditionl classification throughout the theoretical part of this book.

2.7 Problems in a Collimated Beam

In the present section, we will be concerned about the image orientation only. Since two refractions occurring at the entrance and exit faces of a prism do not introduce any change in orientation of the image, the prism acts just like a mirror system in parallel light.

Therefore, the basic formula used to calculate the image vector A' of an arbitrary object vector A for a prism will be

$$(A') = R(A) , \qquad (2.52)$$

where R represents the well-known reflection matrix for the prism as well as for the mirror system.

Since an investigation of the reflection matrix R as well as of the closely related parameters T and 2φ has been developed in full detail in Secs. 2.4 and 2.5, it seems that there is no need to discuss it any more here.

However, we also feel that in some cases it may be convenient to use the most fundamental reflection vector formula

$$A' = A - 2(A \cdot N)N . \qquad (1.1)$$

Example

The collimator, as shown in Fig. 2.29, consists of an objective lens with a reticle and a plane mirror at $45°$. Determine the direction of the line of sight (LOS) projected into the object space corresponding to the mark c on the cross-wire.

Solution

Let the unit vector A represent the direction of the incident parallel rays generated by the objective corresponding to the mark c, N the normal vector, and A' the direction of the LOS; then assume a coordinate system xyz.

Referring to Fig. 2.30, find the components of unit vectors A and N along the axes xyz

$$A_x = 0 , \quad A_y = -\cos v , \quad A_z = \sin v ,$$

$$N_x = \cos 45° = \frac{\sqrt{2}}{2} , \quad N_y = \sin 45° = \frac{\sqrt{2}}{2} , \quad N_z = 0 .$$

Thus,

$$-2(A \cdot N) = \sqrt{2} \cos v .$$

Substituting the above data into Eq. (1.1) yields

$$A_x' = A_x - 2(A \cdot N)N_x = \cos v ,$$

$$A_y' = A_y - 2(A \cdot N)N_y = 0 ,$$

$$A_z' = A_z - 2(A \cdot N)N_z = \sin v .$$

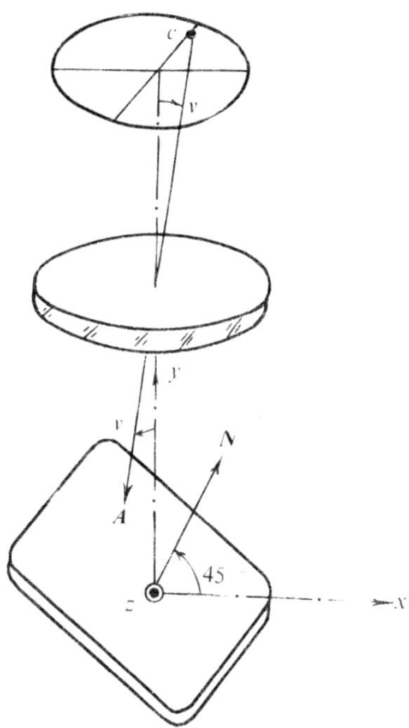

Fig.2.29 .A collimator used for projecting a line of sight

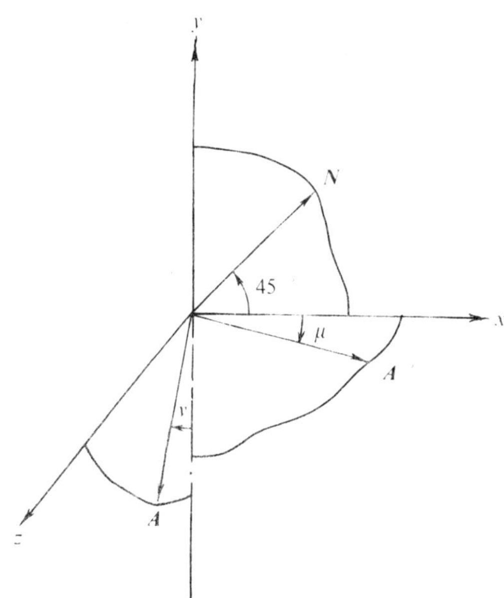

Fig.2.30. A diagram showing the directions of the incident and emergent rays , and normal vector

The component $A_y' = 0$ indicates that the direction of the LOS is parallel to the x-z plane. Let μ be the angle between the vector A' and the x axis. Then,

$$\text{tg } \mu = \frac{A_z'}{A_x'} = \frac{\sin v}{\cos v} = \text{tg } v.$$

Consequently,

$$\mu = v.$$

2.8 Problems in a Convergent Beam

In the present section, we will deduce a general solution to the problem of image formation for a prism.

Referring to Fig. 2.31, an arbitrary reflecting prism is indicated symbolically by a closed curve; its two characteristic parameters R and $\overrightarrow{M_0 M'}$ are given; a known arbitrary object point F is defined by the vector $e = \overrightarrow{FM_0}$ relative to the prism; and F' denotes the conjugate image point to be determined.

Let $o'x'y'z'$ be a right-handed reference coordinate system, and g, g' the position vectors of the points F, F', respectively.

Thus, from Fig.2.31 we have

$$g' = g + e + \overrightarrow{M_0 M'} - R\,(e)_0,$$

(2.53)

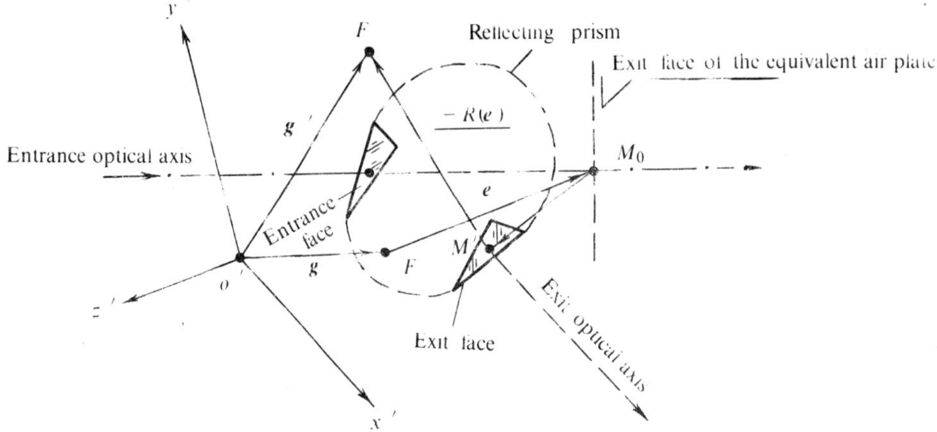

Fig.2.31. A diagram for deriving the formula of the position of the image point corresponding to an arbitrary object point

or in matrix form

$$(g\,')_0\,'=(g\,)_0\,'+(e\,)_0\,'+(\overrightarrow{M_0M\,'})_0\,'-R\,(e\,)_0\,'. \tag{2.54}$$

This is the most general formula used in image formation.

We can readily extend the resulting equation (2.54) to the case of a mirror system by replacing the vector $\overrightarrow{M_0M\,'}$ with $\overrightarrow{MM\,'}$, which represents the vector joining a pair of conjugate points for the mirror system.

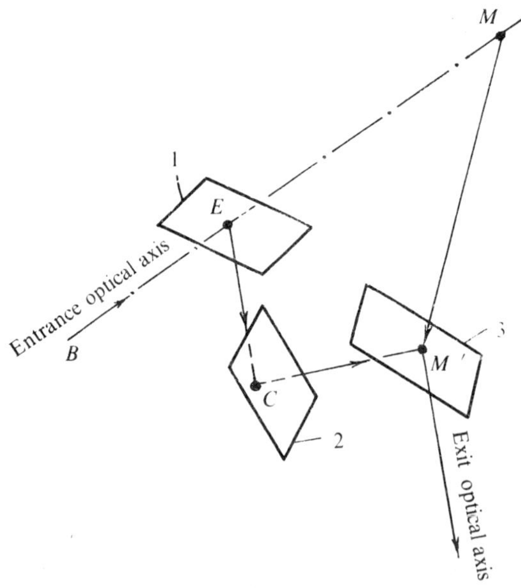

Fig.2.32. A diagram illustrating the method of locating two conjugate points M and $M\,'$ for a general mirror system

The method of locating two conjugate points M and M' in general cases is shown in Fig. 2.32.

The only thing you have to do is to extend the entrance optical axis BE to such a point M that $\overline{EM} = \overline{EC} + \overline{CM}'$, where $E, C,$ and M' are the intersection points of the optical axis with the mirrors.

Therefore, in the case of mirrors system Eq. (2.54) becomes

$$(g')_{0'} = (g)_{0'} + (e)_{0'} + (\overrightarrow{MM'})_{0'} - R\,(e)_{0'} \ . \tag{2.55}$$

The single mirror, double mirror and triple mirror are the most frequently encountered mirror systems. In these cases, we can deal with the problem of choosing a pair of conjugate points in a simple way.

(1) The single mirror

A single plane mirror is indicated in Fig. 2.33. Assume a coordinate system $o'x'y'z'$ and let g and g' be the position vectors of two conjugate points q and q', respectively. The origin o' is

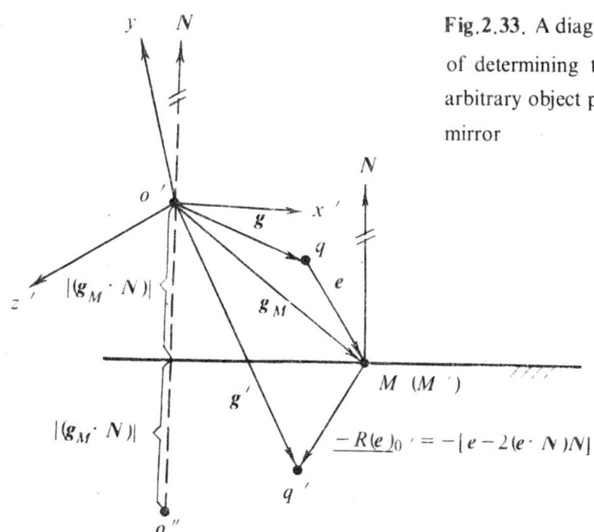

Fig.2.33. A diagram illustrating the method of determining the image point q' of an arbitrary object point q for a single plane mirror

imaged by the mirror at the image point o''.

Now, choose an arbitrary point M in the plane of the mirror and locate it by the position vector g_M. Thus, the point M can be considered as a pair of conjugate points M and M' coincident with each other, i.e., $\overrightarrow{MM'} = 0$. The other symbols in Fig 2.33 keep their original meaning.

Putting $\overrightarrow{MM'} = 0$ into Eq. (2.55) gives

$$(g')_{0'} = (g)_{0'} + (e)_{0'} - R\,(e)_{0'} \tag{2.56}$$

or in vector form

$$g' = g + e - R\,(e)_{0'} \ , \tag{2.57}$$

where according to Eq. (1.1)

$$R(e)_0 = e - 2(e \cdot N)N ,$$ (2.58)

and

$$e = g_M - g .$$ (2.59)

Substituting Eqs. (2.58) and (2.59) into Eq. (2.57) and rearranging it yields

$$g' = g - 2(g \cdot N)N + 2(g_M \cdot N)N$$ (2.60)

or

$$g' = R(g)_0 + 2(g_M \cdot N)N .$$ (2.61)

It is well worthwhile to examine Eq. (2.61) a little further. The first term $R(g)_0$ on the right, depending on the position of an arbitrary object point q, represents a transformation, by which the object point q is initially inverted with respect to the center of symmetry (or center of inversion) o', and then the inverted object point $(-q)$, as shown in Fig. 2.34, is again rotated through 180° about the normal vector N passing o' to the new point q_1. The second term $2(g_M \cdot N)N$ on the right, being independent of the varying input vector g, represents thus a translation, by which the point q_1 will further be displaced by $\overrightarrow{q_1 q'} = 2(g_M \cdot N)N$, reaching the final

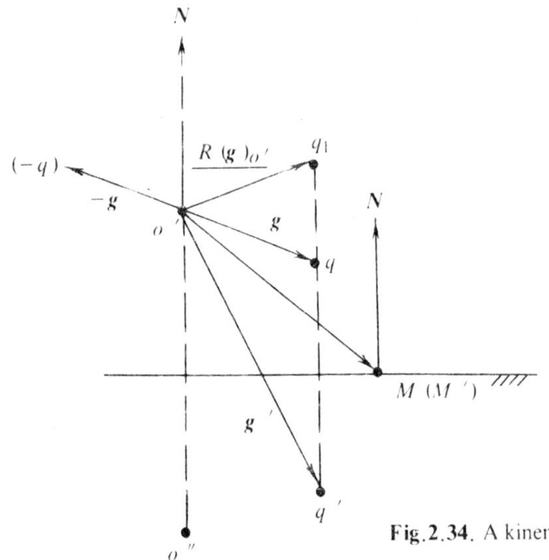

Fig.2.34. A kinematics interpretation of image formation for a single plane mirror

image point q'. Obviously, the magnitude of the linear displacement $2(g_M \cdot N)N$ will be found to be double the distance along the normal to the mirror from the origin o' to the mirror plane, while the sense of the vector $2(g_M \cdot N)N$ will invariably be directed from o' toward the mirror no matter on which side of the mirror the origin o' will be chosen.

The above discussion has interpreted the image formation for a single mirror as a kinematics model of the general finite space motion of a rigid body.

66

Making a comparison between Fig.2.34 and Fig.1.5, you may find that the points o' and o'' in Fig.2.34 correspond to the points o and o' in Fig.1.5, respectively.

(2) The double mirror

A double mirror is indicated in Fig.2.35. The vector P represents the intersecting edge; φ denotes the angle between the mirrors. In this case, any point M on the intersecting edge can be a pair of conjugate points M and M' coincident with each other, i.e., $\overrightarrow{MM'} = 0$. The other symbols in Fig.2.35 have the same meaning as before.

Similarly, we have

$$g' = g + e - R(e)_{0'} \,, \tag{2.62}$$

where according to Eq. (1.9)

$$R(e)_{0'} = e\cos2\varphi + (1 - \cos2\varphi)(e \cdot P)P - \sin2\varphi\,(e \times P)\,, \tag{2.63}$$

and

$$e = g_M - g \,. \tag{2.64}$$

Substituting Eqs. (2.63) and (2.64) into Eq. (2.62) and rearranging it yields

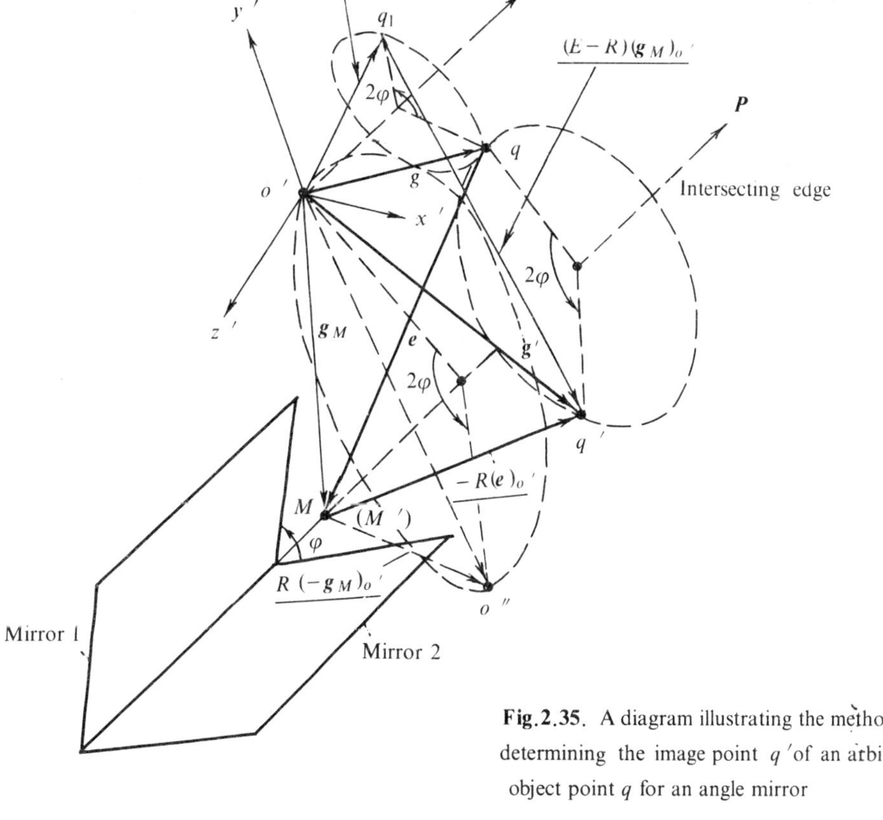

Fig.2.35. A diagram illustrating the method of determining the image point q' of an arbitrary object point q for an angle mirror

$$g' = \{g\cos2\varphi + (1-\cos2\varphi)(g \cdot P)P - \sin2\varphi\,(g \times P)\}$$
$$+ \{g_M - [g_M\cos2\varphi + (1-\cos2\varphi)(g_M \cdot P)P - \sin2\varphi\,(g_M \times P)]\} \qquad (2.65)$$

or

$$g' = R\,(g)_{0'} + (E-R)(g_M)_{0'}. \qquad (2.66)$$

The resulting equation (2.66) also has a kinematics interpretation as stated below.

The first term $R\,(g)_{0'}$ on the right being dependent on the varying input vector g represents a transformation, by which the vector g is rotated through an angle 2φ about the axis P passing through point o' to the new direction $\overrightarrow{o'q_1} = R\,(g)_{0'}$ as shown in the diagram. The second term $(E-R)(g_M)_{0'}$ on the right being independent of the varying input vector g represents thus a translation, by which the point q_1 will further be displaced by $\overrightarrow{q_1q'} = (E-R)(g_M)_{0'} = \overrightarrow{o'o''}$, reaching the final image point q'.

As mentioned in the previous case, the origin o' and its conjugate image point o'' formed by the two-mirror system correspond to the points o and o' in Fig.1.5, respectively.

Obviously, if the origin o' is chosen on the intersecting edge, then $(E-R)(g_M)_{0'} = 0$, and thus $g' = R\,(g)_{0'}$.

(3) The triple mirror

A triple mirror is indicated in Fig.2.36. The vector $P_{1,2}$ represents the intersecting edge of the first two mirrors and the extension of the edge meets the third mirror at point M. Actually, M is the point of intersection of the three mirrors, and in general is the only point which will be imaged by the triple mirror at itself, namely $\overrightarrow{MM'} = 0$. The other symbols in the figure are self-evident.

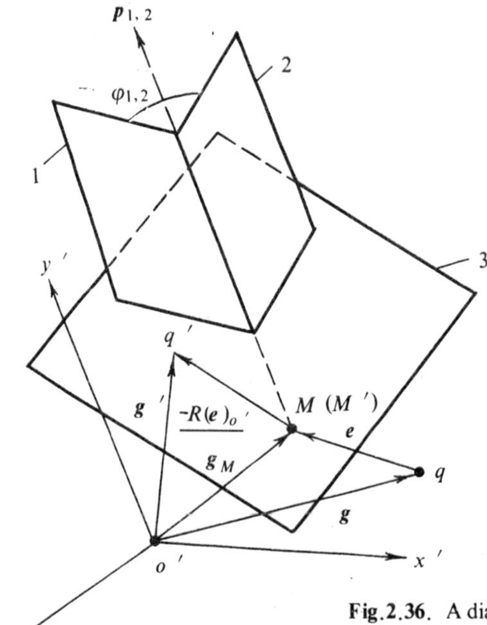

Fig.2.36. A diagram illustrating the method of determining the image point q' of an arbitrary object point q for a triple mirror

From Fig.2.36, we also have

$$g' = g + e - R(e)_{0'},\tag{2.67}$$

$$e = g_M - g.\tag{2.68}$$

Combining Eqs. (2.67) and (2.68) gives

$$g' = R(g)_{0'} + (E-R)(g_M)_{0'}.\tag{2.69}$$

The equation (2.69) has a physical meaning analogous to that in the previous cases.

It should be noted that the second term on the right side of Eq. (2.61) can also be written as

$$2(g_M \cdot N)N = (E-R)(g_M)_{0'}.\tag{2.70}$$

To extend the results obtained for the above three cases to a general mirror system, as shown in Fig.2.37, by using the kinematics model as mentioned in Sect.2.3, we can replace those individual object points with a rigidly unified object body being big enough to contain the origin o' as well. As was done in Sect.1.4, considering the origin o' of the selected coordinate system as a base point of the rigid object body, we then arrive at a generalized formula of determining the image point q' of an arbitrary object point q for a general mirror system as

$$g' = (-1)^t S_{T.2\varphi}(g)_{0'} + \overrightarrow{o'o''},\tag{2.71}$$

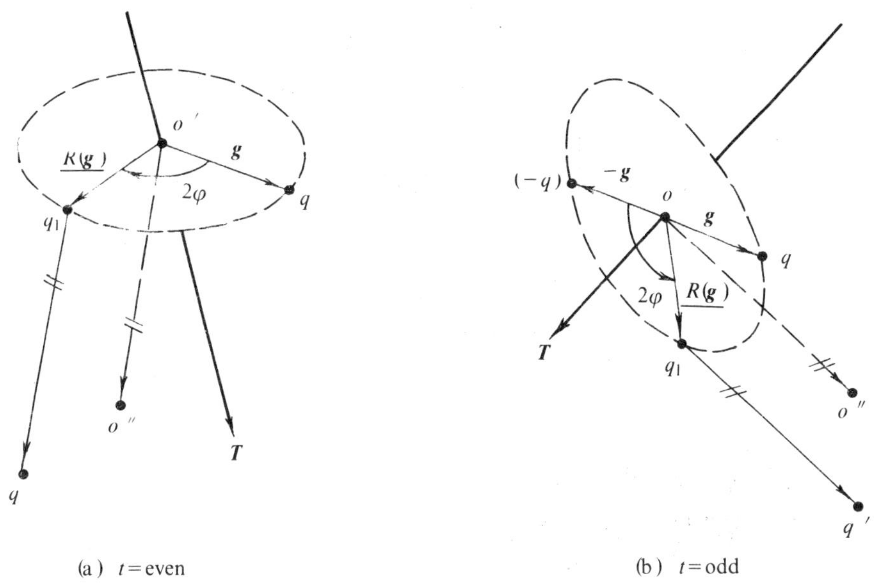

Fig.2.37. A diagram illustrating the method of determining the image point q' of an arbitrary object point q for a general mirror system

where $\overrightarrow{o'o''}$ is the vector joining the origin o' and its image o'' generated by the mirror system.

In general cases,

$$\overrightarrow{o'o''} = (E-R)(g_M)_{0'} + \overrightarrow{MM'}.\tag{2.72}$$

In a special case when $\overrightarrow{MM'} = 0$,

$$\overrightarrow{o'o''} = (E-R)(g_M)_{0'}.\tag{2.73}$$

Furthermore, if the origin o' is chosen to be coincident with such a point M, then $\overrightarrow{o'o''} = 0$, and thus

$$\bar{g}' = R(g)_{0'} \, . \tag{2.74}$$

This situation will be illustrated in the following example.

Example

Determine the image point of an arbitrary object point for a system consisting of three mirrors perpendicular to one another as in Fig. 2.38.

Solution

Assume a reference coordinate system $o'x'y'z'$ with the origin o' coincident with the intersecting point M of the three mirrors.

From Eq. (2.67),

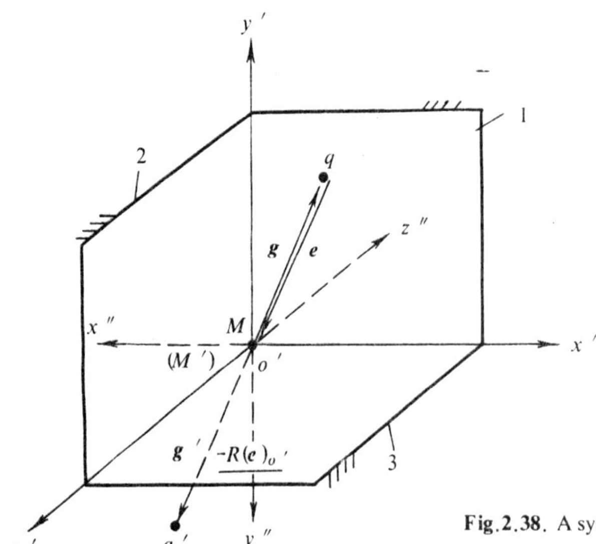

Fig. 2.38. A system consisting of three mutually perpendicular mirrors

$$g' = g + e - R(e)_{0'} \, ,$$

Putting $e = -g$ in the above equation gives

$$g' = R(g)_{0'} \quad \text{or} \quad (g')_0{}' = R(g)_{0'} \, , \tag{2.75}$$

where the reflection matrix R can be found by any one of the well-known methods.

Consider the reference frame $x'y'z'$ as an object coordinate system, and then determine the corresponding image coordinate system $x''y''z''$ formed by the three-mirror system, using graphical method, as shown in Fig. 2.38.

Thus, we obtain the matrix of transformation of coordinates from the image coordinate system $x''y''z''$ to the object coordinate system $x'y'z'$

$$G_{o''o'} = \begin{pmatrix} -1 & 0 & 0 \\ 0 & -1 & 0 \\ 0 & 0 & -1 \end{pmatrix} = -E \, .$$

70

From Eq. (2.36), we also have

$$[R]_{0'} = G_{0''0'} = -E .$$

Substituting the last equation into Eq. (2.75) yields

$$(\boldsymbol{g}')_{0'} = -E\,(\boldsymbol{g})_{0'} = (-\boldsymbol{g})_{0'}$$

or

$$\boldsymbol{g}' = -\boldsymbol{g} .$$

The above resulting formula indicates that for the system consisting of three mirrors normal to one another, its two conjugate spaces are symmetrical with respect to the intersecting point M of the three mirrors as the center of symmetry.

To end this section, it is worth noting that in some books the equation (2.61) is written in matrix form as follows:

$$\begin{pmatrix} 1 \\ g'_{x'} \\ g'_{y'} \\ g'_{z'} \end{pmatrix} = \begin{pmatrix} 1 & 0 & 0 & 0 \\ 2\,(\boldsymbol{g}_M \cdot N)\,N_{x'} & (1-2N_{x'}^2) & -2\,N_{x'}N_{y'} & -2N_{x'}N_{z'} \\ 2\,(\boldsymbol{g}_M \cdot N)\,N_{y'} & -2N_{x'}N_{y'} & (1-2N_{y'}^2) & -2N_{y'}N_{z'} \\ 2\,(\boldsymbol{g}_M \cdot N)\,N_{z'} & -2N_{x'}N_{z'} & -2N_{y'}N_{z'} & (1-2N_{z'}^2) \end{pmatrix} \begin{pmatrix} 1 \\ g_{x'} \\ g_{y'} \\ g_{z'} \end{pmatrix}$$

$$\tag{2.76}$$

or

$$(\boldsymbol{g}')_{0'} = \mathscr{L}\,(\boldsymbol{g})_{0'} , \tag{2.77}$$

where the square matrix of order $n = 4$ represents the reflection matrix for locating the image point of an arbitrary object point for a single plane mirror.

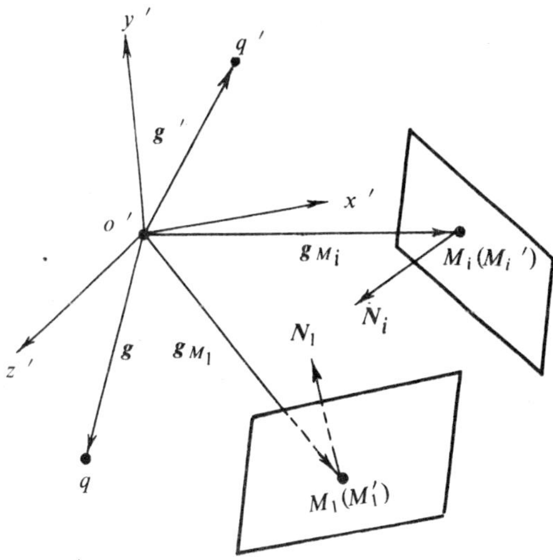

Fig.2.39. A diagram illustrating the method of determining the image point q' of an arbitrary point q for a series of n plane mirrors

The above equation as well as the method of derivation will immediately suggest a generalization for a series of n plane mirrors. In this case, a unified coordinate system $o'x'y'z'$, Fig. 2.39, is necessary. We thus have

$$(g')_{0'} = \mathscr{L}_n \mathscr{L}_{n-1} \cdots \mathscr{L}_i \cdots \mathscr{L}_2 \mathscr{L}_1 (g)_{0'} \, , \tag{2.78}$$

where

$$\mathscr{L}_i = \begin{pmatrix} 1 & 0 & 0 & 0 \\ 2(g_{M_i} \cdot N_i)N_{ix'} & (1 - 2N_{ix'}^2) & -2N_{ix'}N_{iy'} & -2N_{ix'}N_{iz'} \\ 2(g_{M_i} \cdot N_i)N_{iy'} & -2N_{ix'}N_{iy'} & (1 - 2N_{iy'}^2) & -2N_{iy'}N_{iz'} \\ 2(g_{M_i} \cdot N_i)N_{iz'} & -2N_{ix'}N_{iz'} & -2N_{iy'}N_{iz'} & (1 - 2N_{iz'}^2) \end{pmatrix} . \tag{2.79}$$

2.9 The 'Screw Axis' of Image Formation for a Prism

It has already been stated in Sect. 2.3 that the problem of image formation for a prism can be considered as the problem of finite motion of a rigid body, and we are also aware that the latter can be treated ultimately as an equivalent helical motion. Thus, a reasonable question arising from this situation would be if we could find the *screw axis* of image formation for a prism, and the answer is positive.

2.9.1 The Screw Axis of Image Formation for a Prism with an Even Number of Reflecting Surfaces

In order to determine the location of the screw axis r_0 and the axial shift $2d$ of image formation for an e.n. prism, we can directly employ Eqs. (1.40) and (1.42) by replacing the parameters S, P, and θ in those equations with the parameter $\overrightarrow{M_0 M'}, T$, and 2φ, respectively. Thus, we obtain the following two formulas:

$$r_0 = \frac{1}{2} [(\overrightarrow{M_0 M'} \cdot T)T - \overrightarrow{M_0 M'}] + \frac{1}{2} \left(\frac{\sin 2\varphi}{\cos 2\varphi - 1} \right) \overrightarrow{M_0 M'} \times T , \tag{2.80}$$

$$2d = \overrightarrow{M_0 M'} \cdot T . \tag{2.81}$$

In this case, the perpendicular vector r_0 is measured and dropped from the origin M' onto the screw axis $\vec{\lambda}$, which is known as being parallel to the characteristic vector T.

Example

Locate the screw axis $\vec{\lambda}$ and find the axial shift $2d$ for a Penta prism as shown in Fig 2.40, given $\overrightarrow{M_0 M'} = 0.5D i' + \left(-0.5D + 3.414 \dfrac{D}{n} \right) j'$ and $T = k'$ relative to the normalized coordinate system $o'x'y'z'$, and $2\varphi = 90°$.

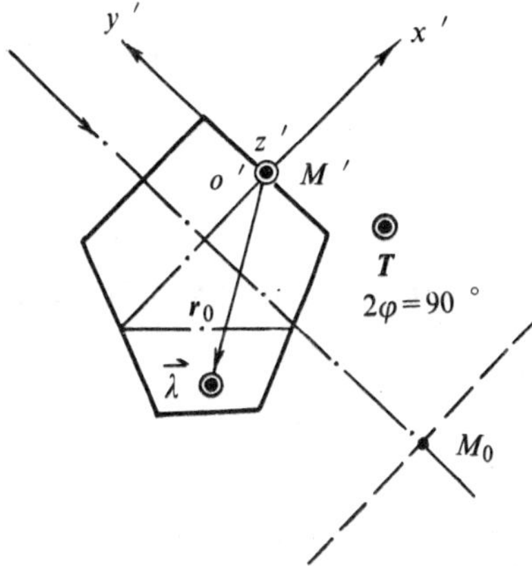

Fig.2.40. A diagram illustrating the location of the screw axis $\vec{\lambda}$ for a Penta prism

Solution

Substituting the given data into Eqs. (2.80) and (2.81) then gives the results

$$r_0 = -1.707\,\frac{D}{n}\,i' + \left(0.5\,D - 1.707\,\frac{D}{n}\right)j'\,,$$

$$2d = 0\,.$$

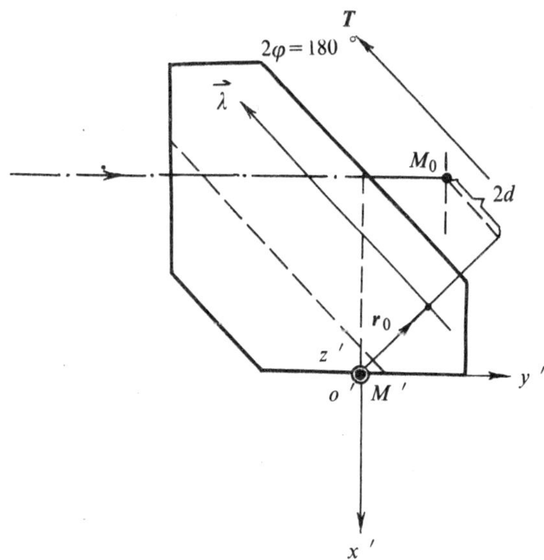

Fig. 2.41. A diagram illustrating the location of the screw axis $\vec{\lambda}$ and the axial shift $2d$ for an Amici prism

Example

Locate the screw axis $\vec{\lambda}$ and find the axial shift $2d$ for an Amici prism as shown in Fig.2.41,

given $\overrightarrow{M_0 M'} = 0.886Di' + \left(0.866D - 1.732 \dfrac{D}{n}\right) j'$ and $T = -0.707\, i' - 0.707\, j'$ relative to

the normalized coordinate system $o'x'y'z'$, and $2\varphi = 180°$.

Solution

Substituting the given data into Eqs. (2.80) and (2.81) then yields the results

$$r_0 = -0.433\, \frac{D}{n}\, i' + 0.433\, \frac{D}{n}\, j',$$
$$2d = -1.225\, D + 1.225\, \frac{D}{n}.$$

Some important properties concerning the screw axis $\vec{\lambda}$ are stated as follows:

"With the object fixed, neither the location nor the orientation of the conjugate image will change no matter through what angle the prism has rotated around its screw axis $\vec{\lambda}$."

"With the object fixed, the conjugate image will not change in location no matter by what distance the prism is displaced along its screw axis $\vec{\lambda}$."

Of course, the above statements are restricted in their full validity to the paraxial region.

2.9.2 The Screw Axis of Image Formation for a Prism with an Odd Number of Reflecting Surfaces

In the present case, for bringing the object to coincide with its conjugate image, an inversion operation prior to the helical motion is necessary.

Since the location of the inverted object depends upon the position of the center of inversion, which may arbitrarily be chosen in the object space, the problem for locating the screw axis of image formation for an o.n. prism thus comes out with an infinite number of answers.

In order to avoid the matter of uncertainty, among the unlimited number of these possible screw axes, mutually parallel, we choose for some reason only such one that it will pass through the correspondingly selected center of inversion.

However, the unique solution for the screw axis $\vec{\lambda}$ according to the above definition still corresponds to an infinite number of centers of inversion which lie on the same screw axis.

For further simplicity, it is reasonable to locate the center of inversion at such a point along the unique screw axis to make the axial shift of the equivalent helical motion of image formation equal to zero, i.e., $2d = 0$.

Based on the above mentioned principle, the diagram in Fig 2.42 is employed to derive the equations for determining the position vector r_i of the center of inversion I measured from the origin $M'(o')$ and the perpendicular vector r_0 dropped from the same origin M' onto the screw axis $\vec{\lambda}$.

To meet the first requirement, the center of inversion I should lie on its corresponding screw axis $\vec{\lambda}$, and to meet furthermore the second requirement the center of inversion I must be located at such a position along the screw axis that the line joining

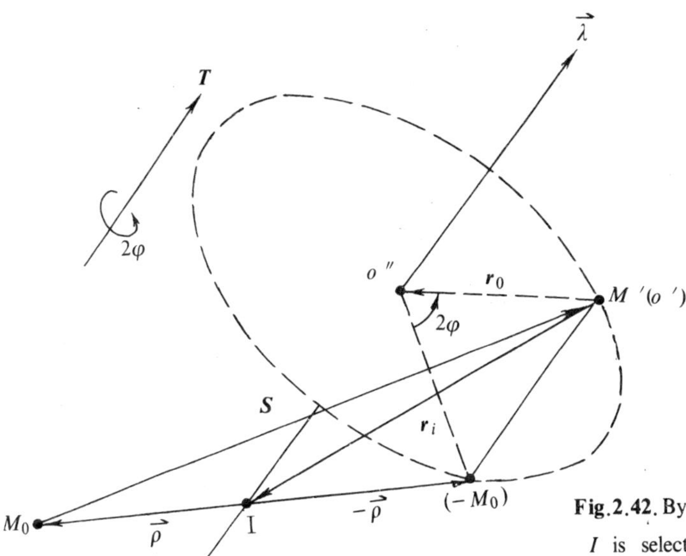

Fig.2.42. By our definition, such a center of inversion I is selected so as to make the corresponding screw axis $\vec{\lambda}$ pass through that center and to convert the helical motion into pure rotation, i.e., $2d = 0$

the inverted point $(-M_0)$ of point M_0, and the point M' will be located in a plane normal to the vector $\vec{\lambda}$. Thus, the point $-M_0$ can be brought to the new position M' simply by a pure rotation through 2φ about $\vec{\lambda}$.

From Fig. 2.42,

$$\overrightarrow{I(-M_0)} = -\vec{\rho} = r_i + s ,$$

$$\overrightarrow{IM'} = -r_i .$$

According to the rotation vector formula (1.9),

$$(r_i + s)\cos2\varphi + (1 - \cos2\varphi)[(r_i + s)\cdot T]T - \sin2\varphi[(r_i + s)\times T]$$

$$= -r_i$$

or

$$r_i(\cos2\varphi + 1) + (1 - \cos2\varphi)(r_i \cdot T)T - \sin2\varphi(r_i \times T)$$

$$= -S\cos2\varphi - (1 - \cos2\varphi)(S \cdot T)T + \sin2\varphi(S \times T) . \tag{2.82}$$

Performing the vector product of both sides of Eq. (2.82) on the right by T yields

$$(\cos2\varphi + 1)r_i \times T - \sin2\varphi[(r_i \cdot T)T - r_i]$$

$$= (-\cos2\varphi)S \times T + \sin2\varphi[(S \cdot T)T - S] . \tag{2.83}$$

Combining Eqs. (2.82) and (2.83) to solve for r_i gives

$$r_i = -\frac{1}{2}S + \frac{1}{2}\left(\frac{\sin2\varphi}{\cos2\varphi + 1}\right)S \times T . \tag{2.84}$$

From the triangle $O''M'I$ in Fig. 2.42,

$$r_0 = r_i - (r_i \cdot T)T , \tag{2.85}$$

where the second term on the right can also be expressed as

$$(r_i \cdot T)T = -\frac{1}{2}(S \cdot T)T . \tag{2.86}$$

Substituting Eq. (2.86) into Eq. (2.85) leads finally to

$$r_0 = \frac{1}{2}[(S \cdot T)T - S] + \frac{1}{2}\left(\frac{\sin 2\varphi}{\cos 2\varphi + 1}\right)S \times T . \tag{2.87}$$

Now, by replacing S by $\overrightarrow{M_0M}'$ Eqs. (2.84) and (2.87) become

$$r_i = -\frac{1}{2}\overrightarrow{M_0M}' + \frac{1}{2}\left(\frac{\sin 2\varphi}{\cos 2\varphi + 1}\right)\overrightarrow{M_0M}' \times T , \tag{2.88}$$

$$r_0 = \frac{1}{2}[(\overrightarrow{M_0M}' \cdot T)T - \overrightarrow{M_0M}'] + \frac{1}{2}\left(\frac{\sin 2\varphi}{\cos 2\varphi + 1}\right)\overrightarrow{M_0M}' \times T . \tag{2.89}$$

Moreover, Eqs. (2.89) and (2.80) can be unified as a single

$$r_0 = \frac{1}{2}[(\overrightarrow{M_0M}' \cdot T)T - \overrightarrow{M_0M}'] + \frac{1}{2}\left[\frac{\sin 2\varphi}{\cos 2\varphi + (-1)^{t-1}}\right]\overrightarrow{M_0M}' \times T . \tag{2.90}$$

Example

Locate the screw axis $\vec{\lambda}$ and the center of inversion I for the prism indicated in Fig. 2.43, given $\overrightarrow{M_0M}' = -0.046D\,i' + \left(-1.445D + 2.981\dfrac{D}{n}\right)j'$ and $T = -k'$ relative to the normalized coordinate system $o'x'y'z'$, and $2\varphi = 90°$, $t = 3$.

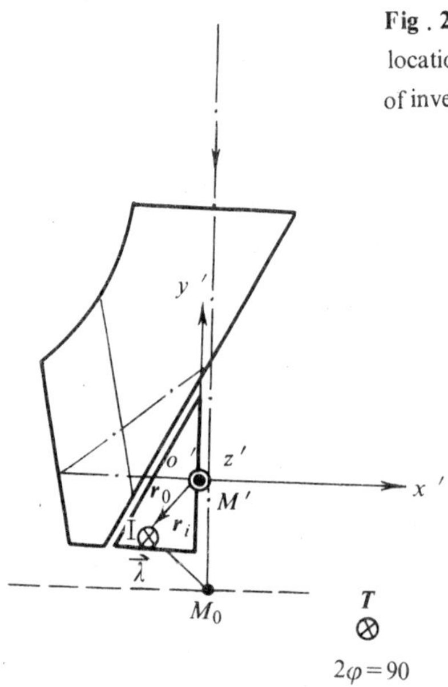

Fig. 2.43. A diagram illustrating the locations of the screw axis $\vec{\lambda}$ and the point of inversion for an o.n. prism with a roof

Solution

Substituting the given data into Eqs. (2.90) and (2.88) gives the results

$$r_0 = \left(0.746D - 1.491\frac{D}{n}\right)i' + \left(0.700D - 1.491\frac{D}{n}\right)j',$$

$$r_i = \left(0.746D - 1.491\frac{D}{n}\right)i' + \left(0.700D - 1.491\frac{D}{n}\right)j'.$$

Since the equivalent helical moton of image formation for the prism has been converted to a pure rotation by choosing such a point of inversion, thus the axial shift $2d = 0$.

Example

Locate the Screw axis $\vec{\lambda}$ and the center of inversion I for a right-angle prism indicated in Fig. 2.44, given $\overrightarrow{M_0M}' = 0.5\,D\,i' + \left(0.5D - \dfrac{D}{n}\right)j'$ and $T = -0.707i' + 0.707j'$ relative to the normalized coordinate system $o'x'y'z'$, and $2\varphi = 180°$, $t = 1$.

Solution

Since $2\varphi = 180°$, we have

$$\frac{1}{2}\left(\frac{\sin 2\varphi}{\cos 2\varphi + 1}\right) = \infty.$$

On the other hand, neither the vector $\overrightarrow{M_0M}'$ nor the vector product $(\overrightarrow{M_0M}' \times T)$ in this case will be equal to zero even if by the vector $\overrightarrow{M_0M}'$ is meant a vector to join any pair of conjugate points, i.e.,

$$\overrightarrow{M_0M}' \neq 0, \quad \overrightarrow{M_0M}' \times T \neq 0.$$

Thus, we obtain

$$r_0 = \infty, \quad r_i = \infty.$$

The result shows that the screw axis of image formation for a right-angle prism is located in-

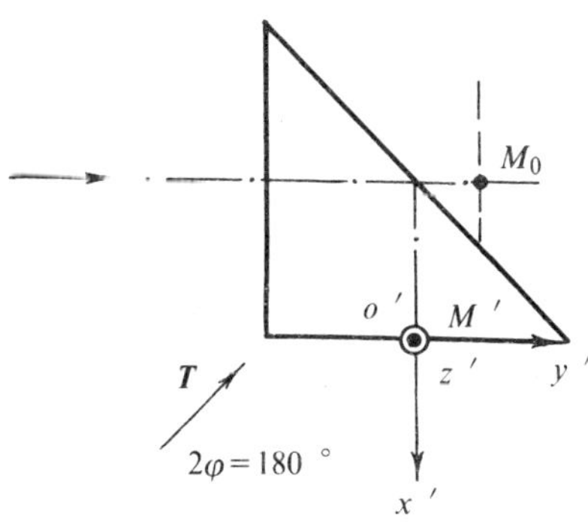

Fig. 2.44. A diagram illustrating the location of the screw axis $\vec{\lambda}$ and the center of inversion I for a right-angle prism with only one reflecting surface

finitely distant from the prism. This conclusion holds true also for those prisms with an odd number of reflecting surfaces, where all the mirror normals are coplanar, or in other words, they are parallel to the same plane, except the cases when $\overrightarrow{M_0M'}=0$ or $\overrightarrow{M_0M'}\times T=0$.

Therefore, a single plane mirror is one of such exceptions. In Fig. 2.45, any point M in the mirror plane can be considered as a pair of conjugate points M and M' which are coincident with each other. Also the origin o' is located at the same point M'.

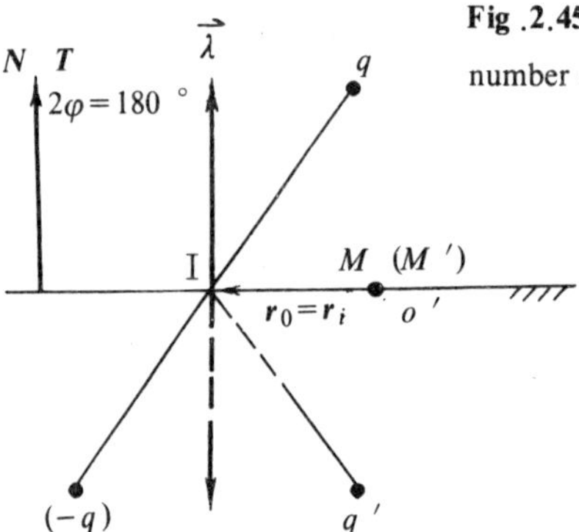

Fig. 2.45. A diagram illustrating an infinite number of screw axes for a single plane mirror

Now, by putting $\overrightarrow{M_0M'}=\overrightarrow{MM'}=0$, $2\varphi=180°$ and $t=1$ into Eqs. (2.88) and (2.90), we obtain

$$r_i = \infty \cdot 0 \,, \quad r_0 = \infty \cdot 0 \,.$$

The above results, together with the vector product in Eqs. (2.88) and (2.90), show that the center of inversion I can be any point in the mirror plane, and the screw axis $\vec{\lambda}$ will be the mirror normal erected at the corresponding center of inversion. Then, any vector normal to the mirror can be the screw axis $\vec{\lambda}$ for that mirror.

Also, some important properties regarding the screw axis $\vec{\lambda}$ for prisms with an odd number of reflecting surfaces are stated below:

"With the object fixed, neither the location nor the orientation of the conjugate image will change no matter through what angle the prism has rotated around its screw axis $\vec{\lambda}$." The reason for this is simply because the screw axis $\vec{\lambda}$ and the center of inversion I both will not change in the course of rotating the prism. That is why we would define such a screw axis for the o.n. prism."

"With the object fixed, as the prism is displaced by a distance Δg along its screw axis $\vec{\lambda}$ in one direction, the conjugate image will displace by twice the distance, namely $2\Delta g$, along the screw axis $\vec{\lambda}$ but in the opposite direction."

"The screw axis $\vec{\lambda}$ for a prism with an odd number of reflecting surfaces and with coplanar mirror normals, according to the previously mentioned definition, is located infinitely distant, except a few cases when $\overrightarrow{M_0M'}$ (or $\overrightarrow{MM'}$) = 0 or $\overrightarrow{M_0M'}$ (or $\overrightarrow{MM'}$) $\times T = 0$."

Finally, it should be noted that the center of inversion I for the o.n. prisms, according to the definition in this text, appears to be a unique point which will be imaged at itself.

Problems

2.1 If you hold a Penta prism and look into the entrance face, what will you see in it? Which phenomenon or picture you will see, one, resulting from the unfolding process only or the other, resulting from the whole process including both the unfolding and the reducing?

2.2 The optical tunnel for a, Penta prism is shown in Fig. 2.46. When an observer is looking into its entrance face ab, at which plane will he see the exit face $g''h''$, the exit face gh of the plane-parallel plate or the exit face $g_0 h_0$ of the equivalent air plate?

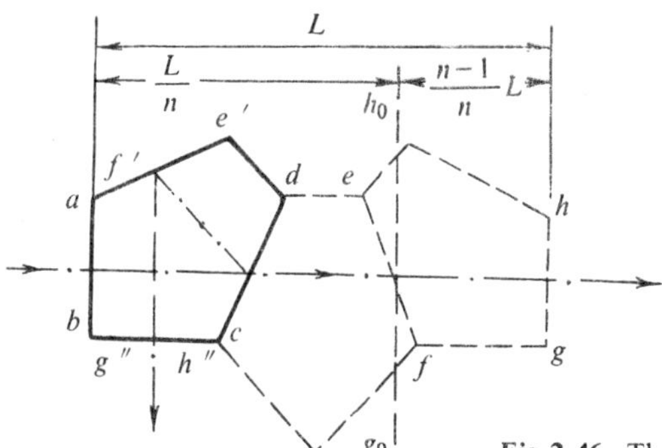

Fig.2.46. The tunnel diagram for a Penta prism

2.3 Since a reflecting prism will disappear after the unfolding and reducing processes. Then how could we investigate the object and image relationship by making use of these two processes through which the prism has become nothing?

2.4 We have derived the idea of the equivalent air plate (or equivalent air thickness) for a prism in our own way. Could you tell the difference and the advantage of doing so as compared with the traditional?

2.5 Could you give an explanation of why the conception of the equivalent air plate can be utilized to determine prism sizes from the point of view of what has been described in the text — the relationship between two conjugate spaces for a prism?

2.6 If you hold a Penta prism and look back into the exit face $g''h''$, then what will you see in it?

2.7 In Fig. 2.47, the objective lens yields the image $A_0 F_0 B_0$ of some infinitely distant object in the posterior focal plane, before the Pechan prism is inserted. Determine how far and in which direction the image $A_0 F_0 B_0$ will move after inserting the Pechan prism into the ray paths at the position defined by a distance d from the lens to the entrance face. Given $f_0' = 120$ mm, $d = 20$ mm, $n = 1.5163$, and the aperture of the prism $D = 20$ mm.

2.8 Do you think that the characteristic vector T and the characteristic angle 2φ for a specific reflecting prism are varying with the input vector A? Why?

2.9. Tell in your own words the significance of locating a pair of basic conjugate points for a reflecting prism.

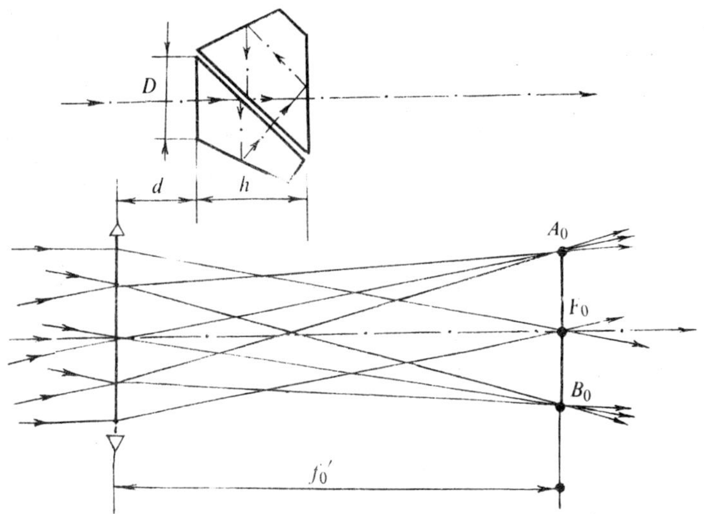

Fig.2.47. Inserting a Pechan prism into a system

Chapter 3
Rotation of Mirror Systems

In this chapter, we shall discuss the image motion caused by a finite angular displacement of a mirror system around a fixed axis with the object remaining stationary.

It should be noted that in the heading the term *mirror system* has replaced the term *reflecting prism*. This is simply because the angle through which a reflecting prism turns is allowed only to be small-valued when the prism is located in a convergent beam. For the problems regarding a reflecting system in a collimated beam, however, there is no need to distinguish between the mirror system and the reflecting prism.

Of course, the theory for treating the problem of finite rotation of a mirror system remains valid in dealing with the case of a small rotation of mirrors as well as of prisms; nevertheless, for certain reasons the problem regarding image motion due to a small rotation of a prism will be discussed in full detail in the later chapters.

The material provided in the present chapter may be used for solving a variety of technical problems regarding mirrors and prisms such as scanning, beam-steering, tracking, angle-measuring, image-stabilizing, aligning, and so on.

3.1 Problems in a Collimated Beam

As mentioned in Sect. 2.6, when the prism is located in a collimated beam, we are concerned about the image orientation (a quantity of three degrees of freedom) only.

Suppose an arbitrary mirror system, Fig. 3.1, has rotated through a finite angle θ about a fixed axis P. Assume an arbitrary input vector A, and let A' and A'_0 be two corresponding output vectors formed by the mirror system prior to and after its rotation, respectively.

Now, we proceed to derive the formula relating the final image vector A'_0 to the conjugate object vector A.

To solve the problem involving a rotating mirror system, transformation of coordinates is traditionally considered as a very effective mathematical tool.

As was done in Sect. 1.3, assume two rectangular coordinate systems: one, the moving coordinate system $x_1' y_1' z_1'$, which is rigidly connected to the rotating mirror system, and the other the fixed coordinate system $x' y' z'$, which is connected to the fixed space. In addition, for the sake of simplicity, the fixed coordinate axes $x' y' z'$ are always assumed to be parallel to the initial orientation of the moving coordinate axes $x_1' y_1' z_1'$, respectively, as shown in Fig. 3.1.

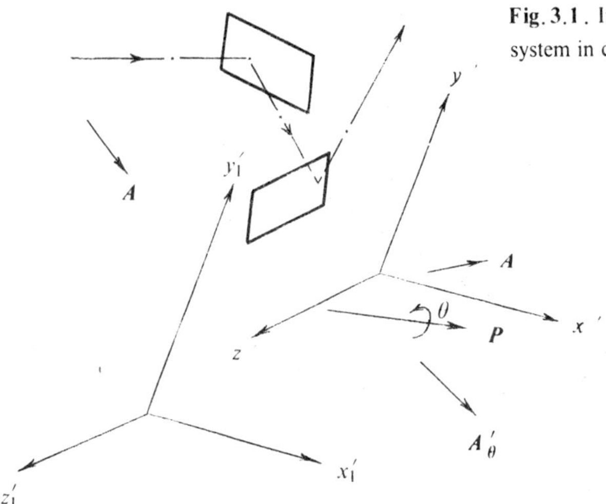

Fig. 3.1. Imaging with a rotating mirror system in collimated light

According to the convention adopted in this book, the symbols $G_{1\,'0\,'}$ and $G_{0\,'1\,'}$ represent the matrices of transformation of coordinates from the moving coordinate system $x_1'\,y_1'\,z_1'$ to the fixed coordinate system $x'\,y'\,z'$ and from the fixed coordinate system $x'\,y'\,z'$ to the moving coordinate system $x_1'\,y_1'\,z_1'$, respectively. No doubt, we have here: $G_{0\,'1\,'} = (G_{1\,'0\,'})^{-1} = (G_{1\,'0\,'})'$.

From the conclusion obtained in Sect. 1.3 and by using Eq. (1.12), we can calculate the matrix $G_{1\,'0\,'}$ in the following way:

$$
G_{1\,'0\,'} = S_{p,0} = \begin{pmatrix}
\cos\theta + 2P_x^2\sin^2\dfrac{\theta}{2} & -P_z\sin\theta + 2P_x P_y\sin^2\dfrac{\theta}{2} & P_y\sin\theta + 2P_x P_z\sin^2\dfrac{\theta}{2} \\[2mm]
P_z\sin\theta + 2P_x P_y\sin^2\dfrac{\theta}{2} & \cos\theta + 2P_y^2\sin^2\dfrac{\theta}{2} & -P_x\sin\theta + 2P_y P_z\sin^2\dfrac{\theta}{2} \\[2mm]
-P_y\sin\theta + 2P_x P_z\sin^2\dfrac{\theta}{2} & P_x\sin\theta + 2P_y P_z\sin^2\dfrac{\theta}{2} & \cos\theta + 2P_z^2\sin^2\dfrac{\theta}{2}
\end{pmatrix}
\quad (3.1)
$$

Let us begin with transforming the coordinates of the input vector A relative to the fixed coordinate system $x'\,y'\,z'$ to the coordinates of the same vector A relative to the moving coordinate system $x_1'\,y_1'\,z_1'$:

$$(A)_{1\,'} = G_{0\,'1\,'}(A)_{0\,'}$$

or

$$(A)_{1\,'} = G_{1\,'0\,'}^{-1}(A)_{0\,'}. \quad (3.2)$$

Then, the output vector A'_0 formed by the mirror system will readily be found primarily relative to the moving coordinate system $x_1'\,y_1'\,z_1'$

$$(A_0')_1 = R\,(A)_{1'} \,, \tag{3.3}$$

where R represents the reflection matrix for the mirror system, expressed relative to the moving coordinate system $x_1'\,y_1'\,z_1'$. However, this reflection matrix R will easily be determined in accordance with the initial position of the mirror system where the moving coordinate axes $x_1'\,y_1'\,z_1'$ appear to be parallel to the orientation of the fixed coordinate axes $x'\,y'\,z'$, respectively, that is to say, the reflection matrix R for a moving mirror system (or a prism) would be calculated in the same manner as that in which the matrix for a fixed mirror system (or prism) was handled in Chapter 2 so long as the relative orientation of the fixed coordinate system with respect to the moving coordinate system has been properly assumed.

Usually, the viewer is located in the fixed space and we ought to transform the coordinates of the output vector A_0' from the moving coordinate system $x_1'\,y_1'\,z_1'$ to the fixed coordinate system $x'\,y'\,z'$

$$(A_0')_{0'} = G_{1'0'}\,(A_0')_{1'} \,. \tag{3.4}$$

Combining Eqs. (3.2), (3.3) and (3.4), we arrive finally at the following equation:

$$(A_0')_{0'} = G_{1'0'}\,RG_{1'0'}^{-1}\,(A)_{0'} \,. \tag{3.5}$$

If introducing a new symbol R_k:

$$R_k = G_{1'0'}\,R\,G_{1'0'}^{-1} \,, \tag{3.6}$$

then Eq. (3.5) can be written in a more compact form:

$$(A_0')_{0'} = R_k\,(A)_{0'} \,, \tag{3.7}$$

where R_k may be called the *kinematical reflection matrix* for the mirror system.

In this case, R and R_k may be considered as two reflection matrices both expressed relative to the fixed coordinate system $x'\,y'\,z'$ for the mirror system in accordance with its initial and final positions, respectively.

Sometimes, only the change in orientation $\Delta A_0'$ of the image vector due to the rotation of the mirror system is of interest. Then, we have

$$(\Delta A_0')_{0'} = (A_0')_{0'} - (A')_{0'} \,. \tag{3.8}$$

Using Eqs. (2.29) and (3.5), Eq. (3.8) can be written as

$$(\Delta A_0')_{0'} = (G_{1'0'}\,RG_{1'0'}^{-1} - R)\,(A)_{0'} \,. \tag{3.9}$$

Example

The optical system of a panoramic telescope used to scan the entire horizon is shown in Fig. 3.2. The elements 3, 4 and 5 forming the basic system of the telescope are held from turning. In the parallel light above the objective, the head prism 1 can be rotated about a vertical axis through $360°$, but when this is done, the image rotates at prism speed. This rotation will be compensated for by the use of a Dove prism 2. The Dove prism must be rotated at half the speed of the head prism. Try to verify it.

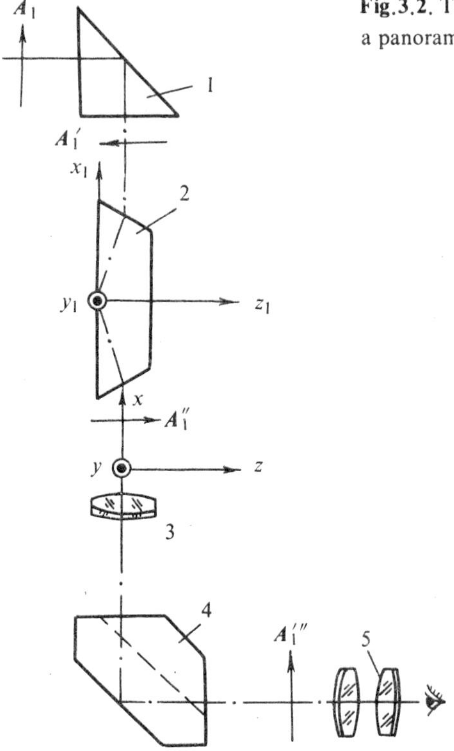

Fig.3.2. The optical system of a panoramic telescope

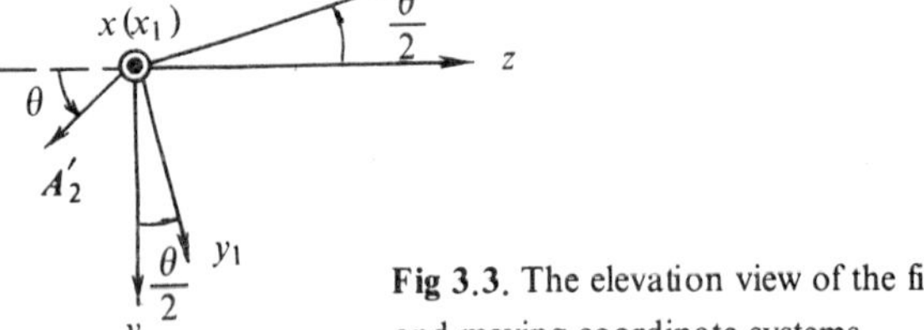

Fig 3.3. The elevation view of the fixed and moving coordinate systems

Solution

Suppose that the diagram in Fig. 3.2 indicates a correct arrangement of three prisms in relation to one another at a certain instant.

Let a distant erect object vector A_1 be imaged through the head prism, then the Dove prism, and then the combination including the objective and the Amici prism; assume that A_1', A_1'', and A_1''' denote its successive image vectors in image spaces of the three prisms, respectively. Thus, the latter one A_1'' within the front focal plane of the eyepiece 5 should be erect too.

We proceed to discuss the object and image relation for the Dove prism. For this purpose, a fixed coordinate system $x\,y\,z$ and a moving coordinate system $x_1\,y_1\,z_1$ are assumed, the latter being rigidly connected to the rotating Dove prism. Since the part of space to be viewed is synchronistically rotated with the head scanning prism, we would assume another distant object vector A_2 to replace A_1, and the vector A_2 will be imaged through the head prism in the orientation designated by A_2' in the elevation view, Fig. 3.3, after the prism has rotated about the vertical axis x through an arbitrary angle θ. As mentioned above, the Dove prism has then rotated about the vertical axis x through an angle $\dfrac{\theta}{2}$.

Now, let $A_{2,0}''$ be the image vector of the object vector A_2' formed by the rotating Dove prism.

From Eq. (3.5), we have

$$(A_{2,0}'')_0 = G_{1,0}\,R\,G_{1,0}^{-1}\,(A_2')_0 \,, \tag{3.10}$$

where according to Eqs. (1.13) and (1.4),

$$G_{1,0} = \begin{pmatrix} 1 & 0 & 0 \\ 0 & \cos\dfrac{\theta}{2} & -\sin\dfrac{\theta}{2} \\ 0 & \sin\dfrac{\theta}{2} & \cos\dfrac{\theta}{2} \end{pmatrix}, \quad G_{1,0}^{-1} = \begin{pmatrix} 1 & 0 & 0 \\ 0 & \cos\dfrac{\theta}{2} & \sin\dfrac{\theta}{2} \\ 0 & -\sin\dfrac{\theta}{2} & \cos\dfrac{\theta}{2} \end{pmatrix},$$

$$R = \begin{pmatrix} 1 & 0 & 0 \\ 0 & 1 & 0 \\ 0 & 0 & -1 \end{pmatrix}.$$

Again, from Fig. 3.3,

$$(A_2')_0 = \begin{pmatrix} 0 \\ \sin\theta \\ -\cos\theta \end{pmatrix}.$$

Putting the above data into Eq. (3.10) gives

$$(A''_{2,0})_0 = \begin{pmatrix} 0 \\ 0 \\ 1 \end{pmatrix}$$

or

$$A''_{2,0} = \boldsymbol{k}.$$

Since $A''_{2,0}$ is parallel to A''_1, the image vector of the object vector A_2 in the image space of the Amici prism will be maintained erect, so that no image lean happens during the scanning.

Example

In Fig. 3.4, is shown a beam- steering collimator to be mounted in an aircraft and used to track a ground point for measuring the relative position of this point with respect to the air-craft. The parallel beam A corresponding to the reticle center c, entering the double mirror system including a gimbaled mirror 1 and a fixed mirror 2, will be projected into the space as the line of sight. The normal N_1 to the rotating mirror in its initial position and the

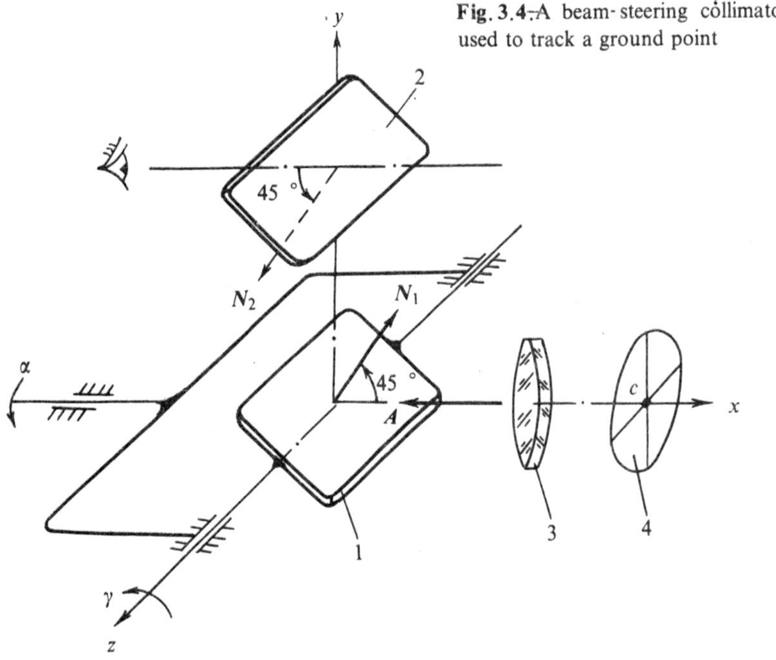

Fig. 3.4. A beam- steering collimator in aircraft used to track a ground point

normal N_2 to the fixed mirror are shown in Fig. 3.5. Determine the direction of the line of sight relative to the fixed coordinate system xyz in terms of the angles of rotation of the gimbaled mirror, namely, the gimbal angles α and γ.

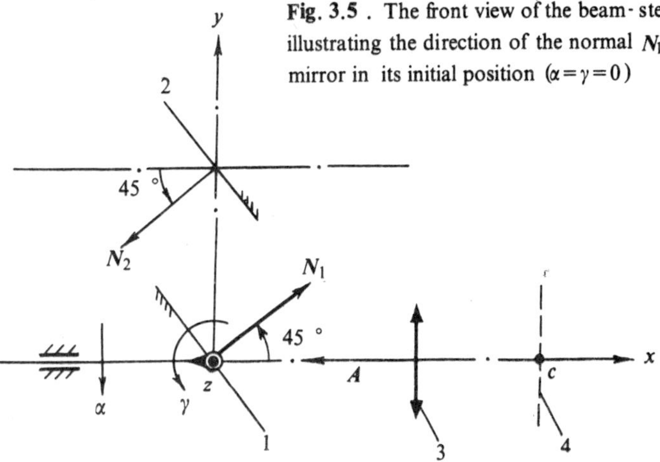

Fig. 3.5 . The front view of the beam-steering collimator illustrating the direction of the normal N_1 to the gimbaled mirror in its initial position $(\alpha = \gamma = 0)$

Solution

Let A' and A'' denote the emerging beams after mirror 1 and mirror 2 , respectively, corresponding to the entering beam A . Customarily, assume a vector $A_l = -A''$ to be the direction of the line of sight.

To simplify the method of deriving the formula relating A' to A, suppose that the mirror 1 is rotated to its final position in two steps, and accordingly an intermediate coordinate system $x_1 y_1 z_1$ is suggested in addition to the fixed and moving coordinate systems xyz and $x_2 y_2 z_2$, the latter being called the final moving coordinate system. Also, for convenience the rotation of the mirror about the outer pivot axis x will be referred to as its motion completed in the first step while the rotation about the inner pivot axis z_1 (not z) will be regarded as that motion completed in the second step.

The two-step motion of the mirror 1 is illustrated in Fig. 3.6 . In the first step, the intermediate coordinate system acting as the moving coordinate system for the time being, starting from its initial orientation xyz, has rotated together with the mirror through an angle α about the axis x to the new orientation $x_1 y_1 z_1$, while in the second step, the final moving coordinate system starting from its home orientation $x_1 y_1 z_1$, which is now reassumed as the relatively fixed coordinate system, will rotate with the mirror through an angle γ about the axis z_1 to the final orientation $x_2 y_2 z_2$.

Let G_{fm} denote the matrix of transformation of coordinates from the fixed coordinate system xyz to the moving coordinate system $x_2 y_2 z_2$, and G_{mf} denote the matrix of transformation of coordinates from the moving coordinate system $x_2 y_2 z_2$ to the fixed coordinate system xyz.

Now, according to Eqs. (1.29) and (1.13) we have

$$G_{fm} = G_{02} = S_{k,-\gamma}\, S_{i,-\alpha} = S_{k,\gamma}^{-1}\, S_{i,\alpha}^{-1} , \tag{3.11}$$

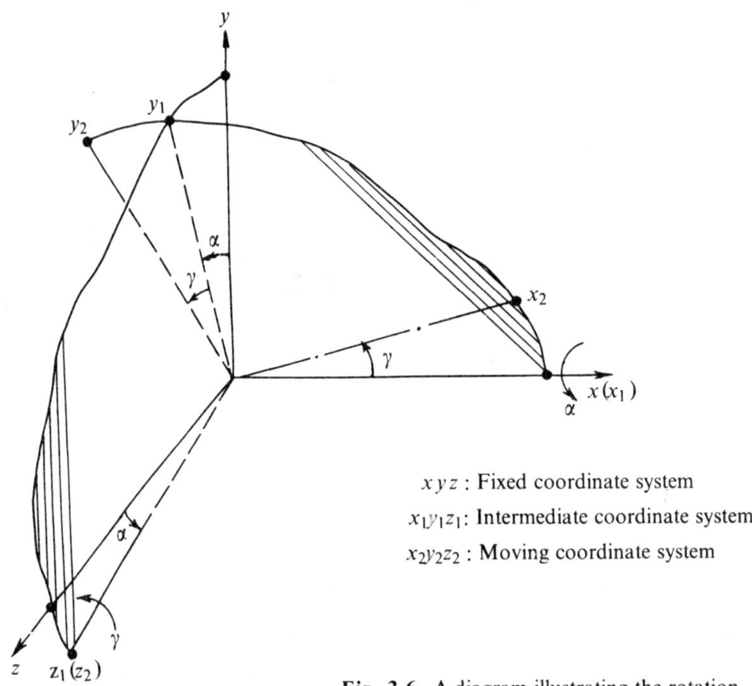

Fig .3.6. A diagram illustrating the rotation
, of the gimbaled mirror completed in two steps

and similarly

$$G_{mf}=G_{20}=S_{i,\alpha}S_{k,\gamma}\ . \tag{3.12}$$

No doubt,

$$S_{k,\gamma}^{-1}\,S_{i,\alpha}^{-1}=(S_{i,\alpha}\,S_{k,\gamma})^{-1}.$$

Therefore, the relationship $G_{02}=G_{20}^{-1}$ will automatically be ensured.

Assume R_1 and R_2 to be the reflection matrices for the mirror 1 and mirror 2 , respectively. Then, from Eqs. (3.5) and (2.29)

$$(A')_0 = S_{i\,\alpha}S_{k,\gamma}\,R_1\,S_{k,\gamma}^{-1}\,S_{i,\alpha}^{-1}\,(A)_0\ ,$$
$$(A'')_0 = R_2\,(A')_0\ .$$

And by definition of the LOS,

$$(A_l)_0 = -\,(A'')_0\ .$$

Combining the above three equations leads to

$$(A_l)_0 = -\,R_2 S_{i,\alpha}S_{k,\gamma}\,R_1 S_{k,\gamma}^{-1}\,S_{i,\alpha}^{-1}\,(A)_0 \tag{3.13}$$

or

$$(A_l)_0 = -R_2 S_{i,\alpha} S_{k,\gamma} R_1 S'_{k,\gamma} S'_{i,\alpha} (A)_0 ,$$

(3.14)

where by use of Eqs. (1.4) and (1.13)

$$R_1 = \begin{pmatrix} 0 & -1 & 0 \\ -1 & 0 & 0 \\ 0 & 0 & 1 \end{pmatrix}, \quad R_2 = \begin{pmatrix} 0 & -1 & 0 \\ -1 & 0 & 0 \\ 0 & 0 & 1 \end{pmatrix},$$

$$S_{i,\alpha} = \begin{pmatrix} 1 & 0 & 0 \\ 0 & \cos\alpha & -\sin\alpha \\ 0 & \sin\alpha & \cos\alpha \end{pmatrix}, \quad S_{k,\gamma} = \begin{pmatrix} \cos\gamma & -\sin\gamma & 0 \\ \sin\gamma & \cos\gamma & 0 \\ 0 & 0 & 1 \end{pmatrix},$$

$$S'_{i,\alpha} = S_{i,-\alpha} = \begin{pmatrix} 1 & 0 & 0 \\ 0 & \cos\alpha & \sin\alpha \\ 0 & -\sin\alpha & \cos\alpha \end{pmatrix}, \quad S'_{k,\gamma} = S_{k,-\gamma} = \begin{pmatrix} \cos\gamma & \sin\gamma & 0 \\ -\sin\gamma & \cos\gamma & 0 \\ 0 & 0 & 1 \end{pmatrix}.$$

In addition,

$$(A)_0 = \begin{pmatrix} -1 \\ 0 \\ 0 \end{pmatrix}.$$

Putting the above matrices into Eq. (3.14) yields

$$(A_l)_0 = \begin{pmatrix} \cos2\gamma\cos\alpha \\ -\sin2\gamma \\ -\cos2\gamma\sin\alpha \end{pmatrix}$$

(3.15)

or

$$A_l = \cos2\gamma\cos\alpha \cdot i - \sin2\gamma \cdot j - \cos2\gamma\sin\alpha \cdot k .$$

(3.16)

To make the result clear, the unit vector A_l is illustrated in Fig. 3.7, representing the direction of the LOS defined by three components in the fixed coordinate system xyz.

3.2 Problems in a Convergent Beam

As mentioned in Sect. 2.6, while the mirror system is located in a convergent beam, we are concerned about both the orientation and location of the image (quantities of six degrees of freedom in total).

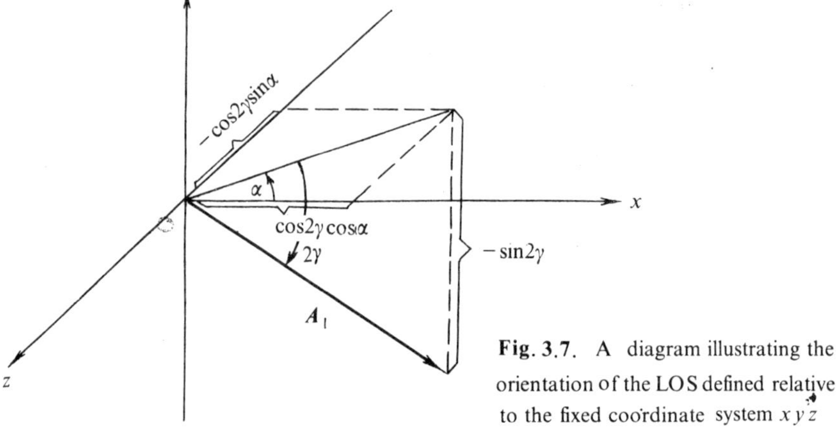

Fig. 3.7. A diagram illustrating the orientation of the LOS defined relative to the fixed coordinate system $x\,y\,z$

The problem presented here is equal to requiring the determination of the image point of an arbitrary, fixed object point formed by a mirror system which has rotated through a given angle about a fixed axis.

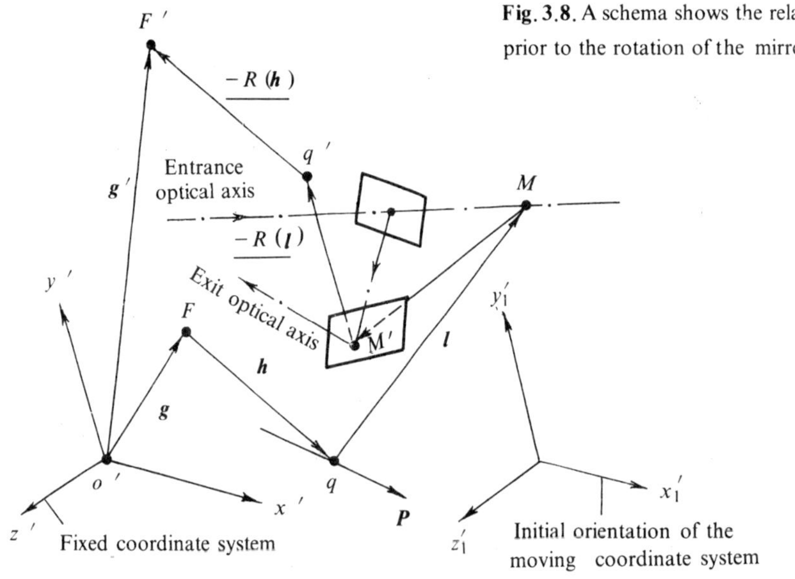

Fig. 3.8. A schema shows the relationship prior to the rotation of the mirror system

In Fig.3. 8, is shown an arbitrary mirror system in its initial position, which will rotate through a finite angle θ about a fixed axis p . The reflection matrix R and a pair of conjugate points $\overrightarrow{MM}\,'$ are given. Assume a fixed coordinate system $o\,'x\,'y\,'z\,'$ as a reference frame, and also a moving coordinate system $x_1{}'y_1{}'z_1{}'$ as shown in its initial orientation. An arbitrary

90

point q on the action line of P is given to locate the axis of rotation.

Assume an arbitrary, stationary object point F, and let F' and F_0' be two corresponding image points formed by the mirror system prior to and after its rotation, respectively.

Since the length between points F and M is changing during the rotation of the mirror system, we will no longer be able to use the vector $\overrightarrow{FM}$ directly to locate the object point F with respect to the mirror system, which is unlike the situation being depicted in Fig. 2.31. Here, the location of the object point F is defined firstly relative to the point q by the vector $h = \overrightarrow{Fq}$, and then the location of the point q in turn will be defined relative to the mirror system by the vector $l = \overrightarrow{qM}$. Consequently, the vector $\overrightarrow{FM}$ has been resolved into two components h and l so that the former remains stationary while the latter will rotate together with the mirror system, and thus these two vectors h and l can easily be handled separately in the following discussion.

From Eq. (2.54), we can readily find the position vector g' of the image point F'

$$(g')_{0'} = (g)_0 + (h)_{0'} + (l)_{0'} + (\overrightarrow{MM'})_0 - R(l)_{0'} - R(h)_{0'}, \tag{3.17}$$

and according to the principle in the previous section we may also easily determine the position vector g_0' of the final image point F_0'

$$(g_0')_0 = (g)_0 + (h)_0 + G_{1'0'}(l)_0 + G_{1'0'}(\overrightarrow{MM'})_0 - G_{1'0'}R(l)_0 - G_{1'0'}RG_{1'0'}^{-1}(h)_0. \tag{3.18}$$

Considering the relationships: $G_{1'0'} = S_{p,0}$ and $G_{1'0'}^{-1} = S_{p,-0}$, Eq. (3.18) can also be written

$$(g_0')_0 = (g)_{0'} + (h)_{0'} + S_{p,0}(l)_{0'} + S_{p,0}(\overrightarrow{MM'})_{0'}$$
$$- S_{p,0}R(l)_{0'} - S_{p,0}RS_{p,-0}(h)_{0'}. \tag{3.19}$$

As the reference frame $x'y'z'$ is arbitrarily selected, the subscript "$0'$" in Eq. (3.19) may be removed

$$(g_0') = (g) + (h) + S_{P,0}(l) + S_{p,0}(\overrightarrow{MM'}) - S_{p,0}R(l) - S_{p,0}RS_{p,-0}(h). \tag{3.20}$$

Sometimes, only the displacement $\Delta g_0' = \overrightarrow{F'F_0'}$ of the image point is of interest. Then, subtracting Eq. (3.17) from Eq. (3.20) gives

$$(\Delta g_0') = (S_{p,0} - E)(l + \overrightarrow{MM'}) - (S_{p,0} - E)R(l) - (S_{p,0}RS_{p,-0} - R)(h). \tag{3.21}$$

Obviously, the position vector g of the object point does not occur in Eq. (3.21), so that the displacement $\Delta g_0'$ of the image point is independent of the choice of the origin o'.

Example

In Fig. 3.9, is shown a collimator to be mounted on a seeker in aircraft, comprising particularly a gyro mirror 2 rigidly attached to the one end of the spin axis of the gyro and being capable of rotating around any axis passing through the gyro center q. The divergent beam starting from the cross point F of the reticle 1 is projected through the collimator to the space, resulting in a parallel beam, and then will be reflected to the viewer by the semisilvered glass 5.

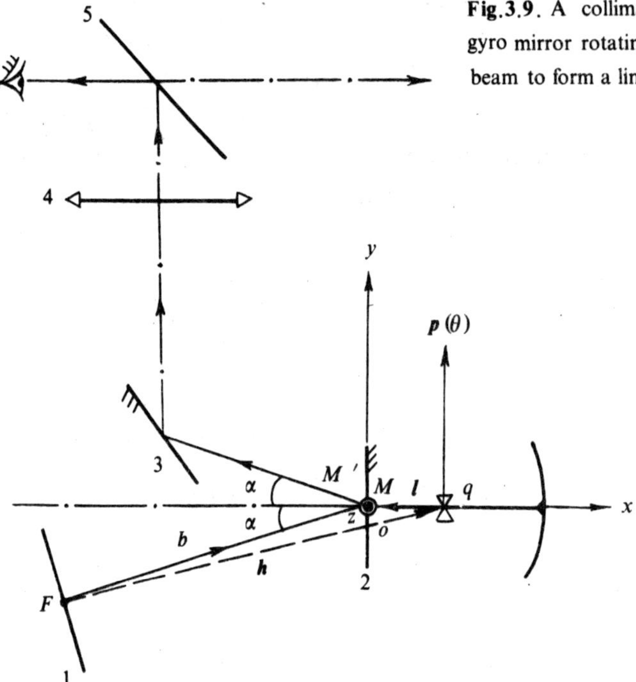

Fig.3.9. A collimator comprising a gyro mirror rotating in a convergent beam to form a line of sight

Assume an aircraft reference frame $oxyz$. The line of sight is said to be in its null orientation parallel to the x direction if the gyro mirror is in its home position, as indicated in Fig.3.9. Now, suppose that the movable mirror has rotated through an angle θ about the axis P passing through q and parallel to the y direction. Determine the direction of the line of sight relative to its null orientation. It should be noted that the origin o representing the initial location of the center of the gyro mirror is a fixed point relative to the aircraft.

Solution

In Fig.3.9, a pair of conjugate points M and M' for the mirror 2 is chosen to be located at point o and coincident with one another. Vectors h and l have the same meaning as that in Fig.3.8.

From the geometrical relation in the diagram, vectors l and h are easily to be found

$$l = \overrightarrow{qM} = -l \cdot i,$$
$$h = \overrightarrow{Fq} = \overrightarrow{FM} - l = (b\cos\alpha + l) \cdot i + b\sin\alpha \cdot j,$$

where $b = \overline{Fo}$ and α being shown in Fig 3.9.

According to Eqs. (1.13)and (1.4),

$$S_{p,0} = \begin{pmatrix} \cos\theta & 0 & \sin\theta \\ 0 & 1 & 0 \\ -\sin\theta & 0 & \cos\theta \end{pmatrix} \;,\qquad R = \begin{pmatrix} -1 & 0 & 0 \\ 0 & 1 & 0 \\ 0 & 0 & 1 \end{pmatrix} \;,$$

and thus

$$(S_{p,0} - E) = \begin{pmatrix} \cos\theta - 1 & 0 & \sin\theta \\ 0 & 0 & 0 \\ -\sin\theta & 0 & \cos\theta - 1 \end{pmatrix} \;,$$

$$(S_{P,0} - E)R = \begin{pmatrix} 1 - \cos\theta & 0 & \sin\theta \\ 0 & 0 & 0 \\ \sin\theta & 0 & \cos\theta - 1 \end{pmatrix} \;,$$

$$(S_{p,0}R\,S_{p,-0}) = \begin{pmatrix} -\cos2\theta & 0 & \sin2\theta \\ 0 & 1 & 0 \\ \sin2\theta & 0 & \cos2\theta \end{pmatrix} \;,$$

$$(S_{P,0}RS_{P,-0} - R) = \begin{pmatrix} -\cos2\theta + 1 & 0 & \sin2\theta \\ 0 & 0 & 0 \\ \sin2\theta & 0 & \cos2\theta - 1 \end{pmatrix} \;.$$

Putting the above matrices and the column matrices (l) and (h) into Eq. (3. 21) and considering $\overrightarrow{MM}' = 0$ yields

$$(\Delta g_0')_0 = \begin{pmatrix} b\cos\alpha \, \cos2\theta - b\cos\alpha + l\,(1 - 2\cos\theta + \cos2\theta) \\ 0 \\ -b\cos\alpha \, \sin2\theta + l\,(2\sin\theta - \sin 2\theta) \end{pmatrix} . \tag{3.22}$$

Obviously, the resulting column matrix gives three components of the displacement $\Delta g_0'$ $= \overrightarrow{F'F_0'}$ of the image point F' along the frame axes xyz. Now, we need to find the corresponding displacement $\Delta g_0'' = \overrightarrow{F''F_0''}$ of the image point F'' in the image space of the whole optical system of the seeker, i.e., the space of the line of sight.

For this purpose, a pair of object and image coordinate systems $x_1 y_1 z_1$ and $x_1' y_1' z_1'$, as indicated in Fig. 3. 10, is constructed for the double mirror system consisting of 3 and 5, in which the objective 4 is excluded for the time being. Also, a pair of conjugate displacements $\Delta g_0' = \overrightarrow{F'F_0'}$ and $\Delta g_0'' = \overrightarrow{F''F_0''}$ for the same mirror system is shown in the figure, and the axes x_1 and x_1' are assumed to pass through the points F' and F'', respectively.

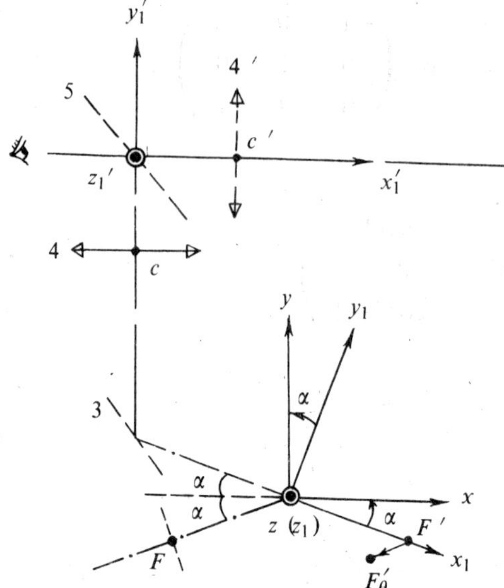

Fig.3.10. The diagram showing the orientations of a pair of object and image coordinate systems $x_1 y_1 z_1$ and $x_1' y_1' z_1'$ for the double mirror system consisting of 3 and 5 (excluding the objective 4) in relation to the reference frame $x\,y\,z$

Now, we shall transform the components (or coordinates) of the displacement $\Delta g_0'$ from the coordinate system xyz to the coordinate system $x_1 y_1 z_1$

$$(\Delta g_0')_1 = G_{01} (\Delta g_0')_0 \ ,$$

where

$$G_{01} = \begin{pmatrix} \cos\alpha & -\sin\alpha & 0 \\ \sin\alpha & \cos\alpha & 0 \\ 0 & 0 & 1 \end{pmatrix} .$$

According to the object and image relationship, we have

$$(\Delta g_0'')_1 = (\Delta g_0')_1 .$$

Combining the above three formulas and considering Eq. (3.22) yields

$$(\Delta g_0'')_1 = \begin{pmatrix} b\cos^2\alpha \cos2\theta - b\cos^2\alpha + l\,(1 - 2\cos\theta + \cos2\theta)\cos\alpha \\ \dfrac{b}{2} \sin2\alpha \cos2\theta - \dfrac{b}{2} \sin2\alpha + l\,(1 - 2\cos\theta + \cos2\theta)\sin\alpha \\ - b\cos\alpha \sin2\theta + l\,(2\sin\theta - \sin2\theta) \end{pmatrix} \qquad (3.23)$$

Referring to Fig. 3.11, the function of the objective can be considered while it is projected onto c' in the image space of the glass, where the viewer is located. The components $\Delta g_{0z_1'}''$ and $\Delta g_{0y_1'}''$ cause deviations $\Delta\beta_\parallel$ and $\Delta\beta_\perp$ which are parallel and perpendicular to the plane of motion of the spin axis, respectively, while the component $\Delta g_{0x_1'}''$ causes convergency or

94

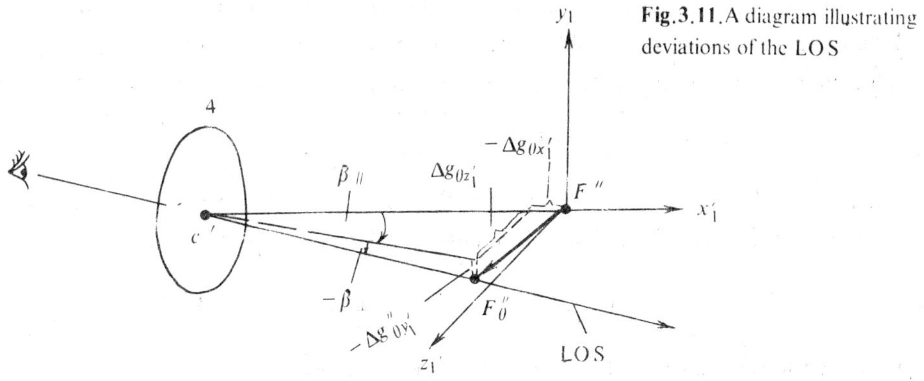

divergency of the emergent beam leading to parallax in the case of viewing an infinitely distant target.

From Fig. 3.11, we have

$$\mathrm{tg}\,\beta_{||} = \frac{\Delta g_{0z_1'}''}{f'+\Delta g_{0x_1'}''} \approx \frac{\Delta g_{0z_1'}''}{f'}\;,$$

$$\sin\beta_\perp = \frac{\Delta g_{0y_1'}''}{f'+\Delta g_{0x_1'}''}\cos\beta_{||} \quad \text{or} \quad \mathrm{tg}\beta_\perp \approx \frac{\Delta g_{0y_1'}''}{f'}\;.$$

Substituting the components from Eq. (3.23) into the above formulas gives

$$\mathrm{tg}\beta_{||} \approx \frac{-b\cos\alpha\,\sin2\theta + l\,(2\sin\theta - \sin2\theta)}{f'}\;,$$

$$\mathrm{tg}\beta_\perp \approx \frac{\dfrac{b}{2}\sin2\alpha\,\cos2\theta - \dfrac{b}{2}\sin2\alpha + l\,(1 - 2\cos\theta + \cos2\theta)\sin\alpha}{f'}\;,$$

where f' denotes the focal length of the objective.

3.3 Theorem of Rotation of a Mirror System

Rewrite Eq. (3.5)

$$(A_0')_{0'} = G_{1'0'}\,RG_{1'0'}^{-1}\,(A)_{0'}\;, \qquad\qquad (3.5)$$

which is thought of as a very fundamental and traditional formula regarding the problem of rotation of a mirror system. Both the physical meaning and accordingly the mathematical process are assumed to be very clear.

However, to gain a better understanding of the behaviour of this formula , particularly its kinematical process, we would replace the matrices $G_{1,0}$ and $G_{1,0}^{-1}$ of transformation of coordinates in Eq. (3.5) with the corresponding rotation matrices $S_{p,0}$ and $S_{p,-0}$, respectively. Thus

$$(A_0') = S_{P,0} R S_{P,-0} (A), \tag{3.24}$$

where the subscript "0" is dropped for the same reason mentioned before.

Now, the equation (3.24) is bracketed in the following manner :

$$(A_0') = S_{p,0} \{ R [S_{p,-0} (A)]\} \tag{3.25}$$

so that the operation of the above formula will be devided into three steps.

Let us investigate each step in detail.

First step: $S_{p,-0} (A)$

The operation involved in this step means that a given object vector A is rotated about the axis P through an angle $-\theta$, by which we have constructed correctly the relative position of the object with respect to the mirror system although the absolute positions of the both are not true. We also realize that the rotation $S_{p,-0}(A)$ is assumed to happen in the object space of the mirror system which is kept stationary temporarily. Introduce a new vector A_1 defined as

$$(A_1) = S_{p,-0} (A). \tag{3.26}$$

Second step: $R [S_{p,-0} (A)]$

According to the physical meaning of the reflection matrix R for a mirror system, the operation performed during the present step is to determine the conjugate part in the image space of the mirror system corresponding to what is involved in the angular brackets .

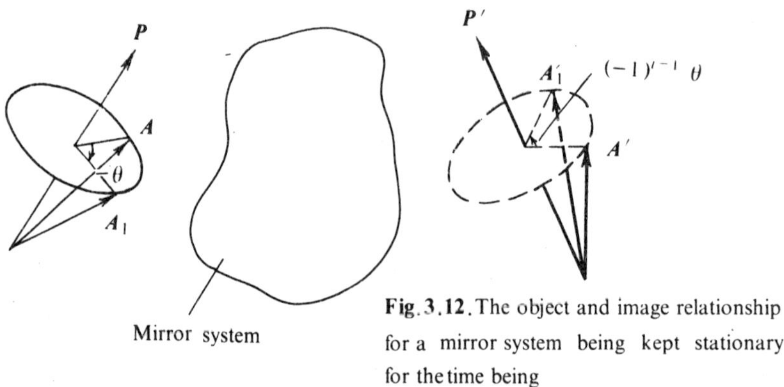

Fig.3.12. The object and image relationship for a mirror system being kept stationary for the time being

To be more clear, we make a schema, Fig.3.12, in which vectors A', P', and A_1' represent the images of the vectors A, P and A_1, respectively, formed by the mirror system which is kept stationary temporarily. Here, P is called the object-axis (of rotation) while P' the image-axis (of rotation).

Since the object vector A has rotated in the object space about the object-axis P through an angle $-\theta$ to the new orientation A_1, the image vector A' will correspondingly rotate in the image space about the image-axis P' through a respective angle $(-1)^l(-\theta)$ to the new orientation A_1' where the coefficient $(-1)^l$ is introduced to take the mirror effect into consideration which will be discussed later.

From what is defined above in Eq. (3.26), we can readily have

$$(A_1') = R[S_{P,-\theta}(A)] . \tag{3.27}$$

Also, the above discussion enables us to express the equation (3.27) in an alternative form, thus stressing on its kinematical feature

$$(A_1') = S_{P',(-1)^l-\theta}(A'). \tag{3.28}$$

It is instructive to notice that the previously obtained vectors A_1 and A_1' are related to the mirror system in its initial position in the same way as the vectors A and A_0' will be related to the mirror system in its final position provided only the respective optical connections and relative positions are of concern.

Third step : $S_{P,\theta}\{R[S_{P,-\theta}(A)]\}$

From Eq. (3.28),

$$S_{P,\theta}\{R[S_{P,-\theta}(A)]\} = S_{P,\theta}(A_1'). \tag{3.29}$$

The mathematical operation indicated in Eq. (3.29) expresses simply that the intermediate image vector A_1' will rotate again through an angle θ about the axis P to a new orientation A_0', i.e., the final image vector.

However, it is well worthwhile to examine the operation during the present step a little further.

Since the relation among the two vectors A_1 and A_1', and the mirror system at its initial place has been established correctly in regard to their relative positions and optical connection, all we have to do in the last step is to adjust their absolute positions without interfering their relative orientation and location with respect to each other.

Now, presume that the vector A_1 and the mirror system rotate jointly through an angle θ about the axis P, then the intermediate image vector A_1' just follows, and as a result the unified whole including A_1, A_1', and the mirror system rotates through θ around P as a single rigid body. So far, the object vector has been recovered to its original orientation: $S_{P,\theta}S_{P,-\theta}(A) = (A)$, and the mirror system has rotated through θ about P to the new position where it should be, therefore the finally obtained image vector A_0' will no doubt be the correct one.

To summarize the whole process, we have a new formula for A_0'

$$(A_0') = S_{P,\theta}S_{P',(-1)^l-\theta}(A'), \tag{3.30}$$

where $(P') = R(P)$, $(A') = R(A)$.

As compared with Eq. (3.5), the new equation (3.30) enables us to begin the computation with the image vector A' as a known value. As a matter of fact, in many cases, particularly for reflecting prisms, where the image vector A' may be obtained by drawing or just by taking a look. Therefore, by using Eq.(3.30)a three-step problem is converted into a two-step problem.

Although the new formula is used to determine the output vector A_0' in regard to its orientation only, yet the idea of this formula can also be extended to dealing with the problems in a convergent beam.

To explore the concept implied in the above equation, we will formulate it as a theorem.

Theorem of rotation of a mirror system :

"With the object fixed, any image motion caused by a rotation of a mirror system through a finite angle θ about an object-axis P may be accomplished by a rotation of the image through an angle $(-1)^{t-1}\theta$ about the image-axis P', followed by another rotation of the intermediate image through an angle θ about the axis P, where t denotes the number of reflections ; P' represents the image of P for the mirror system in its initial position."

The rotation of the image in the first step is referred to as an *optical movement* due to the object and image relationship while the rotation of the image in the second step is regarded as a *mechanical movement*.

We used to call this theorem the *two-step rule*.

Now let us give the reason for introducing the coefficient $(-1)^{t-1}$.

It is shown in Fig.3.13 that an object rotates about an object-axis P through an angle θ while its image in a single plane mirror rotates about the image-axis P' through an angle $-\theta$. In the case of a double mirror system, two minus signs cancel each other, then the angle of rotation of the image will be $+\theta$ again. Reasoning from analog, the angle of rotation of the image will be $(-1)^t\theta$ for a mirror system with t reflections. As we know, the angle of rotation of the

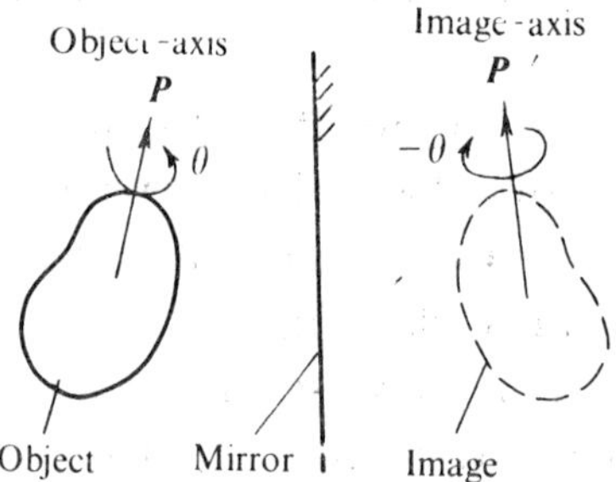

Fig.3.13. A diagram to illustrate the handedness of a mirror image in angular form

object about the object-axis P is $-\theta$, therefore the angle of rotation of the image about the image-axis P' will accordingly be $(-1)^t(-\theta)=(-1)^{t+1}\theta$, and the latter is replaced by $(-1)^{t-1}\theta$ without making any distinction.

3.4 Applications of the Theorem of Rotation of a Mirror System

As mentioned in the previous section, the theorem of rotation of a mirror system can be used to solve problems either in a collimated or in a convergent beam.

3.4.1 Problems in a Collimated Beam

By using the two-step rule, the two basic equations (3.5) and (3.9) for solving problems in a collimated beam in regard to the rotation of mirror systems will become

$$(A_0') = S_{p,0} S_{p',(-1)^{t-1}0} (A') ,\tag{3.30}$$

$$(\Delta A_0') = (S_{p,0} S_{p',(-1)^{t-1}0} - E)(A') ,\tag{3.31}$$

where $(P') = R(P), (A') = R(A).$

These formulas will prove to be useful in cases like the following example.

Example

In Fig. 3.14, is shown an angle mirror for which A and A' represent a pair of input vector and output vector; P denotes a direction parallel to the intersecting edge of the angle mirror. Verify that the output vector A' will not change in direction no matter through what angle θ the mirror system has rotated about the axis P with the input vector A fixed.

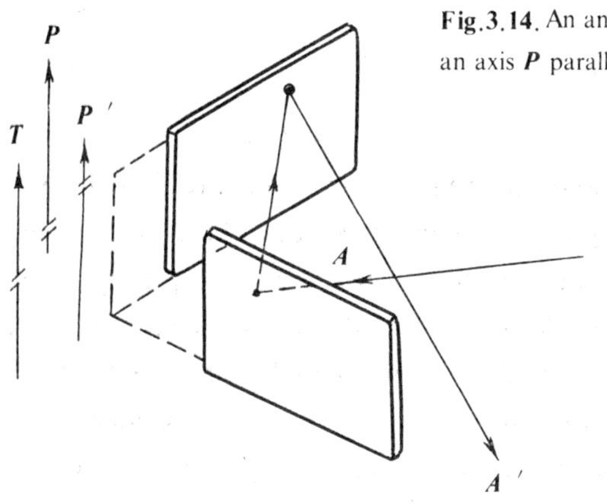

Fig.3.14. An angle mirror rotating about an axis P parallel to its intersecting edge

Proof

From Eq. (2.20), we have

$$(A') = (-1)^t S_{T,2\varphi} (A) .$$

Substituting P and P' for A and A', respectively, in the above equation, and considering $T \parallel P$ yields

$$(P') = (-1)^2 S_{P,2\varphi} (P) .$$

Thus, we have $P' \parallel P$ or $P' = P$.

Now, according to the two-step rule, the image vector A' has been rotating about the axis P' through an angle $(-1)^{2-1}\theta = -\theta$ during the first step, and will then rotate about the axis P

through an angle θ in the second step. Since two axes P' and P are parallel to each other and same in direction, the image vector A' is kept in the original direction.

3.4.2 Image Rotators

From the example in Sect. 3.1, we realize that in some instruments, such as panoramic telescopes, it is necessary to be able to rotate an image about its midpoint to compensate for an unwanted image rotation.

It is well known that the Dove and Pechan prisms, as shown in Fig. 3.15, are two most familiar forms of image rotators, the former being used in collimated light only while the latter in either collimated or convergent light .

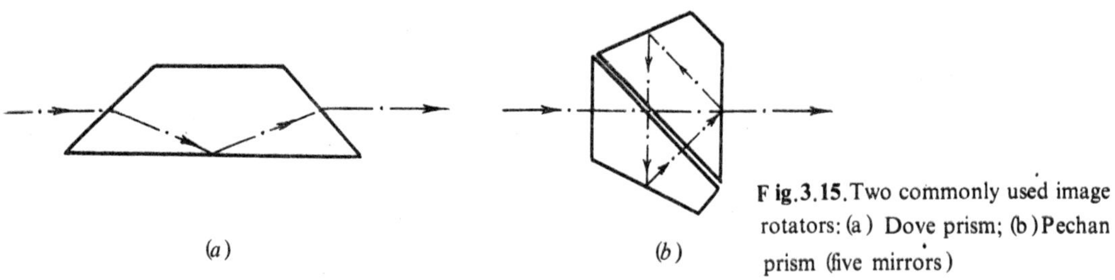

(a)

(b)

F ig.3.15. Two commonly used image rotators: (a) Dove prism; (b)Pechan prism (five mirrors)

The essential and apparent requirements are that the entrance and exit optical axes must lie in the same straight line, which constitutes the axis P of rotation of the prism.

On the other hand, according to the two-step rule, the image motion due to the rotation of the prism through θ about P contains two successive rotations: the preceding one being through $(-1)^{t-1}\theta$ around P' and the following one through θ around P, and as a result the image must rotate constantly about a fixed axis coincident with that around which the prism is turned. Then, to meet this need the image axis P' should also be lying in the same straight line with the object axis P, and thus mathematically we may have either $P' = P$ or $P' = -P$. However, in the latter case the prism having the property of being retrodirective seems impractical.

Now, for the only possible case: $P' = P$, the coefficient $(-1)^{t-1}$ must equal to *plus one*, otherwise the two successive rotations will cancel each other, and the image would not be able to rotate at all. Consequently, there must be an odd number of reflecting surfaces.

To sum up, a model of a prism acting as an image rotator can be defined in terms of its characteristic parameters including t , $T(\vec{\lambda}), 2\varphi$ as well as the entrance and exit optical axes in regard to both the magnitudes and relative positions of them, as indicated in Fig. 3.16.

In the course of rotation of an image rotator about the axis P as in Fig. 3.16, the characteristic vector T or the screw axis $\vec{\lambda}$ is changing in direction all the time, nevertheless the image axis P' will remain coincident with the object axis P due to the unvaried perpendicularity of the vectors $T(\vec{\lambda})$ and P as well as the condition $2\varphi = 180°$.

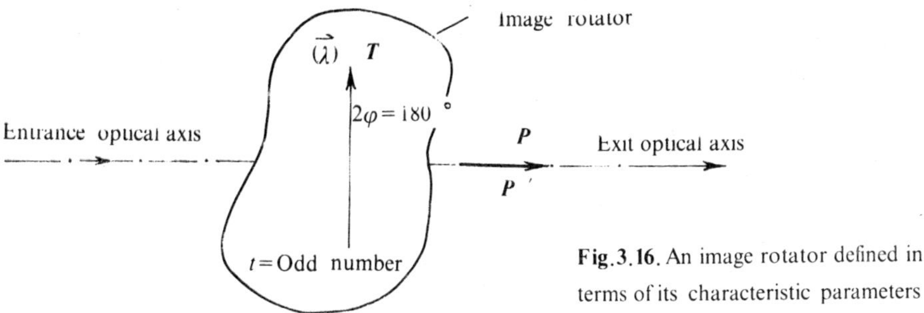

Fig.3.16. An image rotator defined in terms of its characteristic parameters

3.4.3 Problems in a Convergent Beam

When the two-step rule is used to solve the problems in a convergent beam, we should consider not only the orientations but also the locations of the axes P and P'.

Referring to Fig. 3.17, let F be an arbitrary object point in the object space of a given mirror system which will rotate through a finite angle θ about a fixed axis P in the object space, called thus object-axis; F' and P' represent the corresponding images of F and P, respectively, formed by the mirror system in its initial position; q and q' denote a pair of conjugate points

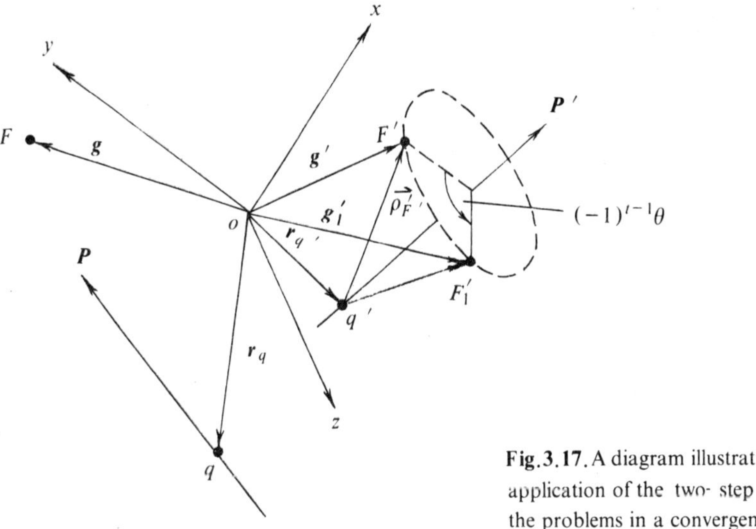

Fig.3.17. A diagram illustrating the application of the two-step rule to the problems in a convergent beam

lying on the axes P and P', respectively, or in other words, the positions of the object- and image-axes in space are located by these two points. Assume a reference coordinate system $oxyz$, and let g, g', g_1', r_q, and $r_{q'}$ represent the position vectors of F, F', F_1', q, and q', respectively; F_1' is the intermediate image point. In order not to make the drawing too complex, we omit the mirror system and the final image point F_0' which will be generated by the object point F through the mirror system in its final position.

Now, we will determine the position vector g_0' of the final image point F_0' by means of the two-step rule.

In the first step, the intermediate image point F_1' can be found by rotating the initial image point F' about the image-axis P' through an angle $(-1)^{l-1}\theta$.

From Fig.3.17, we have

$$g_1' = r_{q'} + \overrightarrow{q'F_1'} \quad , \tag{3.32}$$

$$(\overrightarrow{q'F_1'}) = S_{p',(-1)^{l-1}\theta}(\overrightarrow{\rho_{F'}}) \quad , \tag{3.33}$$

$$\overrightarrow{\rho_{F'}} = g' - r_{q'}. \tag{3.34}$$

Combining the above three equations gives

$$(g_1') = (r_{q'}) + S_{p',(-1)^{l-1}\theta}(g' - r_{q'}) \quad . \tag{3.35}$$

In the second step, the final image point F_0' will be determined by turning the intermediate image point F_1' about the object-axis P through an angle θ. This is equally to use Eq. (3.35) once again by properly substituting. Thus,

$$(g_0') = (r_q) + S_{p,0}(g_1' - r_q). \tag{3.36}$$

Combining Eqs. (3.35) and (3.36), and orderly arranging them yields

$$(g_0') = (E - S_{p,0})(r_q) + S_{p,0}(E - S_{p',(-1)^{l-1}\theta})(r_{q'}) + S_{p,0}S_{p',(-1)^{l-1}\theta}(g') \quad , \tag{3.37}$$

and also

$$\begin{aligned}(\Delta g_0') &= g_0' - g' \\ &= (E - S_{p,0})(r_q) + S_{p,0}(E - S_{p',(-1)^{l-1}\theta})(r_{q'}) + (S_{p,0}S_{p',(-1)^{l-1}\theta} - E)(g'),\end{aligned}$$

$$\tag{3.38}$$

where $(P') = R(P)$; (g') and $(r_{q'})$ can be determined in principle by Eqs. (3.17) or (2.55). However, we prefer to use Eqs. (3.37) and (3.38) only in those cases where the image vectors P', g' and $r_{q'}$ may easily be obtained.

Problems

3.1   Redo the example in Sect.3.1, Fig.3.4, dealing the angles of rotation of the mirror 1 in an order reverse to that in which we did in the text. Then explain which one is more convenient.

3.2   Redo the example in Sect.3.2, Fig.3.9, using the two-step rule.

3.3   Can you give a statement of the theorem of rotation of a mirror system in a different way by assuming that the object and the mirror system first rotate about the object-axis P through an angle θ, then the image follows, and then $\cdots$? What will be told essentially from the difference between these two statements?

3.4 Is it true that the two successive rotations of the image according to the two-step rule are added to each other ? Does each of these two rotations have an independent effect on the resultant motion of the image ?

3.5 Symbols like 'θP' are adopted in some books where θ is a finite angle of rotation and P is a unit vector representing the axis of rotation. Do you feel that these are suitable symbols or not ? Tell why .

3.6 In Fig. 3.8 we have specified two pairs of conjugate points: one, M and M', and the other q and q'. Explain the physical meaning of doing so in an alternative way.

Hint: The rotation axis is the border line at which the fixed space and the moving space intersect with each other, and thus any point q on this line has a dual nature, i. e. , the point q may be regarded as a particle either in the fixed space or in the moving space.

Chapter 4
Small Rotation and Small Translation of Reflecting Prisms

In this chapter, we shall discuss the image motion caused by a small angular displacement plus a small linear displacement of a reflecting prism with the object fixed.

In the case of a small rotation of a prism, we will still be able to use the equations developed specially for a mirror system in the previous chapter to solve these prism-related problems in both collimated and convergent light by simply replacing θ and $\overrightarrow{MM}'$ with $\Delta\theta$ and $\overrightarrow{M_0M}'$, respectively. Such an approach, however, is of little use in this situation because the calculation of a small motion of image by employing these mathematically rigorous and precise formulas is unduly tedious and laborious.

Therefore, we would rather develop, for this purpose, some equations into a new form on the basis of entirely neglecting small quantities of second and higher powers. The approximate result given by these simplified formulas will most likely meet the desired accuracy required for engineering.

Furthermore, these approximate formulas, or generally speaking, such an engineering treatment will still enable us to explore some essential properties and relationships in regard to problems of small tilt or shift of a prism by using several new rules, new theorems and quite a number of characteristic parameters, which might not be discovered due to the complexity of these precise formulas. These generalized natures of a moving reflecting prism will particularly benefit design work.

The material included in the present chapter will prove to be closely related to diverse topics regarding mirror and prism systems such as adjustment, image-stabilizing, error and tolerance calculation, and the like.

4.1 Theorem of Small Rotation of a Reflecting Prism

The theorem of rotation of a mirror system can readily be extended to the case of a reflecting prism if the angle of rotation is assumed to be reasonably small. For some reasons (Ref. 38) we would formulate the latter case as if it were a new theorem.

Theorem of small rotation of a reflecting prism :

"With the object fixed, the image motion caused by a rotation of a reflecting prism through a small angle $\Delta\theta$ about an object-axis P may be accomplished by a rotation of the image through a small angle $(-1)^{t-1}\Delta\theta$ about the image-axis P', followed by another rotation of the intermediate image through a small angle $\Delta\theta$ about the axis P , where t denotes the number of reflections; P' represents the image of P for the prism in its initial position."

The theorem is self-evident and so we need only a few notes on it.

According to the principle discussed in Sect. 1.4.4, we can employ ordinary vectors $(-1)^{t-1} \Delta\theta\,P'$ and $\Delta\theta\,P$ to represent the two successive, small rotations of the image, which may be added up by means of the conventional rules of vector algebra.

The method of doing so implies the fact that all small quantities of higher powers have been neglected and thus the order of two successive small rotations of the image becomes immaterial, that is to say, small errors observe the law of mutual independence.

4.2　Corollary of Small Translation of a Reflecting Prism

Applying the two-step rule to the case of small linear displacement of a reflecting prism leads to the following corollary.

Corollary of small translation of a reflecting prism :

"With the object fixed, the image motion caused by a small linear displacement of a reflecting prism of the magnitude Δg in the direction D (unit vector) may be accomplished by a small linear displacement of the image of the magnitude Δg in the direction $-D'$ (unit vector), followed by another small linear displacement of the intermediate image of the magnitude Δg in the direction D, where D' represents the conjugate image of D through the prism."

In computation, these two small displacements of the image can be represented by $-\Delta g\,D'$ and $\Delta g\,D$.

The corollary is valid for the case of a mirror system whether the linear displacement of the system is infinitely small or finite .

4.3　Equations of Image Motion Caused by a Small Angular Displacement plus a Small Linear Displacement of a Reflecting Prism

4.3.1　Equations of Image Motion Due to a Small Angular Displacement of a Reflecting Prism

Referring to Fig. 4.1, assume an arbitrary prism to be located generally in a convergent beam and rotated through a small angle $\Delta\theta$ about an axis P passing through a given point q. Since both the orientation and the location of the unit vector P will not be varied with the rotation of the prism, it may be considered to be located in the object space and thus P is named the *object-axis* although it has nothing to do with the object. The unit vector P' and the point q' lying on it are the respective images of P and q, and thus P' is called the *image-axis*. The image

106

generated for a given object through the prism is generally assumed to be a body of three dimensions, and hence it may be called the *image body*. A point F' belonging to the image body

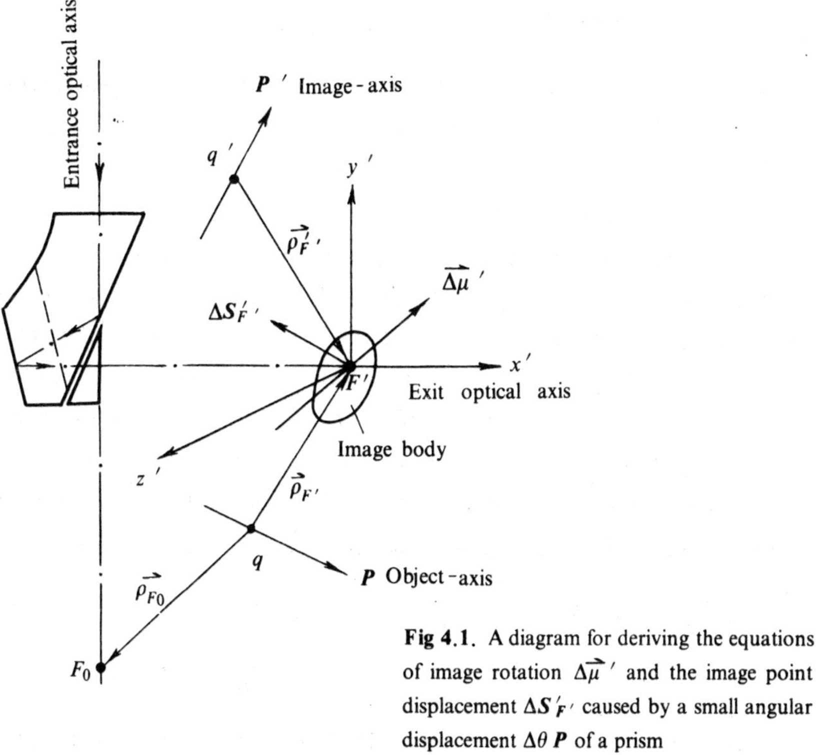

Fig 4.1. A diagram for deriving the equations of image rotation $\vec{\Delta\mu}'$ and the image point displacement $\Delta S_{F'}$ caused by a small angular displacement $\Delta\theta\, P$ of a prism

and usually lying on the exit optical axis of the prism is assigned to represent the location of the image.

In most cases, prisms are used in association with a lens system. Then the image in the strict sense has become a plane picture normal to the exit optical axis of the prism, and thus the point F' will be the center of that image picture. The object point conjugate to F' is denoted by F_0. Draw vectors from q and q' toward F', and label them as $\vec{\rho}_F$ and $\vec{\rho}_F'$, respectively. In addition, $\vec{\rho}_{F_0}$ represents the vector from q toward F_0.

Also, by using the notion of kinematics model as a basic viewpoint in dealing with this problem, the image itself may be thought of as a rigid body. Hence, there is no substantial difference between the motion of an image and that of a rigid body.

Therefore, the image motion, by virtue of being small, may be regarded as consisting of two vector components : one, a small linear displacement of the point F' selected as a base point in the image body, and the other a small angular displacement of the image about an axis through the base point F'. In this text, the two vector components $\Delta S_{F'}'$ and $\Delta\vec{\mu}'$, as shown in Fig.4.1, will particularly be termed the *image point displacement* and the *image rotation*, respectively.

According to the principles discussed in Secs.4.1, 1.4.4 and 1.4.3, we can easily obtain two basic equations of image motion caused by a small angular displacement $\Delta\theta\,\boldsymbol{P}$ (through q) of a prism

$$\Delta\vec{\mu}\,' = \Delta\theta\,\boldsymbol{P} + (-1)^{l-1}\Delta\theta\,\boldsymbol{P}', \tag{4.1}$$

$$\Delta\boldsymbol{S}_{F'}' = \Delta\theta\,\boldsymbol{P}\times\vec{\rho}_{F'} + (-1)^{l-1}\Delta\theta\,\boldsymbol{P}'\times\vec{\rho}_{F'}'. \tag{4.2}$$

With these two equations, we can determine the linear displacement $\Delta\boldsymbol{S}_{E'}'$ of any point E' of the image, not lying on the line of action of $\Delta\vec{\mu}\,'$.

From Fig. 4.2 and Eq. (1.15), we have

$$\Delta\boldsymbol{S}_{E'}' = \Delta\boldsymbol{S}_{F'}' + \Delta\vec{\mu}\,'\times\vec{\rho}_{E'}. \tag{4.3}$$

It is instructive to point out that the idea of the two-step rule is also implied in Eq. (4.2), where the first term $\Delta\theta\,\boldsymbol{P}\times\vec{\rho}_{F'}$ on the right indicates the linear displacement of the image point F' during the rigid rotation of the whole system, including the object, prism, and image as well, through a small angle $\Delta\theta$ around the axis $\boldsymbol{P}$, and thus the corresponding linear displacement of the object point F_0 is equal to $\Delta\theta\,\boldsymbol{P}\times\vec{qF_0} = \Delta\theta\,\boldsymbol{P}\times\vec{\rho}_{F_0}$, while the second term $(-1)^{l-1}\Delta\theta\,\boldsymbol{P}'\times\vec{\rho}_{F'}'$ on the right gives another linear displacement of the image point F' optically conjugate to the corresponding linear displacement of the object point F_0, which equals $(-\Delta\theta\,\boldsymbol{P}\times\vec{\rho}_{F_0})$ because of the following relationship between two conjugate spaces :

$$[(-1)^{l-1}\Delta\theta\,\boldsymbol{P}'\times\vec{\rho}_{F'}'] = -[(-1)^{l}\Delta\theta\,\boldsymbol{P}'\times\vec{\rho}_{F'}']$$

$$= -R\,(\Delta\theta\,\boldsymbol{P}\times\vec{\rho}_{F_0}) = R\,(-\Delta\theta\,\boldsymbol{P}\times\vec{\rho}_{F_0}). \tag{4.4}$$

The result shows that two linear displacements of the object point F_0 cancel each other, therefore the object point F_0 will actually be kept stationary in the two-step process.

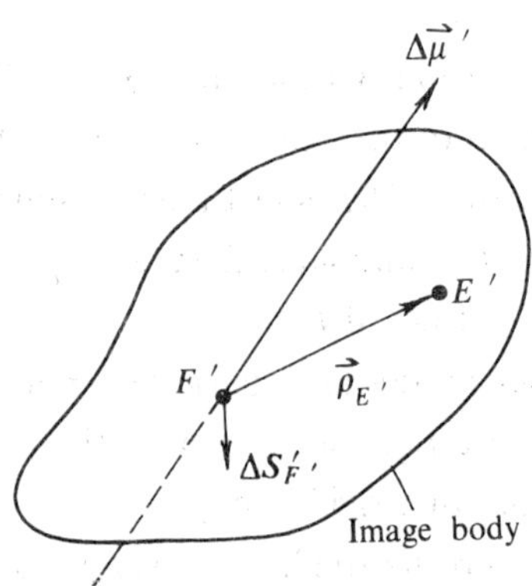

Fig. 4.2. A diagram for determining the linear displacement of any point E of the image body

108

4.3.2 Equation of Image Motion Due to a Small Linear Displacement of a Reflecting Prism

It is evident that a linear displacement of a prism causes only a linear displacement of the image without changing its orientation (or direction).

Assume $\Delta S\ '$ to denote a small linear displacement of the whole image caused by a small linear displacement $\Delta g D$ of the prism.

To show a difference, $\Delta S\ '$ is called the *image displacement*.

According to the corollary in Sect. 4.2 , the equation for determining the image displacement $\Delta S\ '$ caused by a small linear displacement $\Delta g D$ of a prism can be written straightway

$$\Delta S' = \Delta g D - \Delta g D'. \tag{4.5}$$

4.3.3 Equations of Image Motion Due to a Small Angular Displacement plus a Small Linear Displacement of a Reflecting Prism

Combining Eqs. (4.2) and (4.5), and saving Eq. (4.1) will finally give the general equations of image motion caused by both a small angular displacement $\Delta\theta P$ (through q) and a small linear displacement $\Delta g D$ of a prism as follows :

$$\Delta\vec{\mu}\ ' = \Delta\theta P + (-1)^{l-1}\Delta\theta P', \tag{4.1}$$

$$\Delta S'_{F',t} = \Delta\theta P \times \vec{\rho}_{F'} + (-1)^{l-1}\Delta\theta P' \times \vec{\rho}\,'_{F'} + \Delta g D - \Delta g D', \tag{4.6}$$

where $\Delta S'_{F',t}$ denotes the total image point displacement .

4.3.4 Equations of Image Motion in a Normalized Form

In order to make a tabulation of reflecting prisms, we have need to normalize the equations (4.1) and (4.6) by locating the origin $o\ '$ of the coordinate system at point $M\ '$ and replacing most of the image quantities, such as P',D', and $\vec{\rho}'_{F'}$, with their corresponding or adequate object quantities.

Applying the theorem of orientation for reflecting prisms, we have

$$(P\,') = (-1)^l S_{T,2\varphi}\,(P\,), \tag{4.7}$$

$$(D\,') = (-1)^l S_{T,2\varphi}\,(D\,). \tag{4.8}$$

Putting the above formulas into Eqs. (4.1) and (4.5) yields

$$(\Delta\vec{\mu}\,') = \Delta\theta\,(E - S_{T,2\varphi})\,(P\,), \tag{4.9}$$

$$(\Delta S\,') = \Delta g\,[E + (-1)^{l-1}S_{T,2\varphi}]\,(D\,). \tag{4.10}$$

Now, let us deal with the rest of Eq. (4.6), i.e., the equation (4.2), and to make it clear, we will proceed in an orderly way.

First step :

Referring to Fig. 4.3, assume the characteristic point $M\ '$ of an arbitrary prism to be the origin $o\ '$ of the coordinate system, and locate the axis P of rotation of the prism by giving the

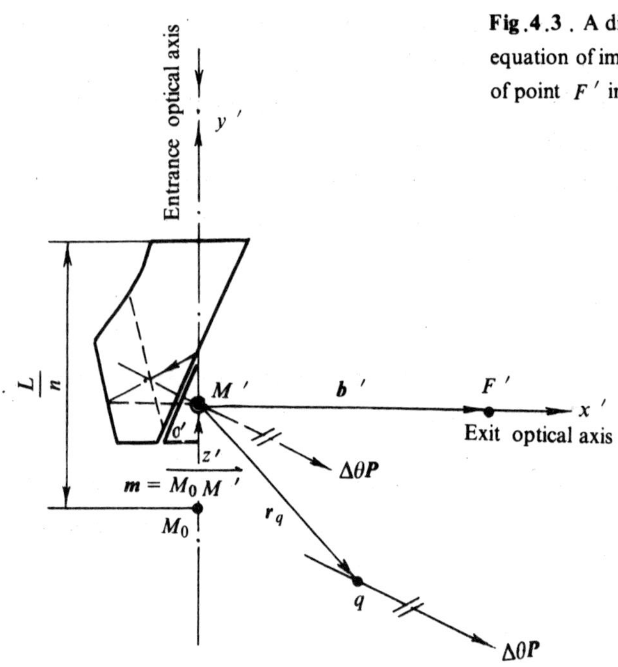

Fig.4.3 . A diagram for deriving the equation of image point displacement of point F' in normalized form

position vector r_q of a point q on the axis P with reference to the origin o'.

Second step :

Choose M' as a base point of the image body and let $\Delta S'_{M'}$ denote the small displacement of the point M' caused by a small angular displacement $\Delta\theta P$ of the prism passing through q. Then, from Eq. (4.3) we can find the image point displacement $\Delta S'_{F'}$ as

$$\Delta S'_F = \Delta S'_{M'} + \Delta\vec{\mu}' \times \overrightarrow{M'F'}$$

or assuming $b' = \overrightarrow{M'F'}$

$$\Delta S'_{F'} = \Delta S'_{M'} + \Delta\vec{\mu}' \times b'. \tag{4.11}$$

Third step :

While determining the displacement $\Delta S'_{M'}$ of the image point M', we may resolve the small angular displacement $\Delta\theta P$ into two components by translating it to a new position passing through the point M', as shown in dotted line in Fig.4.3. Thus, the original, small angular displacement $\Delta\theta P$ of the prism passing through point q is equal to the new one , $\Delta\theta P$ passing through point M', plus another small linear displacement $\Delta\theta P \times \overrightarrow{qM'} = \Delta\theta r_q \times P$ of the prism, and accordingly the small displacement $\Delta S'_{M'}$ of the image point M' can also be devided into two parts : $\Delta S'_{M'1}$ and $\Delta S'_{M'2}$ caused by the angular displacement $\Delta\theta P$ through point M' and the linear displacement $\Delta\theta r_q \times P$ of the prism, respectively.

Consequently,

$$\Delta S'_{M'} = \Delta S'_{M'1} + \Delta S'_{M'2}. \tag{4.12}$$

110

From Eqs. (4.2) and (4.4), we can determine the first part $\Delta S'_{M'\,1}$ of the small displacement of the image point M' by substituting points M', M_0, and again M' for points F', F_0, and q, respectively

$$\Delta S'_{M'\,1} = R\,(-\Delta\theta P \times \overrightarrow{M'M_0}) = (-1)^l\,S_{T.2\varphi}\,(\Delta\theta P \times \overrightarrow{M_0M'}). \tag{4.13}$$

Then, according to Eq. (4.10), we may find the second part $\Delta S'_{M'\,2}$ of the small displacement of the image point M' by replacing the vector $\Delta g D$ with $\Delta\theta r_q \times P$

$$\Delta S'_{M'\,2} = \Delta\theta\,[E + (-1)^{l-1}S_{T.2\varphi}]\,(r_q \times P). \tag{4.14}$$

Finally, combining Eqs. (4.10) through (4.14), and taking account of Eq. (4.9) yields the general equations of image motion in a normalized form

$$(\Delta\vec{\mu}') = \Delta\theta\,(E - S_{T.2\varphi})\,(P), \tag{4.9}$$

$$(\Delta S'_{F'\,,t}) = \Delta\theta\,[E + (-1)^{l-1}S_{T.2\varphi}]\,(r_q \times P) - (-1)^l\Delta\theta S_{T.2\varphi}\,(\overrightarrow{M_0M'} \times P)$$

$$- \Delta\theta\,(b' \times [\,(E - S_{T.2\varphi})\,(P)\,]) + \Delta g\,[E + (-1)^{l-1}S_{T.2\varphi}]\,(D). \tag{4.15}$$

It should be noted that the vector $\overrightarrow{M_0M'}$ in Eq. (4.15) may be replaced by another vector directed from an arbitrary object point toward its corresponding image point, for example $\overrightarrow{oo}'$, and in doing so the equation will remain valid if and only if the new image point o' should be stipulated as the origin of the coordinate system, no matter how the coordinate axes $x'y'z'$ may be orientated. In this case, the vectors r_q and b' keep the primary meaning relative to the new origin o' instead of the point M'. This remarks will be of practical use in the later chapter.

To be more compact, Eqs. (4.9) and (4.15) may be written as

$$(\Delta\vec{\mu}') = \Delta\theta J_{\mu P}\,(P), \tag{4.16}$$

$$(\Delta S'_{F'\,,t}) = \Delta\theta J_{SP}\,(P) + \Delta g J_{SD}\,(D), \tag{4.17}$$

or

$$(\Delta\vec{\mu}') = \Delta\theta J_{\mu P}\,(P), \tag{4.16}$$

$$(\Delta S'_{F'}) = \Delta\theta J_{SP}\,(P), \tag{4.18}$$

$$(\Delta S') = \Delta g J_{SD}\,(D), \tag{4.19}$$

where $J_{\mu P}$, J_{SP}, and J_{SD} are called *the matrix of image rotation*, *the matrix of image point displacement*, and *the matrix of image displacement*, respectively.

From Eqs. (4.9) and (4.10), we have

$$J_{\mu P} = E - S_{T.2\varphi}, \tag{4.20}$$

$$J_{SD} = E + (-1)^{l-1}S_{T.2\varphi}, \tag{4,21}$$

and by using the result from problem 1.3, we obtain

$$J_{SP} = J_{SD}C_q - (-1)^l\,S_{T.2\varphi}\,C_m - C_{b'}\,J_{\mu P}, \tag{4.22}$$

where C_q, C_m, and $C_{b'}$, representing the cross-product-matrices of the vectors r_q, $m = \overrightarrow{M_0M}'$, and b', respectively, will be determined in the following way

$$C_q = \begin{pmatrix} 0 & -z_q' & y_q' \\ z_q' & 0 & -x_q' \\ -y_q' & x_q' & 0 \end{pmatrix},$$

$$C_m = \begin{pmatrix} 0 & -m_{z'} & m_{y'} \\ m_{z'} & 0 & -m_{x'} \\ -m_{y'} & m_{x'} & 0 \end{pmatrix}, \qquad (4.23)$$

$$C_{b'} = \begin{pmatrix} 0 & -b_{z'} & b_{y'} \\ b_{z'} & 0 & -b_{x'} \\ -b_{y'} & b_{x'} & 0 \end{pmatrix}.$$

Finally, by using Eq. (1.12) and putting Eq. (4.23) into Eq. (4.22), the matrices $J_{\mu P}$, J_{SD} and J_{SP} will be given in full detail

$$J_{\mu P} = \begin{pmatrix} 1 - \cos 2\varphi - 2T_x^2 \sin^2\varphi & T_z \sin 2\varphi - 2T_x T_y \sin^2\varphi & -T_y \sin 2\varphi - 2T_x T_z \sin^2\varphi \\[2ex] -T_z \sin 2\varphi - 2T_x T_y \sin^2\varphi & 1 - \cos 2\varphi - 2T_y^2 \sin^2\varphi & T_x \sin 2\varphi - 2T_y T_z \sin^2\varphi \\[2ex] T_y \sin 2\varphi - 2T_x T_z \sin^2\varphi & -T_x \sin 2\varphi - 2T_y T_z \sin^2\varphi & 1 - \cos 2\varphi - 2T_z^2 \sin^2\varphi \end{pmatrix}, \quad (4.24)$$

$$J_{SD} = \begin{pmatrix} 1 + (-1)^{l-1}(\cos 2\varphi + 2T_x^2 \sin^2\varphi) & (-1)^{l-1}(-T_z \sin 2\varphi + 2T_x T_y \sin^2\varphi) & (-1)^{l-1}(T_y \sin 2\varphi + 2T_x T_z \sin^2\varphi) \\[2ex] (-1)^{l-1}(T_z \sin 2\varphi + 2T_x T_y \sin^2\varphi) & 1 + (-1)^{l-1}(\cos 2\varphi + 2T_y^2 \sin^2\varphi) & (-1)^{l-1}(-T_x \sin 2\varphi + 2T_y T_z \sin^2\varphi) \\[2ex] (-1)^{l-1}(-T_y \sin 2\varphi + 2T_x T_z \sin^2\varphi) & (-1)^{l-1}(T_x \sin 2\varphi + 2T_y T_z \sin^2\varphi) & 1 + (-1)^{l-1}(\cos 2\varphi + 2T_z^2 \sin^2\varphi) \end{pmatrix},$$

$$(4.25)$$

$$J_{SP} = \begin{bmatrix}
\begin{aligned}
&(-1)^{t-1}\{-[(z_q{}'+m_{z'})T_{z'}+(y_q{}'\\
&+m_{y'})T_{y'}]\sin2\varphi+2[(z_q{}'+m_{z'})T_{y'}\\
&-(y_q{}'+m_{y'})T_{z'}]T_{x'}\sin^2\varphi\}
\end{aligned}
&
\begin{aligned}
&(-1)^{t-1}\{-(z_q{}'+m_{z'})\cos2\varphi+(x_q{}'\\
&+m_{x'})T_{y'}\sin2\varphi+2[-(z_q{}'+m_{z'})T_{x'}\\
&+(x_q{}'+m_{x'})T_{z'}]T_{x'}\sin^2\varphi\}-z_q{}'
\end{aligned}
&
\begin{aligned}
&(-1)^{t-1}\{(y_q{}'+m_{y'})\cos2\varphi+(x_q{}'\\
&+m_{x'})T_{z'}\sin2\varphi+2[(y_q{}'+m_{y'})T_{x'}\\
&-(x_q{}'+m_{x'})T_{y'}]T_{x'}\sin^2\varphi\}+y_q{}'
\end{aligned}
\\[2em]
\begin{aligned}
&(-1)^{t-1}\{(z_q{}'+m_{z'})\cos2\varphi+(y_q{}'\\
&+m_{y'})T_{x'}\sin2\varphi+2[(z_q{}'+m_{z'})T_{y'}\\
&-(y_q{}'+m_{y'})T_{z'}]T_{y'}\sin^2\varphi\}\\
&-2b{}'T_{x'}T_{z'}\sin^2\varphi+b{}'T_{y'}\sin2\varphi\\
&+z_q{}'
\end{aligned}
&
\begin{aligned}
&(-1)^{t-1}\{-[(z_q{}'+m_{z'})T_{z'}+(x_q{}'\\
&+m_{x'})T_{x'}]\sin2\varphi-2[(z_q{}'+m_{z'})T_{x'}\\
&-(x_q{}'+m_{x'})T_{z'}]T_{y'}\sin^2\varphi\}\\
&-b{}'T_{x'}\sin2\varphi-2b{}'T_{y'}T_{z'}\sin^2\varphi
\end{aligned}
&
\begin{aligned}
&(-1)^{t-1}\{-(x_q{}'+m_{x'})\cos2\varphi\\
&+(y_q{}'+m_{y'})T_{z'}\sin2\varphi+2[(y_q{}'\\
&+m_{y'})T_{x'}-(x_q{}'+m_{x'})T_{y'}]\\
&T_{y'}\sin^2\varphi\}-x_q{}'+b{}'-b{}'\cos2\varphi\\
&-2b{}'T_{z'}^2\sin^2\varphi
\end{aligned}
\\[2em]
\begin{aligned}
&(-1)^{t-1}\{-(y_q{}'+m_{y'})\cos2\varphi\\
&+(z_q{}'+m_{z'})T_{x'}\sin2\varphi+2\\
&[-(y_q{}'+m_{y'})T_{z'}+(z_q{}'\\
&+m_{z'})T_{y'}]T_{z'}\sin^2\varphi\}\\
&-y_q{}'+b{}'T_{z'}\sin2\varphi\\
&+2b{}'T_{x'}T_{y'}\sin^2\varphi
\end{aligned}
&
\begin{aligned}
&(-1)^{t-1}\{(x_q{}'+m_{x'})\cos2\varphi\\
&+(z_q{}'+m_{z'})T_{y'}\sin2\varphi\\
&+2[-(z_q{}'+m_{z'})T_{x'}+(x_q{}'\\
&+m_{x'})T_{z'}]T_{z'}\sin^2\varphi\}\\
&+x_q{}'-b{}'+b{}'\cos2\varphi\\
&+2b{}'T_{y'}^2\sin^2\varphi
\end{aligned}
&
\begin{aligned}
&(-1)^{t-1}\{-[(y_q{}'+m_{y'})T_{y'}\\
&+(x_q{}'+m_{x'})T_{x'}]\sin2\varphi\\
&+2[(y_q{}'+m_{y'})T_{x'}-(x_q{}'\\
&+m_{x'})T_{y'}]T_{z'}\sin^2\varphi\}\\
&-b{}'T_{x'}\sin2\varphi\\
&+2b{}'T_{y'}T_{z'}\sin^2\varphi
\end{aligned}
\end{bmatrix}$$

(4.26)

Problems

4.1 Give the physical meaning for each of the five terms in Eqs. (4.9)and (4.15).

4.2 Derive the expression of the matrix of image point displacement J_{SP} in Eq. (4.26).

4.3 In Fig. 4.4 , is shown a part of telescopic system, including an objective and a roof prism located in convergent light. Determine the components $\Delta S'_{F'y'}$ and $\Delta S'_{F'z'}$ of the image point displacement in the image plane and the component $\Delta\mu'_{x'}$ of the image rotation along the exit optical axis x' of the prism if the latter has rotated through a small angle $\Delta\theta = 50'$ around the axis P lying in its optical-axis section passing through the roof edge mn of it. All the data are given in the diagram with dimensions in millimeters.

Note : It will be known in the next chapter that components $\Delta S'_{F'y'}$ and $\Delta S'_{F'z'}$ are a measure of the optical axis deviation while the component $\Delta\mu'_{x'}$ represents the image lean.

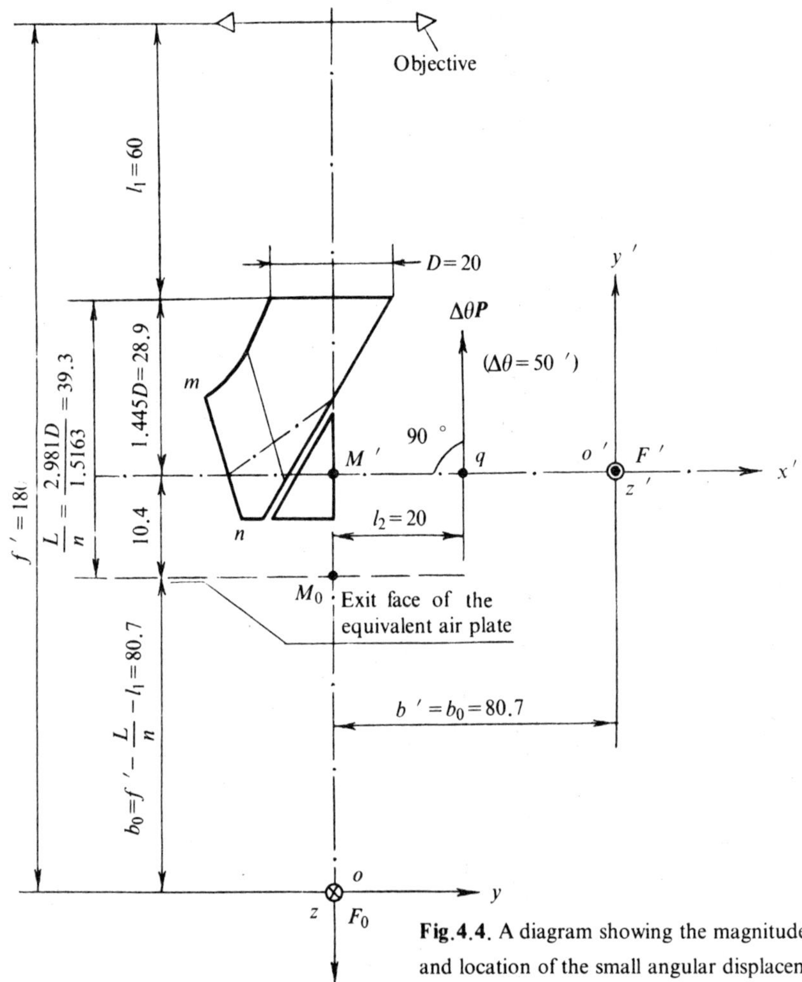

Fig.4.4. A diagram showing the magnitude, direction, and location of the small angular displacement $\Delta\theta\,P$ of the prism, and all dimensions needed

114

4.4 Choose the correct one from the following two expressions:

$$(\boldsymbol{P}\,'\times\vec{\rho}_{F\,'})=S_{T.2\varphi}\;(\boldsymbol{P}\times\vec{\rho}_{F_0})\,,$$

$$(\boldsymbol{P}\,'\times\vec{\rho}_{F\,'})=R\;(\boldsymbol{P}\times\vec{\rho}_{F_0})\,,$$

and tell why the other is wrong (see Fig. 4.1).

Chapter 5
Adjustment of Reflecting Prisms

The adjustment of optical systems is an important step in the manufacture of optical instruments and has to be carried out during the final assembly. Nevertheless , this task needs to be addressed in the designing stage , including system layout and mechanical design. The design , especially of very accurate instruments with compound optical systems , must leave scope for correction during assembly. Optical adjustment , therefore , is essentially to introduce some compensations by adjusting two or three components to balance errors of the whole optical system in order that the system will achieve the required performance.

As compared with lenses , reflecting prisms are more flexible in performing a compensatory function for the reason that they may be either slightly tilted or displaced.

The material in this chapter is intended to discuss the general and basic problems , the most common of all , such as theorems , rules , methods , relationships , parameters , as well as the classification and tabulation of reflecting prisms according to their behaviors in the adjustment aspect. The different and specific topics in this regard for individual optical instruments will be investigated in detail only in Part III .

Such a satisfactory treatment of this specialized area is possible because no matter how different the specific adjustment terms for a variety of optical imaging devices may be , each adjustment term can refer equally to a change in one or a few components of an image motion with a total of six degrees of freedom.

Hence , it is clear that the kinematics approach will still form a strong basis on which we will be able to develop a completely systematized theory of adjustment for reflecting prisms.

5.1 Terminology

Image plane   It refers to the plane of an image formed by a prism , which appears to be a plane picture normal to the exit optical axis of the prism , as shown in Fig. 5.1 .

Image lean   It represents the inclination of image δ in the image plane , or in other words , the rotation of the image around the exit optical axis x' , as indicated in Fig.5.1.

Tilt of image plane   It represents the inclination of image $|\vec{\psi}|$ out of the image plane , or in other words , the rotation of the image around an axis $\vec{\psi}$ lying in the image plane as in Fig.5.2.

The y' and z' tilts of image plane   They represent the components of the tilt of image plane along the y' and z' axes , respectively.

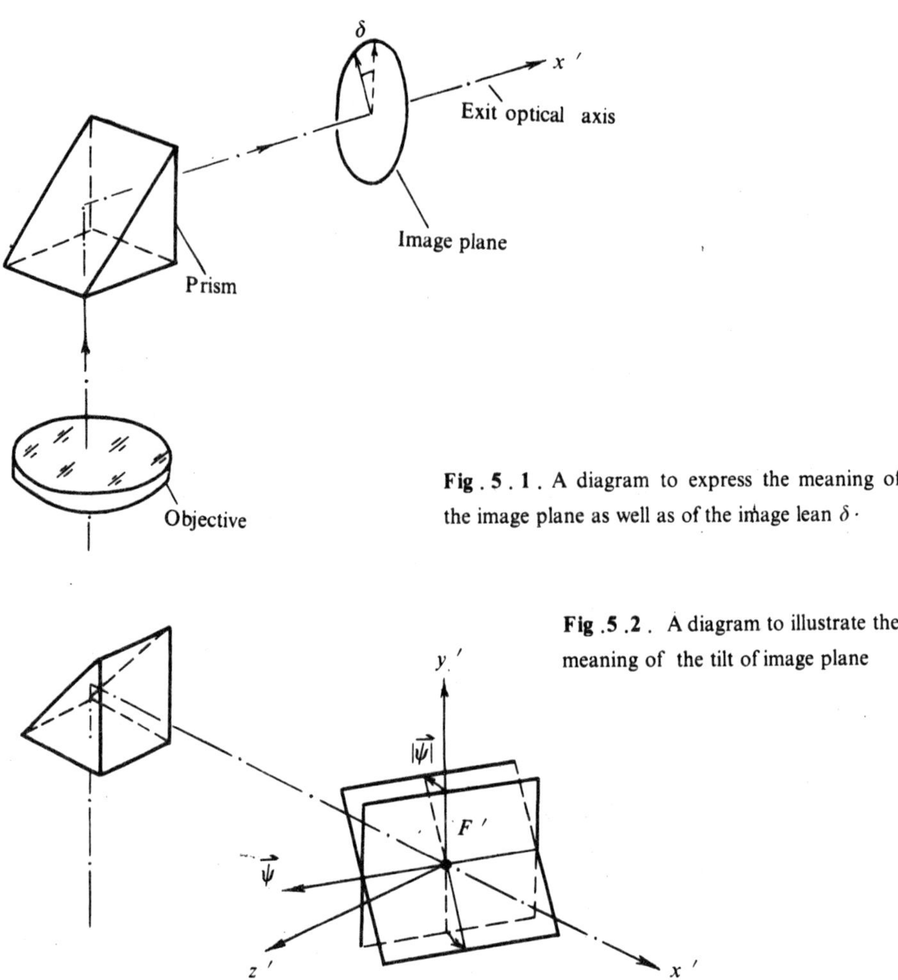

Fig . 5 . 1 . A diagram to express the meaning of
the image plane as well as of the image lean δ .

Fig .5 .2 . A diagram to illustrate the
meaning of the tilt of image plane

Deviation of optical axis (or *optical axis deviation*)   Since the ideal optical axis defined according to the statement in applied optics does not exist in reality , as a working definition , the practical optical axis of an optical device is represented in general by the transverse location of

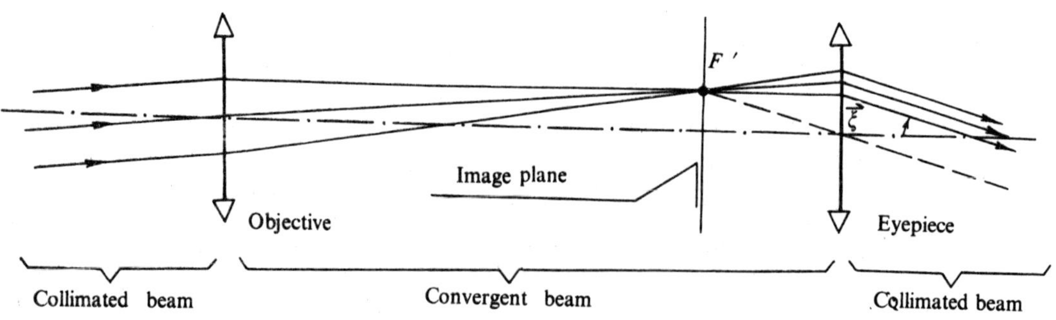

Fig .5 :3 .   A telescopic system , as an example , to show the meaning of
a practical optical axis in terms of the location of the central image point

the central image point , say F' , in the image plane as in Fig. 5. 3 , depending on a certain base-line or base-plane of the whole device , which has been reasonably planned in the designing stage and carefully ensured in different manufacturing phases. For instance , the axis of rotation of hinged binoculars will logically act as a base-line of the device , that is to say , through a process of adjustment the direction of the emergent parallel rays corresponding to the central image point should be parallel to the hinge axis within the limits of tolerance. Then , any visible departure of the direction of the emergent parallel rays from this base-line is called the *optical axis deviation* $\vec{\xi}$. Fig.5.3 shows that the optical axis deviation may appear to be either an incorrect location of the central image point in the image plane in convergent light or equally a wrong direction of the parallel rays , corresponding to the same image point , in collimated light. Therefore , the essential of adjusting the optical axis of an optical instrument is to change the transverse position of the central image point in some way.

The y' and z' optical axis deviations They represent the components of the optical axis deviation along the y' and z' axes , respectively.

Diopter and parallax The former term refers to the convergency or divergency of the emerging pencil of rays due to the departure of the image plane from the anterior focal plane of an eyepiece , while the latter term refers to the phenomenon of a transverse movement of the images of the object and a graticule relative to each other as the viewer's eye is moved over the exit pupil of the instrument. Diopter and parallax are two terms having no bearing on each other conceptually , but both of them depend upon the longitudinal shift of the image plane.

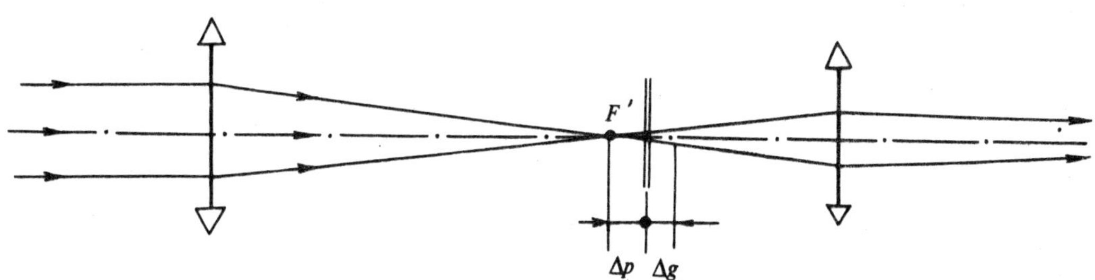

Fig .5 .4 . A telescopic system , as an example , to show the meaning of the diopter of ray pencil as well as of the parallax

The x' image rotation $\Delta\mu'_x$ It is the component of image rotation along the x' axis , or the rotation of image around the x' axis.

The y' image rotation $\Delta\mu'_y$ and the z' image rotation $\Delta\mu'_z$ They are the components of image rotation along the y' , z' axes , or the rotations of image around the y' , z' axes , respectively.

The x' , y' , z' image point displacements $\Delta S_{F'x'}$, $\Delta S'_{F'y'}$, $\Delta S'_{F'z'}$ They represent the components of image point displacement along the x' , y' , z' axes , respectively.

The x' , y' , z' image displacements $\Delta S'_x$, $\Delta S'_y$, $\Delta S'_z$ They denote the components of image displacement along the x' , y' , z' axes , respectively.

5.2 Relation Between the Image Motion and Optical Adjustment

The terms introduced in Sect. 5.1 can be grouped mainly under two categories: one, including the image lean, tilt of image plane, optical axis deviation as well as the diopter, covers a few concepts regarding the optical adjustment, and the other, containing simply six components of two representative vectors $\Delta\vec{\mu}\,'$ and $\Delta S_F'$, of a small image-motion, seems to restate the optical adjustment terms listed above in a kinematics way.

As a matter of fact, they refer to two sides of one problem. Consider an adjusting prism; then the image motion $\Delta\vec{\mu}\,'$ and $\Delta S_F'$, caused by a small movement of the prism is intended for compensating those errors of image position in space, such as δ, $\vec{\psi}$, $\vec{\xi}$ and so forth. One probably sees their correspondence to one another among the terms of the two groups.

However, we still feel that the readers should have a thorough understanding of the relation between the components of image motion and those terms to be adjusted prior to tackling the more difficult parts in this chapter.

Some components like μ_y', and μ_z', have completely different functioning in adjustment depending on whether the adjusting prism is located in a collimated or convergent beam. Thus, our discussion will be divided into the following two cases: collimated beam and convergent beam.

5.2.1 Collimated Beam

Referring to Fig. 5.5, consider an arbitrary prism located in a collimated beam; then the object and the image are both at infinity. Assume a normalized image coordinate system

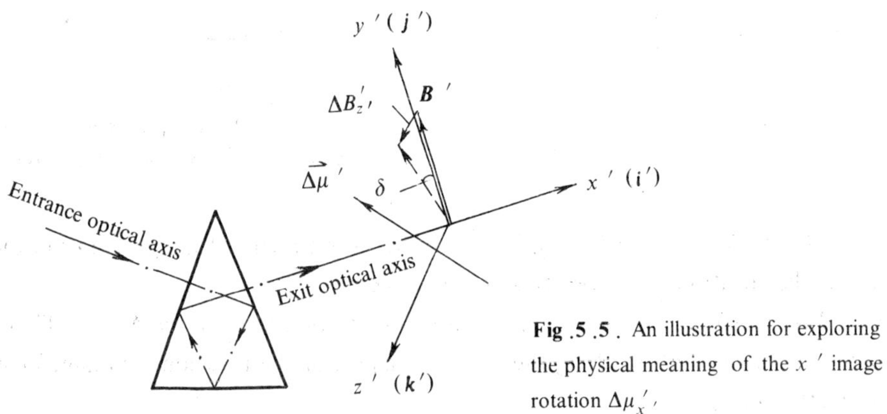

Fig. 5.5. An illustration for exploring the physical meaning of the x' image rotation $\Delta\mu_x'$

$x'y'z'$, and the image plane $y'z'$ can be thought of as being positioned infinitely distant. As indicated in Sect. 2.6, in this case we need to discuss the change in image orientation only.

Now, suppose that an image rotation $\Delta\vec{\mu}\,'$ has occurred due to a small angular displacement of the prism. Firstly, let us determine the magnitude of the image lean δ, which is introduced by the image rotation $\Delta\vec{\mu}\,'$ in order to balance the disagreement of the same type $(-\delta)$.

According to the traditional method, the image lean will be found by assuming a vertical (or horizontal) unit image-vector $B\,'$ along the $y\,'$ axis, and then calculating the change $\Delta B\,'$ of this unit vector due to $\Delta\vec{\mu}\,'$.

From Eq. (1.15),

$$\Delta B' = \Delta\vec{\mu}\,' \times B'. \tag{5.1}$$

The vector $\Delta\vec{\mu}\,'$ of image rotation can be expressed in terms of its three components along the $x\,'$, $y\,'$, $z\,'$ axes

$$\Delta\vec{\mu}\,' = \Delta\mu_{x'}\,' i' + \Delta\mu_{y'}\,' j' + \Delta\mu_{z'}\,' k'. \tag{5.2}$$

From Fig. 5.5,

$$B' = j'. \tag{5.3}$$

Putting Eqs. (5.2) and (5.3) into (5.1) yields

$$\Delta B' = -\Delta\mu_{z'}\,' i' + \Delta\mu_{x'}\,' k'$$

or

$$\Delta B_{x'}\,' = -\Delta\mu_{z'}\,', \qquad \Delta B_{y'}\,' = 0, \qquad \Delta B_{z'}\,' = \Delta\mu_{x'}\,'.$$

From the small right triangle in Fig. 5.5, we have

$$\delta = \frac{\Delta B_{z'}\,'}{|B'|} = \frac{\Delta\mu_{x'}\,'}{1} = \Delta\mu_{x'}\,'. \tag{5.4}$$

The result indicates that the image rotation $\Delta\vec{\mu}\,'$ causes an image lean δ which is just equal to the component of the image rotation along the exit optical axis $\Delta\mu_{x'}\,'$. Thus, the $x\,'$ image rotation will also be called the image lean later on.

Next, determine the optical axis deviation $\vec{\xi}$, which is introduced by the image rotation $\Delta\vec{\mu}\,'$ in order to compensate the disagreement of the same type $(-\vec{\xi})$.

The concept of the practical optical axis as well as of the optical axis deviation has been mentioned in the previous section, which can be enhanced by the illustration in Fig. 5.6, where the central incident parallel rays A is converted by a prism into two emergent parallel beams $A_1\,'$ and $A_2\,'$ prior to and after its rotation, respectively, that will in turn be imaged through a lens at the respective points $F_1\,'$ and $F_2\,'$ in the focal plane. Therefore, in the path of collimated rays before the lens the optical axis deviation appears to be an angular quantity $\vec{\xi}$ formed by vectors $A_2\,'$ and $A_1\,'$, which will thus be related to the image rotation $\Delta\vec{\mu}\,'$.

Similarly, according to the traditional method, the optical axis deviation in collimated light will be found by assuming an axial unit image-vector A' along the $x\,'$ axis, as shown in Fig. 5.7, and then calculating the change $\Delta A'$ of this unit vector due to $\Delta\vec{\mu}\,'$.

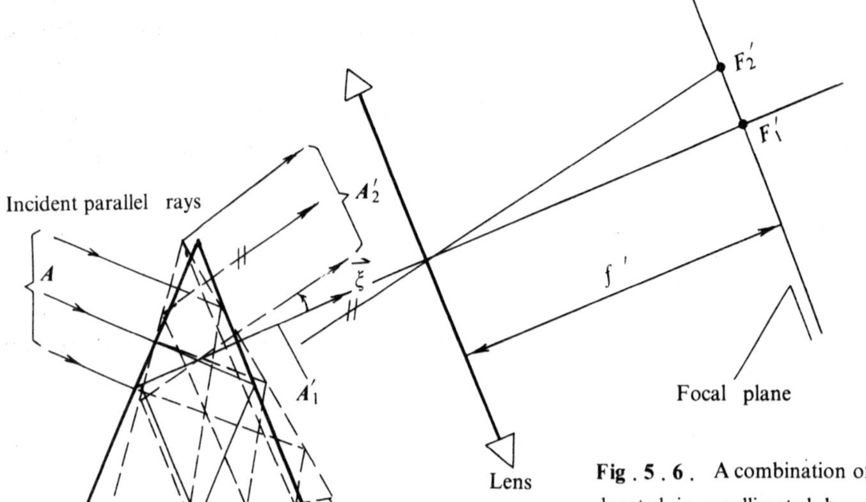

Fig . 5 . 6 . A combination of a prism located in a collimated beam and a lens to illustrate the meaning of the optical axis deviation

From Eq . (1 . 15),

$$\Delta A' = \Delta\vec{\mu}' \times A'.\tag{5.5}$$

Putting Eq . (5 .2) and $A' = i'$ into Eq . (5 .5) yields

$$\Delta A' = \Delta\mu_z' \, j' - \Delta\vec{\mu}_y' \, k'.\tag{5.6}$$

From the geometry in Fig . 5 .7 , we obtain

$$\xi_{y'} = \frac{-\Delta A_{z'}'}{|A'|} = \frac{\Delta\mu_{y'}'}{1} = \Delta\mu_{y'}',\tag{5.7}$$

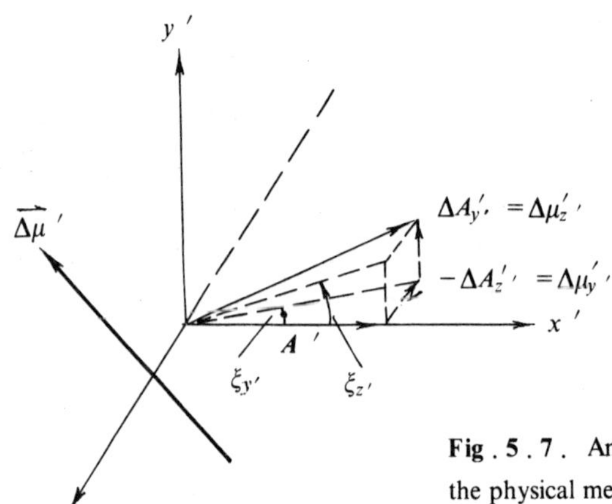

Fig . 5 . 7 . An illustration for exploring the physical meaning of the y', z' image rotations $\Delta\mu_y'$ and $\Delta\mu_z'$

$$\xi_{z'} = \frac{\Delta A_{y'}'}{|A'|} = \frac{\Delta \mu_{z'}'}{1} = \Delta \mu_{z'}'. \tag{5.8}$$

The results show that the components of optical axis deviation $\xi_{y'}$ and $\xi_{z'}$, caused by the image rotation $\Delta \vec{\mu}'$ in collimated light are not different from the y', z' image rotations $\Delta \mu_{y'}'$ and $\Delta \mu_{z'}'$, respectively .

5.2.2 Convergent Beam

While a prism is located in a convergent beam , both the object and image are positioned at finite distances . Under this condition , not only the image rotation $\Delta \vec{\mu}'$ but also the image point displacement $\Delta S_F'$, are meaningful . And the six components $\Delta \mu_{x'}'$, $\Delta \mu_{y'}'$, $\Delta \mu_{z'}'$, $\Delta S_{F'x'}'$, $\Delta S_{F'y'}'$ and $\Delta S_{F'z'}'$ will be related to those terms of optical adjustment correspondingly as stated below .

The x' image rotation $\Delta \mu_{x'}'$ still represents the image lean , while the x' image point displacement is related to the diopter as well as to the parallax . However , it should particularly be noted that the y', z' image rotations $\Delta \mu_{y'}'$ and $\Delta \mu_{z'}'$ cause the y', z' tilts of image plane $\psi_{y'}$ and $\psi_{z'}$, respectively , instead of the y', z' optical axis deviations $\xi_{y'}$ and $\xi_{z'}$, while the latter in convergent light can only be introduced by the z', y' image point displacements $\Delta S_{F'z'}'$ and $\Delta S_{F'y'}'$ correspondingly .

5.3 Problems in a Collimated Beam

The material given in this section is to show how to perform the calculation of adjustment for prisms in a collimated beam mainly by giving some notes and necessary formulas .

It is evident from the above discussion that all we have to do is to calculate the image lean as well as the optical axis deviation , i.e., the three components $\Delta \mu_{x'}'$, $\Delta \mu_{y'}'$, $\Delta \mu_{z'}'$ of the image rotation .

5.3.1 Some Notes on Calculation

a. The angular vector $\Delta \vec{\mu}'$ of image rotation can be translated arbitrarily .

b. The refractions of a prism may be ignored .

c. The translation of a prism may be neglected .

d. The location of the vector $\Delta \theta P$ of a small angular displacement of prisms is meaningless .

5.3.2 Formulas to Be Used

$$\Delta \vec{\mu}' = \Delta \theta P + (-1)^{r-1} \Delta \theta P' \tag{4.1}$$

or

$$(\Delta \vec{\mu}\,') = \Delta\theta\, J_{\mu P}\,(\boldsymbol{P})\,, \tag{4.16}$$

where

$$J_{\mu P} = E - S_{T,2\varphi} \tag{4.20}$$

or

$$J_{\mu P} = E + (-1)^{\iota-1}\, R\,. \tag{5.9}$$

Example

In Fig.5.8, is shown a Taylor V prism, which rotates through a small angle $\Delta\theta$ about the axis $\boldsymbol{P}$ located in the conjugate optical-axis plane and parallel to the base reflecting surface of the prism. Determine the image lean and optical axis deviation due to $\Delta\theta\boldsymbol{P}$.

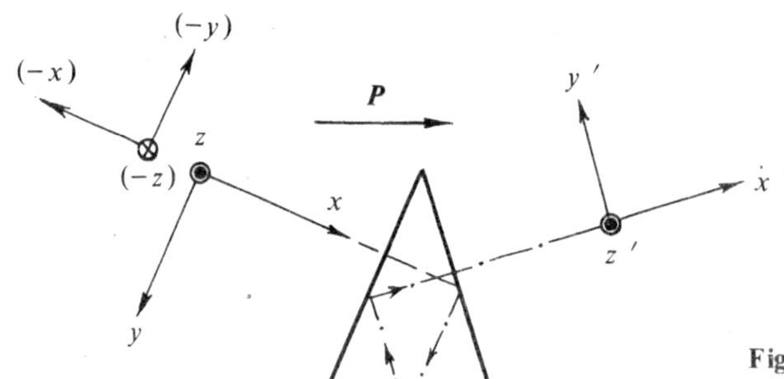

Fig.5.8. A Taylor V prism rotating about $\boldsymbol{P}$ through $\Delta\theta$

Solution

According to the stipulation in Sect.2.3.2, assume a normalized image coordinate system $x'y'z'$ for the given planar prism and correspondingly construct an object coordinate system xyz conjugate to $x'y'z'$, as shown in Fig.5.8.

Applying the principle mentioned in Secs.1.3 and 2.3.2 yields

$$S_{T,2\varphi} = G_{0'-0}\,,$$

where $G_{0'-0}$ is the matrix of transformation of coordinates from the coordinate system $x'y'z'$ to the coordinate system $(-x)(-y)(-z)$.

Thus,

$$S_{T,2\varphi} = \begin{pmatrix} \cos 135^\circ & \cos 45^\circ & \cos 90^\circ \\ \cos 45^\circ & \cos 45^\circ & \cos 90^\circ \\ \cos 90^\circ & \cos 90^\circ & \cos 180^\circ \end{pmatrix} = \begin{pmatrix} -0.707 & 0.707 & 0 \\ 0.707 & 0.707 & 0 \\ 0 & 0 & -1 \end{pmatrix}$$

From Eq. (4.20),

124

$$J_{\mu P}=E-S_{T,2\varphi}=\begin{pmatrix}1.707 & -0.707 & 0\\ -0.707 & 0.293 & 0\\ 0 & 0 & 2\end{pmatrix}.$$

From Fig. 5.8, determine the components of the unit vector P relative to the working coordinate system $x'y'z'$

$$P_{x'}=0.924,\qquad P_{y'}=-0.383,\qquad P_{z'}=0.$$

Putting the known data into Eq. (4.16), we eventually obtain

$$(\Delta\vec{\mu}')=\Delta\theta J_{\mu P}(P)=\Delta\theta\begin{pmatrix}1.707 & -0.707 & 0\\ -0.707 & 0.293 & 0\\ 0 & 0 & 2\end{pmatrix}\begin{pmatrix}0.924\\ -0.383\\ 0\end{pmatrix}=\begin{pmatrix}1.85\\ -0.77\\ 0\end{pmatrix}\Delta\theta$$

or

$$\Delta\mu_{x'}'=1.85\,\Delta\theta,\qquad \Delta\mu_{y'}'=-0.77\,\Delta\theta,\qquad \Delta\mu_{z'}'=0.$$

The results show that the image lean $\Delta\mu_{x'}'$ equals $1.85\Delta\theta$ while the y' optical axis deviation $\Delta\mu_{y'}'$ equals $-0.77\Delta\theta$ and the z' optical axis deviation is zero.

This example can also be solved by using Eq. (4.1).

The equation is here written in its expanded form

$$\Delta\vec{\mu}'=\Delta\theta\{P_{x'}+(-1)^{l-1}P_{x'}'\}i'+\Delta\theta\{P_{y'}+(-1)^{l-1}P_{y'}'\}j'$$

$$+\Delta\theta\{P_{z'}+(-1)^{l-1}P_{z'}'\}k'.$$

From Eq. (2.13),

$$P_{x'}'=P_{x'},\qquad P_{y'}'=P_{y'},\qquad P_{z'}'=P_{z'}.$$

Thus, by substitution, we obtain

$$\Delta\vec{\mu}'=\Delta\theta\{P_{x'}+(-1)^{l-1}P_{x}\}i'+\Delta\theta\{P_{y'}+(-1)^{l-1}P_{y}\}j$$

$$+\Delta\theta\{P_{z'}+(-1)^{l-1}P_{z}\}k'.$$

Again, from Fig. 5.8,

$$P_{x}=0.924,\qquad P_{y}=-0.383,\qquad P_{z}=0.$$

Putting the known values of $P_{x},P_{y},P_{z},P_{x'},P_{y'},P_{z'}$ and t into the expanded equation of $\Delta\vec{\mu}'$ leads to the same result

$$\Delta\vec{\mu}'=1.85\Delta\theta\cdot i'-0.77\Delta\theta\cdot j'.$$

5.4 Problems in a Convergent Beam

The material included in the present section is to indicate how to perform the calculation of adjustment for prisms in a convergent beam mainly by giving some notes and necessary formulas.

Like the image body, a plane image still has six degrees of freedom, therefore we should investigate the effects of all components $\Delta\mu_x'$, $\Delta\mu_y'$, $\Delta\mu_z'$, $S_{F'x'}'$, $\Delta S_{F'y'}'$ and $\Delta S_{F'z'}'$ of the image motion in this case.

However, the image rotation $\Delta\vec{\mu}'$ has previously been treated once in the collimated beam and there will not be anything new in calculating it for this case; particularly, the y', z' image rotations $\Delta\mu_y'$ and $\Delta\mu_z'$ here are only related ⅃ the y', z' tilts of image plane, which will not be very critical in most cases when the field of view of the optical device is not big enough. Besides, the diopter and parallax can usually be put in order simply by shifting the eyepiece and graticule in the axial direction, respectively.

As a matter of fact, we will concentrate our attention on determining the optical axis deviation caused by the y', z' image point displacements $\Delta S_{F'y}'$ and $\Delta S_{F'z'}'$.

5.4.1 Some Notes on Calculation

a. The transverse location of the central image point F' is used directly to represent the optical axis.

b. Both the reflections and the refractions of a prism should be considered.

c. The translation of a prism should also be taken into account.

d. The location of the vector $\Delta\theta P$ of a small angular displacement is also meaningful.

e. The angular vector $\Delta\vec{\mu}'$ of image rotation is referred to as a sliding vector, which is not allowed to be translated laterally.

5.4.2 Formulas to Be Used

$$\Delta\vec{\mu}' = \Delta\theta\,P + (-1)^{i-1}\Delta\theta\,P', \tag{4.1}$$

$$\Delta S_{F'}' = \Delta\theta P \times \vec{\rho}_F' + (-1)^{i-1}\Delta\theta\,P' \times \vec{\rho}_{F'}', \tag{4.2}$$

$$\Delta S' = \Delta g\,D - \Delta g\,D', \tag{4.5}$$

or

$$(\Delta\vec{\mu}') = \Delta\theta\,J_{\mu P}\,(P), \tag{4.16}$$

$$(\Delta S_{F'}') = \Delta\theta\,J_{SP}\,(P), \tag{4.18}$$

$$(\Delta S') = \Delta g\,J_{SD}\,(D), \tag{4.19}$$

126

where the matrices $J_{\mu P}$, J_{SP} and J_{SD} can be determined by Eqs. (4.24), (4.26) and (4.25), respectively.

Since the form of Eq. (4.5) is very similar to that of Eq. (4.1) and also the application of the latter equation has been known in the previous section, it remains to explain the usage of Eq. (4.2) or (4.18).

Very often, the equation (4.2) can be flexibly used to determine the y', z' image point displacements $\Delta S_{F'y'}$ and $\Delta S_{F'z'}$. Also, it was felt that the reader would benefit from the following discussion by having a good grasp of some points as well as the procedure of applying this equation.

Procedure of calculating $\Delta S_{F'y'}$ and $\Delta S_{F'z'}$ (see Fig. 5.9):

(1) Locate the object and image points F_0 and F'.

(2) Construct a pair of completely conjugate coordinate systems $oxyz$ and $o'x'y'z'$.

Since the image point F' is selected as the base point of the whole image, it is reasonable to locate origins o' and o of two completely conjugate coordinate systems $o'x'y'z'$ and $oxyz$ at image point F' and its corresponding object point F_0, respectively.

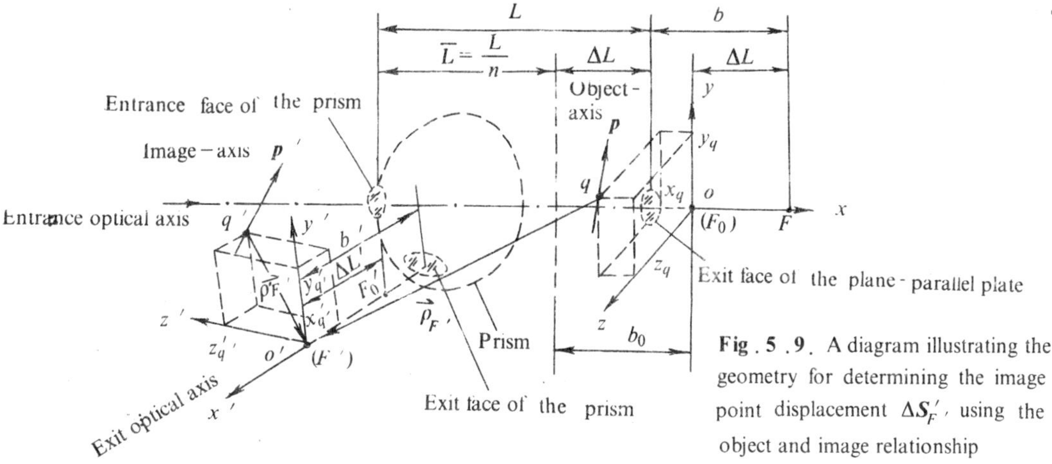

Fig. 5.9. A diagram illustrating the geometry for determining the image point displacement $\Delta S_{F'}$ using the object and image relationship

(3) Locate a point q on the action line of P.

Substantially, this is a step to locate the axis P of rotation in the space. Although q may be an arbitrary point on the action line of the vector P, we should choose a simple one.

(4) Determine the components of P on the axes x, y, z and x', y', z': P_x, P_y, P_z and $P_{x'}$, $P_{y'}$, $P_{z'}$.

(5) Determine the coordinates of q in the coordinate systems $oxyz$ and $o'x'y'z'$: x_q, y_q, z_q and x_q', y_q', z_q'.

(6) Determine the components of P' and the coordinates of q' relative to the image coordinate system $o'x'y'z'$: P_x', P_y', P_z' and x_q', y_q', z_q'.

By using the object and image relationship, we have

$$P_{x'} = P_x, \qquad P_{y'} = P_y, \qquad P_{z'} = P_z, \qquad (2.13)$$

$$x'_{q'} = x_q, \qquad\qquad y'_{q'} = y_q, \qquad\qquad z'_{q'} = z_q. \qquad\qquad (2.17)$$

Although we may choose another point t on the action line of the image axis $\boldsymbol{P'}$ to replace the point q', as shown in Fig. 5.10, yet it is still suggested to utilize the image q' corresponding to a chosen point q in the object space because only in doing so one would be able to have a definite method of determining the required point q' simply by using the relationship between the two conjugate spaces, and this is particularly important for developing the theory a step further. Also, it is in general difficult to locate a point in the image space of a prism without regard to any corresponding point in the object space.

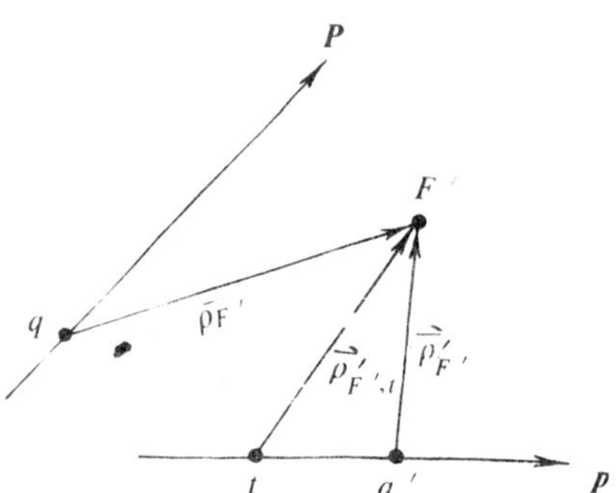

Fig. 5.10. A diagram showing the location of the vector with either point q' or point t

(7) Determine the components of the vectors $\vec{\rho}_{F'}$ and $\vec{\rho}'_{F'}$ along the axes x', y', z'.
From Fig. 5.9,

$$\vec{\rho}_{F'} = -\overrightarrow{o'q}, \qquad\qquad \vec{\rho}'_{F'} = -\overrightarrow{o'q'}.$$

Thus, we have

$$\rho_{F'x'} = -x'_q, \qquad\qquad \rho_{F'y'} = -y'_q, \qquad\qquad \rho_{F'z'} = -z'_q,$$

$$\rho'_{F'x'} = -x'_{q'}, \qquad\qquad \rho'_{F'y'} = -y'_{q'}, \qquad\qquad \rho'_{F'z'} = -z'_{q'}. \qquad (5.10)$$

(8) Determine the y', z' image point displacements $\Delta S'_{F'y'}$ and $\Delta S'_{F'z'}$.

Writing Eq. (4.2) in its expanded form and keeping the y', z' components of the image point displacement, we obtain

$$\Delta S'_{F'y'} = \Delta\theta[\,(P_{z'}\rho_{F'x'} - P_{x'}\rho_{F'z'}) + (-1)^{t-1}(P'_{z'}\rho'_{F'x'} - P'_{x'}\rho'_{F'z'})\,], \qquad (5.11)$$

$$\Delta S'_{F'z'} = \Delta\theta[\,(P_{x'}\rho_{F'y'} - P_{y'}\rho_{F'x'}) + (-1)^{t-1}(P'_{x'}\rho'_{F'y'} - P'_{y'}\rho'_{F'x'})\,]. \qquad (5.12)$$

Example

In Fig. 5.11 is shown a prism mount, which is attached to a stereoscopic telescope. The roof prism rests on the platform and is held in the correct position by a spring strip and some setting device, and the whole prism mount is in turn fixed in the prism

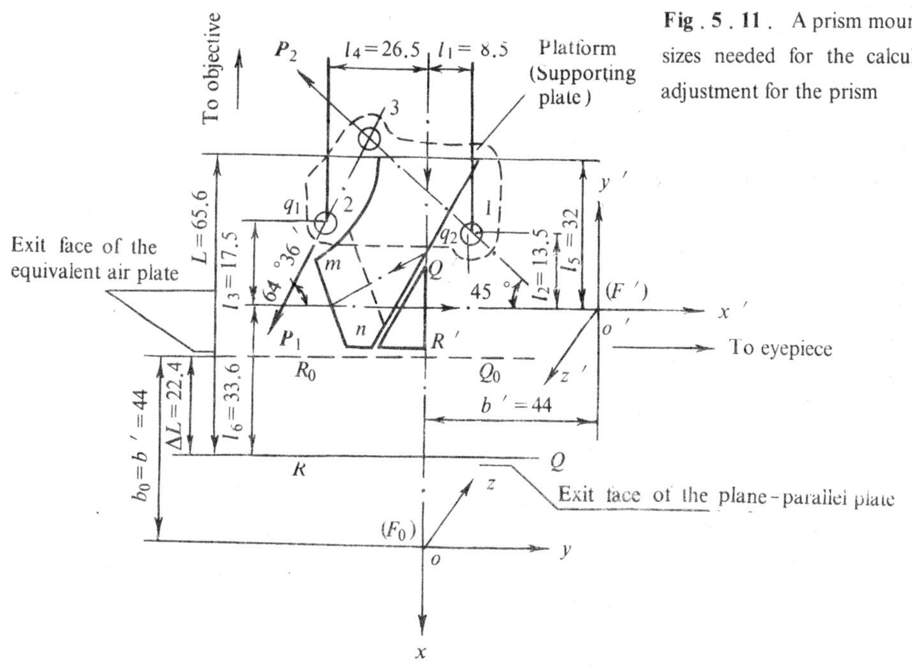

Fig . 5 . 11 . A prism mount with all sizes needed for the calculation of adjustment for the prism

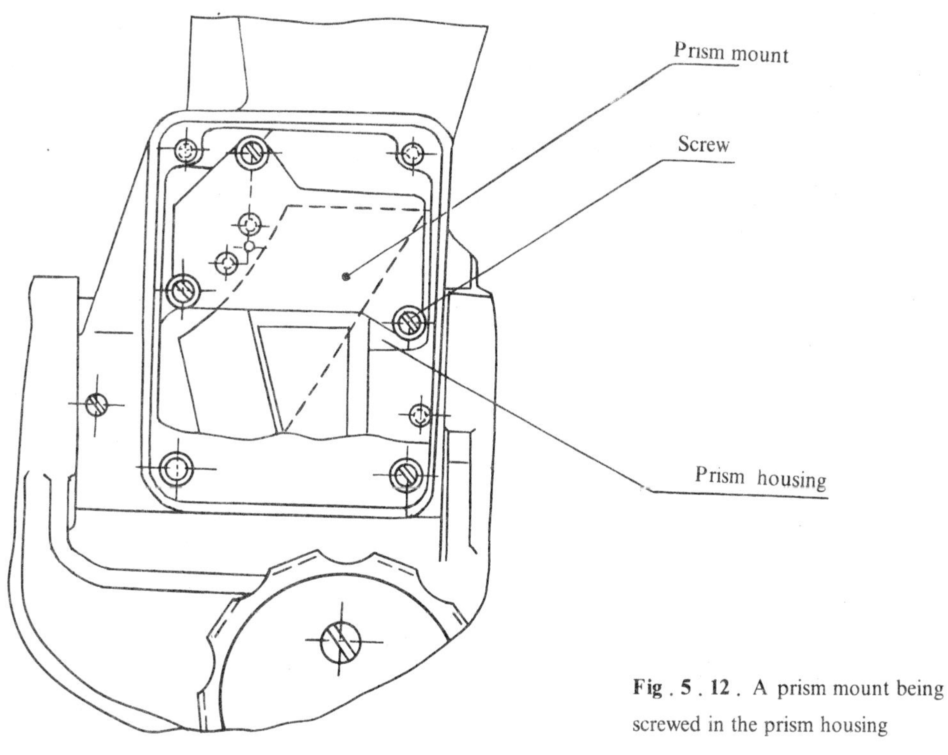

Fig . 5 . 12 . A prism mount being screwed in the prism housing

housing with three screws through the clear holes 1 , 2 , 3 of the platform , under which three raised finishing pads press against the flat spots on the seat in the prism housing , as shown in Fig.5.12.

While adjusting the image lean ,we may put a spacer of different thickness in between the flat spots and the finishing pads under either hole 1 or hole 2 so that the prism may accordingly slightly rotate around either axis P_1 or axis P_2 depending on in which direction the existing image lean might be . Now , it is required to determine the optical axis deviation introduced during the adjustment of the image lean , i.e. , caused by a rotation through a small angle $\Delta\theta$ about the axis P_1 or P_2 .

The relevant data are given below :

Prism aperture : $D = 22$,

Prism index : $n = 1.5163$,

The distance of F' from the exit face of the the prism : $b' = 44$.

The plane of the pad face of the platform , i.e. , the bearing face of the prism mount , is parallel to the optical - axis section passing through the roof edge mn , and at a height of 8.3 mm above the section ,consequently both the axes P_1 and P_2 lying in the bearing face are parallel to the optical - axis section and 8.3 mm above it .

Solution

Let us first consider the case of rotation about the axis P_1 .

(1) Calculate L , ΔL , and b_0 .

$$L = 2.98 D = 65.6 , \qquad \Delta L = \frac{n-1}{n} L = 22.4 ,$$

$$b_0 = b' = 44 .$$

Thus , we can locate the points F_0 and F' along the entrance and exit optical axes .

(2) Construct a pair of conjugate coordinate systems $oxyz$ and $o'x'y'z'$,as shown in Fig.5.11.

(3) Choose the point of intersection of the center line of hole 2 with the pad face as the point q_1 .

(4) Determine the components of P_1 along the axes x , y , z and x',y',z' .

$$P_{1x} = \cos 25°24' = 0.903 , \qquad P_{1y} = -\sin 25°24' = -0.429 , \qquad P_{1z} = 0 ,$$

$$P_{1x'} = -\sin 25°24' = -0.429 , \quad P_{1y'} = -\cos 25°24' = -0.903 , \quad P_{1z'} = 0 .$$

(5) Determine the coordinates of q_1 in the coordinate systems $oxyz$ and $o'x'y'z'$.

$$x_{q_1} = -72.7 , \qquad y_{q_1} = -26.5 , \qquad z_{q_1} = -8.3 ,$$

$$x'_{q_1} = -70.5 , \qquad y'_{q_1} = 17.5 , \qquad z'_{q_1} = 8.3 .$$

(6) Determine the components of P_1' and the coordinates of q_1' relative to the image coordinate system $o'x'y'z'$.

$$P'_{1x'} = P_{1x} = 0.903 , \qquad P'_{1y'} = P_{1y} = -0.429 , \qquad P'_{1z'} = P_{1z} = 0 ,$$

$$x'_{q_1'} = x_{q_1} = -72.7 , \qquad y'_{q_1'} = y_{q_1} = -26.5 , \qquad z'_{q_1'} = z_{q_1} = -8.3 .$$

(7) Determine the components of the vectors $\vec{\rho}_F$ and $\vec{\rho}'_F$ along the axes x' , y' , z' .

130

$$\rho_{F'x'} = -x'_{q_1} = 70.5 , \qquad \rho_{F'y'} = -y'_{q_1} = -17.5 , \qquad \rho_{F'z'} = -z'_{q_1} = -8.3 ,$$

$$\rho'_{F'x'} = -x'_{q_1} = 72.7 , \qquad \rho'_{F'y'} = -y'_{q_1} = 26.5 ; \qquad \rho'_{F'z'} = -z'_{q_1} = 8.3 .$$

(8) Rearrange the relevant data .

$$\boldsymbol{P}_1 : \quad P_{1x'} = -0.429 , \qquad P_{1y'} = -0.903 , \qquad P_{1z'} = 0 ,$$

$$\vec{\rho}_F : \quad \rho_{F'x'} = 70.5 , \qquad \rho_{F'y'} = -17.5 , \qquad \rho_{F'z'} = -8.3 ,$$

$$\boldsymbol{P}'_1 : \quad P'_{1x'} = 0.903 , \qquad P'_{1y'} = -0.429 , \qquad P'_{1z'} = 0 ,$$

$$\vec{\rho}'_F : \quad \rho'_{F'x'} = 72.7 , \qquad \rho'_{F'y'} = 26.5 , \qquad \rho'_{F'z'} = 8.3 .$$

(9) Determine the y', z' image point displacements $\Delta S'_{F'y'}$ and $\Delta S'_{F'z'}$.

Substituting the above data and $t = 3$ into Eqs . (5.11) and (5.12) gives eventually

$$\Delta S'_{F'y'} = -11.1 \Delta\theta , \tag{5.13}$$

$$\Delta S'_{F'z'} = 126.4 \Delta\theta . \tag{5.14}$$

Similarly , while rotating the prism through a small angle $\Delta\theta$ about the axis $\boldsymbol{P}_2$, we will obtain

$$\Delta S'_{F'y'} = 0 , \tag{5.15}$$

$$\Delta S'_{F'z'} = 39.2 \Delta\theta . \tag{5.16}$$

This problem is to be continued in Prat Ⅲ .

5.4.3 The Case of Two Independent Reflecting Prisms in a Convergent Beam

The case of two independent prisms used in a convergent beam in one system is encountered frequently . In Fig . 5.13 is shown an example containing a central prism and a rhomboid prism , which has been employed in certain rangefinders .

While discussing the calculation of adjustment for the central prism , we may suppose that the rhomboid prism is fixed .

Let F'' be the image point in the image space of the rhomboid prism , and assume the data a , b as well as the apertures of two prisms to be given .

Now , locate the image and object points F' and F_0 for the central prism by calculating

$(b + \dfrac{L_2}{n_2})$ and $(b + \dfrac{L_2}{n_2} + a + \dfrac{L_1}{n_1})$, where L_1 , L_2 and n_1 , n_2 are the axial lengths and indices of the prisms 1 and 2 , respectively .

Next , construct three coordinate systems $oxyz$, $o'x'y'z'$, and $o''x''y''z''$ corresponding to one another . Then , the components $\Delta S'_{F'y'}$ and $\Delta S'_{F'z'}$ of displacement of image point F' in the image plane $x'o'y'$ may be determined using Eqs . (5.11) and (5.12) .

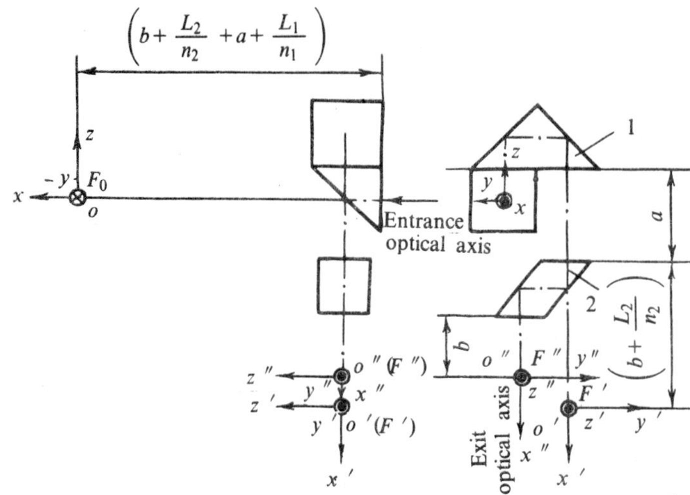

Fig . 5 . 13 . An example illustrating the calculation of adjustment for two prisms located together in a conver-gent beam

Finally , according to the object and image relationship for the rhomboid prism , we obtain

$$\Delta S''_{F''y''}=\Delta S'_{F'y'}, \qquad\qquad \Delta S''_{F''z''}=\Delta S'_{F'z'}. \qquad\qquad (5.17)$$

5.5 Components of Image Rotation and Relevant Characteristic Parameters

As mentioned in the previous sections , each component of the image rotation $\Delta\vec{\mu}\,'$ is meaningful in optical adjustment . The $y\,'$, $z\,'$ image rotations $\Delta\mu_{y}'$ and $\Delta\mu_{z}'$ may represent two components of the optical axis deviation when the prism is located in a collimated beam , or denote two components of the tilt of image plane when the prism is positioned in a convergent beam , and in either case the $x\,'$ image rotation $\Delta\mu_{x}'$ will always refer to the image lean .

For some reasons , chiefly owing to the difficulty in designing the fastening structure for prisms , we have to deal with each component of image rotation individually while adjusting the prism , or in other words , we could hardly treat the image rotation $\Delta\vec{\mu}\,'$ as a single term in one adjusting step . Consider a prism located in collimated light; then it is almost impossible to put both the image lean and the optical axis deviation in order by rotating a certain prism just once .

Therefore , it is worthwhile to derive a general equation to relate any component of the image rotation $\Delta\vec{\mu}\,'$ to the small angular displacement of a prism , by which the image rotation is produced .

5.5.1 Equations of a Component of Image Rotation Due to a Small Angular Displacement of a Reflecting Prism

Let a unit vector r' be an arbitrary direction in the image space of a prism and assume $\Delta\mu_{r'}$ to denote the component of image rotation along the direction r', which for short is called the r' image rotation. Substantially, $\Delta\mu_{r'}$ represents a small rotation of the image about the axis r'.

From Eq. (4.1),

$$\Delta\mu_{r'} = \Delta\theta\,[\,P_{r'} + (-1)^{t-1}\,P'_{r'}\,]\,,\tag{5.18}$$

where $P_{r'}$ and $P'_{r'}$ denote the components of unit vectors P and P' along the direction r', respectively.

Assume $(-1)^t r$ to be a unit vector in the object space and conjugate to r', that is to say, r is said to be conjugate to r' if the number of reflections t is even, whereas r is only a unit vector in the object space with its direction opposite to what is conjugate to r' if the number of reflections t is odd. Obviously, in the latter case $-r$ is conjugate to r'.

Under the condition purposefully stipulated above, we have

$$P'_{r'} = (-1)^t\,P_r\,,\tag{5.19}$$

where P_r denotes the component of the unit vector P along the direction r.

Substituting Eq. (5.19) into Eq. (5.18) gives a unified formula regardless of the number t of reflections

$$\Delta\mu_{r'} = \Delta\theta\,(P_{r'} - P_r)\,.\tag{5.20}$$

According to the meaning of the scalar product, we also have

$$P_r = P\cdot r\,,\qquad\qquad P_{r'} = P\cdot r'\,.$$

Putting the above two formulas into Eq. (5.20) yields

$$\Delta\mu_{r'} = P\cdot\Delta\theta\,(r' - r)\,.\tag{5.21}$$

Now, a new vector $\vec{\delta}_h$ is introduced and defined as follows:

$$\vec{\delta}_h = \Delta\theta\,(r' - r) = \delta_h\,h\,,\tag{5.22}$$

where δ_h is the magnitude of the vector $\vec{\delta}_h$ while h is the unit vector along the direction of $\vec{\delta}_h$.

Thus, we obtain

$$\Delta\mu_{r'} = P\cdot\vec{\delta}_h = \delta_h\cos\,(\widehat{P,\vec{\delta}_h})\,.\tag{5.23}$$

5.5.2 Theorem of Adjustment for Reflecting Prisms

Rewrite the last two equations in one place

$$\begin{cases}\Delta\mu_{r'} = P\cdot\vec{\delta}_h = \delta_h\cos\,(\widehat{P,\vec{\delta}_h})\,,\tag{5.23}\\[2mm]\vec{\delta}_h = \Delta\theta\,(r' - r) = \delta_h\,h\,.\tag{5.22}\end{cases}$$

These two equations will prove to be significant in both the theoretical and practical aspects. To emphasize this, we will formulate them as a theorem.

Theorem of adjustment for reflecting prisms:

" The relation between the component $\Delta\mu_{r'}$ of image rotation along the direction of a unit vector r' in the image space (the r' image rotation) and the small angular displacement $\Delta\theta\, P$ of the prism , which produces the image rotation , is subject to the *law of cosine function* : $\Delta\mu_{r'}' = P \cdot \vec{\delta_h} = \delta_h \cos (P \,\widehat{\cdot}\, \vec{\delta_h})$ while the sole vector parameter $\vec{\delta_h}$ in the function is determined by the *rule of two - vector subtraction*: $\vec{\delta_h} = \Delta\theta\, (r' - r)$, where $(-1)^t\, r$ is a unit vector in the object space and conjugate to r' ; t is the number of reflections ."

The theorem may also be called the *cosine function and two - vector subtraction rule* .

Now , let us further explore the physical meaning of the above theorem .

From Eq . (5 .23) , if the axis P of rotation of a prism is parallel to the parametric vector $\vec{\delta_h}$ of the prism , i .e . , $P \| \vec{\delta_h}$ or $P \| h$, then the r' image rotation reaches its maximum value $\Delta\mu_{r'\,\text{max}}'$:

$$\Delta\mu_{r'}' = \delta_h \cos 0\,° = \delta_h = \Delta\mu_{r'\,\text{max}}' , \tag{5 . 24}$$

in which the maximum value $\Delta\mu_{r'\,\text{max}}'$ of the r' image rotation is none other than the magnitude of the parametric vector itself δ_h .

Obviously , when P is in a direction opposite to $\vec{\delta_h}$, $\Delta\mu_{r'}'$ reaches the negative maximum:

$$\Delta\mu_{r'}' = \delta_h \cos 180\,° = -\delta_h = -\Delta\mu_{r'\,\text{max}}' .$$

Based on the physical meaning , δ_h , h and $\vec{\delta_h}$ are therefore named *the extreme value of r' image rotation* , *extreme - valued axis direction of r' image rotation*, and *extreme - valued characteristic vector of r' image rotation* , respectively .

It is also evident that $\Delta\mu_{r'}'$ will be equal to zero if P is perpendicular to $\vec{\delta_h}$:

$$\Delta\mu_{r'}' = \delta_h \cos 90\,° = 0 .$$

Thus , we have also the *zero - valued axis direction of r' image rotation* , and no doubt all the directions , being perpendicular to the extreme - valued axis direction of r' image rotation , must be the zero - valued axis directions corresponding to that r' image rotation .

In case $\vec{\delta_h} = 0$, $\Delta\mu_{r'}'$ will always be zero no matter about which axis the prism turns . However , this conclusion is true only in the sense of neglecting small quantities of higher powers . In this situation , all the directions belong to the zero - valued axis directions of r' image rotation .

The principle as well as the formulas mentioned above are said to be general to a certain extent since the unit vector r' can be any direction in the image space of a prism .

Comparing the meaning of the extreme - valued axis direction and that of the gradient discussed in Sect .1 .5 , it is worthwhile examining a new term , the gradient of the r' image rotation , designated by $\vec{\eta_h}$ and defined as follows:

$$\vec{\eta_h} = \text{grad}\,\Delta\mu_{r'}' = \frac{\partial\Delta\mu_{r'}'}{\partial\Delta\theta_{x'}}\, i' + \frac{\partial\Delta\mu_{r'}'}{\partial\Delta\theta_{y'}}\, j' + \frac{\partial\Delta\mu_{r'}'}{\partial\Delta\theta_{z'}}\, k' , \tag{5 . 25}$$

where

$$\Delta\theta_{x'} = \Delta\theta\, P_{x'}, \qquad \Delta\theta_{y'} = \Delta\theta\, P_{y'}, \qquad \Delta\theta_{z'} = \Delta\theta P_{z'} \tag{5.26}$$

or

$$\Delta\vec{\theta} = \Delta\theta\, \boldsymbol{P}. \tag{5.27}$$

From, Eqs. (5.23) and (5.22),

$$\Delta\mu_{r'}' = \boldsymbol{P}\cdot\vec{\delta}_h = \Delta\theta\, \boldsymbol{P}\cdot (\boldsymbol{r}' - \boldsymbol{r}).$$

Then,

$$\Delta\mu_{r'}' = \Delta\theta\, P_{x'}(\boldsymbol{r}'-\boldsymbol{r})_{x'} + \Delta\theta\, P_{y'}(\boldsymbol{r}'-\boldsymbol{r})_{y'} + \Delta\theta\, P_{z'}(\boldsymbol{r}'-\boldsymbol{r})_{z'}$$

or

$$\Delta\mu_{r'}' = \Delta\theta_{x'}(\boldsymbol{r}'-\boldsymbol{r})_{x'} + \Delta\theta_{y'}(\boldsymbol{r}'-\boldsymbol{r})_{y'} + \Delta\theta_{z'}(\boldsymbol{r}'-\boldsymbol{r})_{z'}, \tag{5.28}$$

where $(\boldsymbol{r}'-\boldsymbol{r})_{x'}$, $(\boldsymbol{r}'-\boldsymbol{r})_{y'}$, and $(\boldsymbol{r}'-\boldsymbol{r})_{z'}$ are the components of the vector $(\boldsymbol{r}'-\boldsymbol{r})$ along the axes x', y', z'.

Putting Eq. (5.28) into Eq. (5.25) gives

$$\vec{\eta}_h = \mathrm{grad}\,\Delta\mu_{r'}' = (\boldsymbol{r}'-\boldsymbol{r})_{x'}\,\boldsymbol{i}' + (\boldsymbol{r}'-\boldsymbol{r})_{y'}\,\boldsymbol{j}' + (\boldsymbol{r}'-\boldsymbol{r})_{z'}\,\boldsymbol{k}'. \tag{5.29}$$

The above equation can also be written as

$$\vec{\eta}_h = \mathrm{grad}\,\Delta\mu_{r'}' = \boldsymbol{r}' - \boldsymbol{r}. \tag{5.30}$$

Comparing Eqs. (5.30) and (5.22) ultimately yields

$$\vec{\delta}_h = \Delta\theta\,\vec{\eta}_h. \tag{5.31}$$

The resulting equation shows that the extreme‑valued characteristic vector $\vec{\delta}_h$ and the gradient $\vec{\eta}_h$ of r' image rotation are the same in direction but different only in magnitude, therefore the extreme‑valued axis direction $\boldsymbol{h}$ of $\Delta\mu_{r'}'$ may also be determined by operating upon $\Delta\mu_{r'}'$ with the gradient operator $\left(\dfrac{\partial}{\partial\Delta\theta_{x'}},\ \dfrac{\partial}{\partial\Delta\theta_{y'}},\ \dfrac{\partial}{\partial\Delta\theta_{z'}}\right)$, and thus $\vec{\eta}_h$ or $\mathrm{grad}\,\Delta\mu_{r'}'$ will also be called the *gradient axis direction of r' image rotation* later on.

5.5.3 Characteristic Parameters of Components of Image Rotation

One of the important applications of the above mentioned theorem is to help us explore some characteristic parameters of adjustment for reflecting prisms.

Take an arbitrary planar reflecting prism as an example, see Fig. 5.14. Draw a normalized image coordinate system $x'y'z'$. Then, the coordinate system xyz in the object space is constructed based on such a principle that xyz will be related to $x'y'z'$ in the same way as r is related to r'.

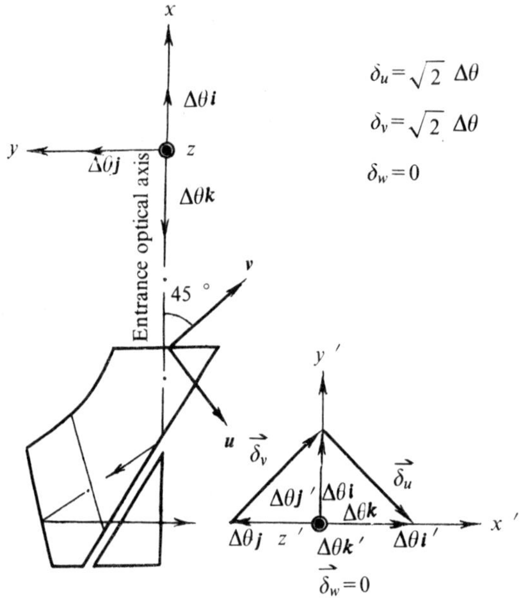

Fig.5.14. Graphical method of obtaining three extreme-valued characteristic vectors for a certain planar reflecting prism

The xyz frame formed in this way will also be called the normalized coordinate system in the object space of reflecting prisms.

Again, let i, j, k and i', j', k' be unit vectors along the axes x, y, z, and x', y', z', respectively.

Since the unit vector r' represents an arbitrary direction in the image space of a prism, we can replace r' successively by i', j' and k' when deducing the theorem. Then, Eqs.(5.23) and (5.22) become the following two sets of equations

$$\left. \begin{aligned} \Delta\mu'_{x'} &= \boldsymbol{P} \cdot \vec{\delta}_u = \delta_u \cos (\boldsymbol{P}^{\frown}\vec{\delta}_u) = \delta_u \cos \alpha \\ \Delta\mu'_{y'} &= \boldsymbol{P} \cdot \vec{\delta}_v = \delta_v \cos (\boldsymbol{P}^{\frown}\vec{\delta}_v) = \delta_v \cos \beta \\ \Delta\mu'_{z'} &= \boldsymbol{P} \cdot \vec{\delta}_w = \delta_w \cos (\boldsymbol{P}^{\frown}\vec{\delta}_w) = \delta_w \cos \gamma \end{aligned} \right\} , \qquad (5.32)$$

$$\left. \begin{aligned} \vec{\delta}_u &= \Delta\theta (i' - i) = \delta_u \boldsymbol{u} \\ \vec{\delta}_v &= \Delta\theta (j' - j) = \delta_v \boldsymbol{v} \\ \vec{\delta}_w &= \Delta\theta (k' - k) = \delta_w \boldsymbol{w} \end{aligned} \right\} , \qquad (5.33)$$

where $\Delta\mu'_{x'}, \Delta\mu'_{y'}, \Delta\mu'_{z'}$ represent the x', y', z' image rotations, respectively; $\vec{\delta}_u, \vec{\delta}_v, \vec{\delta}_w$ represent the extreme-valued characteristic vectors of the x', y', z' image rotations, respectively; $\delta_u, \delta_v, \delta_w$ represent the extreme values of the x', y', z' image rotations, respectively; unit vectors $\boldsymbol{u}, \boldsymbol{v}, \boldsymbol{w}$ represent the extreme-valued axis directions of the x', y', z' image rotations, respectively; $\cos\alpha, \cos\beta, \cos\gamma$ represent the cosine of angles formed by the axis $\boldsymbol{P}$ of rotation with $\boldsymbol{u}, \boldsymbol{v}, \boldsymbol{w}$, respectively.

136

For all the planar reflecting prisms , the three extreme - valued characteristic vectors $\vec{\delta}_u$, $\vec{\delta}_v$ and $\vec{\delta}_w$ can be determined by graphical methods according to the two - vector subtraction rule , namely Eq . (5 .33) , as has been done in Fig .5 .14 .

From Eq . (5 .32) , it is evident that if the three extreme - valued characteristic vectors $\vec{\delta}_u$, $\vec{\delta}_v$ and $\vec{\delta}_w$ for a prism are known* in advance , then the x' , y' , z' image rotations $\Delta\mu'_x$, $\Delta\mu'_y$ and $\Delta\mu'_z$ can be determined as soon as the axis P of rotation is given . Therefore , the theorem or the rule can also be referred to as one of the methods of determining the image rotation .

As a matter of fact , the three extreme - valued characteristic vectors $\vec{\delta}_u$, $\vec{\delta}_v$ and $\vec{\delta}_w$ themselves are significant , because they would be able to provide a clear picture to express the essential behavior of the adjustment for reflecting prisms in collimated light in efficient and transparent language .

For this reason , by canceling the unnecessary part of the drawing in Fig .5 .14 we have obtained the embryonic form of a so - called *adjustment diagram* for the prism , as shown in Fig . 5 .15 , in which three extreme - valued characteristic vectors for the prism are expressed in

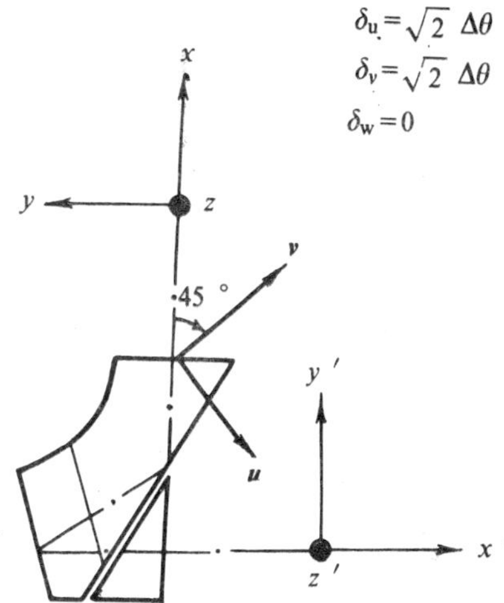

Fig .5 .15 . The embryonic form of an adjustment diagram for a prism

terms of their directions u , v , w and magnitudes δ_u , δ_v , δ_w , and also given is the image coordinate system $x'y'z'$ that corresponds to these vectors .

Similarly , let $\vec{\eta}_u$, $\vec{\eta}_v$ and $\vec{\eta}_w$ be the gradient axis directions of the x' , y' , z' image rotations , respectively , then we have accordingly

---

* It can be restated in a better way: If three gradients $\vec{\eta}_u$, $\vec{\eta}_v$ and $\vec{\eta}_w$ of the x' , y' , z' image rotations are known in advance , then $\Delta\mu'_x$, $\Delta\mu'_y$ and $\Delta\mu'_z$ can be determined once P and $\Delta\theta$ are given . However , this is not very critical , since we may regard $\Delta\theta$ just as a scalar factor .

$$\vec{\eta}_u = \operatorname{grad} \Delta\mu'_{x'} = i' - i$$
$$\vec{\eta}_v = \operatorname{grad} \Delta\mu'_{y'} = j' - j \qquad , \tag{5.34}$$
$$\vec{\eta}_w = \operatorname{grad} \Delta\mu'_{z'} = k' - k$$

$$\vec{\delta}_u = \Delta\theta\,\vec{\eta}_u$$
$$\vec{\delta}_v = \Delta\theta\,\vec{\eta}_v \qquad . \tag{5.35}$$
$$\vec{\delta}_w = \Delta\theta\,\vec{\eta}_w$$

Furthermore, the extreme-valued characteristic vectors $\vec{\delta}_u$, $\vec{\delta}_v$ and $\vec{\delta}_w$ as well as the gradients $\vec{\eta}_u$, $\vec{\eta}_v$ and $\vec{\eta}_w$ can also be expressed in matrix form.

From Eqs. (4.16) and (5.27),

$$(\Delta\vec{\mu}\,') = \vec{J}_{\mu P}\,(\Delta\theta). \tag{5.36}$$

On the other hand, we have

$$\vec{\eta}_u = \frac{\partial\Delta\mu'_{x'}}{\partial\Delta\theta_{x'}}\,i' + \frac{\partial\Delta\mu'_{x'}}{\partial\Delta\theta_{y'}}\,j' + \frac{\partial\Delta\mu'_{x'}}{\partial\Delta\theta_{z'}}\,k'$$
$$\vec{\eta}_v = \frac{\partial\Delta\mu'_{y'}}{\partial\Delta\theta_{x'}}\,i' + \frac{\partial\Delta\mu'_{y'}}{\partial\Delta\theta_{y'}}\,j' + \frac{\partial\Delta\mu'_{y'}}{\partial\Delta\theta_{z'}}\,k' \qquad . \tag{5.37}$$
$$\vec{\eta}_w = \frac{\partial\Delta\mu'_{z'}}{\partial\Delta\theta_{x'}}\,i' + \frac{\partial\Delta\mu'_{z'}}{\partial\Delta\theta_{y'}}\,j' + \frac{\partial\Delta\mu'_{z'}}{\partial\Delta\theta_{z'}}\,k'$$

Comparing Eqs', (5.36) and (5.37) gives

$$\begin{pmatrix} \vec{\eta}_u \\ \vec{\eta}_v \\ \vec{\eta}_w \end{pmatrix} = J_{\mu P} \begin{pmatrix} i' \\ j' \\ k' \end{pmatrix} , \tag{5.38}$$

and

$$\begin{pmatrix} \vec{\delta}_u \\ \vec{\delta}_v \\ \vec{\delta}_w \end{pmatrix} = \Delta\theta\, J_{\mu P} \begin{pmatrix} i' \\ j' \\ k' \end{pmatrix} . \tag{5.39}$$

The equation (5.38) can also be written as

$$(\vec{\eta}_u)' = (1\,0\,0)\,J_{\mu P}, \tag{5.40}$$
$$(\vec{\eta}_v)' = (0\,1\,0)\,J_{\mu P}, \tag{5.41}$$

138

$$(\vec{\eta}_w)' = (0\ 0\ 1)\,J_{\mu P}\,. \tag{5.42}$$

Moreover, it should be noted that the characteristic vector T by virtue of its nature represents a zero-valued axis direction of the component of image rotation along any direction in the image space, therefore T is perpendicular to all extreme-valued axis directions, such as u, v, w, and in general the arbitrary one h (or $\vec{\delta}_h$). Thus, $T \perp u$, $T \perp v$, $T \perp w$, and $T \perp h$.

Consequently, the three extreme-valued axis directions u, v and w for a prism must be parallel to a common plane, which is perpendicular to the characteristic vector T for that prism.

From Eqs. (5.22) and (5.23), it will be seen that none of the components of image rotation can exceed double the angle of rotation of a prism; actually even the image rotation $\Delta\vec{\mu}'$ itself cannot do it either, i.e., we have in all cases

$$|\Delta\vec{\mu}'| \leqslant 2\Delta\theta\,. \tag{5.43}$$

To show the significance of the theorem of adjustment as well as of the characteristic parameters associated with it, let us further investigate some more examples.

Example

To make a comparison, redo the example which has been solved in Sect. 5.3 (Fig. 5.8), but now we have the adjustment diagram of that prism, as shown in Fig. 5.16.

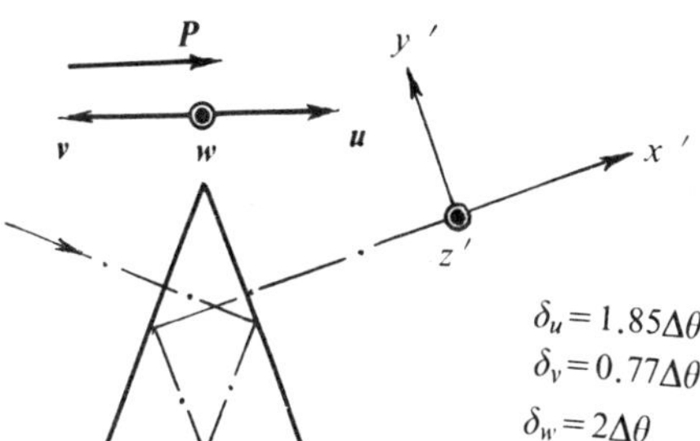

Fig. 5.16. An adjustment diagram is ready for use

Solution

With three parameters $\vec{\delta}_u$, $\vec{\delta}_v$ and $\vec{\delta}_w$ given in the diagram, we can immediately determine the x', y', z' image rotations according to Eq. (5.32)

$$\Delta\mu_{x'}' = \delta_u \cos\,(\widehat{P,\vec{\delta}_u}) = 1.85\,\Delta\theta \cos 0° = 1.85\Delta\theta\,,$$

$$\Delta\mu_{y'}' = \delta_v \cos\,(\widehat{P,\vec{\delta}_v}) = 0.77\Delta\theta \cos 180° = -0.77\Delta\theta\,,$$

$$\Delta\mu_{z'}' = \delta_w \cos\,(\widehat{P,\vec{\delta}_w}) = 2\Delta\theta \cos 90° = 0\,.$$

Example

Investigate the behavior of a roof Pechan concerning its adjustment for image lean, see

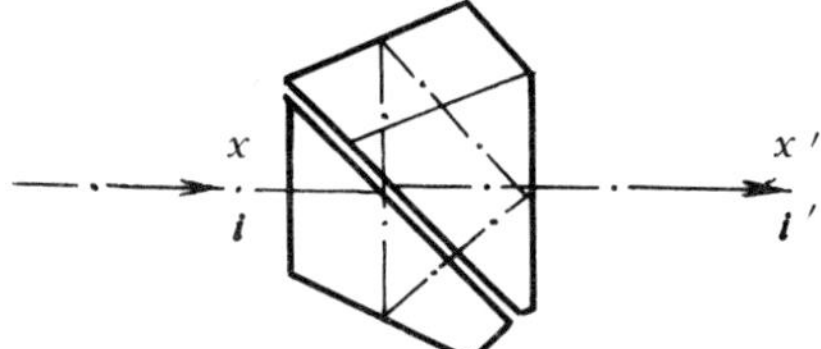

Fig . 5 .17 .   A roof Pechan having $\vec{\delta_u}=0$

Fig .5 .17 .

Discussion

From Fig .5 .17 ,

$$\vec{\delta_u}=\Delta\theta\,(i\,'-i)=0 .$$

Then ,

$$\Delta\mu_{x'}'= P\cdot\vec{\delta_u}=\delta_u\cos(\overset{\frown}{P,\vec{\delta_u}})=0 .$$

The result shows that the $x\,'$ image rotation will always be equal to zero no matter in which direction the axis P of rotation of the roof Pechan might be , therefore the prism is said to be free from image lean . But, we must say that the roof Pechan is not suitable for use in adjusting the image lean .

Example

Make a discussion on how to choose an adequate axis of rotation of a 45 °angle mirror , used as the left end reflector of a stereoscopic rangefinder , for a certain adjustment , see Fig . 5 .18 .

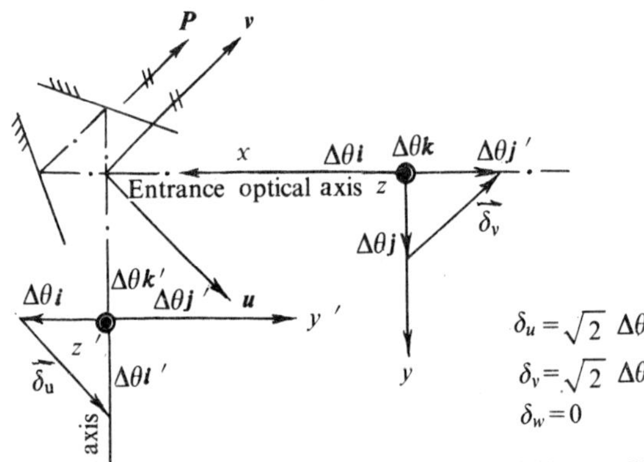

Fig .5 .18 . A diagram illustrating the choice of an axis of rotation of the left end reflector of a rangefinder for compensating the misalignment in height of the measuring mark

Verification

As a rule , we will first find the extreme - valued characteristic vectors for the angle mirror from its adjustment diagram which you may obtain in the Appendix , or just make a drawing of it in a minute if a planar prism as well as a planar mirror system is the case .

As is well known , in this example the y' image rotation $\Delta\mu_{y'}'$ has an influence on the misalignment in height of the mark while the z' image rotation $\Delta\mu_{z'}'$ will cause an error in range , and the x' image rotation is always related to the image lean .

From the adjustment diagram , we have

$$\vec{\delta}_w = 0 , \quad \vec{\delta}_v \perp \vec{\delta}_u .$$

The first result shows that there will be no error in range measurement no matter about which axis the reflector is turned through a small angle . This is supposed to be the chief reason why a pentamirror or a pentaprism should be employed here to replace an ordinary right - angled prism or a single mirror .

The second result , namely $\vec{\delta}_v \perp \vec{\delta}_u$, indicates that the 45 °angle mirror has the advantage of being able to adjust $\Delta\mu_{x'}'$ and $\Delta\mu_{y'}'$, completely independently in two individual steps by selecting an axis P of rotation either parallel to u or to v .

Since the requirement imposed on the reflector is to compensate the misalignment in height of the rangefinder while it it in use , an axis P parallel to v is adopted so as to ensure a high sensitivity of adjustment and to avoid introducing unwanted components of image rotation .

5.6 Components of Image Displacement and Relevant Characteristic Parameters

There is also a need to examine components of image displacement $\Delta S'$ individually .
By comparing the following two equations :

$$\Delta S' = \Delta g\, D - \Delta g\, D' , \tag{4.5}$$

$$\Delta\vec{\mu}' = \Delta\theta\, P + (-1)^{l-1}\Delta\theta\, P', \tag{4.1}$$

we can see a close mathematical resemblance between them , therefore all the respective equations , theorem , as well as characteristic parameters in regard to components of image displacement may be found in a similar way . For this reason , we will give all these counterparts with only a few notes on some new symbols .

5.6.1 Equations of a Component of Image Displacement Due to a Small Linear Displacement of a Reflecting Prism

$$\Delta S_{r'}' = \Delta g\,(D_{r'} - D_{r'}') , \tag{5.44}$$

$$\Delta S_{r'}' = D \cdot \vec{\delta}_e = \delta_e \cos(\widehat{D',\vec{\delta}_e}) , \tag{5.45}$$

$$\vec{\delta}_e = \Delta g\,[\,r' - (-1)^l r\,] = \delta_e e , \tag{5.46}$$

141

where the meaning and the relationship of unit vectors r' and r' remain unchanged; $\Delta S_{r'}'$ denotes the component of image displacement along the direction r' which for short is called the r' image displacement.

Also, δ_e, e, and $\vec{\delta_e}$ represent the *extreme value of r' image displacement*, *extreme-valued shift direction of r' image displacement*, and *extreme-valued characteristic vector of r' image displacement*.

Again, we have

$$\vec{\eta_e} = \text{grad } \Delta S_{r'}' = r' - (-1)^t r, \tag{5.47}$$

$$\vec{\delta_e} = \Delta g \vec{\eta_e}, \tag{5.48}$$

where $\vec{\eta_e}$ represents the *gradient shift direction of r' image displacement*.

5.6.2 Corollary of Adjustment for Reflecting Prisms

The equations (5.45) and (5.46) can accordingly be summarized as a corollary.

Corollary of adjustment for reflecting prisms:

"The relation between the component $\Delta S_{r'}'$ of image displacement along the direction of a unit vector r' in the image space (the r' image displacement) and the small linear displacement $\Delta g D$ of the prism, which produces the image displacement, is subject to the *law of cosine function* : $\Delta S_{r'}' = D \cdot \vec{\delta_e} = \delta_e \cos (\overset{\wedge}{D,\vec{\delta_e}})$ while the sole vector parameter $\vec{\delta_e}$ in the function is determined by the *rule of two-vector subtraction or addition* : $\vec{\delta_e} = \Delta g [r' - (-1)^t r]$, where $(-1)^t r$ is a unit vector in the object space and conjugate to r'; t is the number of reflections."

The corollary may also be called the *cosine function and two-vector subtraction or addition rule*.

5.6.3 Characteristic Parameters of Components of Image Displacement

Correspondingly, we have

$$\left. \begin{array}{l} \Delta S_{x'}' = D \cdot \vec{\delta_a} = \delta_a \cos (\overset{\wedge}{D,\vec{\delta_a}}) = \delta_a \cos \alpha \\[2mm] \Delta S_{y'}' = D \cdot \vec{\delta_b} = \delta_b \cos (\overset{\wedge}{D,\vec{\delta_b}}) = \delta_b \cos \beta \\[2mm] \Delta S_{z'}' = D \cdot \vec{\delta_c} = \delta_c \cos (\overset{\wedge}{D,\vec{\delta_c}}) = \delta_c \cos \gamma \end{array} \right\}, \tag{5.49}$$

$$\left. \begin{array}{l} \vec{\delta_a} = \Delta g [i' - (-1)^t i] = \delta_a a \\[2mm] \vec{\delta_b} = \Delta g [j' - (-1)^t j] = \delta_b b \\[2mm] \vec{\delta_c} = \Delta g [k' - (-1)^t k] = \delta_c c \end{array} \right\}, \tag{5.50}$$

where $\Delta S_{x'}'$, $\Delta S_{y'}'$, $\Delta S_{z'}'$ represent the x', y', z' image displacements, respectively; $\vec{\delta_a}$, $\vec{\delta_b}$, $\vec{\delta_c}$ represent the extreme-valued characteristic vectors of the x', y', z' image displacements, respectively; δ_a, δ_b, δ_c represent the extreme values of the x', y', z' image displacements, respectively; a, b, c represent the extreme-valued shift directions of the x',

y', z' image displacements, respectively.

Here, the relation of the coordinate system xyz in the object space is still kept normalized with respect to the image coordinate system $x'y'z'$ (see Sect. 5.5).

Three adjustment diagrams, containing each six extreme-valued characteristic parameters $\vec{\delta_u}$, $\vec{\delta_v}$, $\vec{\delta_w}$ and $\vec{\delta_a}$, $\vec{\delta_b}$, $\vec{\delta_c}$, are given for three planar reflecting prisms in Figs. 5.19, 5.20 and 5.21, respectively.

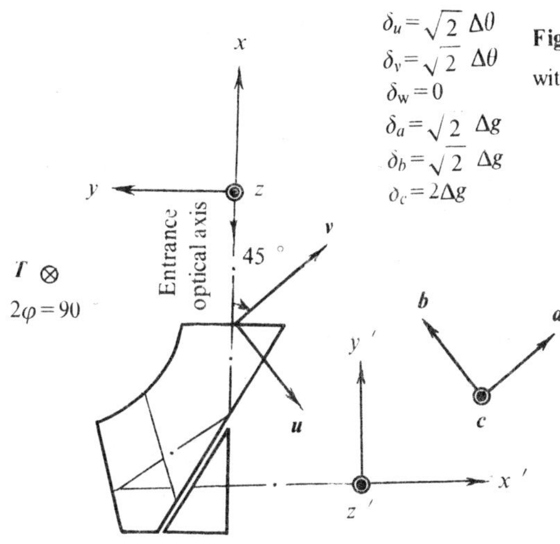

Fig. 5.19. Adjustment diagram for a prism with an odd number of reflections

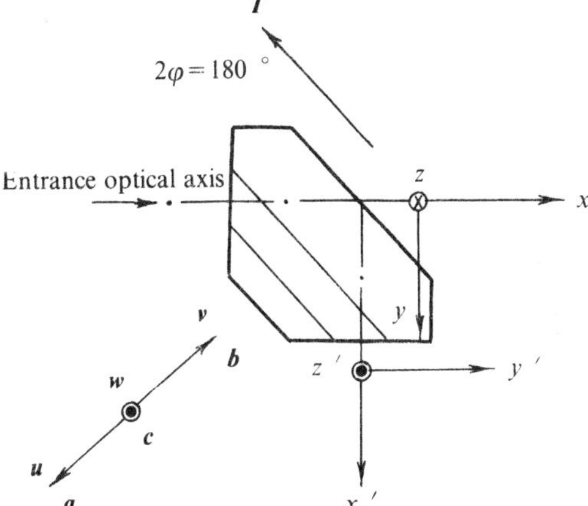

Fig. 5.20. Adjustment diagram for a prism with an even number of reflections

Again, we have

$$\left.\begin{aligned}
\vec{\eta_a} &= \operatorname{grad}\Delta S'_{x'} = i' - (-1)^l i \\
\vec{\eta_b} &= \operatorname{grad}\Delta S'_{y'} = j' - (-1)^l j \\
\vec{\eta_c} &= \operatorname{grad}\Delta S'_{z'} = k' - (-1)^l k
\end{aligned}\right\}, \tag{5.51}$$

$$\begin{aligned}
\vec{\delta}_a &= \Delta g\, \vec{\eta}_a \\
\vec{\delta}_b &= \Delta g\, \vec{\eta}_b \\
\vec{\delta}_c &= \Delta g\, \vec{\eta}_c
\end{aligned} \Bigg\} , \qquad (5.52)$$

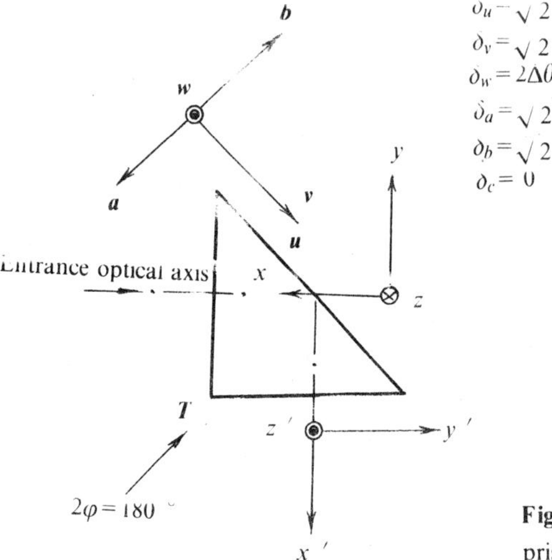

Fig.5.21. Adjustment diagram for a prism with an odd number of reflections

and

$$\begin{pmatrix} \vec{\eta}_a \\ \vec{\eta}_b \\ \vec{\eta}_c \end{pmatrix} = J_{SD} \begin{pmatrix} i' \\ j' \\ k' \end{pmatrix} , \qquad (5.53)$$

$$\begin{pmatrix} \vec{\delta}_a \\ \vec{\delta}_b \\ \vec{\delta}_c \end{pmatrix} = \Delta g J_{SD} \begin{pmatrix} i' \\ j' \\ k' \end{pmatrix} , \qquad (5.54)$$

where $\vec{\eta}_a$, $\vec{\eta}_b$ and $\vec{\eta}_c$ represent the gradient shift directions of the x', y', z' image displacements, respectively.

From Eq.(5.46), we have readily seen that

$$\begin{aligned}
\vec{\delta}_e &= \Delta g\,(r' - r), & t &= \text{even number} \\
\vec{\delta}_e &= \Delta g\,(r' + r), & t &= \text{odd number}
\end{aligned} \Bigg\} . \qquad (5.55)$$

Thus for those prisms with an even number of reflections, the gradients of the x', y', z' image displacements are equal to the gradients of the x', y', z' image rotations, respectively :

$$\left.\begin{array}{l} \vec{\eta}_a = \vec{\eta}_u \\ \vec{\eta}_b = \vec{\eta}_v \\ \vec{\eta}_c = \vec{\eta}_w \end{array}\right\} \ , \quad t = \text{even number}, \tag{5.56}$$

although these two sets of gradients are completely different in meaning. Hence , a , b and c are parallel to u , v and w , respectively , and also T is perpendicular to any one of a , b and c , as shown in Fig.5.20. However ,in this situation T is regarded as the zero- valued shift direction of the component $\Delta S_{r'}'$ of image displacement along an arbitrary direction instead of the zero- valued axis direction of the component $\Delta\mu_{r'}'$ of image rotation along an arbitrary direction .

As to those prisms with an odd number of reflections ,the two sets of gradients are different from one another as they are determined according to different rules .Hence ,a ,b and c also differ from u ,v and w in direction , as shown in Figs.5.19 and 5.21 .Moreover , in this situation the characteristic vector T will no longer represent a zero-valued shift direction common to all components of image displacement , i. e., $\Delta S_{r'}'$, so there is no reason for three extreme- valued shift directions a ,b and c to be normal to T, and thus these three unit vectors are not necessarily parallel to a common plane .That is just the case in Fig.5.19 , where the extreme- valued shift directions a ,b and c constitute a three dimensional frame .

However , there still exists some relation between the two sets of gradients :

$$\left.\begin{array}{l} \vec{\eta}_a = \vec{\eta}_u + 2\,i \\ \vec{\eta}_b = \vec{\eta}_v + 2\,j \\ \vec{\eta}_c = \vec{\eta}_w + 2\,k \end{array}\right\} \ , \quad t = \text{odd number}. \tag{5.57}$$

5.7 Components of Image Point Displacement and Relevant Characteristic Parameters

The basic equations of image motion include three formulas concerning respectively the image rotation $\Delta\vec{\mu}'$, the image point displacement $\Delta S_F'$, and the image displacement $\Delta S'$. A study of $\Delta\vec{\mu}'$ and $\Delta S'$ in terms of a component of them was explored in considerable depth in Secs.5.5 and 5.6.Now , let us proceed to tackle the last , more complicated part $\Delta S_{F'}'$. For this purpose , rewrite the necessary equations:

$$\Delta S_{F'}' = \Delta\theta\,P\times\vec{\rho}_F' + (-1)^{t-1}\Delta\theta\,P'\times\vec{\rho}_{F'}' \tag{4.2}$$

or from Eq.(4.15) ,

$$(\Delta S_F') = \Delta\theta\,[\,E + (-1)^{t-1}S_{T,2\varphi}\,]\,(r_q\times P) - (-1)^t\Delta\theta\,S_{T,2\varphi}\,(\overrightarrow{M_0M}'\times P)$$

$$-\Delta\theta\,(b'\times[\,\underline{(E - S_{T,2\varphi})\,(P)}\,]\,) . \tag{5.58}$$

145

We will also deal with the components $\Delta S'_{F'x'}$, $\Delta S'_{F'y'}$, $\Delta S'_{F'z'}$, and in general $\Delta S'_{F'r'}$ of image point displacement along the axes x', y', z', and an arbitrary direction r' in the image space, which for short are called the x', y', z', and in general r' image point displacements, respectively.

Using the matrix J_{SP} from Eq. (4.26), three components $\Delta S'_{F'x'}$, $\Delta S'_{F'y'}$, and $\Delta S'_{F'z'}$ of image point displacement for an arbitrary prism can be written in a general form as follows:

$$\Delta S'_{F'x'} = \Delta\theta\,[\,P_{x'}\,(l_{11}\,x'_q + l_{12}\,y'_q + l_{13}\,z'_q + l_{14})$$
$$+ P_{y'}\,(l_{21}\,x'_q + l_{22}\,y'_q + l_{23}\,z'_q + l_{24})$$
$$+ P_{z'}\,(l_{31}\,x'_q + l_{32}\,y'_q + l_{33}\,z'_q + l_{34})]\,, \tag{5.59}$$

$$\Delta S'_{F'y'} = \Delta\theta\,[\,P_{x'}\,(m_{11}\,x'_q + m_{12}\,y'_q + m_{13}\,z'_q + m_{14})$$
$$+ P_{y'}\,(m_{21}\,x'_q + m_{22}\,y'_q + m_{23}\,z'_q + m_{24})$$
$$+ P_{z'}\,(m_{31}\,x'_q + m_{32}\,y'_q + m_{33}\,z'_q + m_{34})]\,, \tag{5.60}$$

$$\Delta S'_{F'z'} = \Delta\theta\,[\,P_{x'}\,(n_{11}\,x'_q + n_{12}\,y'_q + n_{13}\,z'_q + n_{14})$$
$$+ P_{y'}\,(n_{21}\,x'_q + n_{22}\,y'_q + n_{23}\,z'_q + n_{24})$$
$$+ P_{z'}\,(n_{31}\,x'_q + n_{32}\,y'_q + n_{33}\,z'_q + n_{34})]\,. \tag{5.61}$$

Then, the component $\Delta S'_{F'r'}$ of image point displacement along an arbitrary direction r' can be expressed as

$$\Delta S'_{F'r'} = \Delta\theta\,[\,P_{x'}\,(d_{11}\,x'_q + d_{12}\,y'_q + d_{13}\,z'_q + d_{14})$$
$$+ P_{y'}\,(d_{21}\,x'_q + d_{22}\,y'_q + d_{23}\,z'_q + d_{24})$$
$$+ P_{z'}\,(d_{31}\,x'_q + d_{32}\,y'_q + d_{33}\,z'_q + d_{34})]\,. \tag{5.62}$$

For any concrete reflecting prism some of the coefficients d_{ij} in Eq. (5.62) may be equal to zero, but such a general expression can still be useful for explaining some of the properties of a component of image point displacement which have no bearing on whether one or the other d_{ij} exists or not.

However, these equations may be written in a more compact form

$$\Delta S'_{F'x'} = \Delta\theta\,[\,l_1\,(q)\,P_{x'} + l_2\,(q)\,P_{y'} + l_3\,(q)\,P_{z'}\,]\,, \tag{5.63}$$

$$\Delta S'_{F'y'} = \Delta\theta\,[\,m_1\,(q)\,P_{x'} + m_2\,(q)\,P_{y'} + m_3\,(q)\,P_{z'}\,]\,, \tag{5.64}$$

$$\Delta S'_{F'z'} = \Delta\theta\,[\,n_1\,(q)\,P_{x'} + n_2\,(q)\,P_{y'} + n_3\,(q)\,P_{z'}\,]\,, \tag{5.65}$$

and in general

$$\Delta S'_{F'r'} = \Delta\theta\,[\,d_1\,(q)\,P_{x'} + d_2\,(q)\,P_{y'} + d_3\,(q)\,P_{z'}\,]\,, \tag{5.66}$$

146

where $l_i(q)$, $m_i(q)$, $n_i(q)$, and $d_i(q)$ represent certain linear combinations of the coordinates of a point q (x_q', y_q', z_q').

The equation (5.66) shows that an arbitrary component $\Delta S_{F'_r}'$ of image point displacement depends upon $\Delta\theta P$ and q, the latter being a sliding point on the action line of P. In other words, $\Delta S_{F'_r}'$ is related to both the parameter R of orientational conjugation and the parameter $\overrightarrow{M_0 M}'$ of locational conjugation for a prism.

Thus, it would be more complicated to gain an insight into the equation (5.58) than it was for Eqs. (4.9) and (4.10).

5.7.1 The Gradient Axis (Extreme-Valued Axis) and the Zero-Valued Axis of a Component of Image Point Displacement

Rewrite the formula of $\Delta S_{F'_r}'$

$$\Delta S_{F'_r}' = \Delta\theta \; [\; d_1(q) P_{x'} + d_2(q) P_{y'} + d_3(q) P_{z'}]\;. \tag{5.66}$$

In Eq. (5.66), $\Delta\theta P$ and r_q are both variable, there would not exist any extreme-valued axis of r' image point displacement, defined for the whole space.

However, by holding r_q fast, we may find the gradient of r' image point displacement $\Delta S_{F'_r}'$ for a fixed point as:

$$\vec{\eta}_{F'hq} = \operatorname{grad} \Delta S_{F'_r}' = \frac{\partial \Delta S_{F'_r}'}{\partial \Delta\theta_{x'}} \; i' + \frac{\partial \Delta S_{F'_r}'}{\partial \Delta\theta_{y'}} \; j' + \frac{\partial \Delta S_{F'_r}'}{\partial \Delta\theta_{z'}} \; k'\;. \tag{5.67}$$

Here, $\vec{\eta}_{F'hq}$ is called the *local gradient axis of r' image point displacement* after its meaning, and the word *local* may be omitted since the notation used is clear.

The partial derivatives can be determined from Eq. (5.66)

$$\frac{\partial \Delta S_{F'_r}'}{\partial \Delta\theta_{x'}} = d_1(q)\;, \qquad \frac{\partial \Delta S_{F'_r}'}{\partial \Delta\theta_{y'}} = d_2(q)\;, \qquad \frac{\partial \Delta S_{F'_r}'}{\partial \Delta\theta_{z'}} = d_3(q)\;. \tag{5.68}$$

By putting Eq. (5.68) back into Eq. (5.66), $\Delta S_{F'_r}'$ can again be expressed as

$$\Delta S_{F'_r}' = \Delta\theta \left(\frac{\partial \Delta S_{F'_r}'}{\partial \Delta\theta_{x'}} \; P_{x'} + \frac{\partial \Delta S_{F'_r}'}{\partial \Delta\theta_{y'}} \; P_{y'} + \frac{\partial \Delta S_{F'_r}'}{\partial \Delta\theta_{z'}} \; P_{z'} \right)\;. \tag{5.69}$$

It should be noted that Eq. (5.69) is related to Eq. (5.67) in the same way as Eq. (5.23) is related to Eq. (5.22).

Also, putting Eq. (5.68) into Eq. (5.67) gives

$$\vec{\eta}_{F'hq} = d_1(q)\; i' + d_2(q) j' + d_3(q) k'\;. \tag{5.70}$$

The above equation shows that there must exist a unique gradient axis of $\Delta S_{F'_r}'$ for a fixed point q so long as any one of the coefficients $d_i(q)$ $(i = 1, 2, 3)$ is not zero at that point.

To examine the zero-valued axis, equate Eq. (5.69) to zero -

$$\Delta S_{F'r'}' = \Delta\theta \left(\frac{\partial \Delta S_{F'r'}'}{\partial \Delta\theta_{x'}} \, P_{x'} + \frac{\partial \Delta S_{F'r'}'}{\partial \Delta\theta_{y'}} \, P_{y'} + \frac{\partial \Delta S_{F'r'}'}{\partial \Delta\theta_{z'}} \, P_{z'} \right) = 0 \qquad (5.71)$$

or

$$- \; \Delta S_{F'r'}' = \Delta\theta \, [\; d_1(q) P_{x'} + d_2(q) P_{y'} + d_3(q) P_{z'} \;] = 0 . \qquad (5.72)$$

Solving Eq. (5.71) or (5.72) for $P_{x'}$, $P_{y'}$ and $P_{z'}$ gives the direction cosines of the zero-valued axes ζ_{dq} of $\Delta S_{F'r'}'$ for a fixed point q.

Obviously, the structure of the scalar product of Eq. (5.71) or (5.72) indicates the fact that the zero-valued axes and the gradient axis of an arbitrary component $\Delta S_{F'r'}'$ of image point displacement for the same point q are orthogonal to each other, as shown in Fig.5.22.

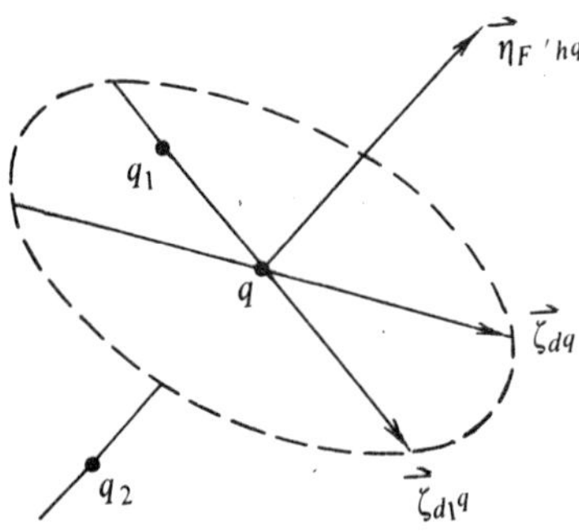

Fig.5.22. The gradient axis $\vec{\eta}_{F'hq}$ and the zero-valued axes $\vec{\zeta}_{dq}$ of $\Delta S_{F'r'}'$ for a fixed point q

Hence, there is an infinte number of zero-valued axes of $\Delta S_{F'r'}'$ for a fixed point q. However, these zero-valued axes are concurrent and coplanar; the intersecting point is q while the plane is normal to the unique gradient axis $\vec{\eta}_{F'hq}$ for the same point q.

Furthermore, according to the physical meaning as well as the mathematical behavior of Eqs. (5.66) and (5.70), the zero-valued axis belongs to a sliding vector while the gradient axis has the property of a fixed vector; the latter is not allowed to be displaced either laterally or longitudinally.

It is shown in Fig.5.22 that zero-valued axis $\vec{\zeta}_{d_1q}$ for a fixed point q can also be considered as a zero-valued axis for another point lying on the action line of $\vec{\zeta}_{d_1q}$, for instance q_1, but the gradient axis $\vec{\eta}_{F'hq}$ for a fixed point q can no longer be kept as a gradient axis for any other point on the action line of $\vec{\eta}_{F'hq}$, for example, q_2 (exceptions may sometimes happen).

Applying Eqs. (5.67) and (5.69) to the x', y', z' image point displacements leads to two corresponding sets of equations.

$$\vec{\eta}_{F\,'uq} = \frac{\partial \Delta S_{F\,'x'}'}{\partial \Delta \theta_{x'}}\, \boldsymbol{i}' + \frac{\partial \Delta S_{F\,'x'}'}{\partial \Delta \theta_{y'}}\, \boldsymbol{j}' + \frac{\partial \Delta S_{F\,'x'}'}{\partial \Delta \theta_{z'}}\, \boldsymbol{k}'$$

$$\vec{\eta}_{F\,'vq} = \frac{\partial \Delta S_{F\,'y'}'}{\partial \Delta \theta_{x'}}\, \boldsymbol{i}' + \frac{\partial \Delta S_{F\,'y'}'}{\partial \Delta \theta_{y'}}\, \boldsymbol{j}' + \frac{\partial \Delta S_{F\,'y'}'}{\partial \Delta \theta_{z'}}\, \boldsymbol{k}'$$

$$\vec{\eta}_{F\,'wq} = \frac{\partial \Delta S_{F\,'z'}'}{\partial \Delta \theta_{x'}}\, \boldsymbol{i}' + \frac{\partial \Delta S_{F\,'z'}'}{\partial \Delta \theta_{y'}}\, \boldsymbol{j}' + \frac{\partial \Delta S_{F\,'z'}'}{\partial \Delta \theta_{z'}}\, \boldsymbol{k}' \qquad (5.73)$$

$$\Delta S_{F\,'x'}' = \Delta \theta \left(\frac{\partial \Delta S_{F\,'x'}'}{\partial \Delta \theta_{x'}}\, P_{x'} + \frac{\partial \Delta S_{F\,'x'}'}{\partial \Delta \theta_{y'}}\, P_{y'} + \frac{\partial \Delta S_{F\,'x'}'}{\partial \Delta \theta_{z'}}\, P_{z'} \right)$$

$$\Delta S_{F\,'y'}' = \Delta \theta \left(\frac{\partial \Delta S_{F\,'y'}'}{\partial \Delta \theta_{x'}}\, P_{x'} + \frac{\partial \Delta S_{F\,'y'}'}{\partial \Delta \theta_{y'}}\, P_{y'} + \frac{\partial \Delta S_{F\,'y'}'}{\partial \Delta \theta_{z'}}\, P_{z'} \right)$$

$$\Delta S_{F\,'z'}' = \Delta \theta \left(\frac{\partial \Delta S_{F\,'z'}'}{\partial \Delta \theta_{x'}}\, P_{x'} + \frac{\partial \Delta S_{F\,'z'}'}{\partial \Delta \theta_{y'}}\, P_{y'} + \frac{\partial \Delta S_{F\,'z'}'}{\partial \Delta \theta_{z'}}\, P_{z'} \right) \qquad (5.74)$$

where $\vec{\eta}_{F\,'uq}$, $\vec{\eta}_{F\,'vq}$, $\vec{\eta}_{F\,'wq}$ represent the gradient axes of the x', y', z' image point displacements for a point q, respectively.

5.7.2 The Gradient Direction of a Component of Image Point Displacement

It is evident that we have another kind of gradient of the r' image point displacement $\Delta S_{F\,'r'}'$ by holding $\Delta \theta P$ constant

$$\vec{\eta}_{F\,'eP} = \mathrm{grad}\,\Delta S_{F\,'r'}' = \frac{\partial \Delta S_{F\,'r'}'}{\partial x_q'}\, \boldsymbol{i}' + \frac{\partial \Delta S_{F\,'r'}'}{\partial y_q'}\, \boldsymbol{j}' + \frac{\partial \Delta S_{F\,'r'}'}{\partial z_q'}\, \boldsymbol{k}', \qquad (5.75)$$

which may be called the *gradient (translation) direction of r' image point displacement*.

The gradient direction $\vec{\eta}_{F\,'eP}$ seems to be less important than the gradient axis $\vec{\eta}_{F\,'hq}$, therefore we will give only some explanation regarding the direction of the vector $\vec{\eta}_{F\,'eP}$.

Referring to Fig.5.23, suppose that a small angular displacement $\Delta \theta P$ of a prism is translated from one position passing through the point q to another position passing through the point q_1. Then $\overrightarrow{qq_1}$, designated by Δr_q, represents the change of the position vector from r_q to r_{q_1}.

Assume the incremental vector Δr_q to be coincident with the gradient direction $\vec{\eta}_{F\,'eP}$ of r' image point displacement $\Delta S_{F\,'r'}'$. Then, on one hand the vector Δr_q should be perpendicular to the small angular displacement $\Delta \theta P$ since any sliding of a small angular displacement $\Delta \theta P$ along its own direction is meaningless; on the other hand, a lateral shift of a small angular displacement $\Delta \theta P$ by Δr_q will equivalently add a small linear displacement $(\Delta r_q \times \Delta \theta P)$ to the prism, therefore the vector Δr_q has also to be perpendicular to the plane formed by the small linear displacement $\Delta\theta P$ and the extreme-valued characteristic vector $\vec{\delta_e}$ of r' image displacement $\Delta S_r'$, because any lateral shift of $\Delta\theta P$ in this plane will be non-significant.

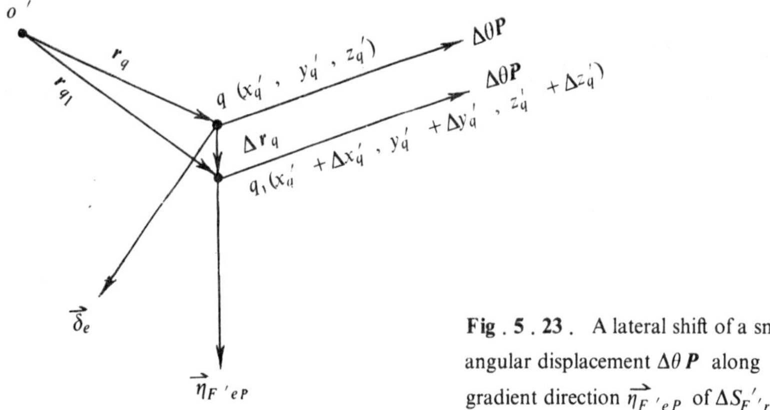

Fig. 5.23. A lateral shift of a small angular displacement $\Delta\theta P$ along the gradient direction $\vec{\eta}_{F'eP}$ of $\Delta S_{F'r'}'$

Thus, the gradient direction $\vec{\eta}_{F'eP}$ of $\Delta S_{F'r'}'$ should meet the following conditions:

$$\vec{\eta}_{F'eP} \cdot P = 0, \tag{5.76}$$

$$\vec{\eta}_{F'eP} \cdot \vec{\delta_e} = 0. \tag{5.77}$$

It should be noted that the gradient direction $\vec{\eta}_{F'eP}$ is independent of the location of the axis P of rotation.

Similarly, applying Eq. (5.75) to the x', y', z' image point displacements we obtain $\vec{\eta}_{F'aP}, \vec{\eta}_{F'bP}, \vec{\eta}_{F'cP}$, which represent the gradient (translation) directions of the x', y', z' image point displacements and have the same properties as indicated in Eqs. (5.76) and (5.77).

5.7.3 Terminology

For the convenience of the following discussion, some new terms will first be defined.

One-dimensional zero-valued axis It represents those axes around which a small rotation of a prism causes no component of image point displacement $\Delta S_F'$ along only one axis, i.e., $\Delta S_{F'x}' = 0$ (simultaneously $\Delta S_{F'y}' \neq 0$, $\Delta S_{F'z}' \neq 0$) or $\Delta S_{F'y}' = 0$ (simultaneously $\Delta S_{F'z}' \neq 0$ and $\Delta S_{F'x}' \neq 0$) or $\Delta S_{F'z}' = 0$ (simultaneously $\Delta S_{F'x}' \neq 0$ and $\Delta S_{F'y}' \neq 0$).

Thus, we have three sorts of such axes, designated by $\vec{\zeta}_{lq}, \vec{\zeta}_{mq}$ and $\vec{\zeta}_{nq}$, which are called the *one-dimensional zero-valued axis of x' image point displacement*, the *one-dimensional zero-valued axis of y' image point displacement*, and the *one-dimensional zero-valued axis of z' image point displacement*, respectively.

Two-dimensional zero-valued axis It represents those axes around which a small rotation of a prism causes no components of image point displacements $\Delta S_F'$ along only two axes, i.e., $\Delta S_{F'x}' = \Delta S_{F'y}' = 0$ (simultaneously $\Delta S_{F'z}' \neq 0$) or $\Delta S_{F'y}' = \Delta S_{F'z}' = 0$

(simultaneously $\Delta S_{F'x'}' \neq 0$) or $\Delta S_{F'z'}' = \Delta S_{F'x'}' = 0$ (simultaneously $\Delta S_{F'y'}' \neq 0$).

There are also three sorts of such axes, designated by $\vec{\zeta}_{lmq}$, $\vec{\zeta}_{mnq}$, and $\vec{\zeta}_{nlq}$, which are called the *two-dimensional zero-valued axis of x'-y' image point displacement*, the *two-dimensional zero-valued axis of y'-z' image point displacement*, and the *two-dimensional zero-valued axis of z'-x' image point displacement*, respectively.

Three-dimensional zero-valued axis $\vec{\zeta}_q$ of image point displacement $\Delta S_F'$. It represents those axes around which a small rotation of a prism causes no displacement of image point F': $\Delta S_F' = 0$, namely, $\Delta S_{F'x'}' = \Delta S_{F'y'}' = \Delta S_{F'z'}' = 0$.

One-dimensional gradient axis and two-dimensional gradient axis. It is easy to understand that $\vec{\eta}_{F'uq}$, $\vec{\eta}_{F'vq}$, $\vec{\eta}_{F'wq}$ are said to be *one-dimensional gradient axes of the x', y', z' image point displacements*, respectively, for a same point q if these three vectors are not coincident with one another. Once two gradient axes referring to a same point q but related to two different components of image point displacement are coincident, then we obtain a *two-dimensional gradient axis*. Theoretically, we should have the three-dimensional gradient axis of image point, but so far we have not actually seen one.

Plane of one-dimensional zero-valued aixs of r' image point displacement It represents that plane in which any line belongs to the one-dimensional zero-valued axis of r' image point displacement. Replacing r' by x', y' and z', we have three sorts of such planes, denoted by Π_l, Π_m and Π_n, which are called the *plane of one-dimensional zero-valued axis of x' image point displacement*, the *plane of one-dimensional zero-valued axis of y' image point displacement*, and the *plane of one-dimensional zero-valued axis of z' image point displacement*, respectively.

Planes of two-dimensional zero-valued axis Π_{lm}, Π_{mn}, Π_{nl} of x'-y', y'-z', z'-x' image point displacements The definitions regarding these planes are omitted since they can be figured out by analogy.

Logically, we should have the plane of three-dimensional zero-valued axis Π_{lmn}.

Planar one-dimensional zero-valued pole and the associated plane Referring to Fig. 5.24, if many* one-dimensional zero-valued axes of r' image point displacement are concurrent at point q $(C_{d\Pi})$, then the point of concurrence $C_{d\Pi}$ is called the *one-dimensional zero-valued pole of r' image point displacement*. Moreover, if these one-dimensional zero-valued axes are coplanar, then we have a *planar one-dimensional zero-valued pole $C_{d\Pi}$ of r' image point displacement*, and the plane Π in which are lying all the axes as well as the pole will thus be called the *associated plane*.

Spatial one-dimensional zero-valued pole of r' image point displacement If the axes $\vec{\zeta}_{d_iq}$ indicated in Fig. 5.24 can not conform with one single plane, the point of concurrence will thus be called the *spatial one-dimensional zero-valued pole of r' image point displacement*.

---

* Actually two is enough.

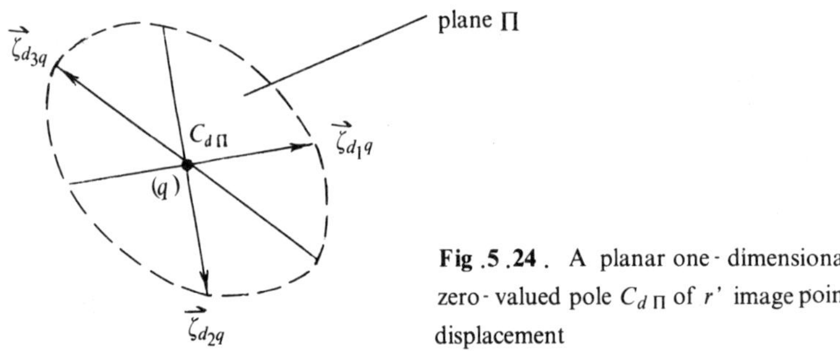

Fig .5 .24 . A planar one - dimensional zero - valued pole $C_{d\Pi}$ of r' image point displacement

Planar and spatial , two- and three- dimensional zero- valued poles   The definitions are omitted for the same reason as above .

Planar three- dimensional zero- valued polar line   For some prisms , there are an infinite number of planar three- dimensional zero- valued poles , which form a continuous straight line , as shown in Fig . 5 . 25 . Thus , we obtain a *planar three- dimensional zero- valued polar line* and all associated planes are therefore perpendicular to the polar line .

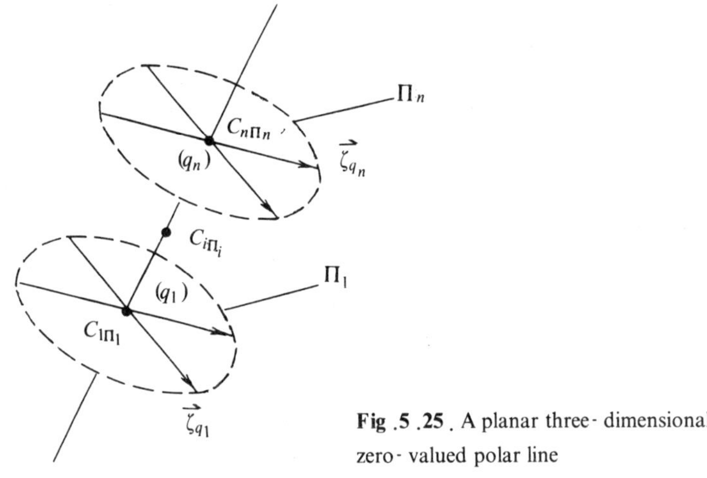

Fig .5 .25 . A planar three- dimensional zero- valued polar line

5 .7 .4 Characteristic Parameters for Coplanar Reflecting Prisms

The definition of a coplanar reflecting prism is given in Sect . 2 . 1 . Prisms of this type are so important   that they comprise over 80% of the prisms included in the Appendix . Carefully examining the three component equations $\Delta S'_{F'x'}$, $\Delta S'_{F'y'}$ and $\Delta S'_{F'z'}$ of image point displacement for those prisms ,you may find that there is a surprising resemblance in the structure of these formulas . Therefore , we feel that it might be necessary and important to make an investigation specially into coplanar reflecting prisms in regard to those characteristic axes , planes , and poles just defined in Sect . 5 . 3 , and hope that some results of such an investigation would lead to a more complete and systematic study of all kinds of reflecting prisms in this aspect .

For this purpose , several typical prisms of this category are shown in Figs . 5 . 26 through 5 . 31 , including , for each , an adjustment diagram and three component equations attached below .

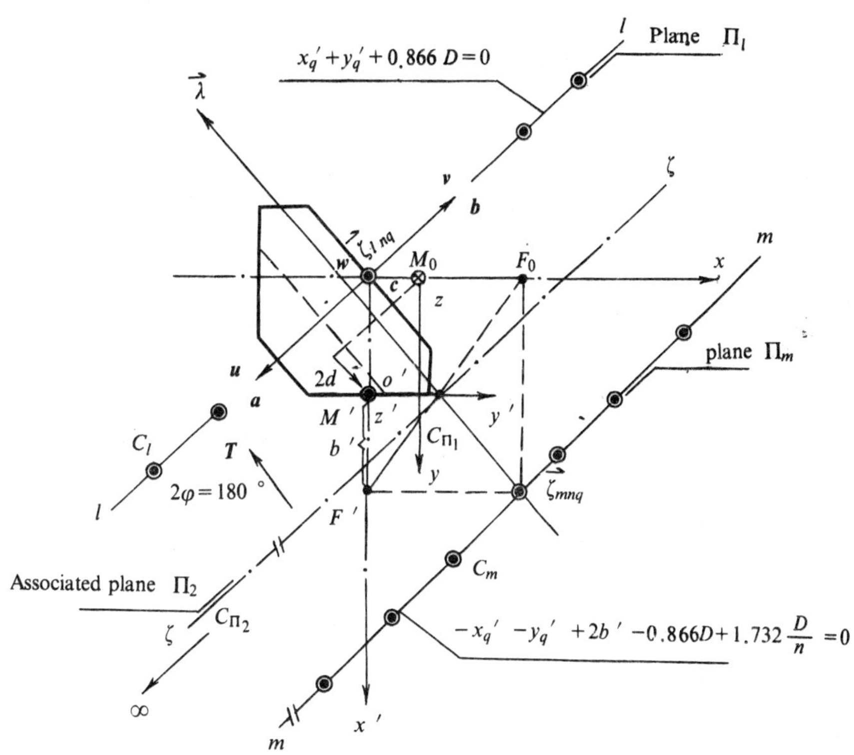

$$\Delta S_{F'x'}' = \Delta\theta[-P_{x'} z_q' - P_{y'} z_q' + P_{z'}(x_q' + y_q' + 0.866\,D)] , \qquad (5.78)$$

$$\Delta S_{F'y}' = \Delta\theta\left[P_{x'} z_q' + P_{y'} z_q' + P_{z'}\left(-x_q' - y_q' + 2b' \right.\right.$$

$$\left.\left. - 0.866\,D + 1.732\,\frac{D}{n} \right)\right] , \qquad (5.79)$$

$$\Delta S_{F'z'}' = \Delta\theta\left[P_{x'}\left(-2y_q' + b' - 0.866\,D + 1.732\,\frac{D}{n} \right) \right.$$

$$\left. + P_{y'}(2x_q' - b' + 0.866\,D) \right] . \qquad (5.80)$$

Fig . 5 . 26 . A representative prism having one planar three - dimensional zero - valued pole C_{Π_1} at finite distance and another one C_{Π_2} at infinity

153

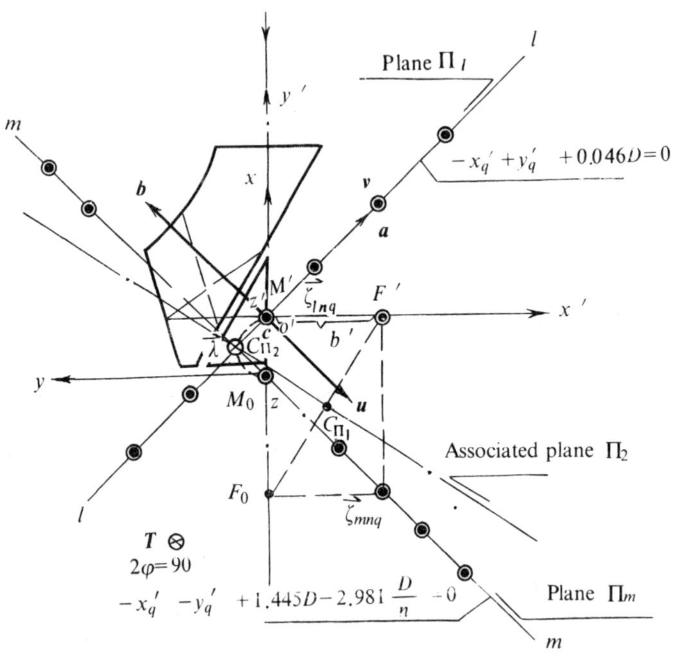

$$\Delta S_{F'x'}' = \Delta\theta \left[P_{x'} z_q' - P_{y'} z_q' + P_{z'}(-x_q' + y_q' + 0.046D) \right],$$

(5.81)

$$\Delta S_{F'y'}' = \Delta\theta \left[P_{x'} z_q' + P_{y'} z_q' + P_{z'}\left(-x_q' - y_q' + 1.445D - 2.981\frac{D}{n}\right)\right],$$

(5.82)

$$\Delta S_{F'z'}' = \Delta\theta \left[P_{x'}\left(-2y_q' - b' + 1.445D - 2.981\frac{D}{n}\right) + P_{y'}(2x_q' - b' - 0.046D)\right].$$

(5.83)

Fig. 5.27. A representative prism having both two planar three-dimensional zero-valued poles C_{Π_1} and C_{Π_2} at finite distance

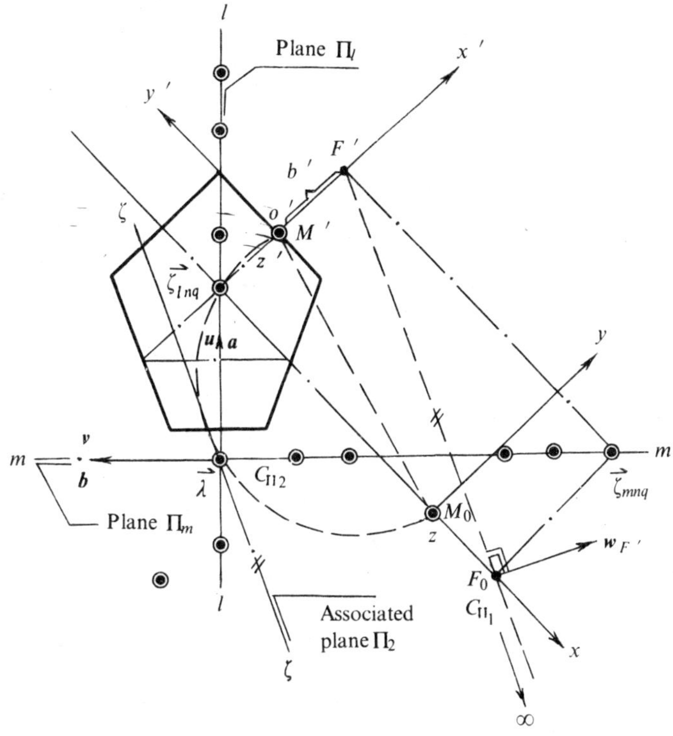

$$\Delta S'_{F'x'} = \Delta\theta\,[\,P_{x'}z'_q - P_{y'}z'_q + P_{z'}(-x'_q + y'_q - 0.5D)]\,, \tag{5.84}$$

$$\Delta S'_{F'y'} = \Delta\theta\left[P_{x'}z'_q + P_{y'}z'_q + P_{z'}\left(-x'_q - y'_q + 0.5D - 3.414\frac{D}{n}\right)\right]\,, \tag{5.85}$$

$$\Delta S'_{F'z'} = \Delta\theta\left[P_{x'}\left(b' - 0.5D + 3.414\frac{D}{n}\right) + P_{y'}(-b' - 0.5D)\right]. \tag{5.86}$$

Fig . 5 .28 . A representative prism having one planar three-dimensional zero-valued pole C_{Π_2} at finite distance and another one C_{Π_1} at infinity

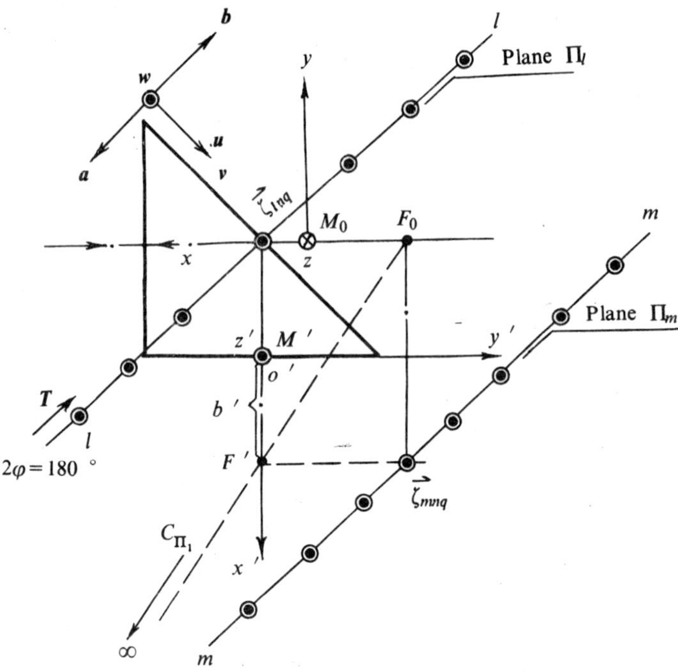

$$\Delta S'_{F'x'} = \Delta\theta\left[-P_{x'}z_q' - P_{y'}z_q' + P_{z'}(x_q' + y_q' + 0.5D)\right],\tag{5.87}$$

$$\Delta S'_{F'y'} = \Delta\theta\left[P_{x'}z_q' + P_{y'}z_q' + P_{z'}\left(-x_q' - y_q' + 2b' - 0.5D + \frac{D}{n}\right)\right],\tag{5.88}$$

$$\Delta S'_{F'z'} = \Delta\theta\left[P_{x'}\left(-b' + 0.5D - \frac{D}{n}\right) + P_{y'}(-b' - 0.5D)\right].\tag{5.89}$$

Fig. 5.29. A representative prism having one planar three-dimensional zero-valued pole C_{Π_1} at infinity only

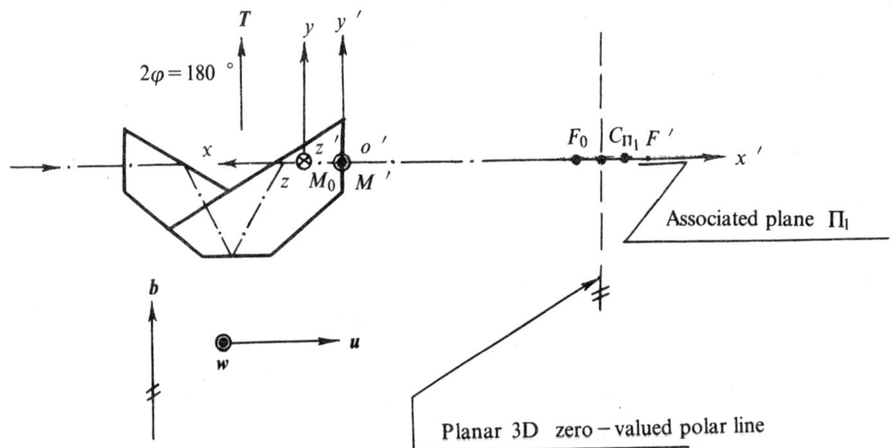

$$\Delta S'_{F'x'} = 0,\tag{5.90}$$

$$\Delta S'_{F'y'} = \Delta\theta\left[2P_{x'}z_q' + P_{z'}\left(-2x_q' + 2b' - 3.464D + 5.196\frac{D}{n}\right)\right],\tag{5.91}$$

$$\Delta S'_{F'z'} = \Delta\theta P_{y'}\left(-3.464D + 5.196\frac{D}{n}\right).\tag{5.92}$$

Fig. 5.30. A representative prism having a planar three-dimensional zero-valued polar line

156

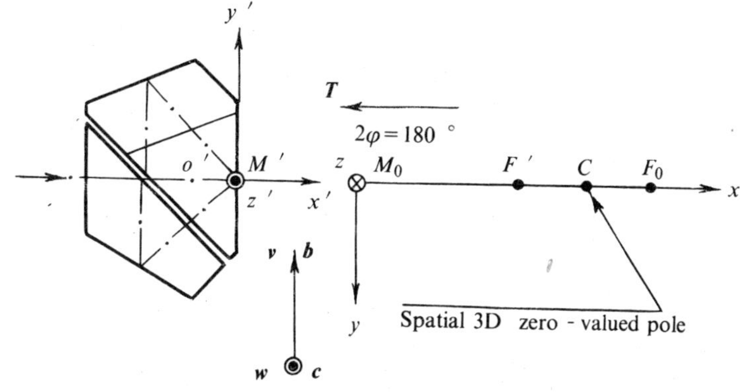

$$\Delta S_{F'x'}' = 0 , \tag{5.93}$$

$$\Delta S_{F'y'}' = \Delta\theta \left[2P_{x'}z_q' + P_z \left(-2x_q' + 2b' - 1.299D + 5.156\frac{D}{n} \right) \right] , \tag{5.94}$$

$$\Delta S_{F'z}' = \Delta\theta \left[-2P_{x'}y_q' + P_{y'} \left(2x_q' - 2b' + 1.299D - 5.156\frac{D}{n} \right) \right] . \tag{5.95}$$

Fig. 5.31. A representative prism having a spatial three-dimensional zero-valued pole C

Some rules which may be useful in finding the zero-valued characteristic parameters will first be discussed below.

a. The vector $\overrightarrow{F_0F'}$ or $\overrightarrow{F'F_0}$ must be a three-dimensional zero-valued axis of image point displacement. This is true simply because the equation of $\Delta S_{F'}'$ will be zero in this case:

$$\Delta S_{F'}' = \Delta\theta\, P \times \overrightarrow{\rho_{F'}} + (-1)^{l-1}\Delta\theta\, P \times \overrightarrow{\rho_{F'}} = 0$$

due to $\overrightarrow{\rho_{F'}} = \overrightarrow{\rho'_F} = 0$.

b. The screw axis $\overrightarrow{\lambda}$ of image formation for a prism is also a three-dimensional zero-valued axis of image point displacement no matter whether the number of reflections of the prism is even or odd.

c. If two three-dimensional zero-valued axes intersect at a point, for instance C_{Π_i}, then the point of intersection C_{Π_i} must be a planar three-dimensional zero-valued pole, and the associated plane Π_i is formed by these two axes. This rule holds also for the cases of one-dimensional and two-dimensional zero-valued axes.

d. Referring to Fig. 5.32, draw two planes 1 and 2 perpendicular to the entrance and exit optical axes x and x', and passing the object and image points F_0 and F', respectively. Now, if these two planes intersect at a finite distance, then the intersecting edge will be a two-dimensional zero-valued axis $\overrightarrow{\zeta_{mnq}}$ of $y'-z'$ image point displacement. That is simply because the four vectors P, P', $\overrightarrow{\rho_F}$, and $\overrightarrow{\rho_{F'}}$ in Eq. (4.2) are all lying in the plane 2 normal to the exit optical axis x'.

By analogy, we can locate a two-dimensional zero-valued axis $\overrightarrow{\zeta_{nlq}}$ as well as a two-dimensional zero-valued axis $\overrightarrow{\zeta_{lmq}}$.

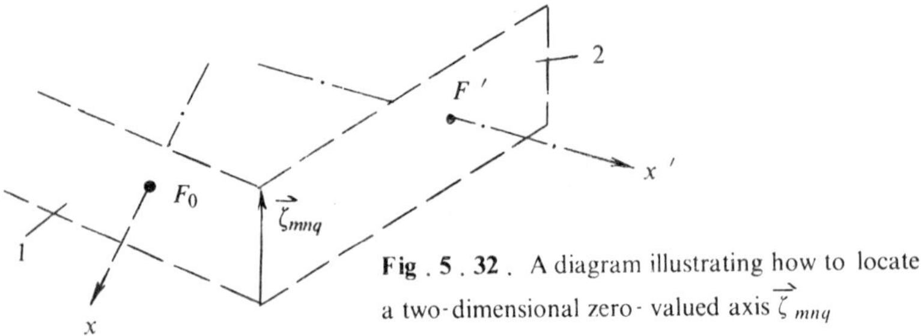

Fig . 5 . 32 . A diagram illustrating how to locate a two-dimensional zero-valued axis $\vec{\zeta}_{mnq}$

However, while locating a two-dimensional zero-valued axis $\vec{\zeta}_{lmq}$ for coplanar reflecting prisms, two planes constructed according to the above mentioned rule will coincide with each other, becoming a single plane, i.e., the optical-axis section of the prism, as shown in Fig.5.33. In this case, any line whatever lying in the sectional plane may be considered as an intersecting edge between the two planes, and will be a two-dimensional zero-valued axis $\vec{\zeta}_{lmq}$. Hence, the optical-axis section for any coplanar reflecting prism must be a plane of two-dimensional zero-valued axis of x'-y' image point displacement. Actually, this result can also be explained by the fact that the four vectors $\boldsymbol{P}$, $\boldsymbol{P'}$, $\vec{\rho}_F$ and $\vec{\rho}_F'$ in Eq. (4.2) are all lying in the optical-axis section normal to the z' axis.

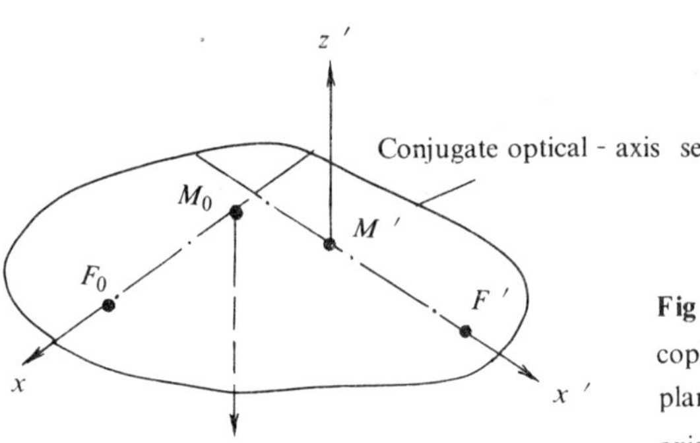

Fig . 5 . 33 . The optical-axis section of a coplanar reflecting prism appears to be a plane Π_{lm} of two dimensional zero-valued axis ζ_{lmq}

e . The three extreme-valued characteristic vectors $\vec{\delta}_a, \vec{\delta}_b$ and $\vec{\delta}_c$ especially their directions $\boldsymbol{a}$, $\boldsymbol{b}$ and $\boldsymbol{c}$ will prove to be of great utility in giving a quick judgement regarding the adjustment property of a small angular displacement $\Delta\theta\boldsymbol{P}$, i.e., the behavior of an adjusting axis $\boldsymbol{P}$ of a prism

The basic idea of using three extreme-valued characteristic vectors $\vec{\delta}_a$, $\vec{\delta}_b$ and $\vec{\delta}_c$ is to resolve a given, small angular displacement $\Delta\theta\boldsymbol{P}$ into two components by translating it to a new position $\Delta\theta\boldsymbol{P}_{\parallel}$. Thus, the original small angular displacement $\Delta\theta\boldsymbol{P}$ of the prism is equal to the new one $\Delta\theta\boldsymbol{P}_{\parallel}$ plus another small linear displacement $\Delta\theta\boldsymbol{P}\times\boldsymbol{S}$ caused by the lateral shift $\boldsymbol{S}$ of the vector $\Delta\theta\boldsymbol{P}$. Of course, the angular vector $\Delta\theta\boldsymbol{P}$ should be displaced laterally in such a manner that the adjustment property of the new one $\Delta\theta\boldsymbol{P}_{\parallel}$ will be visualized much more easily than that of the original one . As to the influence of the additional , small linear displacement of the prism , it can be quickly estimated by its orientation relative to three extreme-valued shift directions $\boldsymbol{a}$, $\boldsymbol{b}$ and $\boldsymbol{c}$.

158

In Fig . 5 . 34 , consider a small angular displacement $\Delta\theta P$ to be parallel to the optical - axis section of a coplanar reflecting prism and passing through point q at a certain height above the

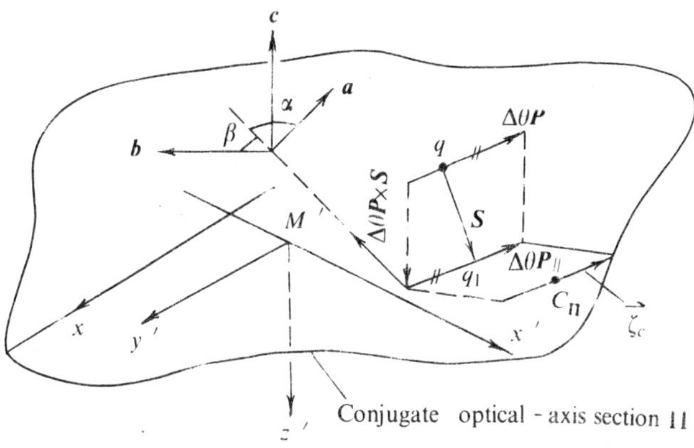

Fig . 5 .34 . A diagram to show how to give a quick judgement of a small angular displacement of a prism in regard to the adjustment property by using a , b and c

section . To judge the adjustment effect of $\Delta\theta P$, it is projected onto the optical - axis section Π , becoming $\Delta\theta P_{\parallel}$, and then the original $\Delta\theta P$ is replaced by the new one $\Delta\theta P_{\parallel}$ plus a small linear displacement $\Delta g = \Delta\theta P \times S$. Now , it can be seen from the diagram that the new angular vector $\Delta\theta P_{\parallel}$ lying in the conjugate optical - axis section , and displaced laterally from a three - dimensional zero - valued axis $\vec{\zeta}_c$ is thus merely a two - dimensional zero - valued axis $\vec{\zeta}_{lmq}$ of x '- y ' image point displacement , while the small linear displacement Δg of the prism forming angles α and β with a and b , respectively , will cause $\Delta S_{F'x'}'$ and $\Delta S_{F'y'}'$. Therefore , the angular displacement $\Delta\theta P$ to be judged will cause all three components of image point displacement .

Very often , the rule can be employed to find some other zero - valued axes from a given zero - valued axis while the number of dimensions of the new axes may be equal to or less than that of the given axis .

In Fig . 5 . 35 , suppose $\vec{\zeta}_{mnq}$ is a given two - dimensional zero - valued axis and it happens to be parallel to both the extreme - valued shift directions b and c . Then , translating $\vec{\zeta}_{mnq}$ in the direction parallel to both b and c , we will obtain a series of other two - dimensional

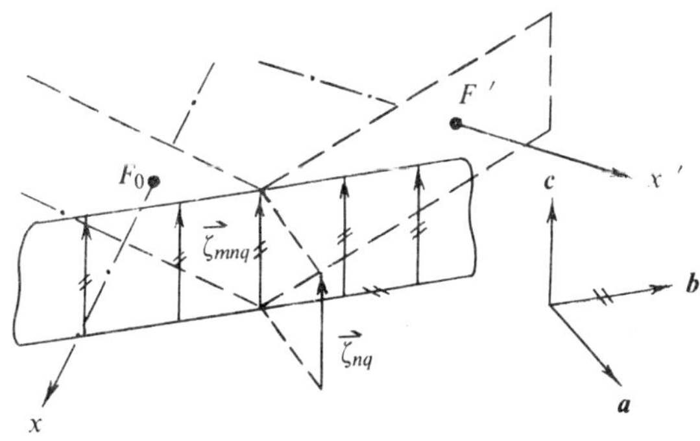

Fig .5 .35 . A diagram showing how to find some new zero - valued axes from a known one

159

zero - valued axes of the same nature while shifting $\vec{\zeta}_{mnq}$ in the direction parallel to c only , we may have a series of one - dimensional zero - valued axes $\vec{\zeta}_{nq}$.

For certain prisms , some of the three extreme - valued shift directions may not exist , for instance $\vec{\delta}_c = 0$, i .e . , c disappears . Now , let Fig . 5 .36 be the case where $\vec{\delta}_c = 0$. In this situation , the z' image point displacement $\Delta S'_{F'_z}$ caused by a given $\Delta\theta P$ will remain unchanged no matter how $\Delta\theta P$ is translated . So , we can determine $\Delta S'_{F'_z}$ by shifting $\Delta\theta P$ laterally to a convenient position passing through the object point F_0 , and then from Eq . (4 .2) , we have $\Delta S'_{F'_z} = (\Delta\theta P \times \overrightarrow{F_0 F'})_{z'}$, i .e . , the scalar projection of the vector product $\Delta\theta P \times \overrightarrow{F_0 F'}$ onto the z' axis .

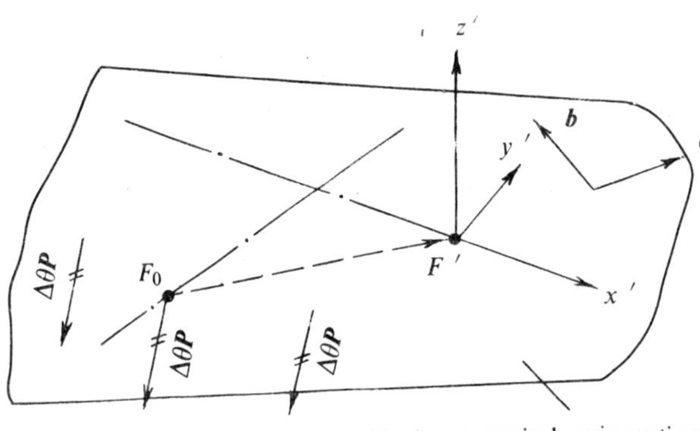

Conjugate optical - axis section

Fig .5 .36 . A diagram to show the case where c does not exist

As a matter of fact , the three extreme - valued characteristic vectors $\vec{\delta}_a$, $\vec{\delta}_b$ and $\vec{\delta}_c$ of x' , y' , z' image displacements for a prism dominate the structure of all the coefficient functions $l_i (q)$, $m_i (q)$ and $n_i (q)$ $(i = 1 , 2 , 3)$ in Eqs . (5 , 59) through (5 . 61) or Eqs . (5 . 63) through (5 .65) , which will be discussed later .

f . In any case , we can directly utilize the three component equations of image point displacement , for example , Eqs . (5 .78) through (5 .80) .

Setting any of the three component equations to zero , for instance , Eq . (5 .63) .

$$\Delta S'_{F'x'} = \Delta\theta [\, l_1 (q) P_{x'} + l_2 (q) P_{y'} + l_3 (q) P_{z'}] = 0 , \qquad (5 .96)$$

and solving Eq . (5 .96) for $P_{x'}$, $P_{y'}$ and $P_{z'}$ with a fixed point q will yield all solutions of the one - dimensional zero - valued axis $\vec{\zeta}_{lq}$ of x image point displacement through that point .

Obviously . setting either two of the three component equations to zero , for instance , Eqs . (5 .64) and (5 .65)

$$\Delta S'_{F'y'} = \Delta\theta [\, m_1 (q) P_{x'} + m_2 (q) P_{y'} + m_3 (q) P_{z'}] = 0 , \qquad (5 .97)$$

$$\Delta S'_{F'z'} = \Delta\theta [\, n_1 (q) P_{x'} + n_2 (q) P_{y'} + n_3 (q) P_{z'}] = 0 , \qquad (5 .98)$$

and solving Eqs . (5 .97) and (5 .98) for $P_{x'}$, $P_{y'}$ and $P_{z'}$ with a fixed point q will possibly give a solution of the two - dimensional zero - valued axis $\vec{\zeta}_{mnq}$ of $y' - z'$ image point displacement through that point .

160

It is also clear that the three-dimensional zero-valued axis must be a solution of the following system of equations :

$$\Delta S_{F'x'}' = \Delta\theta\,[\,l_1\,(q)\,P_{x'} + l_2\,(q)\,P_{y'} + l_3\,(q)\,P_{z'}\,] = 0 , \tag{5.99}$$

$$\Delta S_{F'y'}' = \Delta\theta\,[\,m_1\,(q)\,P_{x'} + m_2\,(q)\,P_{y'} + m_3\,(q)\,P_{z'}\,] = 0 , \tag{5.100}$$

$$\Delta S_{F'z'}' = \Delta\theta\,[\,n_1\,(q)\,P_{x'} + n_2\,(q)\,P_{y'} + n_3\,(q)\,P_{z'}\,] = 0 . \tag{5.101}$$

In this case , if we fix a point q in advance , then the above system of equations (5.99) through (5.101) would generally give no solution , because the number K of independent unkowns $P_{x'}$ and $P_{y'}$ $(P_{x'}^2 + P_{y'}^2 + P_{z'}^2 = 1)$ is less than the number M of independent equations : $K < M$ $(2 < 3)$, that is to say , for a given point q , most likely it is impossible to find a three-dimensional zero-valued axis .

However , if we leave a free choice of location of the three-dimensional zero-valued axis to be determined , then we have four independent unknowns in all for such an axis which is exactly equal to the number of degrees of freedom of a sliding axis in space , as shown in Fig. 5.37 , where two sets of parameters may be adopted: either the coordinates of the two points $q_1(y_1'\,,\,z_1')$ and $q_2\,(x_2'\,,\,y_2')$ of intersection of the sliding line with two coordinate planes or the coordinates of the one point q_1 $(y_1'\,,\,z_1')$ of intersection of the sliding line with one coordinate plane plus two direction cosines $\cos\alpha$ and $\cos\beta$ of the line . Thus $K - M = 4 - 3 = 1$, so that there may exist more than one three-dimensional zero-valued axis in the whole space for a reflecting prism .

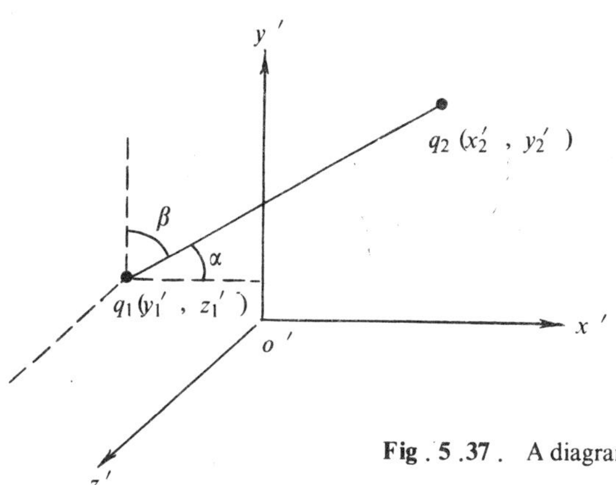

Fig. 5.37. A diagram showing how to specify the direction and location of a sliding axis in space

The system of Eqs. (5.99) , (5.100) and (5.101) can also be expressed in a vector form by setting Eq. (5.58) to zero :

$$(\Delta S_{F'}) = \Delta\theta\,[\,E + (-1)^{l-1}\,S_{T,2\varphi}\,]\,(r_q \times P) - (-1)^l\,\Delta\theta S_{T,2\varphi}\,(\overrightarrow{M_0 M}' \times P)$$

$$- \Delta\theta\,(b' \times [\,E - S_{T,2\varphi}\,]\,(P)) = 0 . \tag{5.102}$$

By using Eq. (4.11) , the above equation (5.102) may be written as

$$\Delta S_{M'}' = -\Delta\vec{\mu}' \times b' , \tag{5.103}$$

where $\Delta S'_M$ represents the displacement of point M' caused by a small angular displacement of a prism $\Delta\theta P$ passing through point q while $\Delta\vec{\mu}'$ represents the image rotation caused by the same $\Delta\theta P$. The action line of $\Delta\vec{\mu}'$ may be considered as passing through point M' provided the latter is chosen as a base point of the image (body) .

The equation (5.103) shows clearly a geometric relation to which the three vectors $\Delta S'_M$, $\Delta\vec{\mu}'$ and b' should conform in order to satisfy the equation : $\Delta S'_F = 0$.

As indicated in Fig. 5.38, the motion of the image due to $\Delta\theta P$ is expressed in terms of a small linear displacement $\Delta S'_M$ of a point M' chosen as a base point in the image (body) plus a small angular displacement $\Delta\vec{\mu}'$ passing thruogh the base point M' as shown in full lines in the diagram .

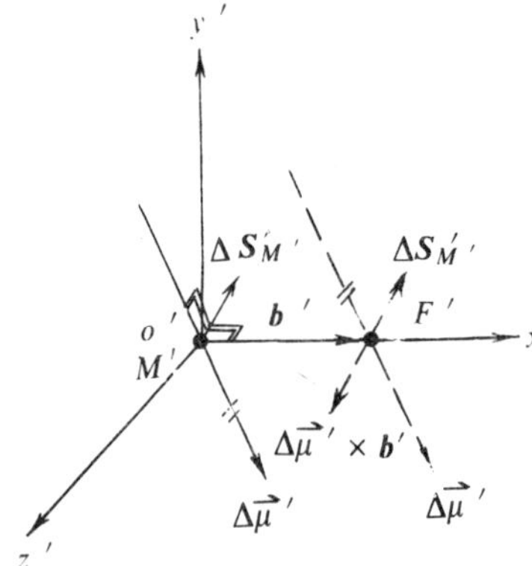

Fig. 5.38. A diagram to illustrate the meaning of setting the equation of $\Delta S'_F$ to zero

The vector $\Delta S'_M$ must first be perpendicular to the vector $\Delta\vec{\mu}'$ so as to be able to convert the image motion further into a special case of helical motion , namely a pure rotation . Then the vector $\Delta S'_M$ should again be perpendicular to the vector b' in order that the axis of the pure rotation be able to pass though the image point F'.

Hence, the physical meaning of the whole equation (5.103) is that while translating the angular vector $\Delta\vec{\mu}'$ from point M' to point F', this additional linear displacement $\Delta\vec{\mu}'\times b'$ brought to the image point F', and the original one $\Delta S'_M$ will cancel each other, resulting in a pure rotation $\Delta\vec{\mu}'$ with its action line passing through point F', as shown in dotted lines in the diagram .

Also , the geometric relation in Eq. (5.103) may serve as a means to make an estimation of the possible directions of three-dimensional zero-valued axes .

Now ,let us apply these individual rules to those prisms listed in Figs. 5.26 through 5.31 .

(1) The Amici prism shown in Fig. 5.26

(i) Using rule d , the conjugate optical-axis section is found to be the plane Π_{lm} of two-dimensional zero-valued axis $\vec{\zeta}_{lmq}$ of $x'-y'$ image point displacement. Actually , this result can also be obtained by substituting $P_{z'}=0$ and $z'_q=0$ into Eqs. (5.78) and (5.79) since the x' and y' image point displacements will thus always be zero , namely $\Delta S'_{F'x'}=0$ and $\Delta S'_{F'y'}=0$, no matter what the direction cosines $P_{x'}$, $P_{y'}$ as well as the coordinates x'_q, y'_q might be .

(ii) Again using rule d , we can first locate a two-dimensional zero-valued axis $\vec{\zeta}_{mnq}$,

162

which happens to be parallel to the plane formed by **b** and **c** in this case. Then, according to rule c, by translating the vector $\overrightarrow{\zeta}_{mnq}$ in a plane parallel to both **b** and **c**, we get further an infinite number of parallels, which remain to be two-dimensional zero-valued axes and intersect the optical-axis section along a straight line mm. In fact, the equation of this foot points line mm can also be found by setting the coefficient of P_z in Eq. (5.79) to zero:

$$- x_q' - y_q' + 2 b' - 0.866 D + 1.732 \frac{D}{n} = 0 .$$

Besides, each foot point, for instance, C_m, is a spatial one-dimensional zero-valued pole of y' image point displacement, and a plane Π_m of one-dimensional zero-valued axis $\overrightarrow{\zeta}_{mq}$ is found as well.

(iii) Similarly, we may locate a two-dimensional zero-valued axis $\overrightarrow{\zeta}_{lnq}$, and by translating it in a plane parallel to both **a** and **c** we get an infinite number of parallels appearing again to be two-dimensional zero-valued axes, whose foot points in the optical-axis section are distributed along a straight line ll, the equation of which can also be found by setting the coefficient of $P_{z'}$ in Eq. (5.78) to zero:

$$x_q' + y_q' + 0.866 D = 0 .$$

Here, we have $ll \parallel mm$ resulting from $a \parallel b$.

Correspondingly, a spatial one-dimensional zero-valued pole C_l as well as a plane Π_l of one-dimensional zero-valued axis $\overrightarrow{\zeta}_{lq}$ is indicated.

(iv) According to rules a, b and c, we have two three-dimensional zero-valued axes $\overrightarrow{F_0 F'}$ and $\overrightarrow{\lambda}$, and their intersection C_{Π_1} appears to be a planar three-dimensional zero-valued pole with the conjugate optical-axis section as its associated plane Π_1.

It is readily seen from Eq. (5.80) that by setting the coefficients of both P_x and $P_{y'}$ to zero we obtain

$$- 2 y_q' + b' - 0.866 D + 1.732 \frac{D}{n} = 0 ,$$

$$2 x_q' - b' + 0.866 D = 0 ,$$

and solving them for x_q' and y_q' gives the coordinates of the pole C_{Π_1} as

$$x_1' = \frac{1}{2} (b' - 0.866 D) ,$$

$$y_1' = \frac{1}{2} \left(b' - 0.866 D + 1.732 \frac{D}{n} \right) .$$

The result shows that this three-dimensional zero-valued pole C_{Π_1} is right at the midpoint of the segment $\overline{F_0 F}'$.

(v) Through the pole C_{Π_1}, draw a line $\zeta \zeta$ parallel to the line mm, and thus parallel to the plane formed by **a**, **b** and **c**. Then, through the line $\zeta \zeta$, draw a plane Π_2 perpendicular to the optical-axis section. It should now be apparent that all the lines lying in the plane Π_2 and parallel to the line $\zeta \zeta$ will also be three-dimensional zero-valued axes. Since the point of intersection of these parallels is at infinity , there is another three-dimensional zero-valued pole C_{Π_2} located infinitely distant along the line $\zeta\zeta$, with Π_2 as its associated plane .

(2) The roof prism shown in Fig . 5 . 27

(i) The meaning of the symbols mm , $\vec{\zeta}_{mnq}$, Π_m , ll , $\vec{\zeta}_{lnq}$ and Π_l remain the same as in Fig . 5 . 26 , and thus they will be given without explanation .

(ii) For this example , the two three-dimensional zero-valued axes $\overrightarrow{F_0F}\,'$ and $\vec{\lambda}$ are non-intersecting; nevertheless , we may treat Eq . (5 . 83) in the same way by setting the coefficients of $P_{x'}$ and $P_{y'}$ to zero :

$$-2\,y_q' - b' + 1.445\,D - 2.981\,\frac{D}{n} = 0\,,$$

$$2\,x_q' - b' - 0.046\,D = 0\,.$$

Then , solving the above equations for x_q' and y_q' will still give the coordinates $(x_1',\,y_1')$ of a three-dimensional zero-valued pole C_{Π_1} :

$$x_1' = \frac{1}{2}\,(b' + 0.046\,D)\,,$$

$$y_1' = \frac{1}{2}\left(-b' + 1.445\,D - 2.981\,\frac{D}{n}\right),$$

with the conjugate optical-axis section as its associated plane Π_1 .

To our surprise , the result shows that this three-dimensional zero-valued pole C_{Π_1} also appears to be at the midpoint of the segment $\overline{F_0F}\,'$.

(iii) Suppose that the screw axis $\vec{\lambda}$ cuts the conjugate optical-axis section at the point C_{Π_2} , then draw a line joining the two points C_{Π_1} and C_{Π_2} , and assume Π_2 to be the plane formed by the line $C_{\Pi_1}C_{\Pi_2}$ and the axis $\vec{\lambda}$. Thus , C_{Π_2} is found to be another three-dimensional zero-valued pole with Π_2 as its associated plane , the validity of which is apparent simply because the line $C_{\Pi_1}C_{\Pi_2}$ itself belongs to a three-dimensional zero-valued axis .

It should be pointed out that the lines ll and mm meet at point C_{Π_2} while the planes Π_l and Π_m intersect each other at the screw axis $\vec{\lambda}$.

Therefore , setting the coefficients of $P_{z'}$ to zero in both the equations (5 . 81) and (5 . 82) we have

$$-x_q' + y_q' + 0.046\,D = 0\,,$$

$$-x_q' - y_q' + 1.445\,D - 2.981\,\frac{D}{n} = 0\,,$$

and solving them for x_q' and y_q' yields the coordinates x_2' and y_2' of the pole C_{Π_2} :

$$x_2' = 0.746\,D - 1.491\,\frac{D}{n}\,,$$

$$y_2' = 0.700\,D - 1.491\,\frac{D}{n}\,.$$

There is a slight difference between the two poles C_{Π_1} and C_{Π_2}; the location of C_{Π_1} is dependent on the position of the image point F' while the location of C_{Π_2} independent of F', however both the associated planes Π_1 and Π_2 are related to F'.

(3) The Penta prism shown in Fig.5.28

our discussion will stress what is particular to this prism.

(i) The most peculiar point for these kinds of prisms is that $\vec{\delta}_c = 0$, i.e., the extreme-valued shift direction c does not exist. According to rule e, the z' image point displacement $\Delta S'_{F'z'}$ has no bearing on the location (i.e. point q) of the axis of rotation, which is well seen from Eq.(5.86), where the coordinates x'_q, y'_q and z'_q all disappear, and the z' image point displacement $\Delta S'_{F'z'}$ depends only upon the direction P of the axis. In this case, there exists an extreme-valued axis direction for $\Delta S'_{F'z'}$, defined in the whole space and designated by $w_{F'}$. As we know, all the axes perpendicular to the optical-axis section appear to be the zero-valued axes of $\Delta S'_{F'z'}$, then the extreme-valued axis, or in this situation the extreme-valued axis direction, $w_{F'}$ must first be parallel to the sectional plane. On the other hand, $\overrightarrow{F_0 F}'$ is a zero-valued axis for all components of $\Delta S'_{F'}$, therefore the line erected from point F_0 perpendicular to $\overrightarrow{F_0 F}'$ and lying in the section will represent such an extreme-valued axis direction $w_{F'}$. If the prism rotates through a small angle $\Delta\theta$ about an axis P parallel to $w_{F'}$, then the z' image point displacement $\Delta S'_{F'z'}$ reaches the extreme value:

$$\Delta S'_{F'z'} = \pm \Delta\theta \, | \overrightarrow{F_0 F}' |$$

or

$$\Delta S'_{F'z'\,\mathrm{max}} = \Delta\theta \, | \overrightarrow{F_0 F}' | .$$

(ii) It should be pointed out in particular that since the structure of the formula of $\Delta S'_{F'z'}$, see Eq.(5.86), has essentially been changed due to the fact that $\vec{\delta}_c = 0$, the midpoint of the segment $\overrightarrow{F_0 F}'$ will no longer be a three-dimensional zero-valued pole.

To see whether there still exists a planar three-dimensional zero-valued pole C_{Π_1} with the optical-axis section Π_1 as its associate plane, we set the right part of Eq.(5.86) to zero:

$$\Delta\theta \left[P_{x'} \left(b' - 0.5 D + 3.414 \frac{D}{n} \right) + P_{y'} \left(-b' - 0.5 D \right) \right] = 0 .$$

Thus,

$$\frac{P_{x'}}{P_{y'}} = \frac{b' + 0.5 D}{b' - 0.5 D + 3.414 \dfrac{D}{n}} .$$

The resulting equation indicates that all the lines lying in the optical-axis section and parallel to the vector $\overrightarrow{F_0 F}'$ appear to be zero-valued axes of $\Delta S'_{F'z'}$. Besides, it is already known that each line in the sectional plane belongs at least to two-dimensional zero-valued axes of x'-y' image point displacement, so these parallels form a family of three‑dimensional zero‑valued axes. Therefore, we obtain a three‑dimensional ze‑ro‑valued pole C_{Π_1} located infinitely distant along the action line of vector $\overrightarrow{F_0 F}\,'$ with the opti‑cal‑axis section Π_1 as its associated plane. The result arrived at is in exact conformity with rule e.

(iii) Suppose the screw axis $\overrightarrow{\lambda}$ cuts the conjugate optical‑axis section at the point C_{Π_2}, and then draw a line $\zeta\,\zeta$ passing through point C_{Π_2} and parallel to vector $\overrightarrow{F_0 F}\,'$. Since the line $\zeta\,\zeta$ and the axis $\overrightarrow{\lambda}$ are both three‑dimensional zero‑valued axes, their intersection C_{Π_2} will thus be a three‑dimensional zero‑valued pole with the plane Π_2 formed by $\zeta\,\zeta$ and $\overrightarrow{\lambda}$ as its associ‑ated plane.

It is also clear from Eqs. (5.84) and (5.85) that setting the coefficients of P_z in both the equations to zero:

$$-x_q'+y_q'-0.5\,D=0,$$

$$-x_q'-y_q'+0.5\,D-3.414\,\frac{D}{n}=0,$$

and solving them for x_q' and y_q' yields then the coordinates x_2' and y_2' of the pole C_{Π_2}:

$$x_2'=-1.707\,\frac{D}{n}\,,$$

$$y_2'=\frac{1}{2}\left(D-3.414\,\frac{D}{n}\right).$$

Note that, the three‑dimensional zero‑valued pole C_{Π_2} also appears to be the inter‑-secting point of the lines $l\,l$ and mm while the two latter lines are parallel to the vectors a and b, respectively.

(4) The right‑angle prism shown in Fig. 5.29

(i) As mentioned in case (3), we have here also a planar three‑dimensional zero‑valued pole C_{Π_1} located infinitely distant along the action line of vector $\overrightarrow{F_0 F}\,'$ with the optical‑axis section Π_1 as its associated plane simply because $\overrightarrow{\delta_c}=0$.

(ii) Another one C_{Π_2} does not exist even at infinity. Actually, the properties mentioned above for the right‑angle prism can also be extended to all the prisms with an odd number of reflecting surfaces and with coplanar mirror normals, except in a few cases where the exit and entrance optical axes are coincident (see next case).

(5) The Abbe K prism shown in Fig. 5.30

The Abbe K prism is possessed of a planar three‑dimensional zero‑valued polar line parallel to the solely existing extreme‑valued shift direction b and passing through the midpoint of the segment $\overline{F_0 F}\,'$. Each point on the polar line is a planar three‑dimensional ze‑ro‑valued pole C_{Π_1} with the associated plane Π_1 normal to the line.

Like the Abbe K prism, all the image rotators, such as the Pechan Z prism, the Uppendahl M prism, and so forth, have a planar three‑dimensional zero‑valued polar line.

(6) The roof Pechan prism shown in Fig. 5.31

From Eqs. (5.93) through (5.95), a spatial three‑dimensional zero‑valued pole C

can be found by setting the coefficient of either $P_{z'}$ or $P_{y'}$ to zero :

$$2\,x_q{'}-2\,b\,'+1.299\,D-5.156\,\frac{D}{n}=0\,,$$

and also

$$y_q{'}=z_q{'}=0\,.$$

Thus , we obtain the coordinates of the pole C :

$$x'=b\,'+\frac{1}{2}\left(-1.299\,D+5.156\,\frac{D}{n}\right),$$

$$y\,'=0\,,$$

$$z\,'=0\,,$$

which appears to be at the midpoint of the segment $\overline{F_0\,F\,'}$.

The roof Pechan has particularly a plane of three-dimensional zero-valued axis Π_{lmn} passing through C and perpendicular to $\vec{\lambda}$.

As will be seen in the following chapter , all the circular beam deflectors (see Sect . 6 . 4) have a spatial three-dimensional zero-valued pole and a plane of three-dimensional zero-valued axis .

To summarize the above discussion , we emphasize the three-dimensional zero-valued axis , since the one- or two-dimensional zero-valued axes may be derived therefrom or easily be found by some other methods . Thus , three-dimensional zero-valued poles are of primary importance ,especially the planar three-dimensional zero-valued pole .

The results obtained for these typical prisms lead to the possibility of one or two planar three-dimensional zero-valued poles for most coplanar reflecting prisms , and a zero-valued polar line or a spatial zero-valued pole of three dimensions in a few cases * .

Now let us prove the above phenomenon in a qualitative way prior to a more strict verification in the next section .

(1) The coplanar reflecting prisms may be roughly divided into two groups according to the relative direction of the screw axis $\vec{\lambda}$ with respect to the prism and the magnitudes of 2φ and $2d$. As shown in Fig . 5 . 39 , for prisms in the first group , the screw axis $\vec{\lambda}$ is perpendicular to the conjugate optical-axis section Π , $2d=0$, and 2φ equals any value while for prisms in the second group , the screw axis $\vec{\lambda}$ is lying in the plane Π , $2\varphi=180$ °,and $2d$ equals any value .

(2) For coplanar reflecting prisms , all the lines lying in the conjugate optical-axis section appear to be two-dimensional zero-valued axes $\vec{\zeta}_{lmq}$ while all the lines perpendicular to that sectional plane appear to be one-dimensional zero-valued axes $\vec{\zeta}_{nq}$. This is evident because of the property indicated in point (1) and due to Eq . (4 .2) .

------

* It will be seen later on that for some reflecting prisms there exist a spatial three-dimensional zero-valued pole and a plane of three-dimensional zero-valued axis of image point displacement .

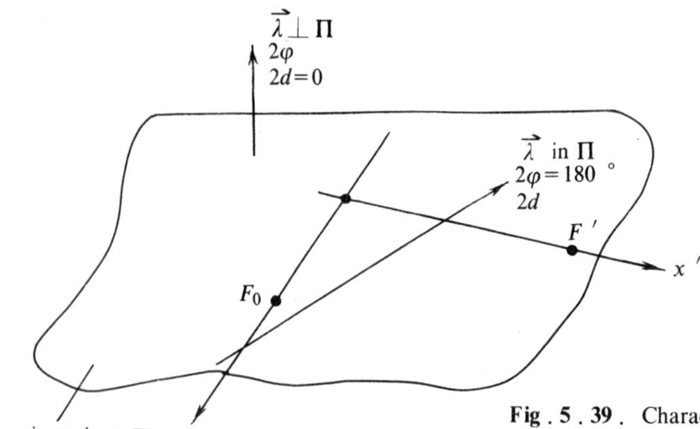

Fig . 5 . 39 . Characteristic parameters for two groups of coplanar reflecting prisms

(3) Based on the second judgement in point (2), the equation (5.61) can be written as

$$\Delta S_{F'z'}' = \Delta\theta\,[\,P_{x'}\,(n_{11}\,x_q'+n_{12}\,y_q'+n_{13}\,z_q'+n_{14})$$

$$+\,P_{y'}\,(n_{21}\,x_q'+n_{22}\,y_q'+n_{23}\,z_q'+n_{24})]\ . \tag{5.104}$$

That is to say , $\Delta S_{F'z'}'$ is independent of $P_{z'}$.

In addition , since P and r_q occur in Eq . (5.58) in the form of vector product , the coefficients n_{11} and n_{22} in Eq . (5.104) must be zero too , thus

$$\Delta S_{F'z'}' = \Delta\theta\,[\,P_{x'}\,(n_{12}\,y_q'+n_{13}\,z_q'+n_{14})$$

$$+\,P_{y'}\,(n_{21}\,x_q'+n_{23}\,z_q'+n_{24})]\ . \tag{5.105}$$

Moreover , for all coplanar reflecting prisms the extreme- valued shift direction c is always perpendicular to the optical- axis section , so long as it exists , i . e ., $\vec{\delta}_c \neq 0$. Under this condition , $\Delta S_{F'z'}'$ will remain unchanged provided P is translated only in the plane perpendicular to the optical- axis section , as shown in Fig . 5 . 40 . This fact indicates that the coefficients n_{13} and n_{23} in Eq . (5.105) should be zero . Then ,

$$\Delta S_{F'z'}' = \Delta\theta\,[\,P_{x'}\,(n_{12}\,y_q'+n_{14})+P_{y'}\,(n_{21}\,x_q'+n_{24})]\ , \tag{5.106}$$

where the coefficients n_{12} and n_{21} are related to T , 2φ and t while the coefficients n_{14} and n_{24} have bearing on b' , D and $\dfrac{D}{n}$.

Still further , if c disappears , then Eq . (5.106) becomes even more simplified

$$\Delta S_{F'z'}' = \Delta\theta\,(n_{14}\,P_{x.}+n_{24}\,P_{y'})\ . \tag{5.107}$$

(4) Combining the above three points , especially comparing the last two equations (5.106) and (5.107) leads to the conclusion that for coplanar reflecting prisms the extreme- valued shift direction c has a decisive effect on one of the planar three-dimensional zero- valued

168

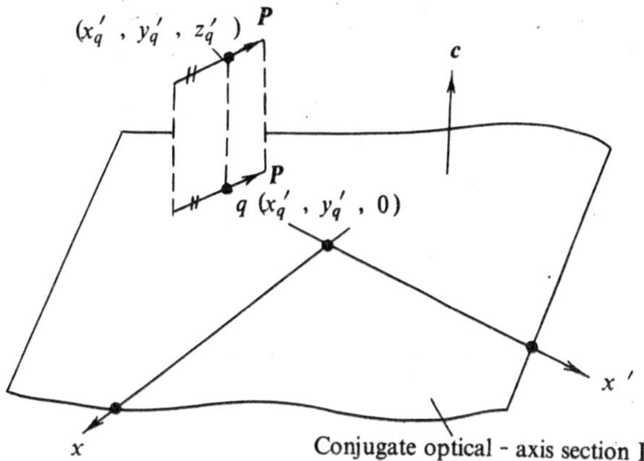

Fig . 5 . 40 . A diagram to show that translating an axis P in the plane perpendicular to the optical - axis section will not affect $\Delta S_{F'z'}'$ if the reflecting prism is coplanar

poles in the following manner: the pole C_{Π_1} appears to be at a finite distance whenever the extreme-value shift direction c exists , while C_{Π_1} will be moved to infinity once c disappears .

(5) To discuss another planar three-dimensional zero-valued pole C_{Π_2}, we start by analyzing the structure of Eqs . (5 , 59) and (5 . 60). Considering $l_{11}=l_{22}=l_{33}=m_{11}=m_{22}=m_{33}=0$,

$$\Delta S_{F'x'}' = \Delta\theta\,[\,P_{x'}\,(l_{12}\,y_q'+l_{13}\,z_q'+l_{14}) + P_{y'}\,(l_{21}\,x_q'+l_{23}\,z_q'+l_{24})$$

$$+ P_{z'}\,(l_{31}\,x_q'+l_{32}\,y_q'+l_{34})]\,, \tag{5.108}$$

$$\Delta S_{F'y'}' = \Delta\theta\,[\,P_{x'}\,(m_{12}\,y_q'+m_{13}\,z_q'+m_{14}) + P_{y'}\,(m_{21}\,x_q'+m_{23}\,z_q'+m_{24})$$

$$+ P_{z'}\,(m_{31}\,x_q'+m_{32}\,y_q'+m_{34})]\,. \tag{5.109}$$

Besides , the extreme-valued shift directions a and b for coplanar reflecting prisms are always parallel to the conjugate optical-axis section . Now , let an axis P be parallel to the sectional plane , and thus parallel to both a and b . Then , $\Delta S_{F'x'}'$ as well as $\Delta S_{F'y'}'$ will remain unchanged if P is translated only in the plane parallel to the optical-axis section . Hence , the coefficients of x_q' and y_q' in the round brackets for both Eqs . (5 . 108) and (5 . 109) have to be zero , that is , $l_{12}=l_{21}=m_{12}=m_{21}=0$. Therefore ,

$$\Delta S_{F'x'}' = \Delta\theta\,[\,P_{x'}\,(l_{13}\,z_q'+l_{14}) + P_{y'}\,(l_{23}\,z_q'+l_{24})$$

$$+ P_{z'}\,(l_{31}\,x_q'+l_{32}\,y_q'+l_{34})]\,, \tag{5.110}$$

$$\Delta S_{F'y'}' = \Delta\theta\,[\,P_{x'}\,(m_{13}\,z_q'+m_{14}) + P_{y'}\,(m_{23}\,z_q'+m_{24})$$

$$+ P_{z'}\,(m_{31}\,x_q'+m_{32}\,y_q'+m_{34})]\,. \tag{5.111}$$

Again , based on the first judgement in point (2) , the coefficients l_{14} , l_{24} , m_{14} and m_{24} must all be zero . Thus ,

$$\Delta S'_{F'_x'} = \Delta\theta \left[l_{13} z'_q P_{x'} + l_{23} z'_q P_{y'} + (l_{31} x'_q + l_{32} y'_q + l_{34}) P_{z'} \right] = 0 , \qquad (5.112)$$

$$\Delta S'_{F'_y'} = \Delta\theta \left[m_{13} z'_q P_{x'} + m_{23} z'_q P_{y'} \right.$$

$$\left. + (m_{31} x'_q + m_{32} y'_q + m_{34}) P_{z'} \right] = 0 . \qquad (5.113)$$

(6) From Eqs. (5.112) and (5.113) , considering again the second judgement in point (2) we can see that all the perpendiculars to the sectional plane , erected at the foot points along the straight line ll determined by setting $l_{31} x'_q + l_{32} y'_q + l_{34} = 0$, belong to the two-dimensional zero-valued axes $\vec{\zeta}_{lnq}$. , as shown in Fig. 5.26 or 5.27 , while all the per-pendiculars to the sectional plane erected at the foot points along the straight line mm deter-mined by setting $m_{31} x'_q + m_{32} y'_q + m_{34} = 0$, belong to the two-dimensional zero-valued axes $\vec{\zeta}_{mnq}$.

It is already clear that we have $ll \parallel a$ and $mm \parallel b$.

Now , if a and b are nonparallel , the two straight lines ll and mm meet at a finite distance and the point of intersection C_{Π_2} will be a planar three-dimensional zero-valued pole; if $a \parallel b$, the two straight lines ll and mm become parallel to each other and thus the pole C_{Π_2} will be considered as being at infinity , except for those prisms with an odd number of reflections and with coplanar mirror normals (see Fig. 5.29) .

(7) To combine the last two points , we may draw also the conclusion that the relative orientation of two extreme-valued shift directions a and b has close bearing on another planar three-dimensional zero-valued pole C_{Π_2} for coplanar reflecting prisms (with exceptions): when a and b are nonparallel , C_{Π_2} is located at a finite distance while C_{Π_2} goes to infinity as long as a and b become parallel to each other .

Finally we may say : if the three extreme-valued axis directions u , v , and w are said to be a decisive factor in governing the nature of the three components $\Delta\mu'_{x'}$, $\Delta\mu'_{y'}$ and $\Delta\mu'_{z'}$ of image rotation then the three extreme-valued shift directions a , b and c may be considered as a dominant factor in affecting the properties of the three components $\Delta S'_{F'_{x'}}$, $\Delta S'_{F'_{x'}}$ and $\Delta S'_{F'_{z'}}$ of image point displacement .

The above verification is neither rigorous nor complete , however it does help us solve the problem in a strict way (see next section) .

5.7.5 Existence Conditions of Planar Three-Dimensional Zero-Valued Poles for Reflecting Prisms and General Method of Finding These Poles

We will start by seeking a method of determining the planar three-dimensional zero-valued poles for a reflecting prism and thus we may readily know what the existence conditions would be for planar three-dimensional zero-valued poles .

In doing so , we prefer to utilize the concept of the gradient axis of a component of image point displacement (see Sect. 5.7.1) because the method based on this idea seems to be the simplest one among a few alternative approaches which we have tried .

It is obvious that if you want the point q shown in Fig . 5 . 41 to be a planar three - dimensional zero - valued pole for a prism , then it is certainly possible to find one three - dimensional zero - valued axis passing through this point , for example $\vec{\zeta}_q$. Next , suppose $\vec{\eta}_{F'uq}$, $\vec{\eta}_{F'vq}$ and $\vec{\eta}_{F'wq}$ are the gradient axes of the x' , y' and z' image point displacements , respectively , for the point q . According to the relationship of orthogonality between the gradient axis and zero - valued axis for the same point q , $\vec{\eta}_{F'uq}$, $\vec{\eta}_{F'vq}$ and $\vec{\eta}_{F'wq}$ are all perpendicular to $\vec{\zeta}_q$ and will thus be lying in one and the same plane . For the time being , the point q is a one - dimensional zero - valued pole shared by the x' , y' and z' image point displacements with three associated planes Π_{lq} , Π_{mq} and Π_{nq} intersecting at the action line of $\vec{\zeta}_q$ and normal to $\vec{\eta}_{F'uq}$, $\vec{\eta}_{F'vq}$ and $\vec{\eta}_{F'wq}$, respectively .

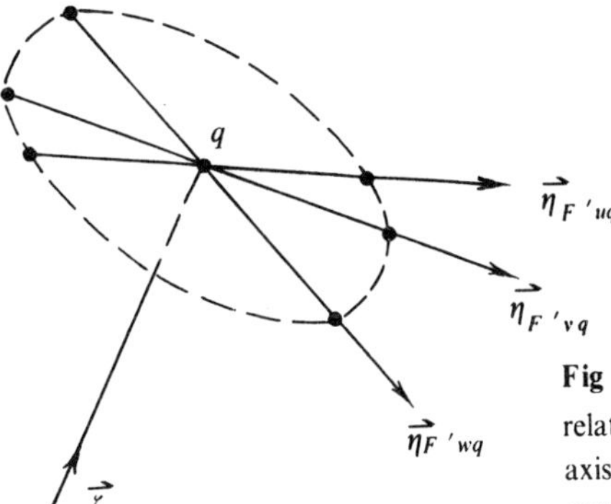

Fig . 5 . 41 . A diagram showing the relationship between a zero - valued axis and three gradient axes for the same point q

Furthermore , suppose that the three gradient axes $\vec{\eta}_{F'uq}$, $\vec{\eta}_{F'vq}$ and $\vec{\eta}_{F'wq}$ happen to be coincident with one another and accordingly the three associated planes Π_{lq} , Π_{mq} and Π_{nq} will also be united to form a single plane Π , as shown in Fig . 5 . 42 . Under this condition , the point

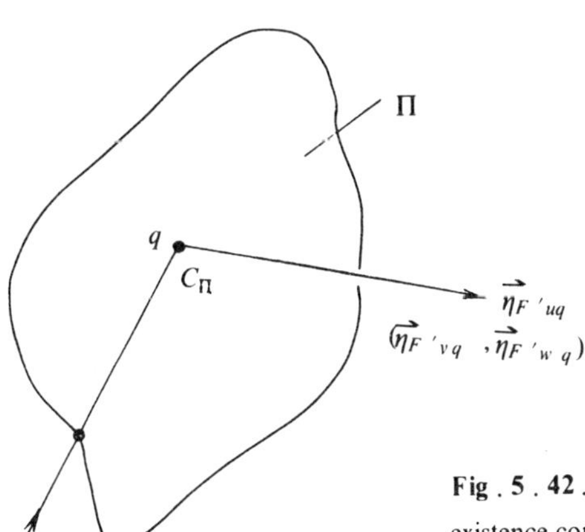

Fig . 5 . 42 . A diagram to show the existence condition of a planar three - dimensional zero - valued pole

q will appear to be a planar three-dimensional zero-valued pole C_Π with the associated plane Π, the latter being perpendicular to $\vec{\eta}_{F'uq}$ or $\vec{\eta}_{F'vq}$ or $\vec{\eta}_{F'wq}$.

Rewrite Eq. (5.73)

$$\vec{\eta}_{F'uq} = \frac{\partial \Delta S'_{F'x'}}{\partial \Delta \theta_{x'}}\, i' + \frac{\partial \Delta S'_{F'x'}}{\partial \Delta \theta_{y'}}\, j' + \frac{\partial \Delta S'_{F'x'}}{\partial \Delta \theta_{z'}}\, k' \left.\begin{array}{c} \\ \\ \end{array}\right\}$$

$$\vec{\eta}_{F'vq} = \frac{\partial \Delta S'_{F'y'}}{\partial \Delta \theta_{x'}}\, i' + \frac{\partial \Delta S'_{F'y'}}{\partial \Delta \theta_{y'}}\, j' + \frac{\partial \Delta S'_{F'y'}}{\partial \Delta \theta_{z'}}\, k' \left.\right\} \qquad (5.73)$$

$$\vec{\eta}_{F'wq} = \frac{\partial \Delta S'_{F'z'}}{\partial \Delta \theta_{x'}}\, i' + \frac{\partial \Delta S'_{F'z'}}{\partial \Delta \theta_{y'}}\, j' + \frac{\partial \Delta S'_{F'z'}}{\partial \Delta \theta_{z'}}\, k' \left.\right\}$$

According to the above discussion, the three gradient axes $\vec{\eta}_{F'uq}$, $\vec{\eta}_{F'vq}$ and $\vec{\eta}_{F'wq}$ should first of all be parallel to one anothor, which may be expressed mathematically as follows:

$$\frac{\dfrac{\partial \Delta S'_{F'x'}}{\partial \Delta \theta_{x'}}}{\dfrac{\partial \Delta S'_{F'y'}}{\partial \Delta \theta_{x'}}} = \frac{\dfrac{\partial \Delta S'_{F'x'}}{\partial \Delta \theta_{y'}}}{\dfrac{\partial \Delta S'_{F'y'}}{\partial \Delta \theta_{y'}}} = \frac{\dfrac{\partial \Delta S'_{F'x'}}{\partial \Delta \theta_{z'}}}{\dfrac{\partial \Delta S'_{F'y'}}{\partial \Delta \theta_{z'}}}\,, \qquad (5.114)$$

$$\frac{\dfrac{\partial \Delta S'_{F'y'}}{\partial \Delta \theta_{x'}}}{\dfrac{\partial \Delta S'_{F'z'}}{\partial \Delta \theta_{x'}}} = \frac{\dfrac{\partial \Delta S'_{F'y'}}{\partial \Delta \theta_{y'}}}{\dfrac{\partial \Delta S'_{F'z'}}{\partial \Delta \theta_{y'}}} = \frac{\dfrac{\partial \Delta S'_{F'y'}}{\partial \Delta \theta_{z'}}}{\dfrac{\partial \Delta S'_{F'z'}}{\partial \Delta \theta_{z'}}}\,, \qquad (5.115)$$

and then these three gradient axes must refer to the same point q, or in other words, they are calculated for the same position vector $r_q(x'_q, y'_q, z'_q)$.

To sum up the above two parts, we realize that all the factors to make it possible for the simultaneous equations (5.114) and (5.115) to have at least a real solution of the position vector $r_q(x'_q, y'_q, z'_q)$ of point q will form such necessary and sufficient conditions that there exists one planar three-dimensional zero-valued pole. Such conditions will be discussed later in detail.

Here, we have $K - M = 3 - 4 = -1$, which indicates one degree of constraint of the simultaneous equations (5.114) and (5.115). Thus, in general it is not ensured that an arbitrary reflecting prism will be able to have one planar three-dimensional zero-valued pole. However, as known from the examples already investigated, many reflecting prisms do have even more than one planar three-dimensional zero-valued poles. Roughly speaking we may distinguish two types of these particular cases.

For a better understanding rewrite Eq. (5.74)

172

$$
\left.\begin{aligned}
\Delta S'_{F'x'} &= \Delta\theta\left(\frac{\partial\Delta S'_{F'x'}}{\partial\Delta\theta_{x'}}\,P_{x'} + \frac{\partial\Delta S'_{F'x'}}{\partial\Delta\theta_{y'}}\,P_{y'} + \frac{\partial\Delta S'_{F'x'}}{\partial\Delta\theta_{z'}}\,P_{z'}\right) \\[2ex]
\Delta S'_{F'y'} &= \Delta\theta\left(\frac{\partial\Delta S'_{F'y'}}{\partial\Delta\theta_{x'}}\,P_{x'} + \frac{\partial\Delta S'_{F'y'}}{\partial\Delta\theta_{y'}}\,P_{y'} + \frac{\partial\Delta S'_{F'y'}}{\partial\Delta\theta}\,P_{z'}\right) \\[2ex]
\Delta S'_{F'z'} &= \Delta\theta\left(\frac{\partial\Delta S'_{F'z'}}{\partial\Delta\theta_{x'}}\,P_{x'} + \frac{\partial\Delta S'_{F'z'}}{\partial\Delta\theta_{y'}}\,P_{y'} + \frac{\partial\Delta S'_{F'z'}}{\partial\Delta\theta_{z'}}\,P_{z'}\right)
\end{aligned}\right\}, \quad (5.74)
$$

For the particular cases of the first type, one of the three gradient axes, for instance $\vec{\eta}_{F'wq}$, does not exist, i.e.,

$$
\frac{\partial\Delta S'_{F'z'}}{\partial\Delta\theta_{x'}} = \frac{\partial\Delta S'_{F'z'}}{\partial\Delta\theta_{y'}} = \frac{\partial\Delta S'_{F'z'}}{\partial\Delta\theta_{z'}} = 0. \tag{5.116}
$$

Accordingly, from Eq. (5.74), we have

$$
\Delta S'_{F'z'} = 0.
$$

The above results tell that the gradient axis $\vec{\eta}_{F'wq}$ of $\Delta S'_{F'z'}$ for a certain point q does not exist. Thus, any line passing through this point q in space appears to be a one-dimensional zero-valued axis of z' image point displacement, or in other words, the point q is a spatial one-dimensional zero-valued pole of $\Delta S'_{F'z'}$. Under this condition, q is said to be a planar three-dimensional zero-valued pole if the remaining two gradient axes $\vec{\eta}_{F'uq}$ and $\vec{\eta}_{F'vq}$ coincide with each other at this point. The pole C_{Π_1} for the prism shown in Fig. 5.27 is just such a case.

For the particular cases of the second type, two of the three gradient axes, for instance, $\vec{\eta}_{F'uq}$ and $\vec{\eta}_{F'vq}$ do not exist, i.e.,

$$
\left.\begin{aligned}
\frac{\partial\Delta S'_{F'x'}}{\partial\Delta\theta_{x'}} &= \frac{\partial\Delta S'_{F'x'}}{\partial\Delta\theta_{y'}} = \frac{\partial\Delta S'_{F'x'}}{\partial\Delta\theta_{z'}} = 0 \\[2ex]
\frac{\partial\Delta S'_{F'y'}}{\partial\Delta\theta_{x'}} &= \frac{\partial\Delta S'_{F'y'}}{\partial\Delta\theta_{y'}} = \frac{\partial\Delta S'_{F'y'}}{\partial\Delta\theta_{z'}} = 0
\end{aligned}\right\}. \quad (5.117)
$$

Accordingly, from Eq. (5.74), we have

$$
\Delta S'_{F'x'} = 0, \qquad \Delta S'_{F'y'} = 0.
$$

The above results indicate that the gradient axes $\vec{\eta}_{F'uq}$ and $\vec{\eta}_{F'vq}$ of $\Delta S'_{F'x'}$ and $\Delta S'_{F'y'}$, respectively, do not exist for a certain point q. Thus, any line passing through this point q in space appears to be a two-dimensional zero-valued axis of x'-y' image point displacement, or in other words, the point q is a spatial two-dimensional zero-valued pole of both

$\Delta S_F{'}_x{'}$ and $\Delta S_F{'}_y{'}$, i.e., a planar three-dimensional zero-valued pole at least.

Now, let us develop Eqs. (5.114) and (5.115) for reflecting prisms in full detail. Since this part of the work seems to require a lengthy article, it will be carried out in an orderly manner.

(1) Formula to be used

We should here employ the real expression of the image point displacement, namely Eq. (5.58), instead of Eqs. (5.59) through (5.61).

Rewrite Eq. (5.58)

$$(\Delta \boldsymbol{S}_F{'}) = \Delta\theta\,[\,E + (-1)^{l-1}\,S_{T.2\varphi}\,]\,(\boldsymbol{r}_q \times \boldsymbol{P})$$
$$- (-1)^l\,\Delta\theta\,S_{T.2\varphi}\,(\overrightarrow{M_0 M}{'} \times \boldsymbol{P}) - \Delta\theta\,(\boldsymbol{b}{'} \times [\,(E - S_{T.2\varphi})\,(\boldsymbol{P})\,]\,). \qquad (5.58)$$

(2) Choice of coordinate system

Unlike the one- or two-dimensional zero-valued characteristic parameters, the properties of the three-dimensional zero-valued axes and poles are unaffected by the rotations and shifts of coordinate axes, therefore to simplify the derivation of the resulting formula as well as to unify the coordinate systems for both the planar and spatial reflecting prisms we will for the time being adopt a new coordinate system by relating it to the characteristic parameters of image formation for reflecting prisms, i.e., by mainly assuming the screw axis $\overrightarrow{\lambda}$ to be the $z{'}$ axis of the coordinate system.

As mentioned before, for reflecting prisms with an odd number of reflections and $2\varphi = 180°$ the screw axis $\overrightarrow{\lambda}$ according to the definition in this text appears to be at infinity. Then, in these cases we should assign an arbitray point I of inversion and assume the corresponding screw axis $\overrightarrow{\lambda}_I$ at finite distance to be the $z{'}$ axis.

Thus, although a unified coordinate system has been suggested, we still have to treat it individually for two types of reflecting prisms.

(i) The coordinate system for all reflecting prisms with an even number of reflections plus the reflecting prisms with an odd number of reflections excluding those with $2\varphi = 180°$

Referring first to Fig. 5.43(a) where the prisms have an even number of reflections, assume the screw axis $\overrightarrow{\lambda}$ to be the $z{'}$ axis, then from the image point $F{'}$ drop a perpendicular to the $z{'}$ axis, thus with the foot point $o{'}$ to represent the origin and with the vector $\overrightarrow{o{'}F}{'}$ to represent the $x{'}$ axis, and finally assign the $y{'}$ axis to form a right-handed coordinate system. Besides, from the origin $o{'}$ lay off a segment $\overline{o{'}o}$ along the screw axis equal to the axial shift $2d$. It is evident that points o and $o{'}$ are a conjugate pair. The $x{'}$ axis here will no longer be the exit optical axis of prisms, therefore the vector $\overrightarrow{o{'}F}{'}$ is designated by $\boldsymbol{b}{''}$ so as to be different from $\boldsymbol{b}{'}$.

Now, referring to Fig. 5.43(b) where the prisms have an odd number of reflections, we have here $2d = 0$. However, we need to consider the inversion operation. Since the center of inversion I for this kind of reflecting prisms with an odd number of reflections is located on the screw axis $\overrightarrow{\lambda}$, namely the $z{'}$ axis, and in general at a distance of e from the origin $o{'}$, the conjugate point o in the object space corresponding to point $o{'}$ in the image space can be found at a distance of $2e$ from the origin $o{'}$. The rest of the diagram is the same as that in Fig. 5.43(a).

174

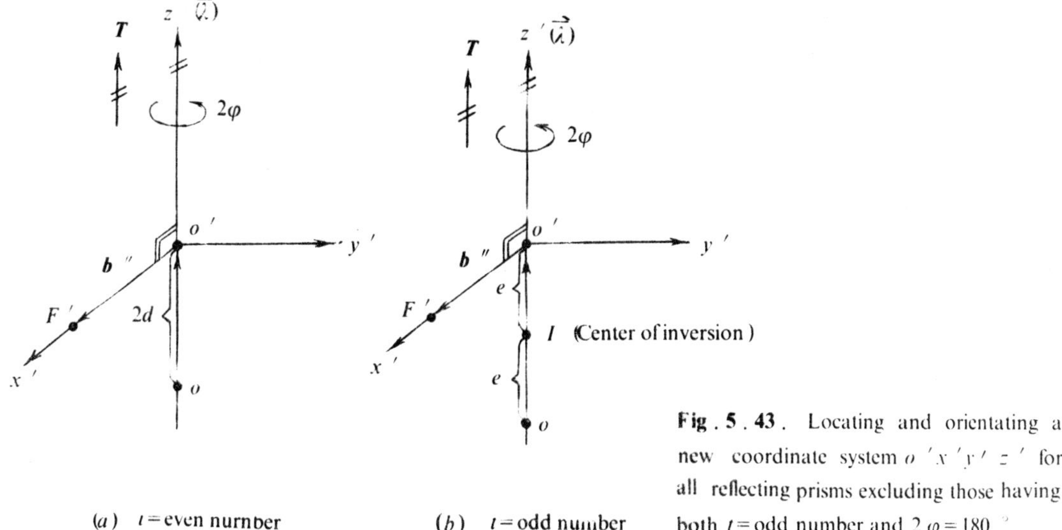

Fig . 5 . 43 . Locating and orientating a new coordinate system $o'x'y'z'$ for all reflecting prisms excluding those having both t = odd number and $2\varphi = 180°$

To unify these two cases mathematically ,we employ a single symbol 2ε to take the place of $2d$ as well as of $2e$.

It should be pointed out that you may keep using the equation (5.58) with the new coordinate system shown in Fig . 5 . 43; if so , only the vectors $\overrightarrow{M_0M'}$ and b' in Eq . (5.58) should be replaced by the vectors $\overrightarrow{oo'}$ and b'' indicated in Fig . 5 . 43 ,respectively. Besides , the matrix $S_{T.2\varphi}$ will be significantly simplified for the characteristic vector T is parallel to the z' axis .

Consequently ,for this type of reflecting prisms and the coordinate system associated with them ,we have for Eq . (5.58) the following expressions of substitution :

$$\left.\begin{array}{l} \overrightarrow{M_0M'} = \overrightarrow{oo'} = 2\varepsilon k' \\[1em] b' = \overrightarrow{o'F'} = b'' = b''i' \\[1em] S_{T.2\varphi} = \begin{pmatrix} \cos 2\varphi & -\sin 2\varphi & 0 \\ \sin 2\varphi & \cos 2\varphi & 0 \\ 0 & 0 & 1 \end{pmatrix} \end{array}\right\} \qquad (5.118)$$

(ii) The coordinate system for reflecting prisms with an odd number of reflections and $2\varphi = 180°$

Reflecting prisms of this type can also be divided into two parts: one , which includes those prisms with coplanar mirror normals ,and the other which covers the rest of this category .

First of all ,for any prism of this type choose such a point I as the center of inversion that the corresponding screw axis $\vec{\lambda}_t$ will be found at a finite distance and the axial shift will just be zero $;2d = 0$. Then, the vector $\vec{\lambda_t}$ is assumed to be the z' axis, as shown in Fig. 5.44, and the whole coordinate system $o'x'y'z'$ is established in the same way as before.

The vector $\overrightarrow{oo'}$ here is no longer coincident with the z' axis, and is thus designated by another symbol $\vec{\xi}$ merely to be readily distinguishable from that in Fig. 5.43.

Now, return to the two parts of reflecting prisms in this category.

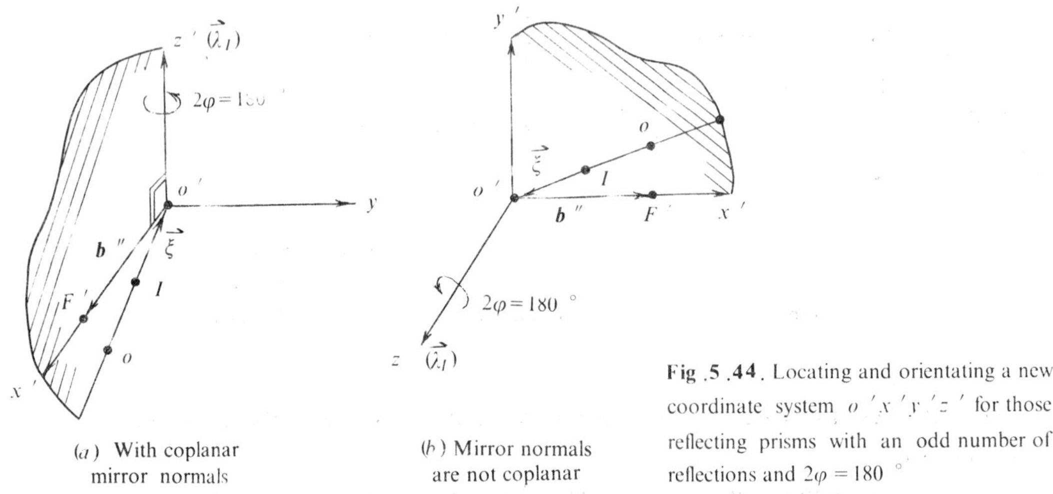

(a) With coplanar mirror normals

(b) Mirror normals are not coplanar

Fig. 5.44. Locating and orientating a new coordinate system $o'x'y'z'$ for those reflecting prisms with an odd number of reflections and $2\varphi = 180°$

In Fig. 5.44(a) referring to those prisms with coplanar mirror normals, the vector $\vec{\xi}$ is lying in the $x'z'$ plane. Then, $\vec{\xi} = \xi_{x'} \boldsymbol{i}' + \xi_{z'} \boldsymbol{k}'; \xi_{y'} = 0$.

In Fig. 5.44 (b) referring to some of the rest, whose mirror normals are not coplanar, the vector $\vec{\xi}$ is lying in the $x'y'$ plane. Then, $\vec{\xi} = \xi_{x'} \boldsymbol{i}' + \xi_{y'} \boldsymbol{j}'; \xi_{z'} = 0$. The roof rhomboid prism, as shown in Fig. 5.45, can be considered as an example in point.

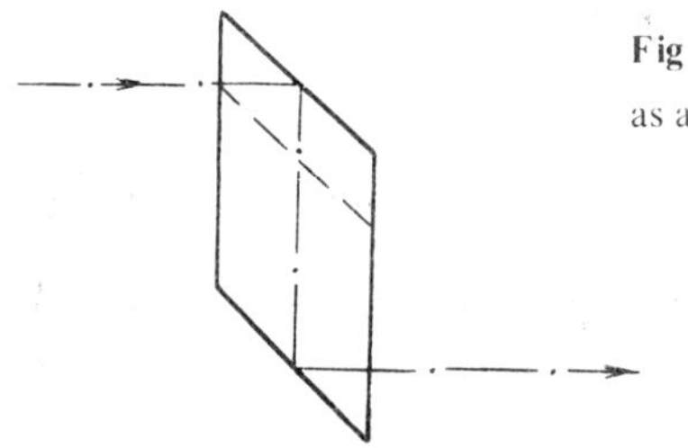

Fig. 5.45. A roof rhomboid prism, as an example to show $\xi_{z'} = 0$

To unify these two cases mathematically, we assume $\vec{\xi} = \xi_{x'} \boldsymbol{i}' + \xi_{y'} \boldsymbol{j}' + \xi_{z'} \boldsymbol{k}'$ while deriving a general fomula, and then the results for the two cases mentioned above can be obtained from the general formula by putting $\xi_{y'} = 0$ and $\xi_{z'} = 0$, respectively.

Similarly, for this type of reflecting prisms and the coordinate system associated with them, we have for Eq. (5.58) the following expressions of substitution:

$$\overrightarrow{M_0 M'} = \overrightarrow{o\,o'} = \vec{\xi} = \xi_{x'}\,\boldsymbol{i}' + \xi_{y'}\,\boldsymbol{j}' + \xi_{z'}\,\boldsymbol{k}'$$

$$\boldsymbol{b}' = \overrightarrow{o\,'F'} = \boldsymbol{b}'' = b''\boldsymbol{i}'$$

$$S_{T,2\varphi} = \begin{pmatrix} -1 & 0 & 0 \\ 0 & -1 & 0 \\ 0 & 0 & 1 \end{pmatrix}$$

$$(5.119)$$

(3) Derivation of a general formula for determining the planar three-dimensional zero-valued poles as well as for judging the existence conditions of such poles

Here, reflecting prisms of two different types will be discussed separately.

(i) For all reflecting prisms with an even number of reflections plus the reflecting prisms with an odd number of reflections excluding those with $2\varphi = 180°$

Putting Eq. (5.118) into Eq. (5.58), and then determining the partial derivatives of the three components $\Delta S'_{F'x'}$, $\Delta S'_{F'y'}$ and $\Delta S'_{F'z'}$ of image point displacement with respect to $\Delta\theta_{x'}$, $\Delta\theta_{y'}$ and $\Delta\theta_{z'}$ yields the actual form of Eqs. (5.114) and (5.115) as

$$\frac{-(-1)^{l-1}\sin2\varphi \cdot z'_q - 2\varepsilon(-1)^{l-1}\sin2\varphi}{[1+(-1)^{l-1}\cos2\varphi]z'_q + 2\varepsilon(-1)^{l-1}\cos2\varphi}$$

$$= \frac{-[1+(-1)^{l-1}\cos2\varphi]z'_q - 2\varepsilon(-1)^{l-1}\cos2\varphi}{-(-1)^{l-1}\sin2\varphi \cdot z'_q - 2\varepsilon(-1)^{l-1}\sin2\varphi}$$

$$= \frac{(-1)^{l-1}\sin2\varphi \cdot x'_q + [1+(-1)^{l-1}\cos2\varphi]y'_q}{-[1+(-1)^{l-1}\cos2\varphi]x'_q + (-1)^{l-1}\sin2\varphi \cdot y'_q}, \qquad (5.120)$$

$$\frac{[1+(-1)^{l-1}\cos2\varphi]z'_q + 2\varepsilon(-1)^{l-1}\cos2\varphi}{-[1+(-1)^{l-1}]y'_q + b''\sin2\varphi}$$

$$= \frac{-(-1)^{l-1}\sin2\varphi \cdot z'_q - 2\varepsilon(-1)^{l-1}\sin2\varphi}{[1+(-1)^{l-1}]x'_q - b''(1-\cos2\varphi)}$$

$$= \frac{-[1+(-1)^{l-1}\cos2\varphi]x'_q + (-1)^{l-1}\sin2\varphi \cdot y'_q}{0}. \qquad (5.121)$$

The discussion of the above equations may be easier to understand if the corresponding formulas of the three gradient axes $\vec{\eta}_{F'uq}$, $\vec{\eta}_{F'vq}$ and $\vec{\eta}_{F'wq}$ are given below

$$\vec{\eta}_{F'uq} = \{-(-1)^{l-1}\sin2\varphi \cdot z'_q - 2\varepsilon(-1)^{l-1}\sin2\varphi\} \cdot \boldsymbol{i}' + \{-[1+(-1)^{l-1}\cos2\varphi]z'_q$$
$$- 2\varepsilon(-1)^{l-1}\cos2\varphi\} \cdot \boldsymbol{j}' + \{(-1)^{l-1}\sin2\varphi \cdot x'_q + [1+(-1)^{l-1}\cos2\varphi]y'_q\} \cdot \boldsymbol{k}',$$

$$(5.122)$$

$$\vec{\eta}_{F'_{vq}} = \{ [1 + (-1)^{l-1}\cos2\varphi] z'_q + 2\varepsilon (-1)^{l-1}\cos 2\varphi \} \cdot i'$$
$$+ \{ -(-1)^{l-1}\sin2\varphi \cdot z'_q - 2\varepsilon (-1)^{l-1} \sin 2\varphi \} \cdot j'$$
$$+ \{ -[1 + (-1)^{l-1}\cos2\varphi] x'_q + (-1)^{l-1}\sin 2\varphi \cdot r'_q \} \cdot k', \tag{5.123}$$
$$\vec{\eta}_{F'_{wq}} = \{ -[1 + (-1)^{l-1}] y'_q + b'' \sin2\varphi \} \cdot i' + \{ [1 + (-1)^{l-1}] x'_q$$
$$- b'' (1 - \cos 2\varphi) \} \cdot j'. \tag{5.124}$$

Prior to the following discussion, it is worthwhile to restate that the image point F', to which the image point displacement $\Delta S'_F$ is related, must be considered as being an arbitrary point lying on the exit optical axis of the prism.

By comparing Eqs. (5.122) and (5.123), we find

$$\frac{\partial \Delta S'_{F'y'}}{\partial \Delta \theta_{y'}} = \frac{\partial \Delta S'_{F'x'}}{\partial \Delta \theta_{x'}}, \tag{5.125}$$

$$\frac{\partial \Delta S'_{F'y'}}{\partial \Delta \theta_{x'}} = -\frac{\partial \Delta S'_{F'x'}}{\partial \Delta \theta_{y'}}. \tag{5.126}$$

Let

$$K = \frac{-(-1)^{l-1}\sin2\varphi \cdot z'_q - 2\varepsilon (-1)^{l-1}\sin2\varphi}{[1 + (-1)^{l-1}\cos2\varphi] z'_q + 2\varepsilon (-1)^{l-1}\cos2\varphi}. \tag{5.127}$$

Then, from the left equation of Eq. (5.120), we have

$$K = -\frac{1}{K} \quad \text{or} \quad K = \sqrt{-1}. \tag{5.128}$$

The imaginary root of K indicates that in general cases, simultaneous Eqs. (5.120) and (5.121) do not have a real solution, i.e., it is not sure whether reflecting prisms of this type may have one planar three-dimensional zero-valued pole referring to an arbitrary image point F' lying on the exit optical axis.

According to Eqs. (5.122), (5.123), and (5.125) through (5.128), the result $K = \sqrt{-1}$ shows essentially that if either the numerator or the denominator of K is not equal to zero, then it is not possible to have the two gradient axes $\vec{\eta}_{F'_{uq}}$ and $\vec{\eta}_{F'_{vq}}$ parallel to each other, and still less possible to make them coincident.

Therefore, the equations (5.120) and (5.121) for reflecting prisms of this type may have real solutions if and only if both the numerator and the denominator of K are equal to zero, that is to say, the first two coefficients of both $\vec{\eta}_{F'_{uq}}$ and $\vec{\eta}_{F'_{vq}}$ should be zero:

$$-(-1)^{l-1}\sin2\varphi \cdot z'_q - 2\varepsilon (-1)^{l-1}\sin2\varphi = 0, \tag{5.129}$$

$$[1 + (-1)^{l-1}\cos2\varphi] z'_q + 2\varepsilon (-1)^{l-1}\cos2\varphi = 0. \tag{5.130}$$

Now, the above two conditions will be discussed for several different combinations

178

of 2φ and 2ε.

   a . $2\varphi \neq 180°$ and $2\varepsilon \neq 0$

This is the most general case which includes all the spatial reflecting prisms and the ordinary planar reflecting prisms in this category .

   From Eq . (5.129) , we have

$$z'_q = -2\varepsilon . \tag{5.131}$$

Then , putting Eq . (5.131) into Eq . (5.130) leads to a contradictory result: '1 = 0' . This fact tells that both the spatial and ordinary planar reflecting prisms of this type do not even have one planar three‑dimensional zero‑valued pole .

   Now , only the coplanar reflecting prisms of this type are left .

   b . $2\varphi \neq 180°$ and $2\varepsilon = 0$

Putting these data into Eqs . (5.120) through (5.124) yields

$$\frac{-(-1)^{l-1}\sin2\varphi \cdot z'_q}{[1+(-1)^{l-1}\cos2\varphi]\,z'_q} = \frac{-[1+(-1)^{l-1}\cos2\varphi]z'_q}{-(-1)^{l-1}\sin2\varphi \cdot z'_q}$$

$$= \frac{(-1)^{l-1}\sin2\varphi \cdot x'_q + [1+(-1)^{l-1}\cos2\varphi]y'_q}{-[1+(-1)^{l-1}\cos2\varphi]x'_q + (-1)^{l-1}\sin2\varphi \cdot y'_q} , \tag{5.132}$$

$$\frac{[1+(-1)^{l-1}\cos2\varphi]z'_q}{-[1+(-1)^{l-1}]y'_q + b''\sin2\varphi} = \frac{-(-1)^{l-1}\sin2\varphi \cdot z'_q}{[1+(-1)^{l-1}]x'_q - b''(1-\cos2\varphi)}$$

$$= \frac{-[1+(-1)^{l-1}\cos2\varphi]x'_q + (-1)^{l-1}\sin2\varphi \cdot y'_q}{0} , \tag{5.133}$$

$$\vec{\eta}_{F'uq} = \{-(-1)^{l-1}\sin2\varphi \cdot z'_q\}\boldsymbol{i}' + \{-[1+(-1)^{l-1}\cos2\varphi]z'_q\}\boldsymbol{j}'$$

$$+ \{(-1)^{l-1}\sin2\varphi \cdot x'_q + [1+(-1)^{l-1}\cos2\varphi]y'_q\}\boldsymbol{k}', \tag{5.134}$$

$$\vec{\eta}_{F'vq} = \{[1+(-1)^{l-1}\cos2\varphi]z'_q\}\boldsymbol{i}' + \{-(-1)^{l-1}\sin2\varphi \cdot z'_q\}\boldsymbol{j}'$$

$$+ \{-[1+(-1)^{l-1}\cos2\varphi]x'_q + (-1)^{l-1}\sin2\varphi \cdot y'_q\}\boldsymbol{k}', \tag{5.135}$$

$$\vec{\eta}_{F'wq} = \{-[1+(-1)^{l-1}]y'_q + b''\sin2\varphi\}\boldsymbol{i}' + \{[1+(-1)^{l-1}]x'_q$$

$$- b''(1-\cos2\varphi)\}\boldsymbol{j}'. \tag{5.136}$$

   Considering the conditions of Eqs . (5.129) and (5.130) , we should put $z'_q = 0$. Then , we have

$$\vec{\eta}_{F'uq} = \{(-1)^{l-1}\sin2\varphi \cdot x'_q + [1+(-1)^{l-1}\cos2\varphi]y'_q\}\boldsymbol{k}', \tag{5.137}$$

$$\vec{\eta}_{F\,'vq} = \{ -[1 + (-1)^{t-1}\cos2\varphi]\, x_q' + (-1)^{t-1}\sin2\varphi \cdot y_q' \}\, \boldsymbol{k}', \tag{5.138}$$

$$\vec{\eta}_{F\,'wq} = \{ -[1 + (-1)^{t-1}]\, y_q' + b''\sin2\varphi \}\, \boldsymbol{i}'$$
$$+ \{ [1 + (-1)^{t-1}]\, x_q' - b''(1 - \cos2\varphi) \}\, \boldsymbol{j}'. \tag{5.139}$$

We still can distinguish two cases.

If $t = $ odd number, then Eqs. (5.137) through (5.139) become

$$\vec{\eta}_{F\,'uq} = \{ \sin2\varphi \cdot x_q' + [1 + \cos2\varphi]\, y_q' \}\, \boldsymbol{k}', \tag{5.140}$$

$$\vec{\eta}_{F\,'vq} = \{ -[1 + \cos2\varphi]\, x_q' + \sin2\varphi \cdot y_q' \}\, \boldsymbol{k}', \tag{5.141}$$

$$\vec{\eta}_{F\,'wq} = \{ -2y_q' + b''\sin2\varphi \}\, \boldsymbol{i}' + \{ 2x_q' - b''(1 - \cos2\varphi) \}\, \boldsymbol{j}'. \tag{5.142}$$

From the resulting equations (5.140) through (5.142), it is clear that there may exist two planar three-dimensional zero-valued poles C_{Π_1} and C_{Π_2}.

The coordinates of C_{Π_1} are determined by the following equations:

$$-2y_q' + b''\sin2\varphi = 0, \tag{5.143}$$

$$2x_q' - b''(1 - \cos2\varphi) = 0, \tag{5.144}$$

resulting in

$$\left.\begin{aligned} x_1' &= \frac{b''}{2}(1 - \cos2\varphi) \\[2mm] y_1' &= \frac{b''}{2}\sin2\varphi \\[2mm] z_1' &= 0 \end{aligned}\right\}, \tag{5.145}$$

and the associated plane Π_1 is thus passing through the pole C_{Π_1} and perpendicular to the gradient axes $\vec{\eta}_{F\,'uC_1}$ or $\vec{\eta}_{F\,'vC_1}$, i.e., $\boldsymbol{k}'$.

The coordinates of C_{Π_2} are determined by the following equations:

$$\sin2\varphi \cdot x_q' + [1 + \cos2\varphi]\, y_q' = 0, \tag{5.146}$$

$$-[1 + \cos2\varphi]\, x_q' + \sin2\varphi \cdot y_q' = 0, \tag{5.147}$$

giving

$$\left.\begin{aligned} x_2' &= 0 \\ y_2' &= 0 \\ z_2' &= 0 \end{aligned}\right\}, \tag{5.148}$$

and the associated plane Π_2 is thus passing through the pole C_{Π_2} and perpendicular to the gradient axis $\vec{\eta}_{F\,'wC_2}$. Putting Eq. (5.148) into Eq. (5.142) yields

$$\vec{\eta}_{F\,'wC_2} = b\,''\sin2\varphi \cdot \boldsymbol{i}\,' - b''(1-\cos2\varphi) \cdot \boldsymbol{j}\,' . \qquad (5.149)$$

we may remember that the prism shown in Fig.5.27 belongs to this case.

If $t =$ even number, then Eqs.(5.137) through (5.139) become

$$\vec{\eta}_{F\,'uq} = \{-\sin2\varphi \cdot x_q' + [1-\cos2\varphi]\, y_q'\}\boldsymbol{k}\,', \qquad (5.150)$$

$$\vec{\eta}_{F\,'vq} = \{-[1-\cos2\varphi]\, x_q' - \sin2\varphi \cdot y_q'\}\boldsymbol{k}\,', \qquad (5.151)$$

$$\vec{\eta}_{F\,'wq} = b\,''\sin2\varphi \cdot \boldsymbol{i}\,' - b\,'(1-\cos2\varphi) \cdot \boldsymbol{j}\,' . \qquad (5.152)$$

From the resulting equations (5.150) through (5.152), it is also clear that there may exist two planar three-dimensional zero-valued poles C_{Π_1} and C_{Π_2}.

The coordinates of C_{Π_2} is determined by the following equations:

$$-\sin2\varphi \cdot x_q' + [1-\cos2\varphi]y_q' = 0 , \qquad (5.153)$$

$$-[1-\cos2\varphi]\, x_q' - \sin2\varphi \cdot y_q' = 0 , \qquad (5.154)$$

yielding

$$\left.\begin{array}{l} x_2' = 0 \\ y_2' = 0 \\ z_2' = 0 \end{array}\right\} , \qquad (5.155)$$

and the associated plane Π_2 is thus passing through the pole C_{Π_2} and perpendicular to the gradient axis $\vec{\eta}_{F\,'wC_2}$, the latter being

$$\vec{\eta}_{F\,'wC_2} = b\,''\sin2\varphi \cdot \boldsymbol{i}\,' - b''(1-\cos2\varphi) \cdot \boldsymbol{j}\,' . \qquad (5.156)$$

The other pole C_{Π_1} is moved to infinity along the direction of a certain unit vector $\boldsymbol{P}_1$ whose direction cosines may be determined by the following equations:

$$\left.\begin{array}{l} b\,''\sin2\varphi\, P_{x'} - b''(1-\cos2\varphi)P_{y'} = 0 \\ P_{z'} = 0 \end{array}\right\} , \qquad (5.157)$$

giving

$$\left.\begin{array}{l} \dfrac{P_{1y'}}{P_{1x'}} = \dfrac{1-\cos2\varphi}{\sin2\varphi} \\[2mm] P_{1z'} = 0 \end{array}\right\} \qquad (5.158)$$

The associated plane Π_1 is thus perpendicular to the gradient axes $\vec{\eta}_{F\,'uC_1}$ or $\vec{\eta}_{F\,'vC_1}$, i.e., $\boldsymbol{k}\,'$ and coincident with the $x\,'y\,'$ plane.

The prism shown in Fig.5.28 belongs to this case.

c.$2\varphi = 180\,°$ and $2\varepsilon \neq 0$

Here, we will discuss the reflecting prism with an even number of reflections only.

Putting these data into Eqs. (5.122) through (5.124) yields

$$\vec{\eta}_{F'uq} = (-2z'_q - 2\varepsilon)\cdot \boldsymbol{j}' + 2y'_q \cdot \boldsymbol{k}', \tag{5.159}$$

$$\vec{\eta}_{F'vq} = (2z'_q + 2\varepsilon)\cdot \boldsymbol{i}' + (-2x'_q)\cdot \boldsymbol{k}', \tag{5.160}$$

$$\vec{\eta}_{F'wq} = -2b''\cdot \boldsymbol{j}'. \tag{5.161}$$

Now, the conditions of Eqs. (5.129) and (5.130) should be applied to Eqs. (5.161) and (5.159) by putting $y'_q = 0$, and then rearrange the three gradient axes in the following order:

$$\vec{\eta}_{F'wq} = -2b''\cdot \boldsymbol{j}', \tag{5.162}$$

$$\vec{\eta}_{F'uq} = (-2z'_q - 2\varepsilon)\cdot \boldsymbol{j}', \tag{5.163}$$

$$\vec{\eta}_{F'vq} = (2z'_q + 2\varepsilon)\cdot \boldsymbol{i}' + (-2x'_q)\cdot \boldsymbol{k}'. \tag{5.164}$$

From the resulting equations (5.162) through (5.164), it is easier to see the pole C_{Π_1} the coordinates of which may be determined by the following equations:

$$2z'_q + 2\varepsilon = 0, \tag{5.165}$$

$$-2x'_q = 0, \tag{5.166}$$

resulting in

$$\left. \begin{array}{l} z'_1 = -\varepsilon \\ x'_1 = 0 \\ y'_1 = 0 \end{array} \right\}. \tag{5,167}$$

The associated plane Π_1 is thus passing through the pole C_{Π_1} and perpendicular to the gradient axis $\vec{\eta}_{F'wC_1}$, the latter being

$$\vec{\eta}_{F'wC_1} = -2b''\cdot \boldsymbol{j}'. \tag{5.168}$$

To find the other pole C_{Π_2} it is advisable to derive the equations of the three components of image point displacement corresponding to the three gradient axes expressed in Eqs. (5.162) through (5.164):

$$\Delta S'_{F'z'} = \Delta\theta\{-2b''P_{y'}\}, \tag{5.169}$$

$$\Delta S'_{F'x'} = \Delta\theta\{(-2z'_q - 2\varepsilon)P_{y'}\}, \tag{5.170}$$

$$\Delta S'_{F'y'} = \Delta\theta\{(2z'_q + 2\varepsilon)P_{x'} + (-2x'_q)P_{z'}\}. \tag{5.171}$$

The above equations show that the pole C_{Π_2} is moved to infinity along the x' direction and the associated plane Π_2 will be passing through a certain point $(0,0,-\varepsilon)$ and perpendicular to $\boldsymbol{k}'$.

The roof right-angle prism shown in Fig. 5.29 is just the case.

Moreover , for some prisms of this category , the screw axis $\vec{\lambda}$ is coincident with the exit optical axis as well as with the entrance optical axis . Then , we have $b'' = 0$, and thus from Eq . (5.169)

$$\Delta S'_{F'_{z'}} = 0 .$$ (5.172)

Combining Eq . (5.172) with Eqs . (5.170) and (5.171) we find a spatial three- dimensional zero- valued pole C for the prism of this class with coordinates as

$$\left. \begin{array}{l} x'_c = 0 \\ y'_c = 0 \\ z'_c = -2\varepsilon \end{array} \right\} .$$ (5.173)

The roof Pechan prism as in Fig .5.31 may be an example in point .
d . $2\varphi = 0°$, $2\varepsilon = 0$, and $t = $ odd number
Putting these data into Eqs . (5.122) through (5.124) yields

$$\vec{\eta}_{F'_{uq}} = -2z'_q \cdot j' + 2y'_q \cdot k' ,$$ (5.174)

$$\vec{\eta}_{F'_{vq}} = 2z'_q \cdot i' - 2x'_q \cdot k' ,$$ (5.175)

$$\vec{\eta}_{F'_{wq}} = -2y'_q \cdot i' + 2x'_q \cdot j' .$$ (5.176)

It is clear from Eqs . (5.174) through (5.176) that a spatial three- dimensional zero- valued pole C is found to be at the origin o' :

$$\left. \begin{array}{l} x'_c = 0 \\ y'_c = 0 \\ z'_c = 0 \end{array} \right\} .$$ (5.177)

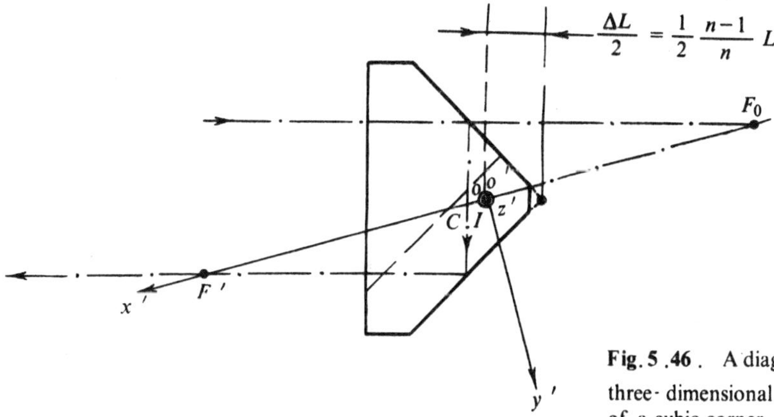

Fig. 5.46 . A diagram to show a spatial three- dimensional zero- valued pole C of a cubic corner

A prism , as shown in Fig . 5 . 46 , called the *cubic corner* is just such a case .

e . $2\varphi = 0$ ° , $2\varepsilon \neq 0$, and $t =$ even number

Putting these data into Eqs . (5 . 122) through (5 . 124) gives

$$\vec{\eta}_{F'uq} = 2\varepsilon \cdot \boldsymbol{j}' \, , \tag{5.178}$$

$$\vec{\eta}_{F'vq} = -2\varepsilon \cdot \boldsymbol{i}' \, , \tag{5.179}$$

$$\vec{\eta}_{F'wq} = 0 \, . \tag{5.180}$$

By Eqs . (5 . 74) , we have accordingly

$$\Delta S'_{F'x'} = \Delta\theta \{ 2\varepsilon P_{y'} \} \, , \tag{5.181}$$

$$\Delta S'_{F'y'} = \Delta\theta \{ -2\varepsilon P_{x'} \} \, , \tag{5.182}$$

$$\Delta S'_{F'z'} = 0 \, . \tag{5.183}$$

From either Eqs . (5 . 178) through (5 . 180) or Eqs . (5 . 181) through (5 . 183) , it is evident that all the lines parallel to the z' direction will be three - dimensional zero - value axes no matter where they are located . Thus , we obtain a spatial three - dimensional zero - valued pole C located infinitely distant along the z' axis with a rotating associated plane about the z' axis .

A rhomboid prism is given in Fig . 5 . 47 as an example for those prisms of this class .

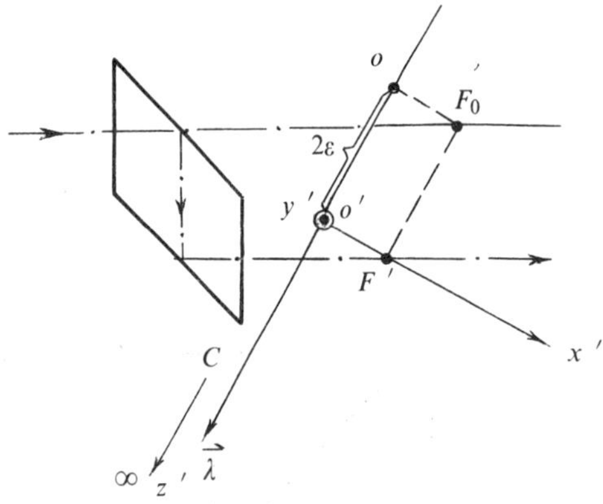

Fig . 5 . 47 . A diagram to show a spatial three - dimensional zero - valued pole C at infinity of a rhomboid prism

(ii) For reflecting prisms with an odd number of reflections and $2\varphi = 180$ °

Putting Eq . (5 . 119) into Eq . (5 . 58) , and then determining the partial derivatives of three components $\Delta S'_{F'x'}$, $\Delta S'_{F'y'}$ and $\Delta S'_{F'z'}$ of image point displacement with respect to $\Delta\theta_{x'}$, $\Delta\theta_{y'}$ and $\Delta\theta_{z'}$ yields the actual form of Eqs . (5 . 73) and (5 . 74) as

$$\vec{\eta}_{F'uq} = \xi_z \cdot \boldsymbol{j}' - \xi_y \cdot \boldsymbol{k}' \, , \tag{5.184}$$

$$\vec{\eta}_{F'vq} = -\xi_z \cdot \boldsymbol{i}' + \xi_x \cdot \boldsymbol{k}' \, , \tag{5.185}$$

184

$$\eta_{F'wq} = (-2y'_q - \xi_{y'}) \cdot i' + (2x'_q + \xi_{x'} - 2b'') \cdot j', \qquad (5.186)$$

$$\Delta S'_{F'x'} = \Delta\theta \{\xi_{z'} P_{y'} - \xi_{y'} P_{z'}\}, \qquad (5.187)$$

$$\Delta S'_{F'y'} = \Delta\theta \{-\xi_{z'} P_{x'} + \xi_{x'} P_{z'}\}, \qquad (5.188)$$

$$\Delta S'_{F'z'} = \Delta\theta \{(-2y'_q - \xi_{y'})P_{x'} + (2x'_q + \xi_{x'} - 2b'')P_{y'}\}. \qquad (5.189)$$

Consider also some different cases.

a. $\xi_{x'} \neq 0$, $\xi_{y'} \neq 0$ and $\xi_{z'} \neq 0$

All the ordinary planar reflecting prisms of this type refer to the present case. A prism corresponding to a three-mirror system, as shown in Fig. 5.48, is given as an example for those prisms of this category.

From either Eqs. (5.184) through (5.186) or Eqs. (5.187) through (5.189), it is obvious that the ordinary planar reflecting prisms of this type do not have any planar three-dimensional zero-valued pole at all.

b. $\xi_{x'} \neq 0$, $\xi_{z'} \neq 0$ and $\xi_{y'} = 0$

The data combined in such a way refer to those prisms of this type with coplanar mirror normals. Putting $\xi_{y'} = 0$ into Eqs. (5.184) through (5.189) gives

$$\vec{\eta}_{F'uq} = \xi_{z'} \cdot j', \qquad (5.190)$$

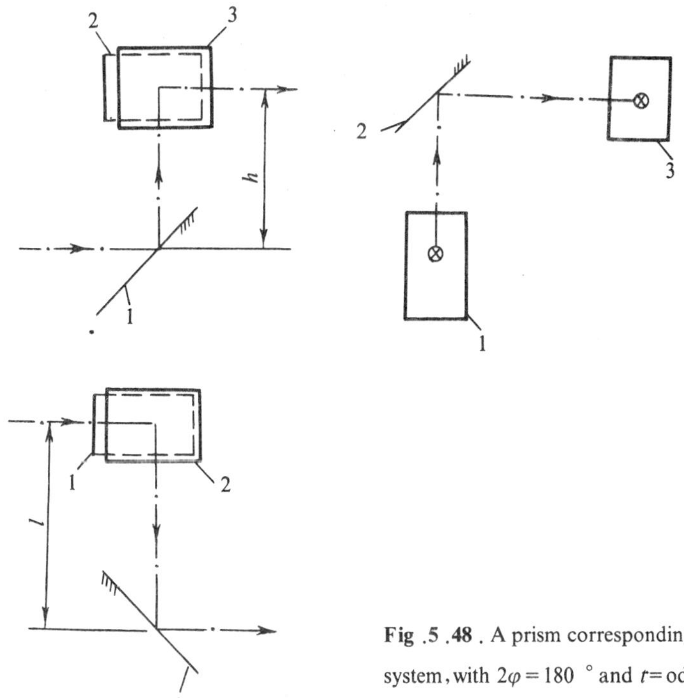

Fig. 5.48. A prism corresponding to a three-mirror system, with $2\varphi = 180°$ and $r = $ odd number represents an ordinary planar reflecting prism while $h \neq l$

$$\vec{\eta}_{F'vq} = -\xi_{z'} \cdot \boldsymbol{i}' + \xi_{x'} \cdot \boldsymbol{k}' \ , \tag{5.191}$$

$$\vec{\eta}_{F'wq} = -2y'_q \cdot \boldsymbol{i}' + (2x'_q + \xi_{x'} - 2b'') \cdot \boldsymbol{j}' , \tag{5.192}$$

$$\Delta S'_{F'x'} = \Delta\theta \left\{ \xi_{z'} P_{y'} \right\} \ , \tag{5.193}$$

$$\Delta S'_{F'y'} = \Delta\theta \left\{ -\xi_{z'} P_{x'} + \xi_{x'} P_{z'} \right\} \ , \tag{5.194}$$

$$\Delta S'_{F'z'} = \Delta\theta \left\{ -2y'_q P_{x'} + (2x'_q + \xi_{x'} - 2b'')P_{y'} \right\} \ . \tag{5.195}$$

After analyzing Eqs. (5.190) through (5.195) carefully, we find that there exists a planar three-dimensional zero-valued pole C_{Π_1} located infinitely distant along the direction determined by the following equations :

$$\Delta S'_{F'x'} = \Delta\theta \left\{ \xi_{z'} P_{y'} \right\} = 0 \ ,$$

$$\Delta S'_{F'y'} = \Delta\theta \left\{ -\xi_{z'} P_{x'} + \xi_{x'} P_{z'} \right\} = 0 \ ,$$

giving

$$\left. \begin{array}{l} P_{1y'} = 0 \\[2ex] \dfrac{P_{1x'}}{P_{1z'}} = \dfrac{\xi_{x'}}{\xi_{z'}} \end{array} \right\} \ , \tag{5.196}$$

while the conjugate optical-axis section is the associated plane Π_1, i.e., the $x'z'$ plane with $y'_q = 0$. The right-angle prism shown in Fig. 5.29 is just such a case.

Moreover, if the entrance optical axis and the exit optical axis for prisms of this category are coincident with each other, then we have also

$$\xi_{z'} = 0 \ . \tag{5.197}$$

Thus, Eqs. (5.190) through (5.195) become

$$\vec{\eta}_{F'uq} = 0 \ , \tag{5.198}$$

$$\vec{\eta}_{F'vq} = \xi_{x'} \cdot \boldsymbol{k}' , \tag{5.199}$$

$$\vec{\eta}_{F'wq} = -2y'_q \cdot \boldsymbol{i}' + (2x'_q + \xi_{x'} - 2b'') \cdot \boldsymbol{j}' , \tag{5.200}$$

$$\Delta S'_{F'x'} = 0 \ , \tag{5.201}$$

$$\Delta S'_{F'y'} = \Delta\theta \left\{ \xi_{x'} P_{z'} \right\} \ , \tag{5.202}$$

$$\Delta S'_{F'z'} = \Delta\theta \left\{ -2y'_q P_{x'} + (2x'_q + \xi_{x'} - 2b'')P_{y'} \right\} \ . \tag{5.203}$$

From Eqs. (5.198) through (5.203), we can find first in the $x'y'$ plane a planar three-dimensional zero-valued pole C_{Π_1} whose coordinates will be determined by the following equations :

$$-2y'_q = 0 \\ 2x'_q + \xi_{x'} - 2b'' = 0 \\ z'_q = 0 \Bigg\} \quad , \tag{5.204}$$

yielding

$$x'_1 = b'' - \frac{\xi_{x'}}{2} \\ y'_1 = 0 \\ z'_1 = 0 \Bigg\} \quad . \tag{5.205}$$

The associated plane Π_1 is thus the $x'z'$ plane. Note that the pole C_{Π_1} bisects the segment $\overline{F_0 F'}$.

Now, if we move the pole C_{Π_1} along the y' direction, we find that it continues to be a planar three-dimensional zero-valued pole with its associated plane perpendicular to the y' axis. The loci of C_{Π_1} is called the *planar three-dimensional zero-valued polar line*. :

The Abbe prism shown in Fig. 5.30 may serve as an example. It should be noted that the planar three-dimensional zero-valued polar line is parallel to the uniquely existing gradient shift direction b.

c. $\xi_{x'} \neq 0$, $\xi_{y'} \neq 0$ and $\xi_{z'} = 0$

The data combined in such a way refer to those prisms of this type whose mirror normals are not coplanar, for instance, the roof rhomboid prism shown in Fig. 5.49.

Putting $\xi_{z'} = 0$ into Eqs. (5.184) through (5.189) yields

$$\vec{\eta}_{F'uq} = -\xi_{y'} \cdot k' , \tag{5.206}$$

$$\vec{\eta}_{F'vq} = \xi_{x'} \cdot k' , \tag{5.207}$$

$$\vec{\eta}_{F'wq} = (-2y'_q - \xi_{y'}) \cdot i' + (2x'_q + \xi_{x'} - 2b'') \cdot j' , \tag{5.208}$$

$$\Delta S'_{F'x'} = \Delta\theta \{ -\xi_{y'} P_{z'} \} , \tag{5.209}$$

$$\Delta S'_{F'y'} = \Delta\theta \{ \xi_{x'} P_{z'} \} , \tag{5.210}$$

$$\Delta S'_{F'z'} = \Delta\theta \{ (-2y'_q - \xi_{y'}) P_{x'} + (2x'_q + \xi_{x'} - 2b'') P_{y'} \} . \tag{5.211}$$

Similarly, there exists a planar three-dimensional zero-valued polar line which is perpendicular to the $x'y'$ plane and cuts the segment $\overline{F_0 F'}$ at its midpoint C_{Π_1}, i.e., one of the planar three-dimensional zero-valued poles, the coordinates of which may be determined by the following equations :

$$-2y'_q - \xi_{y'} = 0 \\ 2x'_q + \xi_{x'} - 2b'' = 0 \\ z'_q = 0 \Bigg\} \quad , \tag{5.212}$$

giving

$$x'_1 = b'' - \frac{\xi_{x'}}{2}$$

$$y'_1 = - \frac{\xi_{y'}}{2}$$

$$z'_1 = 0$$

$$(5.213)$$

All the associated planes Π_1 are perpendicular to the polar line .

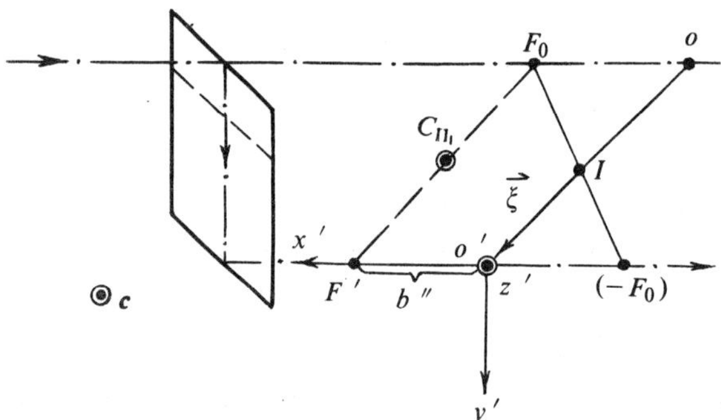

Fig.5.49. A diagram to show a planar three-dimensional zero-valued polar line of a roof rhomboid prism where $\xi_{z'} = 0$

5.8 Classification of Reflecting Prisms

Now we are able to present a new classification of reflecting prisms since we have investigated in depth their behavior in regard to both image formation and adjustment .

As a matter of fact , the behavior of image formation describes the static nature of a prism while the character of adjustment depicts the instantaneously moving property of the prism .

It will be seen from the classification chart (Fig.5.50) that the reflecting prisms are first classified according to their character of image formation roughly into three categories: spatial , ordinary planar , and coplanar reflecting prisms . Then , the last category , namely , the coplanar reflecting prisms will further be divided according to the character of adjustment in full detail into many small groups .

A reasonable or a logical classification as usual may always serve the purpose of exploring something new .

5.9 Tabulation of Reflecting Prisms

In the previous chapters and sections , we have touched upon twenty or so characteristic parameters , which may be basically grouped under two categories , as listed below :

188

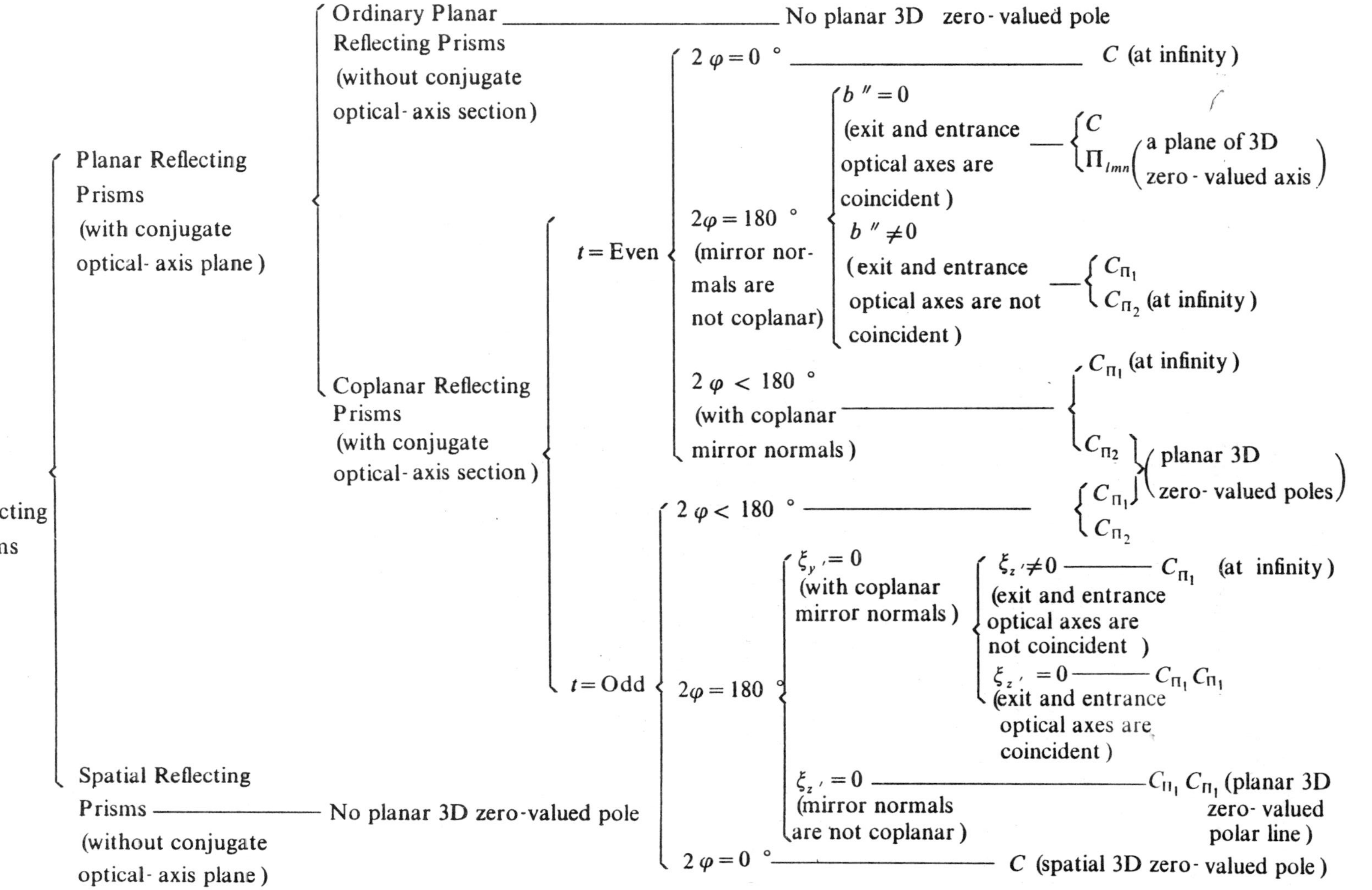

Fig . 5 .50 . Classification chart of reflecting prisms

189

Characteristic parameters of image formation:

Characteristic vector T

Characteristic angle 2φ

Vector of a pair of basic conjugate points $\overrightarrow{M_0M}\,'$

Position vector locating the screw axis $\vec{\lambda}$ of image formation r_0

Position vector of the center I of inversion r_i

Axial shift of the helical motion of image formation $2d$

Characteristic parameters of adjustment:

Extreme-valued axis direction of x' image rotation $(\Delta\mu_{x'}')\boldsymbol{u}$

Extreme-valued axis direction of y' image rotation $(\Delta\mu_{y'}')\boldsymbol{v}$

Extreme-valued axis direction of z' image rotation $(\Delta\mu_{z'}')\boldsymbol{w}$

Extreme value of x' image rotation $\Delta\mu_{x'\,max}' = \delta_u$

Extreme value of y' image rotation $\Delta\mu_{y'\,max}' = \delta_v$

Extreme value of z' image rotation $\Delta\mu_{z'\,max}' = \delta_w$

Extreme-valued shift direction of x' image displacement $(\Delta S_{x'}')\boldsymbol{a}$

Extreme-valued shift direction of y' image displacement $(\Delta S_{y'}')\boldsymbol{b}$

Extreme-valued shift direction of z' image displacement $(\Delta S_{z'}')\boldsymbol{c}$

Extreme value of x' image displacement $\Delta S_{x'\,max}' = \delta_a$

Extreme value of y' image displacement $\Delta S_{y'\,max}' = \delta_b$

Extreme value of z' image displacement $\Delta S_{z'\,max}' = \delta_c$

Planar three-dimensional zero-valued pole $C_{\Pi_1}(x_1',y_1',z_1';\Pi_1)$

Planar three-dimensional zero-valued pole $C_{\Pi_2}(x_2',y_2',z_2';\Pi_2)$

Planar three-dimensional zero-valued polar line $C_{\Pi_1}C_{\Pi_1}(x_1',y_1',z_1';\Pi_1;C_{\Pi_1}C_{\Pi_1}\perp\Pi_1)$

Spatial three-dimensional zero-valued pole $C(x_c',y_c',z_c')$

These characteristic parameters enable us to express the effects of the internal structure of prisms on the input-output relationships in efficient and transparent language .

For engineering use , we prepare quite a complete set of tables of reflecting prisms , which covers 54 commonly used prisms , and for each of them an individual table is given with 1 adjustment diagram , 1 matrix , 20 characteristic parameters , and 6 equations of image motion .

To show an example , derive and calculate all the necessaries in making the table for a spatial reflecting prism shown in Fig 5 . 51 .

Assume $o'x'y'z'$ to be the reference coordinate system .

(1) Check the length (thickness) L of the plane-parallel plate

From Fig 5 . 51 .

$$L = \overline{M'a} + \overline{ab} + \overline{bc}\,,$$

and

$$\overline{M'a} = \frac{D}{2}\,\mathrm{tg}\,60° = 0.866\,D\,,\quad \overline{ab} = \frac{D}{2}\,\frac{1}{\cos 60°} = D\,,$$

$$\overline{bc} = 0.5D\,.$$

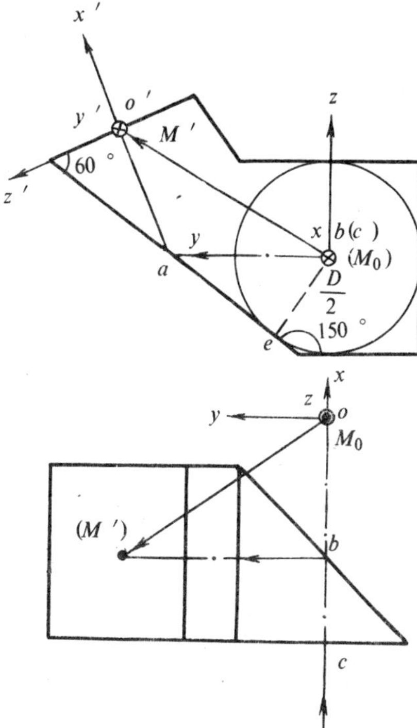

Fig . 5 . 51 . A spatial reflecting prism , serving as an example to show how to make a table of a prism

Therefore ,

$$L = 2.366D .$$

(2) $\overrightarrow{M_0 M}\,'$

From Fig . 5 .51 ,

$$(\overrightarrow{M_0 M}\,')_{x'} = (\overline{M'a} + \overline{ab}\cos 60°) = 1.366 D ,$$

$$(\overrightarrow{M_0 M}\,')_{y'} = \overline{bc} - \frac{L}{n} = \left(0.5 - \frac{2.366}{n}\right)D ,$$

$$(\overrightarrow{M_0 M}\,')_{z'} = \overline{ab}\sin 60° = 0.866D .$$

Therefore ,

$$\overrightarrow{M_0 M}\,' = 1.366D \cdot i' + \left(0.5D - 2.366\frac{D}{n}\right) \cdot j' + 0.866D \cdot k' .$$

(3) T

While determining T , it is advisable to assume the coordinate system xyz in object space as the reference frame .

From Fig . 5 .51 ,

$$i' = 0 \cdot i + \sin 30° \cdot j + \cos 30° \cdot k ,$$

$$j' = i + 0 \cdot j + 0 \cdot k ,$$
$$k' = 0 \cdot i + \cos 30° \cdot j - \sin 30° \cdot k .$$

Suppose

$$T = \cos\alpha \cdot i + \cos\beta \cdot j + \cos\gamma \cdot k .$$

As an axis of rotation , the characteristic vector T should satisfy the following two equations :

$$T \cdot i' = T \cdot i ,$$
$$T \cdot j' = T \cdot j .$$

Thus ,

$$\sin 30° \cos\beta + \cos 30° \cos\gamma = \cos\alpha , \qquad (5.214)$$

$$\cos\alpha = \cos\beta . \qquad (5.215)$$

Besides ,

$$\cos^2\alpha + \cos^2\beta + \cos^2\gamma = 1 . \qquad (5.216)$$

Combining Eqs . (5.214) through (5.216) gives

$$\cos\alpha = \pm\sqrt{\frac{3}{7}} .$$

Correspondingly , we have two solutions of T :

$$T_1 = \sqrt{\frac{3}{7}}\,i + \sqrt{\frac{3}{7}}\,j + \sqrt{\frac{1}{7}}\,k ,$$

$$T_2 = -\sqrt{\frac{3}{7}}\,i - \sqrt{\frac{3}{7}}\,j - \sqrt{\frac{1}{7}}\,k .$$

(4) 2φ

Rewrite Eq . (2.41) :

$$\cos 2\varphi = \frac{(T \times j') \cdot (T \times j)}{|(T \times j)|^2} .$$

Thus ,

$$\cos 2\varphi = \frac{-\cos^2\beta}{\cos^2\beta + \cos^2\gamma} .$$

Putting the known values of $\cos\beta$ and $\cos\gamma$ into the above equation yields

$$\cos 2\varphi = -0.75 .$$

Then ,

$$2\varphi = 138°36' \quad \text{or} \quad 221°24' .$$

According to the fact that the coordinate axes xyz are rotated through an angle 2φ about T to new directions $x'y'z'$, we can easily see that T_1 corresponds to $2\varphi_1 = 221°24'$, while T_2 to $2\varphi_2 = 138°36'$.

(5) r_0

Replacing the parameters S , P and θ in Eq . (1.40) with the parameters $\overrightarrow{M_0 M'}$, T and 2φ , respectively , yields

$$r_0 = \frac{1}{2}\left[\,(\overrightarrow{M_0M'} \cdot T)T - \overrightarrow{M_0M'}\,\right] + \frac{1}{2}\left(\frac{\sin 2\varphi}{\cos 2\varphi - 1}\right)(\overrightarrow{M_0M'} \times T).$$

From the preceding steps, we have obtained

$$T = 0.655\,i' + 0.655\,j' + 0.378\,k',$$

$$2\varphi = 221^\circ\,24',$$

$$\overrightarrow{M_0M'} = 1.366D \cdot i' + \left(0.5D - 2.366\frac{D}{n}\right)\cdot j' + 0.866D \cdot k'.$$

Putting these data into the formula of r_0 gives

$$r_0 = \left(-0.247D - 0.677\frac{D}{n}\right)\cdot i' + \left(0.267D + 0.675\frac{D}{n}\right)\cdot j'.$$

(6) $2d$

Similarly, from Eq. (1.42), we have

$$2d = \overrightarrow{M_0M'} \cdot T.$$

By substituting, we obtain

$$2d = 1.550D - 1.550\frac{D}{n}.$$

(7) $S_{T.2\varphi}$, $J_{\mu P}$, J_{SD} and J_{SP}

Now, by putting the obtained quantities T, 2φ, $\overrightarrow{M_0M'}$ and t into Eqs. (1.12), (4.24), (4.25) and (4.26), and considering $m = \overrightarrow{M_0M'}$, we have

$$S_{T.2\varphi} = \begin{pmatrix} 0 & 1 & 0 \\ 0.5 & 0 & 0.866 \\ 0.866 & 0 & -0.5 \end{pmatrix},$$

$$J_{\mu p} = \begin{pmatrix} 1 & -1 & 0 \\ -0.5 & 1 & -0.866 \\ -0.866 & 0 & 1.5 \end{pmatrix},$$

$$J_{SD} = \begin{pmatrix} 1 & -1 & 0 \\ -0.5 & 1 & -0.866 \\ -0.866 & 0 & 1.5 \end{pmatrix},$$

$$
J_{SP} = \begin{pmatrix}
-z_q' - 0.866D & -z_q' & x_q' + y_q' + 1.366D \\[2ex]
\begin{aligned}0.866y_q' + z_q' - 0.866b' \\ +0.433D - 2.049\,\dfrac{D}{n}\end{aligned} & \begin{aligned}-0.866x_q' \\ +0.5z_q' - 0.75D\end{aligned} & \begin{aligned}-x_q' - 0.5y_q' + 1.5b' \\ -0.25D + 1.183\,\dfrac{D}{n}\end{aligned} \\[3ex]
\begin{aligned}-1.5y_q' + 0.5b' \\ -0.25D + 1.183\,\dfrac{D}{n}\end{aligned} & \begin{aligned}1.5x_q' + 0.866z_q' \\ -b' + 1.433D\end{aligned} & \begin{aligned}-0.866y_q' + 0.866b' \\ -0.433D + 2.049\,\dfrac{D}{n}\end{aligned}
\end{pmatrix}
$$

(8) u, v, w, δ_u, δ_v and δ_w

From Eq. (5.39),

$$
\begin{pmatrix} \vec{\delta_u} \\ \vec{\delta_v} \\ \vec{\delta_w} \end{pmatrix} = \Delta\theta\, J_{\mu P} \begin{pmatrix} i' \\ j' \\ k' \end{pmatrix} = \Delta\theta \begin{pmatrix} 1 & -1 & 0 \\ -0.5 & 1 & -0.866 \\ -0.866 & 0 & 1.5 \end{pmatrix} \begin{pmatrix} i' \\ j' \\ k' \end{pmatrix}.
$$

Thus,

$$
\delta_u = \sqrt{(1)^2 + (-1)^2 + (0)^2} \cdot \Delta\theta = 1.414\,\Delta\theta,
$$

$$
\delta_v = \sqrt{(-0.5)^2 + (1)^2 + (-0.866)^2} \cdot \Delta\theta = 1.414\,\Delta\theta,
$$

$$
\delta_w = \sqrt{(-0.866)^2 + (0)^2 + (1.5)^2} \cdot \Delta\theta = 1.732\,\Delta\theta.
$$

Then, dividing the three rows of elements of the matrix $J_{\mu P}$ by the coefficients 1.414, 1.414, 1.732 of δ_u, δ_v, δ_w, respectively, yields

$$
\begin{pmatrix} u \\ v \\ w \end{pmatrix} = \begin{pmatrix} 0.707 & -0.707 & 0 \\ -0.354 & 0.707 & -0.612 \\ -0.5 & 0 & 0.866 \end{pmatrix} \begin{pmatrix} i' \\ j' \\ k' \end{pmatrix}.
$$

(9) a, b, c, δ_a, δ_b and δ_c

Similarly, from Eq. (5.54),

$$
\begin{pmatrix} \vec{\delta_a} \\ \vec{\delta_b} \\ \vec{\delta_c} \end{pmatrix} = \Delta g\, J_{SD} \begin{pmatrix} i' \\ j' \\ k' \end{pmatrix}.
$$

Thus,

$$
\delta_a = 1.414\,\Delta g,
$$

$$
\delta_b = 1.414\,\Delta g,
$$

$$
\delta_c = 1.732\,\Delta g.
$$

Then,

194

$$\begin{pmatrix} a \\ b \\ c \end{pmatrix} = \begin{pmatrix} 0.707 & -0.707 & 0 \\ -0.354 & 0.707 & -0.612 \\ -0.5 & 0 & 0.866 \end{pmatrix} \begin{pmatrix} i' \\ j' \\ k' \end{pmatrix}.$$

$(10)\,\Delta\mu'_x,\,\Delta\mu'_y,\,\Delta\mu'_z,\,\Delta S'_{F'x'},\,\Delta S'_{F'y'}$ and $\Delta S'_{F'z'}$

Rewrite Eqs. (4.16) and (4.17)

$$(\Delta\vec{\mu}') = \Delta\theta\, J_{\mu p}\,(\boldsymbol{P}), \tag{4.16}$$

$$(\Delta S'_F) = \Delta\theta\, J_{SP}\,(\boldsymbol{P}) + \Delta g\, J_{SD}\,(\boldsymbol{D}). \tag{4.17}$$

The subscript 't' of $\Delta S'_{F't}$ is omitted since the meaning is clear, and this will also be done in the Appendix.

Substituting the known matrices $J_{\mu p}$, J_{SD} and J_{SP} into Eqs. (4.16) and (4.17), we find

$$\Delta\mu'_x = \Delta\theta\,(P_{x'} - P_{y'}),$$

$$\Delta\mu'_y = \Delta\theta\,(-0.5P_{x'} + P_{y'} - 0.866P_{z'}),$$

$$\Delta\mu'_z = \Delta\theta\,(-0.866P_{x'} + 1.5P_{z'}),$$

$$\Delta S'_{F'x'} = \Delta\theta[\,P_{x'}(-z'_q - 0.866D) - P_{y'}z'_q + P_{z'}(x'_q + y'_q + 1.366D)] + \Delta g\,(D_{x'} - D_{y'}),$$

$$\Delta S'_{F'y'} = \Delta\theta\left[P_{x'}\left(0.866y'_q + z'_q - 0.866b' + 0.433D - 2.049\,\frac{D}{n}\right) + P_{y'}\,(-0.866x'_q\right.$$

$$\left. + 0.5z'_q - 0.75D) + P_{z'}\left(-x'_q - 0.5y'_q + 1.5b' - 0.25D + 1.183\,\frac{D}{n}\right)\right]$$

$$+ \Delta g\,(-0.5D_{x'} + D_{y'} - 0.866D_{z'}),$$

$$\Delta S'_{F'z'} = \Delta\theta\left[P_{x'}\left(-1.5y'_q + 0.5b' - 0.25D + 1.183\,\frac{D}{n}\right) + P_{y'}\,(1.5x'_q + 0.866z'_q\right.$$

$$\left. - b' + 1.433D) + P_{z'}\left(-0.866y'_q + 0.866b' - 0.433D + 2.049\,\frac{D}{n}\right)\right]$$

$$+ \Delta g\,(-0.866D_{x'} + 1.5D_{z'}).$$

5.10 Formulas of Transformation — Theorem of Exchange of Object-Image Spaces for Reflecting Prisms

All the equations in the tables of reflecting prisms are given in accordance with such a direction of the ray path that the x' axis in the adjustment diagram of the prism will indicate the direction of the exit optical axis. For the convenience of the following discussion, presume that the direction of ray path indicated in the adjustment diagram for a prism in the table refers to the forward direction while the opposite the reverse direction.

Prisms may sometimes happen to be used with a reverse direction of the ray path . Then a problem will arise by asking whether we need to derive new equations or the original equations can still be available by taking some measures to adapt them .

To answer this question , let us try to find some formulas of transformation .

Referring to Fig . 5 . 52 , the left part (a) shows the usage of a prism with a forward ray path while the right part (b) the usage of the prism with a reverse ray path . Let F and F' be a pair of conjugate points belonging correspondingly to a pair of conjugate bodies . Again assume $(-1)^t r$ and r' to be a pair of conjugate directions (unit vectors) .

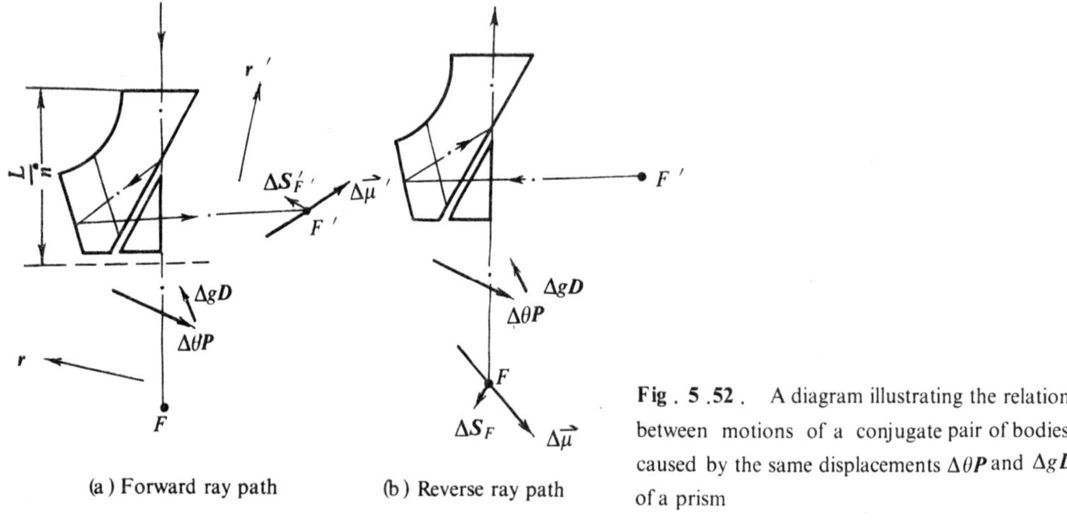

Fig . 5 .52 . A diagram illustrating the relation between motions of a conjugate pair of bodies caused by the same displacements $\Delta\theta P$ and $\Delta g D$ of a prism

In case the ray path is in the forward direction , the image motion caused by a small angular displacement $\Delta\theta P$ plus a small linear displacement $\Delta g D$ with the object (F) fixed is represented by $\Delta S_F'$ and $\Delta\vec{\mu}'$ whereas in case the ray path is in the reverse direction the image motion caused by the same displacements $\Delta\theta P$ plus $\Delta g D$ with the object (F') fixed is represented by ΔS_F and $\Delta\vec{\mu}$.

Consider the case of the forward ray path , see Fig . 5 . 52 (a) . Having completed the motion of $\Delta\theta P$ and $\Delta g D$, the prism becomes fixed , then the object (F) is given a motion of $\Delta S_{F,1}$ and $\Delta\vec{\mu}_1$ so as to give the image (F') a motion of $-\Delta S_F'$ and $-\Delta\vec{\mu}'$. Thus , the image (F') will be returned to its original position as if it had been kept fixed all the time. Now, comparing this final state of the left-side figure (a) with the state of the right-side figure (b), we will readily arrive at a conclusion that $\Delta S_{F,1}$ and $\Delta\vec{\mu}_1$ will be equal to ΔS_F and $\Delta\vec{\mu}$, respectively, or in other words, $\Delta\vec{\mu}$ $(\Delta\vec{\mu}_1)$ is conjugate to $(-1)^{t-1}\Delta\vec{\mu}'$ while ΔS_F $(\Delta S_{F,1})$ is conjugate to $-\Delta S_F'$ no matter whether t is an even or an odd number. Since r is conjugate to $(-1)^t r'$, we have finally

$$\Delta\mu_r = -\Delta\mu'_{r'} , \tag{5.217}$$

$$\Delta S_{Fr} = (-1)^{t-1}\Delta S'_{F'r'} . \tag{5.218}$$

196

These are the so- called *formulas of transformation* ,with which each scalar quantity of image motion for a prism used in the forward direction can be converted into the corresponding quantity of image motion for the same prism used in the reverse direction .

To emphasize the above result , we shall formulate it as a theorem .

Theorem of exchange of object- image spaces for reflecting prisms :

"Let points F and F' represent a conjugate pair of bodies for a reflecting prism in the sense of a limit * , and assume that each of these two conjugate bodies will in turn be taken as the image while the same prism is applied one time in the forward direction of the ray path and the other time in the reverse direction of the ray path. Then, the two image- motions ΔS_F plus $\Delta \vec{\mu}$ and $\Delta S_F'$ plus $\Delta \vec{\mu}'$, caused by the same displacement $\Delta \theta P$ plus $\Delta g D$ of the prism with the object fixed , will be related to each other in such a manner that $\Delta \vec{\mu}$ and ΔS_F (or $\Delta \vec{\mu}'$ and $\Delta S_F'$) will be conjugate to $(-1)^{t-1} \Delta \vec{\mu}'$ and $-\Delta S_F'$ (or $(-1)^{t-1} \Delta \vec{\mu}$ and $-\Delta S_F$) , respectively , for the prism , where t represents the number of reflections , and moreover the projections of $\Delta \vec{\mu}$, ΔS_F and $\Delta \vec{\mu}'$, $\Delta S_F'$ on the direction of two unit vectors r and r' , respectively , will be subject to two formulas of transformation : $\Delta \mu_r = -\Delta \mu_{r'}'$; $\Delta S_{Fr} = (-1)^{t-1} \Delta S_{Fr'}'$, if $(-1)^t r$ and r' are assumed to be a conjugate pair of unit vectors located in spaces containing F and F' , respectively ."

Example

In Fig .5 . 53 , a prism FP- 90 $^\circ$ is assumed to be located in a convergent beam . Determine the image lean $\Delta \mu_{x_1}'$ and two components $\Delta S_{Fy_1}'$ and $\Delta S_{Fz_1}'$ of the image point (F) displacement caused by a small angular displacement $\Delta \theta P$ plus a small linear displacement $\Delta g D$ of the prism with the object fixed , given $D = 20 ; n = 1 . 5163 ; t = 3 ; b = 30 ; \Delta \theta = 30' ; \Delta g = 0 . 1$; P and D are both parallel to the x_1 direction , and the action line of P passes through the center q , as shown in the diagram .

Solution

For convenience , the adjustment diagram of the prism FP - 90° taken from the tables of reflecting prisms in the Appendix is indicated in Fig .5 . 54 . Then , the given axes and points P , q , F and D are located and orientated in relation to the adjustment diagram .

Comparing Fig .5 . 53 with Fig .5 . 54 , we see that the prism illutrated in this example is applied in the reverse direction of ray path .

Now , consider F as an object point for the prism in the adjustment diagram , and then we can determine the location b' of its image point F' relative to the exit facc of the prism by the following equation :

$$b' = -\left(\frac{L}{n} + b\right) = -\left(\frac{3D}{n} + b\right) .$$

---

* An image body will become an image point as its volume approaches zero as a limit .

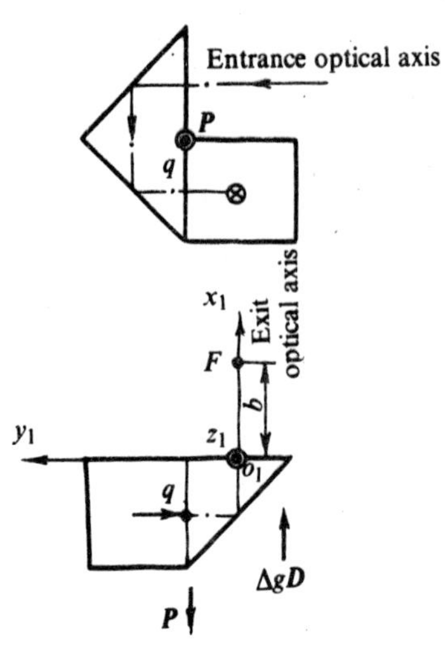

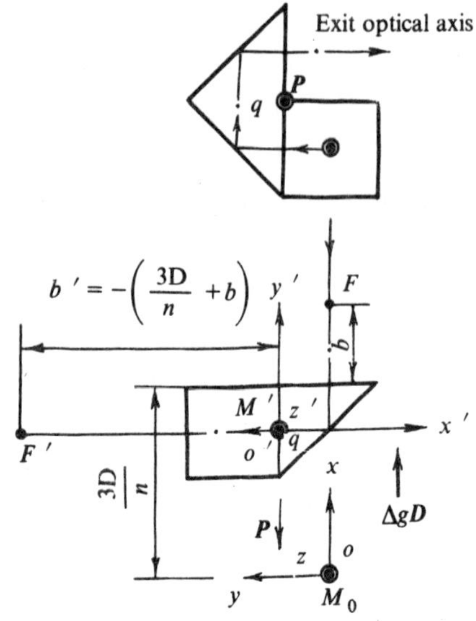

Fig . 5 . 53 . A prism FP - 90 ° in
the reverse direction of ray path

Fig . 5 . 54 . A prism FP - 90 ° in
the forward direction of ray path

· First , we need to find the x' image rotation $\Delta\mu_{x'}'$ and the two components $\Delta S_{F'y'}'$ and $\Delta S_{F'z'}'$ of the image point (F') displacement caused by the given motion of the prism shown in the adjustment diagram (Fig . 5 . 54).

From Fig . 5 . 54 ,

$$P_{x'} = 0 , \qquad P_{y'} = -1 , \qquad P_{z'} = 0 ,$$

$$x_q' = 0 , \qquad y_q' = 0 , \qquad z_q' = -0.5D = -10 ,$$

$$D_{x'} = 0 , \qquad D_{y'} = 1 , \qquad D_{z'} = 0 ,$$

$$b' = -\left(\frac{3D}{n} + b\right) = -\left(\frac{60}{1.5163} + 30\right) = -69.57 .$$

Then , putting these known data into the relevant equations , and considering $\Delta\theta = 30\,'$ and $\Delta g = 0.1$ yields

$$\Delta S_{F'y'}' = \Delta\theta\left[P_{x'} z_q' + P_{y'}(z_q' + D) \right.$$
$$\left. - P_{z'}\left(x_q' + y_q' - 0.5D + 3\frac{D}{n}\right) \right] + \Delta g\,(-D_{x'} + D_{y'})$$
$$= \frac{30}{60}\,\frac{\pi}{180}\,(-1)(-0.5\times 20 + 20) + 0.1 = 0.01 ,$$

$$\Delta S_{F'z'}' = \Delta\theta\left[P_{x'}\left(-2y_q' - b' + 0.5D - 3\frac{D}{n}\right) \right.$$
$$\left. + P_{y'}(2x_q' - b' - 0.5D) \right] + 2\Delta g\,D_{z'}$$

$$= \frac{30}{60} \frac{\pi}{180} [(-1)(69.57) + 0.5 \times 20] = -0.52 ,$$

$$\Delta \mu_{x'}' = \Delta \theta (P_{x'} - P_{y'})$$

$$= 30'(0+1) = 30' .$$

According to Eqs. (5.217) and (5.218), if the prism in Fig. 5.54 is used in the reverse direction, the x image rotation $\Delta \mu_x$ and the two components ΔS_{Fy} and ΔS_{Fz} of the image point (F) displacement will be equal to

$$\Delta S_{Fy} = (-1)^{t-1} \Delta S_{F'y'}' = \Delta S_{F'y'}' = 0.01 ,$$

$$\Delta S_{Fz} = (-1)^{t-1} \Delta S_{F'z'}' = \Delta S_{F'z'}' = -0.52 ,$$

$$\Delta \mu_x = -\Delta \mu_{x'}' = -30' .$$

Finally, by comparing the two coordinate systems $x_1 y_1 z_1$ and $x y z$, we obtain

$$\Delta S_{Fy_1}' = \Delta S_{Fy} = 0.01 ,$$

$$\Delta S_{Fz_1}' = \Delta S_{Fz} = -0.52 ,$$

$$\Delta \mu_{x_1}' = \Delta \mu_x = -30' .$$

To verify the validity of the above results, we give the equations derived specially for the case shown in Fig. 5.53

$$\Delta S_{Fy_1}' = \Delta \theta \left[P_{x1} z_{1q} + P_{y1} (-z_{1q} + D) \right.$$

$$\left. + P_{z1} \left(-x_{1q} + y_{1q} - 0.5D - 3 \frac{D}{n} \right) \right] + \Delta g (D_{x1} - D_{y1}) , \tag{5.219}$$

$$\Delta S_{Fz_1}' = \Delta \theta \left[P_{x1} \left(-2 y_{1q} + b + 0.5D + 3 \frac{D}{n} \right) \right.$$

$$\left. + P_{y1} (2 x_{1q} - b + 0.5D) \right] + 2\Delta g D_{z1} , \tag{5.220}$$

$$\Delta \mu_{x_1}' = \Delta \theta (P_{x1} + P_{y1}) . \tag{5.221}$$

Determine the direction cosines of P and D as well as the coordinates of q and F relative to the coordinate system $o_1 x_1 y_1 z_1$

$$P_{x1} = -1 , \qquad P_{y1} = 0 , \qquad P_{z1} = 0 ,$$

$$x_{1q} = -0.5D = -10 , \quad y_{1q} = 0.5D = 10 , \quad z_{1q} = 0.5D = 10 ,$$

$$D_{x1} = 1 , \qquad D_{y1} = 0 , \qquad D_{z1} = 0 ,$$

$$b = 30 .$$

Then , putting these data into Eqs. (5.219) through (5.221) and considering $\Delta\theta = 30\,'$ and $\Delta g = 0.1$ yields

$$\Delta S'_{Fy_1} = \frac{30}{60}\;\frac{\pi}{180}\;[\,(-1)\,(10)\,] + 0.1 = 0.01\,,$$

$$\Delta S'_{Fz_1} = \frac{30}{60}\;\frac{\pi}{180}\;[\,(-1)(-2)(10) - 10 - 69.57] = -0.52\,,$$

$$\Delta\mu'_{x_1} = 30\,'\,(-1) = -30\,'\,,$$

which are exactly the same as the previously obtained results.

Problems

5.1   Redo the example in Sect. 5.3 using first the principle for the finite rotation of mirror systems, and then ignoring the small quantities of second and higher powers, and make comments on those relevant methods.

5.2   Determine the three components of image rotation $\Delta\vec{\mu}'$ due to a small angular displacement $\Delta\theta P$ of a spatial reflecting prism shown in Fig. 5.55 where $x\,y\,z$ and $x'y'z'$ represent a conjugate pair of coordinate systems. Given that the axis of rotation P is directed along the z' axis and parallel to the plane of the diagram.

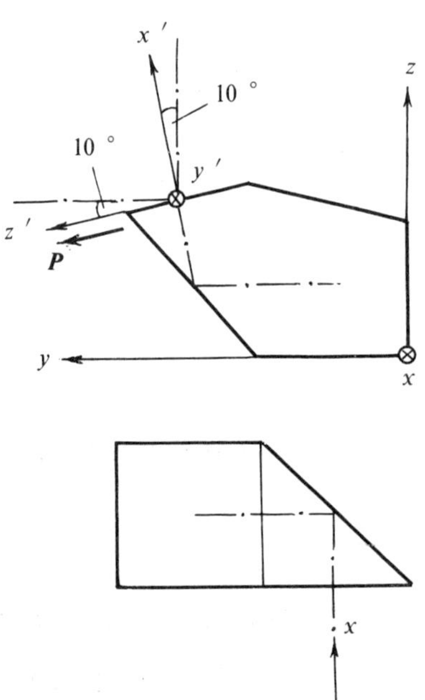

Fig. 5.55. A diagram to illustrate a spatial reflecting prism rotating through a small angle $\Delta\theta$ around P

5.3 Confirm the solutions given in Eqs. (5.15) and (5.16) by calculation.

5.4 Make a complete table for a Penta prism used in the forward direction of the ray path and check it with that presented in the Appendix.

5.5 Generally speaking spatial reflecting prisms do not even have one planar three-dimensional zero-valued pole, but exceptionally the spatial reflecting prisms with an odd number of reflections do have planar three-dimensional zero-valued poles *for a certain fixed image point* F_c' lying on their exit optical axes. Can you tell why and how to locate F_c'?

Hint : This special image point F_c' is the point of intersection of the exit optical axis of that prism with the plane passing through its center of inversion I and perpendicular to its screw axis of image formation $\vec{\lambda}$.

Chapter 6
Image-Stabilizing Reflecting Prisms

Very often, optical instruments are mounted on unstable platforms.

For an optical viewing device used in this situation, the apparent image appears to be oscillating and blurred. Measures should be taken to provide image compensation which maintains a steady image. This presents a very clear concept of what the term *image stabilization* directly means.

However, for an optical measuring (angular measurements)device used under this condition, the word *image stabilization* implies that the instrument is autocontrolled to ensure a stabilized reference frame relative to which the angles are measured.

There are two forms of image stabilization: (i)platform stabilization and (ii)subsystem stabilization. We will discuss the second class where the image, or sometimes only the line of sight, is indirectly controlled by moving only a portion of the optical system. The components of this portion are called *compensating elements*.

Lenses, mirrors, prisms, optical wedges, and plane-parallel plates can all serve as such compensation elements, but our investigation will be directed to mirrors and reflecting prisms only.

The sensor used for detecting an angular perturbation may be one of the following types: a gyroscope, pendulum, level bubble, and a combination of them. The selection of a proper approach depends mainly on the nature of the vibration, the object to be stabilized, and the desired accuracy of stabilization.

It is worthwhile to point out that the problem regarding image stabilizing prisms is essentially the self-adjustment of prisms.

6.1 Relative Image Stabilization and Absolute Image Stabilization

A telescope shown in Fig. 6.1 has rotated from the initial position 1 to its final position 2 due to a small angular perturbation $\Delta \varepsilon \, P$ of the platform on which the telescope rests.

In this case, as a rule, assume one fixed coordinate system $x_1 \, y_1 \, z_1$ and another moving coordinate system $x_2 \, y_2 \, z_2$, the latter being rigidly connected to the telescope.

If the image is stabilized relative to the moving coordinate system $x_2 \, y_2 \, z_2$, namely a target in object space will be viewed relatively stationary through the vibrated telescope, then this type

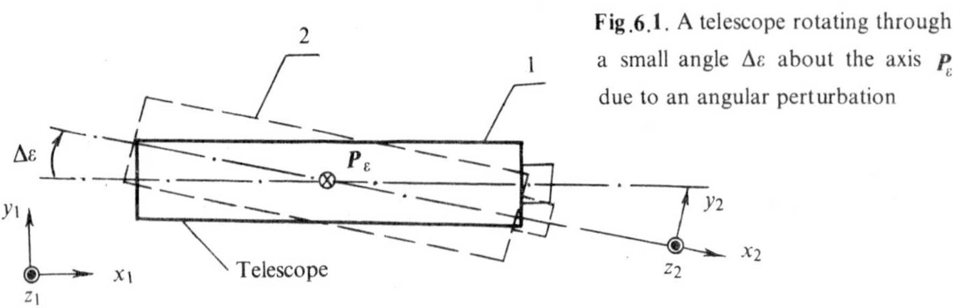

Fig.6.1. A telescope rotating through a small angle $\Delta\varepsilon$ about the axis P_ε due to an angular perturbation

of stabilization method is called the *relative image stabilization*.

Correspondingly, if the image is stabilized relative to the fixed coordinate system $x_1y_1z_1$, then the type of stabilization method is called the *absolute image stabilization*.

As a matter of fact, no coordinate system in space is absolutely fixed. A so called fixed coordinate system can also be referred to as a special case of a moving coordinate system; thus, there is no material difference between these two types of image stabilization.

Therefore, we will discuss the principles of relative image stabilization only.

6.2 Problems in a Collimated Beam

6.2.1 Degrees of Freedom of Image Rotation Due to a Small Angular Displacement of a Reflecting Prism

There might be some confusion in distinguishing the degrees of freedom of the image rotation $\Delta\vec{\mu}'$ due to a small angular displacement $\Delta\theta P$ of a reflecting prism from the degrees of freedom of the angular displacement of the prism $\Delta\theta P$ itself.

No doubt, an arbitrary small rotation $\Delta\theta P$ of a prism has three degrees of freedom, but one may thus make a mistake by simply saying that a small image rotation $\Delta\vec{\mu}'$ caused by an arbitrary small angular displacement $\Delta\theta P$ of a prism also has three degrees of freedom.

The reason for this will be seen from Fig.6.2 , where a unit vector P of the axis of rotation of a prism is resolved into components along and perpendicular to the characteristic vector T of the prism, getting $P_{//}$ and $P_\perp$.

Everyone knows that the component $\Delta\theta P_{//}$ parallel to T will cause no image rotation at all, and only the component $\Delta\theta P_\perp$ perpendicular to T is effective. Thus, we arrive at the following conclusion : although the mechanical motion of a small angular displacement $\Delta\theta P$ of a prism itself has three degrees of freedom , yet the small image rotation $\Delta\vec{\mu}'$ caused by such a small rotation of the prism has two degrees of freedom only.

If you are still not yet convinced of this point, then we can prove it directly, using the formula of image rotation $\Delta\vec{\mu}'$:

$$(\Delta\vec{\mu}') = \Delta\theta \ (E - S_{T.2\varphi}) \ (P) \tag{4.9}$$

204

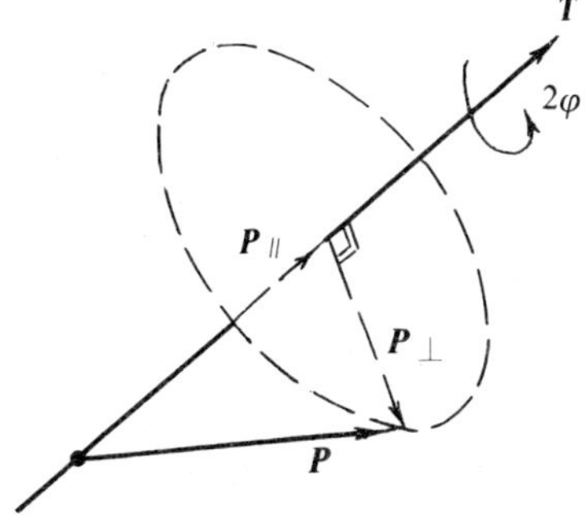

Fig .6 .2 . A resolution of a unit vector P of an axis of rotation into components along and perpendicular to the characteristic vector T

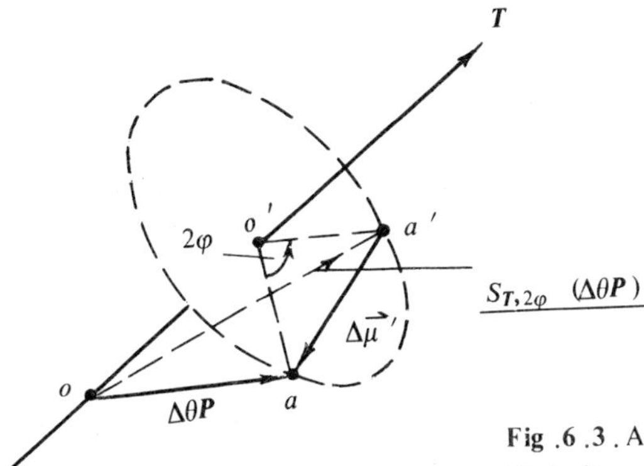

Fig .6 .3 . A diagram illustrating another kind of two-vector subtraction rule

or

$$(\Delta\vec{\mu}') = \Delta\theta\, P - S_{T.2\varphi}\,(\Delta\theta P) .\qquad(6 .1)$$

The second term $S_{T.2\varphi}(\Delta\theta P)$ on the right side of Eq .(6.1)may be thought to be a vector obtained by turning the vector $\Delta\theta P$ about the axis T through an angle 2φ. Thus, the equation (6.1) shows that the image rotation $\Delta\vec{\mu}'$ can also be calculated according to a new kind of two-vector subtraction rule, as illustrated in Fig. 6.3, which leads to the result that the vector of image rotation $\Delta\vec{\mu}'$ caused by a small angular displacement $\Delta\theta P$ of a reflecting prism will always be perpendicular to the characteristic vector T, no matter what rotation axis P as well as what number of reflections t might be.

So, we have arrived at the same conclusion as we did previously.

A stabilized prism used to introduce an image rotation $\Delta\vec{\mu}'$ of two degrees of freedom for compensation should be gimbaled for rotation about two properly disposed axes.

Very often, a reflecting prism rotates around a fixed axis, and thus can offer an angular compensation of only one degree of freedom.

On the other hand, a complete solution of the problem for image stabilization in a collimated beam needs an angular compensation of three degrees of freedom, so a prismatic stabilizing system intended for a complete solution should include two or three correction prisms.

However, for some optical sighting and viewing devices, the image stabilization problem is restricted to rotations around two axes perpendicular to the line of sight since the image rotation around the line of sight is not important in terms of retaining a correct direction of the line of sight only. The image stabilization of this kind is accordingly called the *LOS stabilization*.

Obviously, a LOS stabilization problem needs an angular compensation of two degrees of freedom, and one or two correction prisms are necessary.

6.2.2 Formula for Image Stabilization with Small Angular Perturbation $\Delta\vec{\varepsilon}=\Delta\varepsilon\,\boldsymbol{P}_\varepsilon$

A part of a telescopic system is shown in Fig. 6.4, where the elements 1 and 2 represent two correction prisms located in a collimated beam before an objective lens. The preceding system may also be a prism assembly or a lens group; however, this part is rigidly connected to the telescope housing no matter for what purpose it is intended.

Now, suppose the whole telescope is rotated through a small angle $\Delta\varepsilon$ about an arbitrary axis $\boldsymbol{P}_\varepsilon$ in space due to an unavoidable vibration imparted to the platform on which the telescope housing is mounted.

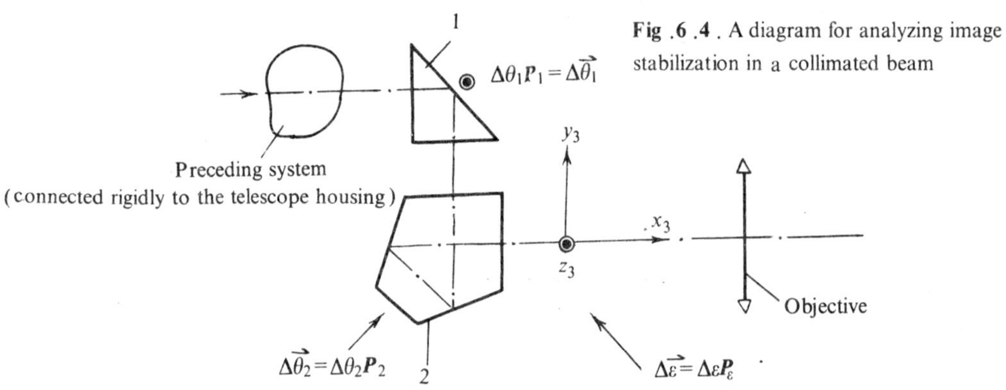

Fig. 6.4. A diagram for analyzing image stabilization in a collimated beam

To compensate the image motion caused by the angular perturbation $\Delta\vec{\varepsilon}=\Delta\varepsilon\boldsymbol{P}_\varepsilon$ mentioned above, prisms 1 and 2 are assumed to have completed small angular compensations $\Delta\theta_1\boldsymbol{P}_1$ and $\Delta\theta_2\boldsymbol{P}_2$, respectively, to change the direction of the light beams leaving the last compensation prism and travelling then towards the objective lens.

The problem will be discussed in such a manner as if the telescope housing is fixed in space; then the object space (or just the object) seems to experience a relatively small angular displacement $-\Delta\vec{\varepsilon}$; similarly all the small angular displacements $\Delta\theta_1\boldsymbol{P}_1$ and $\Delta\theta_2\boldsymbol{P}_2$ of the compensating prisms are measured relative to the fixed instrument.

For this purpose, assume $x_3\,y_3\,z_3$ to be a moving coordinate system connected rigidly to the telescope housing. According to our regulation, the x_3 axis is coincident with the exit optical axis of the last compensation prism, in this case, the prism 2.

206

In the following discussion, the moving coordinate system $x_3 y_3 z_3$ will serve as a unified reference frame all the time, and this will give us a very clearly mental picture particularly while we are deriving the formula for relative image stabilization.

Let $[R_p]_3$ denote the operator matrix of the preceding system, and $[R_1]_3$ and $[R_2]_3$ denote the reflection matrices of prisms 1 and 2, respectively. The operator matrix has a meaning similar to that of a reflecting matrix in the sense that they both modify the direction of an input vector, but the system corresponding to the former may not contain any reflecting surface.

The subscript '3' means that all the matrices are expressed relative to the coordinate system $x_3 y_3 z_3$.

Now, we proceed to handle the problem in the following two steps:

(1) The whole system, including two compensating prisms as well, has completed a small angular displacement $\Delta\vec{\varepsilon}$ due to an angular perturbation, and then relatively consider that the object in the object space of the whole system (or just of the preceding system) has completed a small angular displacement $-\Delta\vec{\varepsilon}$ with respect to the fixed instrument.

Now, consider the optical system of that part containing the preceding system, prism 1, and prism 2, and let $(-\Delta\vec{\varepsilon})_3$ and $(\Delta\vec{\mu}_d'')_3$ be a conjugate pair of small angular displacements for this optical system, expressed relative to the same coordinate system $x_3 y_3 z_3$.

Then, according to the object and image relationship, we have

$$(\Delta\vec{\mu}_d'')_3 = (-1)^{t_e} [R_2]_3 [R_1]_3 [R_p]_3 (-\Delta\vec{\varepsilon})_3 , \tag{6.2}$$

where t_e = even number when the system forms images of same handedness, and t_e = odd number when the system generates images of different handedness. If the system in question is prismatic, then t_e will be the total number of reflections.

The subscripts '3' and the angular brackets in Eq. (6.2) may be omitted since the meaning is clear. Thus,

$$(\Delta\vec{\mu}_d'') = (-1)^{t_e} R_2 R_1 R_p (-\Delta\vec{\varepsilon}) . \tag{6.3}$$

According to the physical meaning, $\Delta\vec{\mu}_d''$ represents the image rotation of disturbance occurring in the image space of the prism 2 due to an angular perturbation $\Delta\vec{\varepsilon}$ of the instrument if no compensation measure has been taken.

(2) Suppose that the compensating prisms start to work, and they are automatically controlled to provide small angular displacements $\Delta\vec{\theta}_1 = \Delta\theta_1 P_1$ and $\Delta\vec{\theta}_2 = \Delta\theta_2 P_2$ for image compensation, respectively. It should be kept in mind that $\Delta\vec{\theta}_1$ and $\Delta\vec{\theta}_2$ are all measured relative to the reference frame $x_3 y_3 z_3$.

Let $\Delta\vec{\mu}_{1,2}''$ and $\Delta\vec{\mu}_2''$ be the image rotations occurring in the image space of prism 2 due to $\Delta\vec{\theta}_1$ and $\Delta\vec{\theta}_2$, respectively, and $\Delta\vec{\mu}_c''$ be the total image rotation for compensation:

$$\Delta\vec{\mu}_c'' = \Delta\vec{\mu}_{1,2}'' + \Delta\vec{\mu}_2'' . \tag{6.4}$$

From Eqs (4.9) and (2.20),

$$(\Delta\vec{\mu}_2'') = [E + (-1)^{t_2-1} R_2] (\Delta\vec{\theta}_2) , \tag{6.5}$$

$$(\Delta\vec{\mu}_{1,2}'') = (-1)^{t_2} R_2 [E + (-1)^{t_1-1} R_1] (\Delta\vec{\theta}_1) , \tag{6.6}$$

where t_1 and t_2 represent the numbers of reflections of prisms 1 and 2, respectively:

Thus,

$$(\Delta\vec{\mu}_c'') = (-1)^{t_2}R_2[E+(-1)^{t_1-1}R_1]\,(\Delta\vec{\theta}_1) + [E+(-1)^{t_2-1}R_2]\,(\Delta\vec{\theta}_2)$$

or

$$(\Delta\vec{\mu}_c'') = \sum_{i=1}^{2} K_{i,2}[E+(-1)^{t_i-1}R_i]\,(\Delta\vec{\theta}_i). \tag{6.7}$$

At most, we may have three compensating prisms in one system. Then, Eqs. (6.3) and (6.7) become

$$(\Delta\vec{\mu}_d''') = (-1)^{t_c}R_3R_2R_1R_p\,(-\Delta\vec{\varepsilon}), \tag{6.8}$$

$$(\Delta\vec{\mu}_c''') = \sum_{i=1}^{3} K_{i,3}[E+(-1)^{t_i-1}R_i]\,(\Delta\vec{\theta}_i), \tag{6.9}$$

where $K_{i,3}$ represents the operator matrix of the system located between the image spaces of the i compensating prism and the last (third) compensating prism for transferring an angular vector, in which $K_{3,3}=1$.

For a complete solution of a problem of image stabilization, we have to meet the following condition:

$$(\Delta\vec{\mu}_c''') = -(\Delta\vec{\mu}_d'''). \tag{6.10}$$

Putting Eqs. (6.8) and (6.9) into Eq. (6.10) gives

$$\sum_{i}^{3} K_{i,3}[E+(-1)^{t_i-1}R_i]\,(\Delta\vec{\theta}_i) = -(-1)^{t_c}R_3R_2R_1R_p\,(-\Delta\vec{\varepsilon}). \tag{6.11}$$

While using two compensating prisms to solve an optical-axis stabilization problem, we have

$$\begin{pmatrix} 0 & 1 & 0 \\ 0 & 0 & 1 \end{pmatrix}\sum_{i}^{2} K_{i,2}[E+(-1)^{t_i-1}R_i]\,(\Delta\vec{\theta}_i) = -\begin{pmatrix} 0 & 1 & 0 \\ 0 & 0 & 1 \end{pmatrix}(-1)^{t_c}R_2R_1R_p\,(-\Delta\vec{\varepsilon}). \tag{6.12}$$

In the last case (see Fig. 6.4), if the axes of rotation P_1 and P_2 are both fixed, then in designing such a prismatic compensation system and arranging the axes of rotation, some requirements should be met as far as possible in order that the vectors of image rotation $\Delta\mu_{1,2}''$ and $\Delta\vec{\mu}_2''$ for the LOS stabilization be perpendicular to each other and both perpendicular to the exit optical axis x_3 of the prism 2 as well.

In the three-prism system, the total image rotation $\Delta\vec{\mu}_c''$ can be written

$$\Delta\vec{\mu}_c'' = \Delta\vec{\mu}_{1,2,3}''' + \Delta\vec{\mu}_{2,3}'' + \Delta\vec{\mu}_3''. \tag{6.13}$$

Similar requirements should be met as far as possible in order that the three vectors of image rotations $\Delta\vec{\mu}_{1,2,3}'''$, $\Delta\vec{\mu}_{2,3}''$, and $\Delta\vec{\mu}_3''$ be perpendicular to one another.

Consideration should also be given to the main function of these compensating prisms in the system. Conflicts may arise among those requirements and the system designer has to try to find the optimal result.

6.2.3 Formula for Image Stabilization with Finite Angular Perturbation

In some individual cases, the angular perturbation $\Delta\varepsilon$ and thus the angular compensation

$\Delta\theta$ may reach such visible magnitudes that their second and higher power terms could no longer be neglected. Then, we have to go back to the finite rotation problem.

Referring to the same figure 6.4, retain the original meanings of the axes P_ε, P_1 and P_2, and replace the small angles $\Delta\varepsilon$, $\Delta\theta_1$, and $\Delta\theta_2$ by finite angles ε, θ_1, and θ_2, respectively. The moving coordinate system $x_3 y_3 z_3$ will still be regarded as the reference frame all the time.

The problem is also to be handled in two steps but in a slightly different way :

(1) Let A be a certain object vector in the object space of the preceding system, and then find the corresponding image vector A'' formed by the system in front of the objective lens prior to rotations of the whole system as well as of any individual parts of it.

From the known formula,

$$(A'') = R_2 R_1 R_p (A),\tag{6.14}$$

where each matrix is expressed relative to the coordinate system $x_3 y_3 z_3$.

(2) Now, the motion is considered to be completed in two steps. First, assume that the telescope as a single unit has rotated about an axis P_ε through an angle ε due to an angular perturbation of the instrument and this process will be regarded as if the object has rotated about the same axis P_ε through an angle $-\varepsilon$ with respect to the 'fixed' telescope. Then, this is followed by compensation rotations of prisms 1 and 2 about axes P_1 and P_2 through angles θ_1 and θ_2, respectively. Here, we have the primary moving coordinate system $x_3 y_3 z_3$ connected to a telescope and the secondary moving coordinate systems attached to each prisms, and in this process the coordinate system of the telescope is regarded as the 'fixed' reference frame while the coordinate system of each prism the moving frame. Also, it should be noted that the angular compensation θ_1 and θ_2 are measured relative to the coordinate system $x_3 y_3 z_3$.

Let A_r'' be the image vector conjugate to the same object vector A formed by the same system at the end of the perturbation and compensation process.

From Eqs.(1.29) and (3.5), we obtain

$$(A_r'') = S_{P_2, \theta_2} R_2 S_{P_2, -\theta_2} S_{P_1, \theta_1} R_1 S_{P_1, -\theta_1} R_p S_{P_\varepsilon, -\varepsilon} (A).\tag{6.15}$$

To maintain the relative direction of the outgoing beam from the last compensation prism with respect to the moving reference frame, it is necessary that

$$(A_r'') = (A'').\tag{6.16}$$

Thus, putting Eqs. (6.14) and (6.15) into Eq. (6.16) yields

$$S_{P_2, \theta_2} R_2 S_{P_2, -\theta_2} S_{P_1, \theta_1} R_1 S_{P_1, -\theta_1} R_p S_{P_\varepsilon, -\varepsilon} (A) = R_2 R_1 R_p (A).\tag{6.17}$$

It should be noted that the input column vector (A) in the above equation (1.17) is not necessarily to be considered as an arbitrary one.

When we solve an optical-axis stabilization or a LOS stabilization problem, by assuming appropriate coordinate systems (see Sect.10.3) the vector A may be considered as being equal to

$$(A) = \begin{pmatrix} 1 \\ 0 \\ 0 \end{pmatrix}.$$

And at most, when we completely solve an image stabilization problem, by handling it in the same way the vector A in Eq.(1.17) will successively be assumed to be an axial one A_1 and a transversal one A_2 :

$$(A) = \begin{pmatrix} 1 & 0 \\ 0 & 1 \\ 0 & 0 \end{pmatrix}.$$

In most cases, a compensating prism acts simultaneously as a scanning element. Besides, the formula to be used in designing the steering and stabilizing prism assembly is closely related to the structure of mounting the prisms on the gimbal rings of the gyroscope as well as to the arrangement of gimbal axes relative to the housing of the instrument.

Therefore, the actual formula for image stabilization with finite angular perturbation could be much more complicate than Eq.(6.17). However, the material discussed in the present section 6.2.3 has provided a basic idea which will be combined with the principle to be developed in Chapter 10 to enable us to tackle the practical engineering problems in this regard.

Sometimes, a combination of the platform stabilization with the subsystem stabilization is used; then the former belongs to the primary stage of compensation while the latter the secondary stage of compensation. In this case, the angular perturbation left for the secondary stabilization becomes much smaller and we can thus employ the approximate equation (6.11).

6.3 Problems in a Convergent Beam

6.3.1 Type of Prisms to Be Used for Image Stabilization in a Convergent Beam

For the same reason mentioned before, the discussion of the image compensation problem will be restricted to a LOS stabilization only, and because of the limited space the stabilized prism is usually attached directly to the inner gimbal ring of a gyroscope. This is the case that we call *the inertial stabilization* in which the angular compensation prism remains stabilized with respect to the inertial space. Hence, the angular compensation $\Delta\vec{\theta}$ of the stabilized prism will always be equal to the angular perturbation $\Delta\vec{\varepsilon}$ in magnitude but opposite in direction, i.e., $\Delta\vec{\theta} = -\Delta\vec{\varepsilon}$, and the action line of $\Delta\vec{\theta}$ passes the center of the gyroscope. Therefore, reflecting prisms used in this situation should meet some requirements and belong to certain types of prisms.

This question will be illustrated by an example. An objective 1 of a photographic system is shown in Fig.6.5. The optical axis of the system is deflected through 90 °by the mirror 2 which serves also as a compensation element for stabilizing the optical axis.

Consider first the optical-axis stabilization around the z axis.

Draw for the mirror 2 a pair of completely conjugate coordinate systems $oxyz$ and $o'x'y'z'$ with two origins o and o' located at a pair of conjugate points F and F', respectively, where F is the posterior focus of the objective lens.

Referring to Fig.6.6, in the first step assume that the system including the mirror rotates

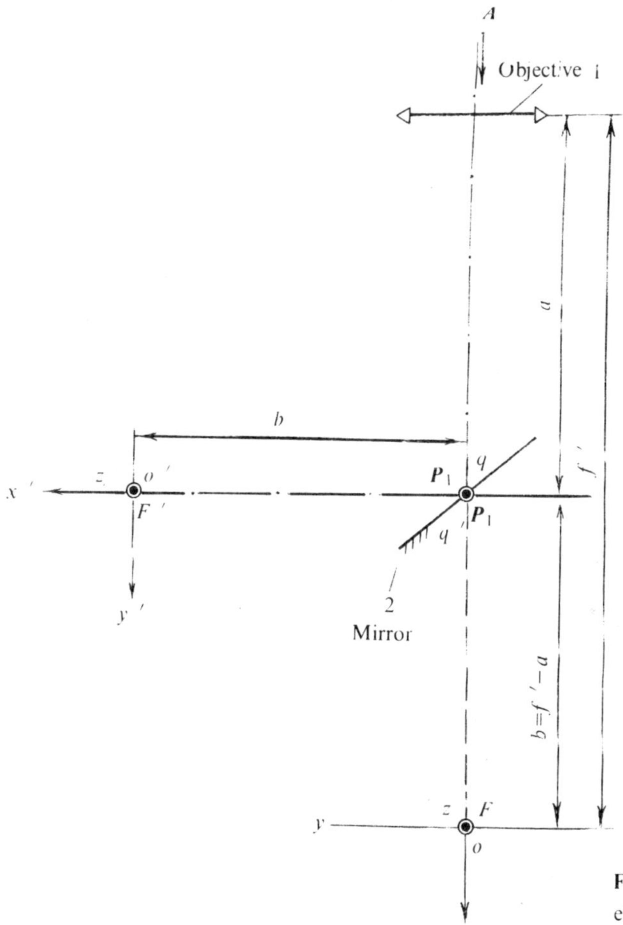

Fig .6 .5. A mirror used as a stabilized element in the convergent beam of a photographic system

as a single unit about the z axis through a small angle $\Delta\varepsilon$ due to an angular perturbation. Thus ,the vertically incident parallel beam A entering the objective lens at a small obliquity $\Delta\varepsilon$ relative to the optical axis in the plane of rotation will be brought to a focus at F_1' , resulting in an image point displacement of disturbance $\Delta S_{F'd}' = \overline{F'F_1'}$ whose conjugate in the object space of the mirror is $\overline{FF_1}$.

From Fig .6 .6 .

$$\Delta S_{F'd}' = \Delta\varepsilon f'. \tag{6.18}$$

Then , in the second step the mirror 2 is returned to its o iginal orientation by the gyroscope to introduce an image point displacement of compensation $\Delta S_{F'c}'$ so that the image point will be moved back from F_1' to F' (at least along the y' direction) .

Now let us determine $\Delta S_{F'c}'$. In the same figure 6 .6 ,assume that the mirror 2 has rotated through a small angle $\Delta\theta_1$ about the axis P_1 whose action line passes through the point of the optical axis corner q . Here , P_1 is parallel to the z axis ,and $\Delta\theta_1 = -\Delta\varepsilon$.

The component of the image point displacement along the y' axis $\Delta S_{p'y'c}'$ is calculated , using Eq .(5 .11) by letting $t = 1$

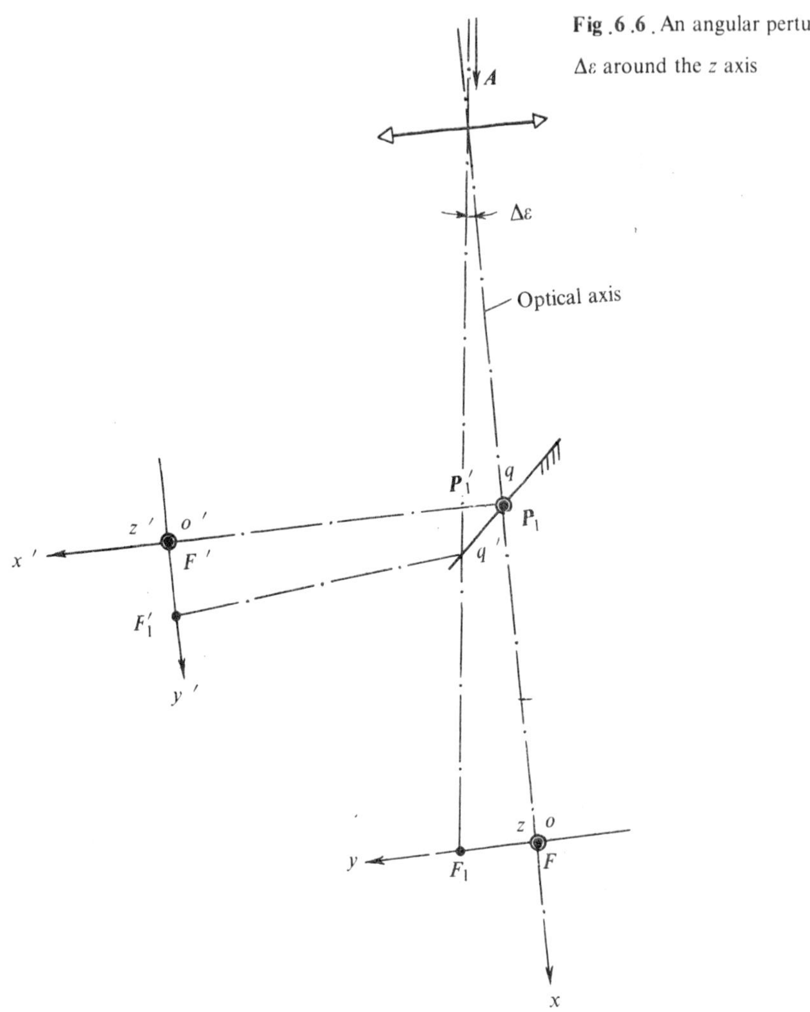

Fig.6.6. An angular perturbation $\Delta\varepsilon$ around the z axis

$$\Delta S_{F'_{y'c}} = -\Delta\varepsilon\,(P_{1z'}\rho_{F'x'} - P_{1x'}\rho_{F'z'}) - \Delta\varepsilon\,(P_{1z'}'\rho_{F'x'}' - P_{1x'}'\rho_{F'z'}') . \qquad (6.19)$$

From Fig.6.5,

$$P_{1x'} = P_{1x'}' = 0 , \qquad\qquad P_{1z'} = P_{1z'}' = 1 ,$$

$$\rho_{F'x'} = \rho_{F'x'}' = f'-a , \qquad \rho_{F'z'} = \rho_{F'z'}' = 0 .$$

Putting these data into Eq.(6.19) yields

$$\Delta S_{F'_{y'c}} = -2\Delta\varepsilon(f'-a) . \qquad\qquad (6.20)$$

On the other hand, for the implementation of the optical-axis stabilization, the following condition should be met :

$$\Delta S_{F'_{y'c}} = -\Delta S_{F'_d}' . \qquad\qquad (6.21)$$

Substituting Eqs. (6.18) and (6.20) into Eq. (6.21) gives

$$2\Delta\varepsilon(f'-a) = \Delta\varepsilon \cdot f' .$$

212

Thus ,

$$a = \frac{f'}{2} \; .$$ (6.22)

This is the condition equation that should be observed for the stabilization of the optical axis around the z axis . Thus , the mirror 2 must be placed at half the focal length f' .

Now , consider the optical-axis stabilization around the y axis .

Referring to Fig . 6 . 7 , in the first step assume that the whole system rotates about the y axis through a small angle Δv due to an angular perturbation . Thus , the vertically incident parallel beam A entering the objective lens at a small obliquity Δv relative to the optical axis in the plane of rotation will be focused on F_2' whose conjugate in the object space of the mirror is F_2 .

From Fig . 6 . 7 ,

$$\Delta S_{F'_2}' = - \Delta v \cdot f' \; .$$ (6.23)

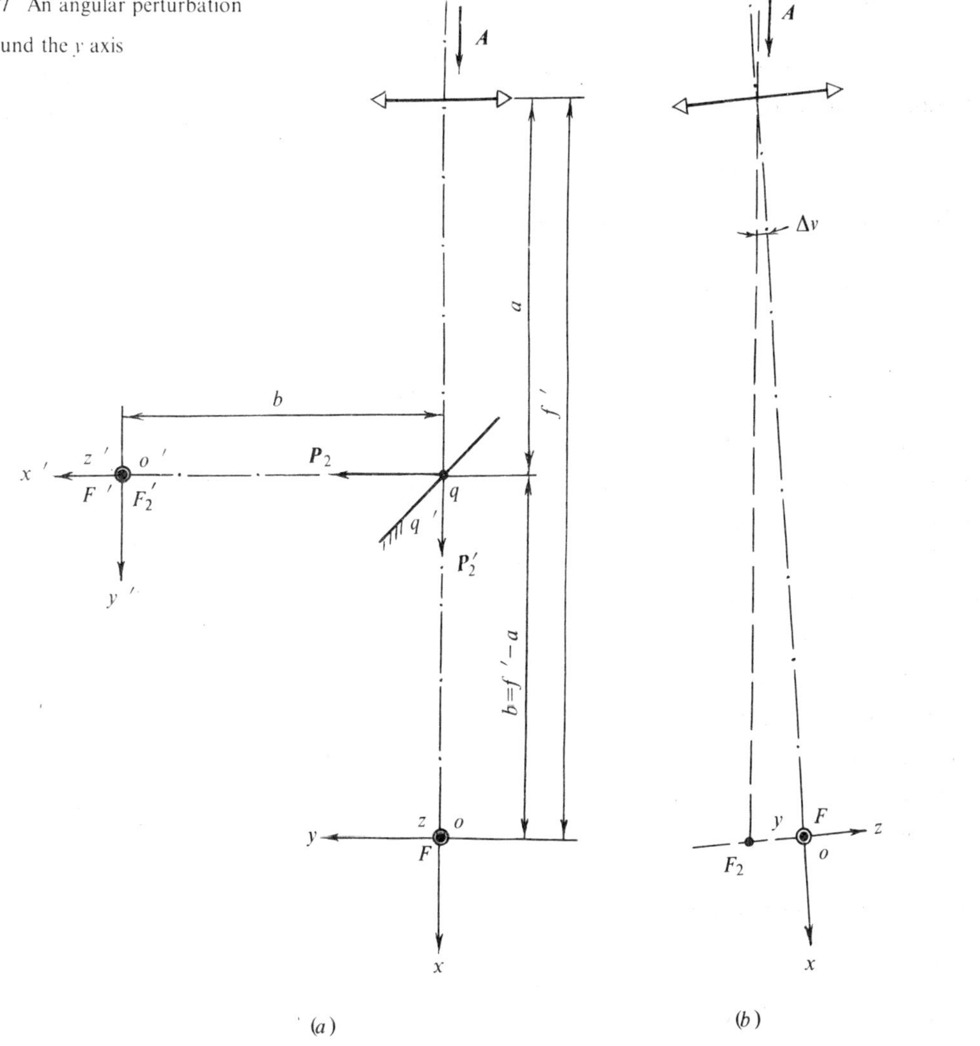

Fig . 6 . 7 An angular perturbation Δv around the y axis

Then ,in the second step the mirror 2 will rotate through an angle $-\Delta v$ about the axis P_2 parallel to the y axis and passing through the center q of the gimbal ,so as to introduce an image point displacement of compensation $\Delta S'_{F'c}$.

Similarly ,the component of the image point displacement along the z' axis $\Delta S'_{F'z'c}$ is calculated ,using Eq . (5.12) by letting $t=1$

$$\Delta S'_{F'y'c} = -\Delta v\ (P_{2x}\,'\rho_{F'y'} - P_{2y}\,'\rho_{F'x'}) - \Delta v\ (P'_{2x}\,'\rho'_{F'y'} - P'_{2y}\,'\rho_{F'x'}). \qquad (6.24)$$

From Fig. 6.7 (a) ,

$$P_{2y'} = P'_{2x'} = 0 , \qquad P_{2x'} = P'_{2y'} = 1 ,$$

$$\rho_{F'y'} = \rho'_{F'y'} = 0 , \qquad \rho_{F'x'} = \rho'_{F'x'} = f' - a .$$

Putting these data into Eq . (6.24) yields

$$\Delta S'_{F'z'c} = \Delta v(f' - a). \qquad (6.25)$$

On the other hand ,for the implementation of the optical-axis stabilization ,the following condition should be met :

$$\Delta S'_{F'z'c} = -\Delta S'_{F'd}. \qquad (6.26)$$

Substituting Eqs . (6.23) and (6.25) into Eq . (6.26) gives

$$\Delta v(f' - a) = \Delta v{\cdot}f' .$$

Thus ,

$$a = 0 . \qquad (6.27)$$

This is the condition equation that should be observed for the stabilization of the optical axis around the y axis. The zero value of such a structural length cannot possibly be carried out in practice .

The difference between the two condition equations (6.22) and (6.27) means that a single mirror cannot properly be used for optical-axis stabilization in a convergent beam .

It is also apparent from the above example that the reason for this is the unsymmetrical structure of the two formulas of $\Delta S'_{F'y'}$ and $\Delta S'_{F'z'}$ for a single mirror ,or in other words ,the situation is due to the fact that the characteristic of the adjustment of the single mirror around the y axis is not idendical to that of the adjustment of the same mirror around the z axis .

A further investigation will show that reflecting prisms like the roof Pechan , roof Abbe , and so on ,all have symmetrical behavior in this regard .

6.3.2 Fromula for Optical-Axis Stabilization

Now ,let us derive a general formula for optical-axis stabilization by using a single reflecting prism in a convergent beam .

214

Referring to Fig $.6.8.$ define a vector f', which is directed along the exit optical axis of the prism, i.e., the x' direction, and has a magnitude f' being always equal to the focal length of the objective lens regardless of the actual position of the focal point F' relative to the objective lens. Also, C_f represents the cross-product-matrix of the vector f'.

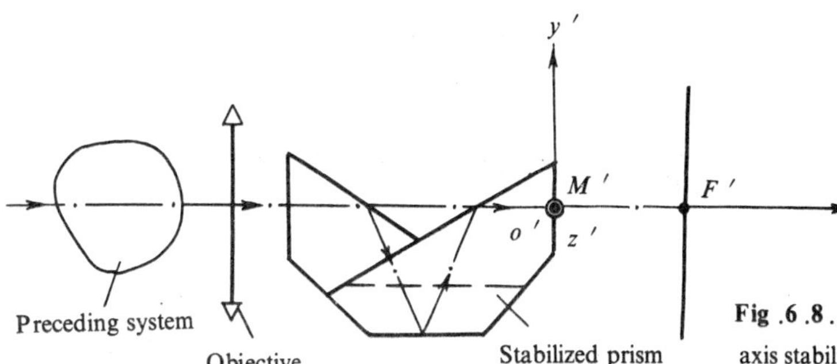

Fig $.6.8.$ A diagram for analyzing optical-axis stabilization in a convergent beam

Thus, we have

$$f' = f' \cdot i', \tag{6.28}$$

$$C_f = \begin{pmatrix} 0 & -f_z' & f_y' \\ f_z' & 0 & -f_x' \\ -f_y' & f_x' & 0 \end{pmatrix} = \begin{pmatrix} 0 & 0 & 0 \\ 0 & 0 & -f' \\ 0 & f' & 0 \end{pmatrix}. \tag{6.29}$$

The meaning of the previously used symbols remains unchanged.
This problem will also be handled in a two-step way.

(1) As the whole system has completed a small angular displacement $\Delta\vec{\varepsilon}$ due to a perturbation, the image point displacement $\Delta S_{F'd}'$ caused therefrom is determined as follows:

$$\Delta S_{F'd}' = [\,(-1)^{l_e} R\, R_p\,(-\Delta\vec{\varepsilon})\,] \times f' \tag{6.30}$$

or by writing the cross product in matrix form Eq. (6.30) becomes

$$(\Delta S_{F'd}') = -(-1)^{l_e} C_f\, R\, R_p\,(-\Delta\vec{\varepsilon}), \tag{6.31}$$

where R represents the reflection matrix of the stabilized prism.

(2) Then, the stabilized prism has completed a small angular displacement $\Delta\vec{\theta}$ so as to introduce an image point displacement of compensation $\Delta S_{F'c}'$.
From Eq. (4.18),

$$(\Delta S_{F'c}') = J_{SP}\,(\Delta\vec{\theta}), \tag{6.32}$$

where J_{SP} represents the matrix of image point displacement of the stabilized prism.

According to the requirement of optical-axis stabilization , we should have

$$\begin{pmatrix} 0 & 1 & 0 \\ & & \\ 0 & 0 & 1 \end{pmatrix} (\Delta S_{F'c}') = - \begin{pmatrix} 0 & 1 & 0 \\ & & \\ 0 & 0 & 1 \end{pmatrix} (\Delta S_{F'd}') . \tag{6.33}$$

Putting Eqs . (6.31) and (6,32) into Eq. (6.33) yields

$$\begin{pmatrix} 0 & 1 & 0 \\ & & \\ 0 & 0 & 1 \end{pmatrix} J_{SP} (\Delta \vec{\theta}) = \begin{pmatrix} 0 & 1 & 0 \\ & & \\ 0 & 0 & 1 \end{pmatrix} (-1)^{le} C_f R R_p (-\Delta \vec{\varepsilon}) . \tag{6.34}$$

Example

In Fig .6 .9 , a roof Pechan prism is inserted into the convergent light path after an objective lens to serve as both an image erector and an optical-axis compensator . Now we discuss if a

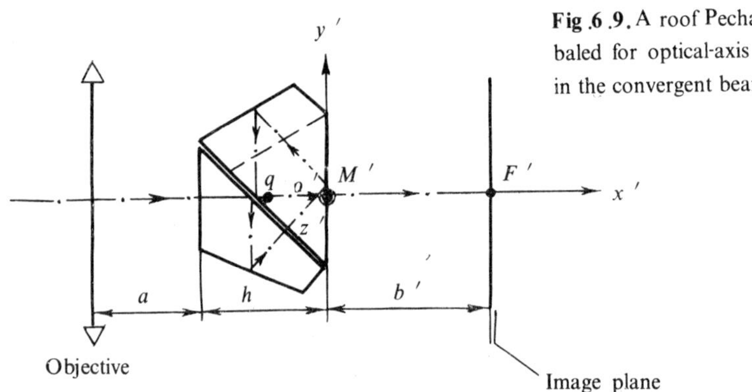

Fig .6 .9. A roof Pechan prism gimbaled for optical-axis stabilization in the convergent beam

reflecting prism of this type can be used for the inertial stabilization condition . If the result is favorable , then determine the location of the center q of the gimbal relative to the prism ,i .e ., the intersection of the mechanical rotation axis with the optical axis .

· From the Appendix , we can find the necessary matrices J_{SP} and $S_{T.2\varphi}$ for the roof Pechan prism :

$$J_{SP} = \begin{pmatrix} 0 & 0 & 0 \\ 2z_q' & 0 & -2x_q' + 2b' - 1.299D + 5.156\dfrac{D}{n} \\ -2y_q' & 2x_q' - 2b' + 1.299D - 5.156\dfrac{D}{n} & 0 \end{pmatrix} , \tag{6.35}$$

216

$$R = (-1)^6 S_{T,2\varphi} = \begin{pmatrix} 1 & 0 & 0 \\ 0 & -1 & 0 \\ 0 & 0 & -1 \end{pmatrix}. \tag{6.36}$$

Since the stabilized prism is directly mounted on the gimbal of a gyroscope, we have

$$\Delta\vec{\theta} = -\Delta\vec{\varepsilon}. \tag{6.37}$$

Putting Eqs. (6.35) through (6.37) and Eq. (6.29) into Eq. (6.34) and considering $R_p = E$ and $t_e = 6$, we obtain

$$\begin{pmatrix} 2z_q' & 0 & -2x_q' + 2b' - 1.299D + 5.156\dfrac{D}{n} \\ -2y_q' & 2x_q' - 2b' + 1.299D - 5.156\dfrac{D}{n} & 0 \end{pmatrix}$$

$$= \begin{pmatrix} 0 & 0 & f' \\ 0 & -f' & 0 \end{pmatrix}. \tag{6.38}$$

Equating the corresponding elements of the first row on both sides of Eq. (6.38) yields

$$\left. \begin{array}{l} 2z_q' = 0 \\ -2x_q' + 2b' - 1.299D + 5.156\dfrac{D}{n} = f' \end{array} \right\}. \tag{6.39}$$

Equating the corresponding elements of the second row on both sides of Eq. (6.38) gives

$$\left. \begin{array}{l} -2y_q' = 0 \\ 2x_q' - 2b' + 1.299D - 5.156\dfrac{D}{n} = -f' \end{array} \right\}. \tag{6.40}$$

Then, from Eq. (6.39), we have

$$z_q' = 0 \quad, \quad x_q' = \frac{-f' + 2b' - 1.299D + 5.156\dfrac{D}{n}}{2} \quad,$$

and from Eq. (6.40) we obtain

$$y_q' = 0 \quad, \quad x_q' = \frac{-f' + 2b' - 1.299D + 5.156\dfrac{D}{n}}{2} \quad.$$

Fortunately, there is no conflict between the two sets of results deduced from the two respective equations (6.39) and (6.40); particularly two solutions for x_q' are identical.

This fact verifies the validity of our prediction that reflecting prisms like the roof Pechan can properly be used as an inertial stabilization method for stabilizing the optical axis in a convergent beam simply because the characteristics of the adjustment of these prisms around the y' axis are identical to those of the adjustment of the same prisms around the z' axis.

Considering that

$$f' = b' + \overline{M_0M'} + h + a ,$$

we have

$$f' = b' + (5.156\frac{D}{n} - 1.299D) + h + a .$$

Then,

$$x'_q = \frac{b' - a - h}{2} .$$

Furthermore, letting $b' = a$, we obtain

$$x'_q = -\frac{h}{2} .$$

The resulting coordinates of point $q : x'_q = \frac{h}{2} , y'_q = z'_q = 0$, show that the center of the gimbal appears to be coincident with the geometrical center of the roof Pechan.

6.4 Circular Beam Deflectors

Reflecting prisms like the roof Pechan, such as the roof Dove, roof Abbe, and so forth, will be investigated a little further.

Again, take the roof Pechan prism as an example. An adjustment diagram of the roof Pechan is shown in Fig. 6.10.

Suppose that a unit vector A represents a parallel beam entering the prism along its entrance optical axis and another unit vector A' represents the conjugate parallel beam emerging from the prism along its exit optical axis.

Now, let us examine how the emerging parallel beam A' will change in direction when the prism is rotated about an axis P parallel to the y' axis.

If the angle of rotation of the prism is assumed to be very small, say $\Delta\theta$, then the change of direction of the emerging parallel beam A', i.e., the image rotation $\Delta\vec{\mu}'$ can readily be determined, using the adjustment diagram of the prism.

Considering the fact that $P\,/\!/\,v$, $P\perp w$, and $\vec{\delta}_u = 0$, we have

$$\Delta\mu'_{x'} = 0 ,$$

$$\Delta\mu'_{y'} = \Delta\mu'_{y'\,max} = 2\Delta\theta ,$$

$$\Delta\mu'_{z'} = 0 .$$

218

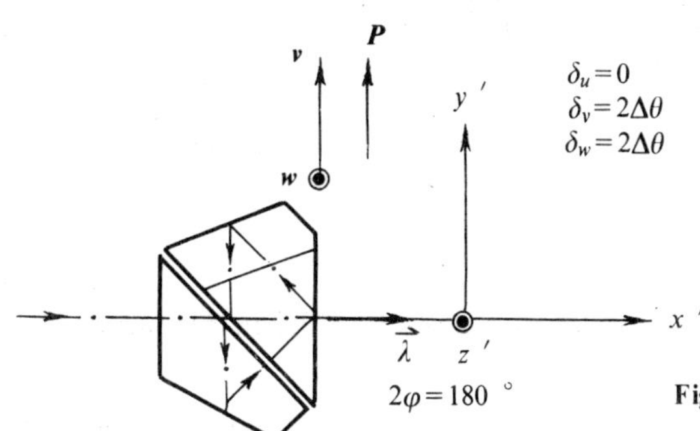

Fig .6 .10 . An adjustment diagram of a roof Pechan prism

Thus ,

$$\Delta\vec{\mu}\,' = \Delta\mu'_x \cdot i\,' + \Delta\mu'_y \cdot j\,' + \Delta\mu'_z \cdot k\,' = 2\Delta\theta \cdot j\,'.$$

The result shows that the image beam $A\,'$ rotates in the same direction as the prism is turned , but at double the angle of rotation of the prism .

Obviously , the same result will be obtained while the prism is rotated through a small angle $\Delta\theta$ about the $z\,'$ axis .

Actually, if only the rotation axis P of the roof Pechan prism is perpendicular to its optical axis , then it will always act as a double speed beam deflector.

By the same reasoning which was used in Sect. 3.4.2 to verify the property of an image rotator, we can prove that the conclusion arrived at will also be extended to the case of a finite rotation.

Therefore, if a roof Pechan rotates through a finite angle θ about an arbitrary axis P perpendicular to its optical axis, the emerging parallel beam $A\,'$ will rotate through a double angle 2θ about the same axis P.

Then, a reflecting prism of this type may be called the circular beam deflector.

To sum up, a circular beam deflector has the property of circular symmetry about its optical axis in regard to the characteristics of image formation as well as of adjustment in both parallel and convergent light.

A model of a prism acting as a circular beam deflector can also be defined in terms of its characteristic parameters including t , $T\,(\vec{\lambda})$, 2φ as well as the relative position of the entrance and exit axes , as shown in Fig . 6 .11 .

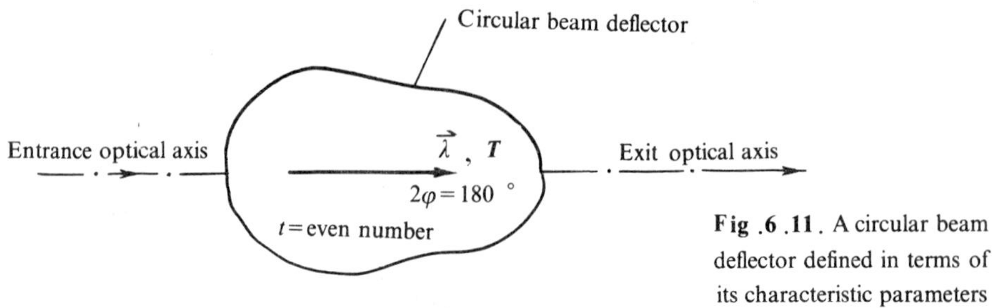

Fig .6 .11. A circular beam deflector defined in terms of its characteristic parameters

6.5   Three-Axis Image-Stabilizing Reflecting Prism Assembly

So far as we know, an image rotator can be used for image-lean compensation by turning it about its optical axis while a circular beam deflector can be used for optical-axis stabilization by turning it about two axes perpendicular to each other and both perpendicular to its optical axis.

Now, two image rotators are combined in such a manner so as to form a circular beam deflector, as shown in Fig. 6.12.

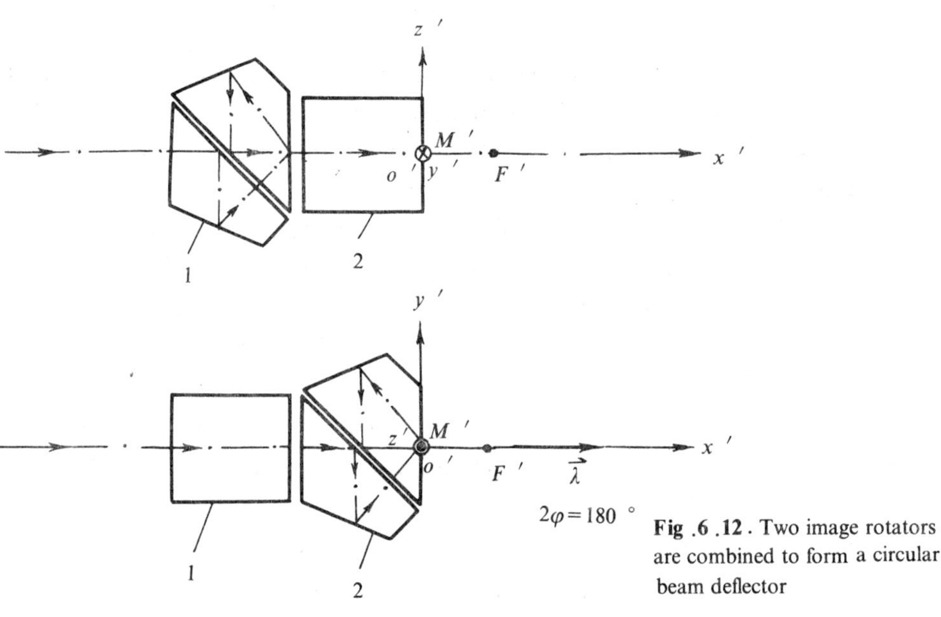

Fig .6 .12. Two image rotators are combined to form a circular beam deflector

One Pechan prism is allowed to rotate alone about the optical axis, acting as an image rotator to provide image compensation around the optical axis while two Pechan prisms as a unit will rotate about two axes perpendicular to each other and again both perpendicular to the optical axis, acting as a circular beam deflector to provide optical-axis stabilization to retain the central image position in the image plane.

Thus, totally the prism assembly provides three-axis image stabilization. This prism assembly can be used in both collimated and convergent light.

Note that the angular compensation $\Delta\theta$ around the optical axis of one Pechan prism relative to the other should be sufficiently small in order that the property of the prism assembly as a circular beam deflector would not be changed too much.

Like a roof Pechan prism, this prism assembly can also be employed as an erecting system for compact binoculars.

Problems

6.1 Consider a three-axis image-stabilizing reflecting prism assembly, as shown in Fig.6.12, and assume A to be an incident parallel beam entering along the entrance optical axis. Suppose that one Pechan, say prism 1 , has rotated relative to another through an angle θ about the x' axis. Derive a formula to represent the emerging parallel beam A_{ψ}' if the assembly as a unit has rotated through an angle ψ about an axis P perpendicular to the x' axis, using first the precise equations and again the approximate equations.

6.2 Give some other examples of circular beam deflectors.

6.3 Do the example in Sect.6.3.2 for a roof Abbe prism.

Part III

Adjustment and Image Stabilization of Optical Instruments

Chapter 7
Adjustment of Hinged Binoculars

The present chapter is devoted to the study of a common problem of adjustment for those optical instruments within a more specialized area , i.e. , the individual adjustment of optical axis of hinged binoculars.

This material may be regarded as a complement to Part Ⅱ in certain aspects since part of the basic knowledge can relate to more than the theory of mirror and prism systems.

7.1 Concept of Individual Adjustment of Optical Axis

There are two types of hinged binoculars: single - hinged and double- hinged. Typical examples for these two types will be illustrated later on in Figs.8.1 and 9.1; the former is a most popular type of hand- held binoculars where two telescope- members have a common hinge while the latter is a stereoscopic telescope where each telescope- member has a hinge of its own.

No matter to which type the hinged binoculars belong , the whole process of its adjustment may usually be regarded as consisting of two major steps: the first , the adjustment of optical axis for each telescope- member , and the second the adjustment of optical axes for two tele-scope- members together. For short , we call them the *individual adjustment of optical axis* , and the *joint adjustment of optical axes* , respectively.

The purpose of individual adjustment of optical axis is to set the optical axis of the individ-ual telescope- member parallel to the mechanical rotation axis of its own hinge , or sometimes to the common hinge axis.

Referring to Fig. 7. 1 , the optical axis of one telescope- member is said to be parallel to its mechanical rotation axis (hinge axis) if a collimated beam B which is parallel to the hinge axis $O_1 O_1$ enters the telescope and leaves it along the direction $B_p{}'$ which is again parallel to the axis $O_1 O_1$

On the contrary , if the collimated beam B which is parallel to the mechanical rotation axis $O_1 O_1$ enters the telescope ,and after passing through it ,will deviate from the axis $O_1 O_1$,for in-stance ,emerging along the direction $B_0{}'$,then the angular deviation of the emergent parallel beam $B_0{}'$ from its original direction B or $B_p{}'$ would thus reasonably be defnded as the opti-cal axis deviation , and the latter may be designated by a vector $\vec{\xi}$ for being of small value. Obvi-ously , the optical axis deviation $\vec{\xi}$ is an angular vector quantity with two degrees of free-dom (two components). So , the goal of the individual adjustment of optical axis is to eliminate

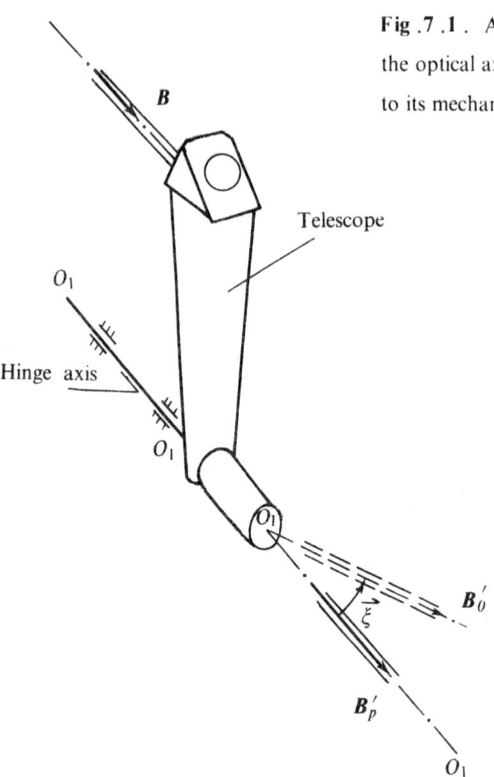

Fig .7 .1 . A diagram to illustrate if the optical axis of a telescope is parallel to its mechanical axis or not

the optical axis deviation $\vec{\xi}$.

The aim of the joint adjustment of optical axes of binoculars is to ensure that the two collimated beams emerging from two respective telescope - members , which correspond to the same incident collimated beam B travelling in the direction along the mechanical axis , should run parallel to each other , or with the departure from parallelism within a given tolerance. The situation must be kept while one telescope rotates with respect to the other around the hinge axis to any or at least some specified positions.

Since the individual adjustment of optical axis will always serve as an essential part of the adjustment of optical axes of hinged binoculars , it is worthwhile to undertake a study of this procedure in full detail .

7.2 Equipment and Method of Individual Adjustment of Optical Axis

The optical device for the individual adjustment of optical axis of one telescope - member in general includes three parts: collimators , a holder , and a measuring telescope, as shown in Fig .9.1.

Referring to Fig. 7.2 , the optical axes O_2O_2 of the two fixed collimators 1 and 2 are parallel to each other. Their centers of cross - lines represent two object - points located infinitely distant , from which two collimated beams emerge parallel to each other along the direction of axes O_2O_2 .

226

The binoculars to be adjusted is placed in the holder between the collimators and the measuring telescope. In the figure, only one telescope-member of the binoculars is shown. For adjusting the popular hand-held binoculars, the holder is equipped with two control knobs, by which the azimuth and elevation of the product fixed to the holder can be changed so that

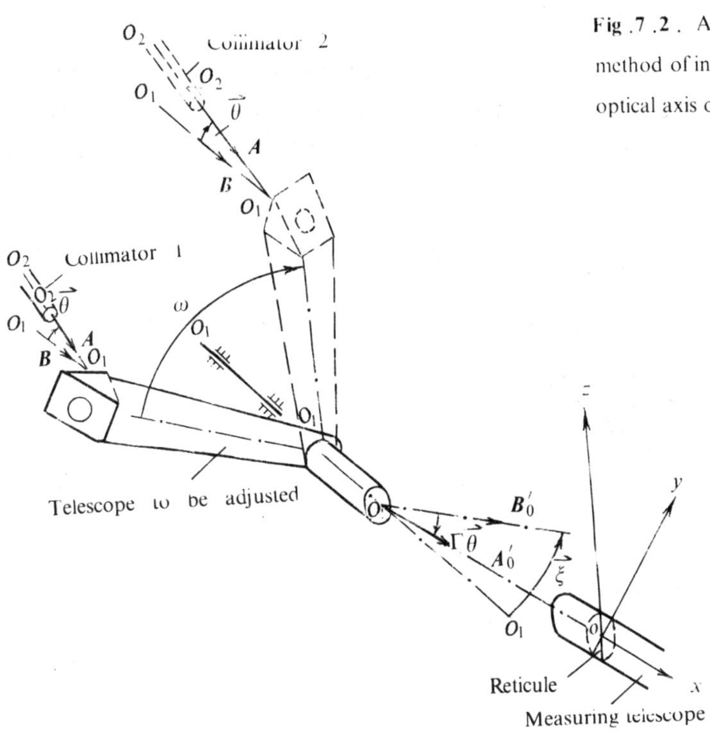

Fig.7.2. A diagram to show the method of individual adjustment of optical axis of one telescope-member

its hinge axis may approximately be orientated along the axes of the collimators. The set-up, as shown in Fig.8.2, represents the case just mentioned. For adjusting the stereoscopic telescopes, as shown in Fig.9.1, where the product itself is possessed of both a horizontal and a vertical rotation knobs, the holder thus becomes much simpler and in this case appears to be a ball support with a handle for fastening the product.

The measuring telescope is used to detect the direction of the parallel beam emerging from the product-telescope. It is usually a Keplerian telescope without an image erector, having only a cross-hair within the focal plane of its objective.

There may also be two versions of the mounting manner of measuring telescopes according to whether they are fixed immovably or free on the frame of the set-up (adjusting device). Two representative examples are again shown in Figs.8.2 and 9.1. In the former case we have a fixed measuring telescope, where its optical axis is aligned exactly parallel to the optical axes of collimators, while in the latter case we have a movable measuring telescope.

Since the fixed version of a measuring telescope can be referred to as a special case of the movable version, the following discussion is directed to the movable version of a measuring telescope, and the relationship obtained will be valid for the fixed version as well.

227

Again referring to Fig. 7.2 , it is easy to understand that the individual adjustment of optical axis should be made at two positions of the telescope-member around its hinge axis , as shown in full and dotted lines , respectively , in the diagram.

Accordingly , this adjusting process involves two main operations: one , called the *aiming* , and the other the *reading*.

(1) Aiming operation

Having fastened the product-telescope in the holder , turn it by hand around the hinge axis O_1O_1 to a position that is roughly directed to the collimator 1 , and then rotate the driving knobs on the holder or on the product-telescope itself to bring the image o' of collimator

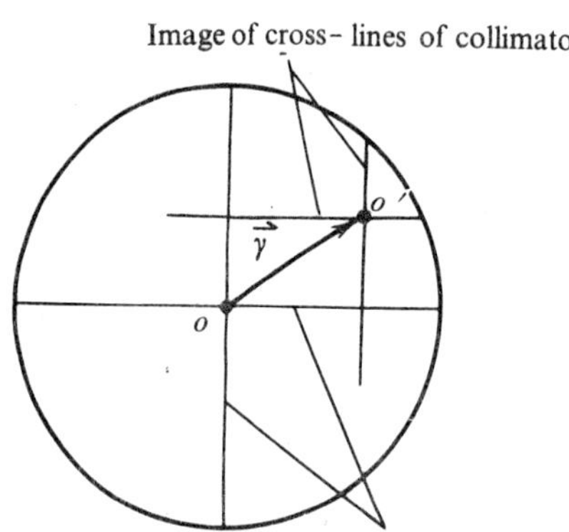

Fig.7.3. The image o' of the center of collimator cross-lines is shifted away from the center o of measuring telescope cross-lines

cross-lines (object cross-lines), formed through the succeeding optical system, to coincide with the measuring telescope cross-lines o (measuring cross-lines). During this operation, the measuring telescope can also be displaced and tilted slightly if it belongs to the movable version. This operation and its final result are called the *aiming* and the *aiming state*, respectively.

(2) Reading operation

Here again , turn the product-telescope by hand around its hinge axis O_1O_1 to another position that is directed to the collimator 2. Because of the existing optical axis deviation , the image of the collimator cross-lines o' formed through the succeeding optical system will be shifted from the measuring cross-lines o , as shown in Fig.7.3 . The vector $\vec{\gamma}$ denoting a shift from o to o' is defined as the *image shift* (measured in angular value) which is mainly dependent on the optical axis deviation $\vec{\xi}$ of that telescope-member , thus providing a basis , depending on which one will be able to make a proper adjustment. The image shift $\vec{\gamma}$ is also a vector quantity having two components. This operation and the result thereof are named the *reading* and the *reading state* , respectively. It should be noticed that during this operation neither the driving knobs on the holder or on the product nor the orientation of the optical axis of the measuring telescope should be disturbed.

In order to complete the individual adjustment of optical axis of one telescope-member in

228

one cycle including only one aiming and one reading operations , we have to derive first the relationship between the image shift $\vec{\gamma}$ and the optical axis deviation $\vec{\xi}$.

7.3 Relationship Between the Image Shift and Optical Axis Deviation

Referring again to Fig . 7 . 2 , all the lines O_1O_1 represent the directions parallel to the mechanical rotation axis of the product-telescope. Let A represent the direction of parallel beams entering the telescope along the axes of collimators and $\vec{\theta}$ represent the angle between the hinge axis O_1O_1 and collimator axis O_2O_2 . In short , this angle $\vec{\theta}$ may be called the *deviation of hinge axis*. For the movable version of the measuring telescope , both the magnitude and direction of the deviation of hinge axis are random , but in any case it is a quantity of small value. Besides , assume that a vector B represents the direction of parallel beams entering the telescope along the hinge axis O_1O_1 .

Suppose the product -telescope is in the aiming state , as shown in full lines in the diagram , and assume $A_0{'}$ and $B_0{'}$ to be the directions of two parallel beams emerging from the telescope and corresponding to the two incident parallel beams A and B , respectively.

Because of the existence of optical axis deviation of the detected telescope , the emergent parallel beam $B_0{'}$ should make an angle $\vec{\xi}$ with the hinge axis O_1O_1 . Also , due to the object and image relationship , two emergent parallel beams $A_0{'}$ and $B_0{'}$ make an angle $\Gamma\vec{\theta}$ with each other , where Γ represents the magnification of the telescope. According to the principle of telescopic optical system , the magnified deviation $\Gamma\vec{\theta}$ will be swung in a plane parallel to that plane in which the deviation of hinge axis $\vec{\theta}$ is done.

When the product-telescope is transferred to the reading state , as shown in dotted lines in the figure , the direction of the incident parallel beam coming from the center of cross-lines of the collimator remains unchanged and may thus still be denoted by the vector A . However , the two parallel beams emerging from the telescope and corresponding to the two incident parallel beams A and B , respectively , will now become $A{'}$ and $B{'}$, which are not shown in Fig. 7.2. Since the angle between A and B remains $\vec{\theta}$, the two emergent parallel beams $A{'}$ and $B{'}$ will intersect at the same angle $\Gamma\vec{\theta}$.

It should be pointed out that during the aiming operation the optical axis of the measuring telescope becomes parallel to the direction $A_0{'}$ of the emergent parallel beam.

All the parallel beams emerging from the product-telescope in both the aiming and reading states will be imaged through the objective of the measuring telescope in its focal plane , so let us locate the image points in this plane for each of the emergent parallel beams just described.

Fig .7.4 shows the field of view of the measuring telescope . Assume $oxyz$ to be a coordinate system attached to the measuring telescope for evaluating the image shift $\vec{\gamma}$, where o represents the center of cross-lines of the measuring telescope , x represents its optical axis ,

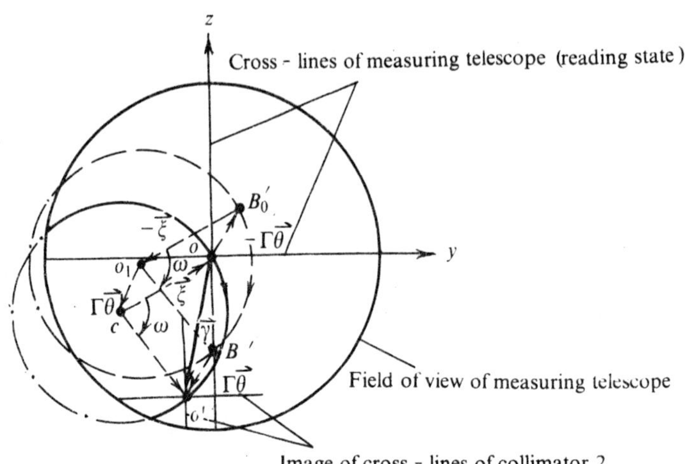

Fig .7 .4. Field of view of the measuring telescope being in reading state

and y and z represent the cross- lines of the measuring telescope in reading state .[*].

As mentioned before , the parallel beam A_0' emerging from the product- telescope in aiming state has been imaged at the origin o of the reference coordinate system yoz , but now the parallel beam A' emerging from the same telescope in reading state is imaged through the objective of the measuring telescope at o' in its focal plane , so that the vector $\overrightarrow{oo}'$ represents the image shift $\vec{\gamma}$. The procedures for locating the image point o' are shown in dotted lines in the diagram. First , we can find the image point B_0' corresponding to the emerging parallel beam B_0' by drawing from o the vector $-\overrightarrow{\Gamma\theta}$, and locate another point o_1 by drawing from B_0' the vector $-\vec{\xi}$. Comparing Figs. 7.4 and 7.2 , one will readily see that if a bundle of parallel rays is directly travelling along the hinge axis O_1O_1 into the measuring telescope , then it will be imaged at the point o_1 . Now , while being transferred from the aiming state to the reading state , the product- telescope rotates through an angle ω about its hinge axis. During this process , we may easily form a picture in our mind where the parallel beam B_0' specially defined above will turn around the fixed axis O_1O_1 to produce a conical scan , and accordingly in the focal plane of the measuring telescope the imaginary image point o_1 remains fixed , like a center , around which the image point B_0' will rotate through the same angle ω in the same direction to the image point B' , which corresponds to the parallel beam B' emerging from the telescope in reading state. Thus , while turning the telescope around its hinge axis , the locus of B' is a circle about o_1 with the magnitude of the optical axis deviation ξ as radius. For the reason mentioned above , since the two parallel beams A' and B' will intersect at constant angle $\overrightarrow{\Gamma\theta}$ regardless of the rotation of the product- telescope around its hinge axis , the image point o' can thus be located by drawing from B' the vector $\overrightarrow{\Gamma\theta}$. Then , it is evident that during the process of rotating the product- telescope , the image point o' will also move in a

____

*We made a particular mention of the operation state , aiming or reading , with which the coordinate system representing the cross- lines of the measuring telescope is associated because for some versions of the set- up the orientation of the cross- lines is different in various states.

230

circular locus , which may be obtained by displacing the circular locus of B' by a linear vector $\overrightarrow{\Gamma\theta}$. Thus , the center c of the circular locus of o' is accordingly determined by drawing from o_1 the vector $\overrightarrow{\Gamma\theta}$. Alternatively , the image point o' can be located by turning the image point o about c through an angle ω in the same direction.

From the parallelogram $oB_0'o_1c$, $\overrightarrow{co}$ is equal to the optical axis deviation. Hence the center c may also be found by drawing from o the vector $-\overrightarrow{\xi}$.

From Fig. 7.4 ,

where
$$\overrightarrow{\gamma} = \overrightarrow{oo}\,' = \overrightarrow{co}\,' - \overrightarrow{co} , \tag{7.1}$$
$$\overrightarrow{co} = \overrightarrow{\xi} . \tag{7.2}$$

The vector $\overrightarrow{co}\,'$ may be regarded as a result obtained by turning the vector $\overrightarrow{co}$ about the center c through an angle ω. Such a planar vector-rotation problem can easily be performed by using complex numbers in their polar form. Thus , we have

$$\overrightarrow{co}\,' = \overrightarrow{co}\, e^{i\omega} = \overrightarrow{\xi}\, e^{i\omega} . \tag{7.3}$$

We shall take the positive sense for ω to be in the counterclockwise direction .

Putting Eqs . (7.2) and (7.3) into Eq. (7.1) gives

$$\overrightarrow{\gamma} = \overrightarrow{\xi}\, e^{i\omega} - \overrightarrow{\xi} ,$$

and thus
$$\overrightarrow{\gamma} = \overrightarrow{\xi}\,(e^{i\omega} - 1) . \tag{7.4}$$

The resulting equation (7.4) is a general formula relating the image shift $\overrightarrow{\gamma}$ to the optical-axis deviation $\overrightarrow{\xi}$. Note that the image shift $\overrightarrow{\gamma}$ depends only on the optical axis deviation $\overrightarrow{\xi}$ and the rotation angle ω , but has no bearing on the deviation $\overrightarrow{\theta}$ of the hinge axis. Some people find it hard to accept the latter fact. Actually , the effect of $\overrightarrow{\theta}$ on $\overrightarrow{\gamma}$ is removed from Eq. (7.4) by the aiming operation. It also postulates certain condition that both $\overrightarrow{\theta}$ and $\overrightarrow{\xi}$ must be of small magnitudes so that small quantities of second and higher powers may be neglected in order to obtain this approximate formula. Note again that in Eqs. (7.4) and (7.1) the vector $\overrightarrow{\xi}$ or $\overrightarrow{co}$ represents the optical axis deviation of the product-telescope in aiming state while the vector $\overrightarrow{\xi}\,e^{i\omega}$ or $\overrightarrow{co}\,'$ represents the optical axis deviation of the same telescope in reading state.

To find the components of the image shift $\overrightarrow{\gamma}$, replacing the vectors $\overrightarrow{\gamma}$ and $\overrightarrow{\xi}$ in Eq. (7.4) by their expressions in complex number yields

$$\gamma_y + i\gamma_z = (\xi_y + i\xi_z)\,(e^{i\omega} - 1) .$$

From Euler's formula ,

Then
$$e^{i\omega} = \cos\omega + i\sin\omega .$$

$$\gamma_y + i\gamma_z = (\xi_y + i\xi_z)\,(\cos\omega + i\sin\omega - 1)$$
$$= (\xi_y + i\xi_z)\,(-2\sin^2\frac{\omega}{2} + i\sin\omega) .$$

Considering that $i^2 = -1$, the above formula becomes

$$\gamma_y + i\gamma_z = (-2\sin^2\frac{\omega}{2} \cdot \xi_y - \sin\omega \cdot \xi_z)$$
$$+ i\,(\sin\omega \cdot \xi_y - 2\sin^2\frac{\omega}{2} \cdot \xi_z) .$$

Equating the real components and the imaginary components from both sides of the above formula gives finally

$$\left.\begin{array}{l} \gamma_y = -2\sin^2\dfrac{\omega}{2}\cdot\xi_y - \sin\omega\cdot\xi_z \\[3mm] \gamma_z = \sin\omega\cdot\xi_y - 2\sin^2\dfrac{\omega}{2}\cdot\xi_z \end{array}\right\} . \qquad (7.5)$$

In fact, the optical axis deviation $\vec{\xi}$ of the product-telescope is an unknown for the time being and during the individual adjustment of optical axis the optical axis deviation $\vec{\xi}$ may be determined from the detected image shift $\vec{\gamma}$.

From Eq. (7.4),

$$\vec{\xi} = \frac{\vec{\gamma}}{e^{i\omega}-1} = \frac{\vec{\gamma}}{(\cos\omega-1)+i\sin\omega} .$$

Multiplying both the numerator and denominator of the term on the right side by $[(\cos\omega-1)-i\sin\omega]$ and replacing vectors by complex numbers leads to

$$\left.\begin{array}{l} \xi_y = -\dfrac{1}{2}\gamma_y + \dfrac{1}{2}\cot\dfrac{\omega}{2}\cdot\gamma_z \\[3mm] \xi_z = -\dfrac{1}{2}\cot\dfrac{\omega}{2}\cdot\gamma_y - \dfrac{1}{2}\gamma_z \end{array}\right\} . \qquad (7.6)$$

The above equations (7.4) through (7.6) are developed for the adjusting device with a movable version of the measuring telescope. For the reason mentioned before, they will also be valid for the adjusting device with a fixed version of the measuring telescope.

7.4 Implementation of Individual Adjustment of Optical Axis

Now, the problem is how to adjust the relevant elements, including the mirror or reflecting prism in our case, so that the optical axis deviation $\vec{\xi}$ of the telescope to be adjusted will vanish.

The manner in which this essential step of adjustment is to be carried out could be quite different depending mainly on the adjusting construction of the product. If the adjusting structure (or mechanism) is accessible, then one will be able to continue adjusting the product-telescope while it is being fixed in the holder.

In this case, our discussion will be followed by seeking the answer to a highly important question: 'To which place should the image point o' corresponding to the center of cross-lines of collimator 2 be brought by adjusting a relevant element of the telescope in order to eliminate its optical axis deviation $\vec{\xi}$?'

As mentioned before, the optical axis deviation of the detected telescope in reading state appears to be $\overrightarrow{co}\,'$ or $\overrightarrow{\xi e^{i\omega}}$ in the field of view of the measuring telescope instead of $\overrightarrow{co}$ or $\overrightarrow{\xi}$,

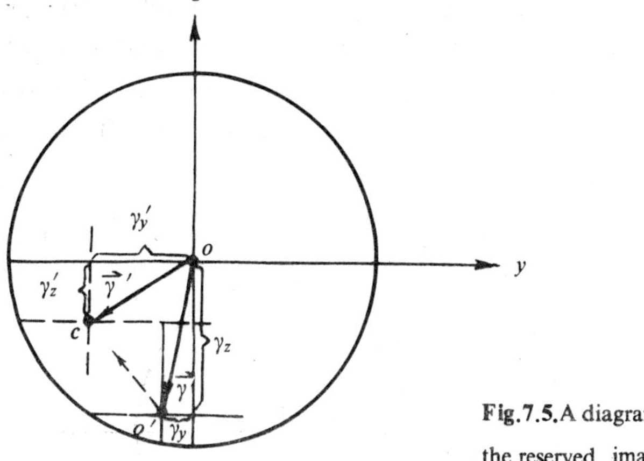

Fig.7.5.A diagram illustrating the reserved image shift $\overrightarrow{\gamma}\,'$

as shown in Figs.7.4 and 7.5. Thus , the adjustment should be made to produce an óptical axis deviation $\overrightarrow{o'c}$ or $-\overrightarrow{\xi e^{i\omega}}$ to balance the original $\overrightarrow{co}\,'$, that is to say , viewing through the measuring telescope the operator has to continue to shift a relevant compensator until the moving image point $o\,'$ appears to be coincident with point c.

The physical meaning of doing so can also be explained in the following way. As described in Sect.7.3 , while the product- telescope turns around its hinge axis , the image point $o\,'$ moves accordingly in a circle about the point c with radius ξ; now , once the image point $o\,'$ becomes coincident with the center c , it will no longer move due to the zero-valued radius , i.e , $\xi = 0$. Therefore , such a phenomenon that the image point $o\,'$ will not vary in position with the rotation of the product- telescope around its hinge axis may be taken as a symbol to indicate that the optical axis of that telescope has been parallel to its mechanical rotation axis. At this moment , although the conjugate pair of parallel beams for the product- telescope A and $A\,'$ corresponding to the image point o' (being coincident with c) both are not parallel to the mechanical rotation axis O_1O_1 , yet they do intersect this axis at angles θ and $\Gamma\theta$, respectively. The latter fact implies that the direction of hinge axis O_1O_1 itself represents a conjugate pair of parallel beams for the adjusted telescope B and $B\,'$ (being coincident with o_1)under this condition.

Referring to Fig.7.5 , let us find the position vector $\overrightarrow{\gamma}\,'$ of the center c , which may be called the *reserved image shift* according to its physical meaning.

Comparing Figs.7.4 and 7.5 , we have

$$\overrightarrow{\gamma}\,' = -\overrightarrow{\xi}\ . \tag{7.7}$$

Thus , from Eq. (7.6) we obtain

$$\left. \begin{aligned} \gamma_y' &= \frac{1}{2}\,\gamma_y - \frac{1}{2}\cot\frac{\omega}{2}\cdot\gamma_z \\[2mm] \gamma_z' &= \frac{1}{2}\cot\frac{\omega}{2}\cdot\gamma_y + \frac{1}{2}\,\gamma_z \end{aligned} \right\} \cdot \tag{7.8}$$

From the geometry in Fig. 7.3 , we may estimate the position of the center c according to the following rule (see Fig . 7 .6):

'The center c appears to be the vertex of a isosceles triangle with the image shift $\overrightarrow{oo}$ ' as its base and with the angle at the vertex equal to the angle ω through which the product-telescope has rotated from the aiming state to the reading state, and relative to the base oo ' the center c is located in that direction in which the telescope rotates. Of course, the center c is also lying on the perpendicular bisector of $\overline{oo}$ '.'

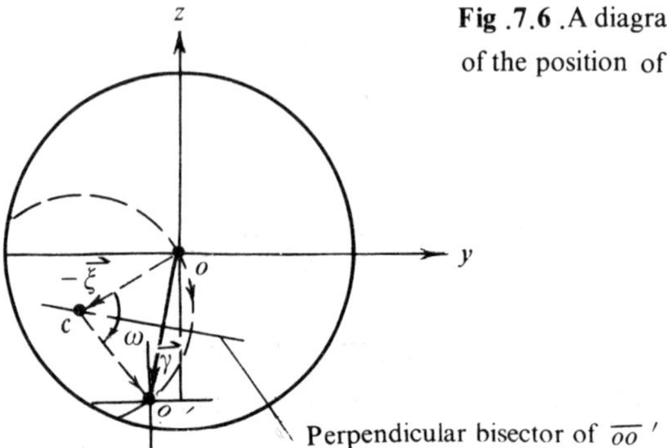

Fig .7.6 .A diagram to show an estimation of the position of the center c

Perpendicular bisector of $\overline{oo}$ '

Very often , the adjusting structure (or mechanism) is not accessible. Then , after the product-telescope has been detected in the foregoing way resulting in the known data of image shift , it should be dismounted from the holder and even disassembled. In this case , a great deal of calculation of adjustment is needed.

Two typical examples for the above two forms of implementation of individual adjustment of optical axis will be illustrated in the next two chapters .

problems

7.1 Can you handle the individual adjustment of optical axis in a different way without perform-ing the aiming and reading operation ?

7.2 Referring to Fig .7.4, what will happen if the adjustment is made to pull the image point o ' back to the origin o ?

Chapter 8
Calculation of Adjustment of Prism Binoculars

The adjustment of a pair of prism binoculars discussed in this chapter is a typical example which illustrates the individual adjustment of optical axis in those cases where the calculation of adjustment is needed after a reading operation.

8.1 Construction of the Hinge of a Pair of Binoculars

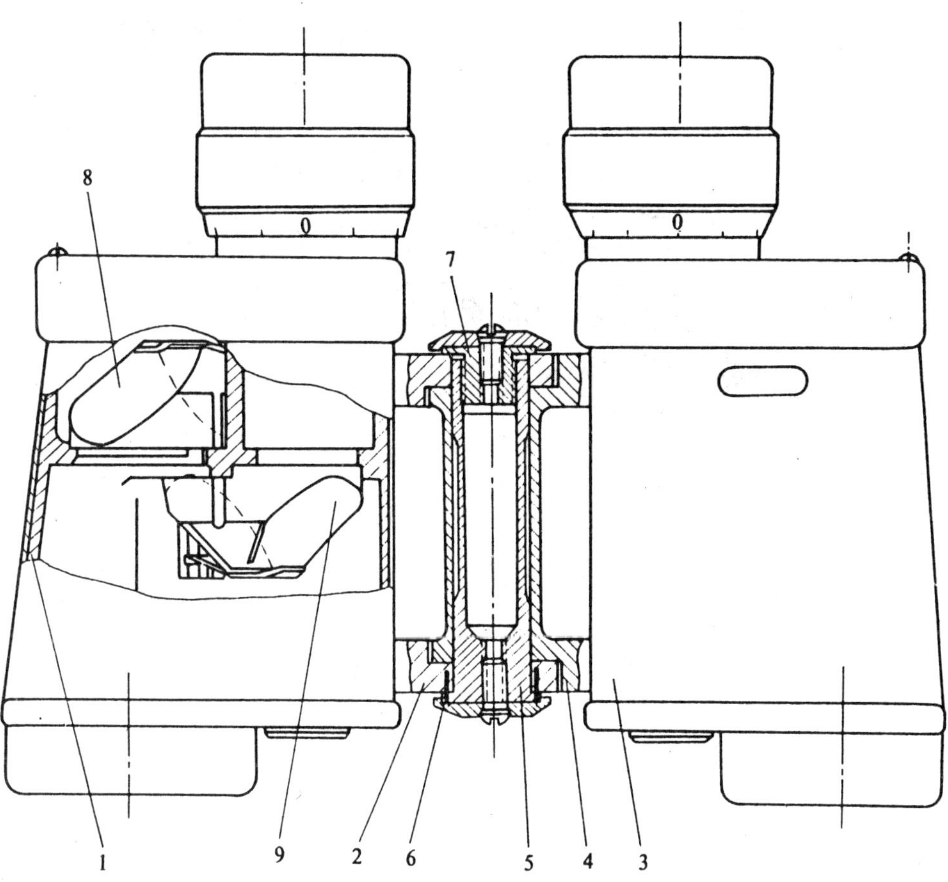

Fig. 8.1. Structure of the hinge of a pair of binoculars

Referring to Fig 8.1, the hinge of a pair of binoculars usually includes three parts:lug 2 as a part of the left-side prism housing 1, sleeve 4 as a part of the right-side prism housing 3, and a hollow tapered shaft 5. The shaft is inserted into the lug and the tapered hole of the sleeve, and is rigidly attached to the lug with the help of a tightening plug screw 7 and pins 6. Thus, the right-side and left-side prism housings may rotate relative to each other around the hinge axis for setting a proper distance between the two eyepieces.

8.2 Equipment for Adjusting Binoculars

Referring to Fig. 8.2, the set-up for adjusting binoculars consists of three parts: collimators (1, 2, 3), a product holder 4, and a measuring telescope 12 of the fixed version.

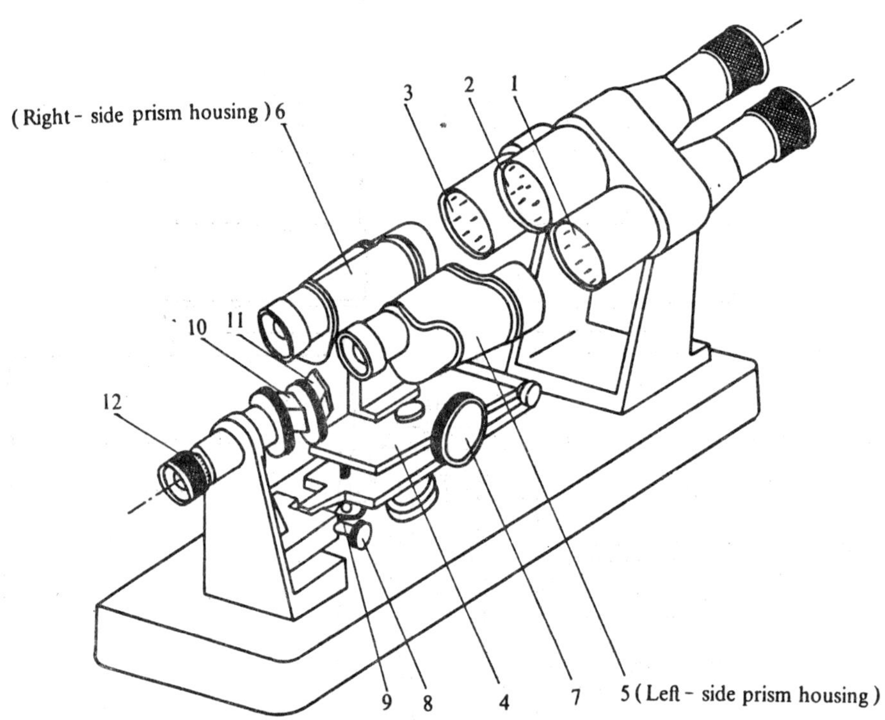

Fig.8.2. A set-up for adjusting binoculars

The binoculars to be adjusted is placed upside down in the holder so that relative to the operator (observer) the left-side prism housing 5 appears to be on the right side while the right-side prism housing 6 on the left side. The sleeve part (number 4 in Fig.8.1) is fastened in the seat of the holder by turning the clamping knob 7. Thus, the right side prism housing 6 is immovable with its objective axis always directed to the collimator 3, while the left -side prism

236

housing 5 will be able to rotate about the hinge axis so that its objective axis may be directed to either collimator 1 or collimator 2. For short, we would later on use accordingly the following expression: the left - side housing is at position 1 or at position 2. By turning the knobs 8 and 9 , we can change both the azimuth and elevation of the binoculars.

The reticule in the focal plane of the measuring telescope is shown in Fig. 8. 3, where the small rectangle with data is used to control the errors in parallelism of the binocular axes.

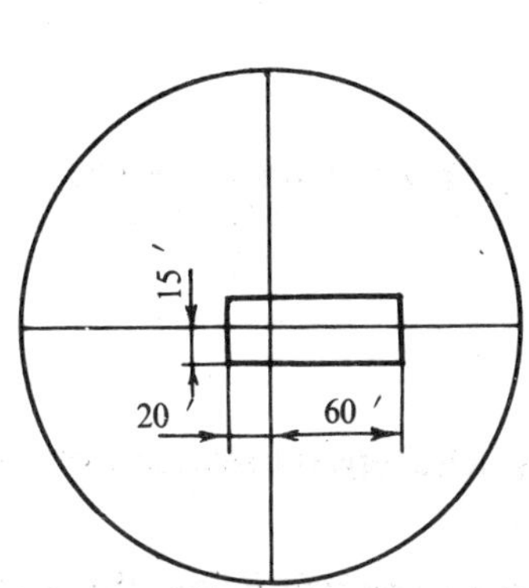

Fig. 8.3. Reticule of measuring telescope

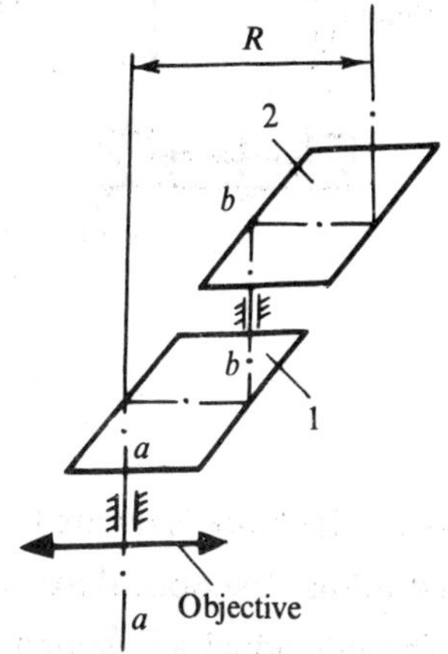

Fig. 8.4 . Double rhomboid prism head

The measuring telescope is equipped with a double rhomboid prism head before the objective, as shown in Fig. 8. 4. The two rhomboid prisms 1 and 2 are able to turn together around the axis aa, which is coincident with the optical axis of the objective, while the rhomboid prism 2 may rotate alone about the axis bb. Consequently, with the help of these two hinges 10 and 11 (see Fig. 8. 2) the optical axis of the measuring telescope will reach any place   within a cylindrical space about the axis aa with radius R, retaining its original direction, so that the operator can see the image of cross - lines either from the collimator 1 or from the collimator 2.

As mentioned in Sect .7. 2 , the present measuring telescope is mounted immovably on the frame and its optical axis has been aligned parallel to the optical axes of collimators.

8.3 Method and Procedure of Adjustment of Optical Axis

Again referring to Fig. 8. 1, for convenience of discussion the prism 8 with its hypotenuse face directed to the objective is called the *big Porro* while the other one 9 the *small Porro*.

After a detail analysis of the adjustment behavior of a Porro prism by using the previously developed principle, we learn that the adjustment of optical axis deviation may be carried out by inserting spacers of different thickness in between the hypotenuse faces of prisms and the flat spots on the seats in the prism housing, as shown in Fig. 8. 5 (a), and the spacers will be placed at those properly selected spots marked with a point

'.' and designated by the abbreviations S.I.,S.O.,B.I.,and B.O., which denote the inner and outer sides of the small and big prisms.

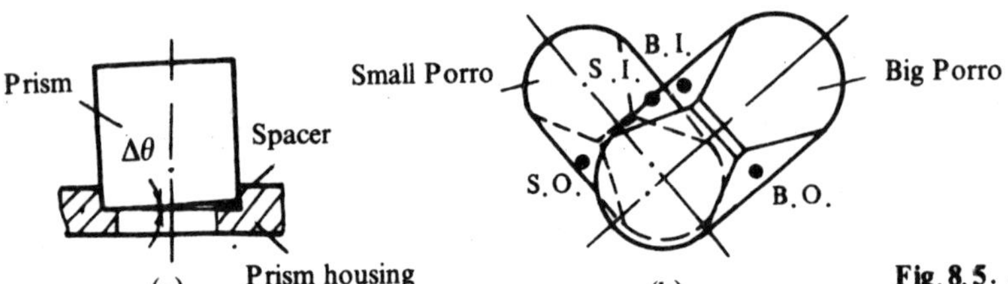

Fig.8.5. A diagram showing where the spacers will be placed

After the binoculars are fastened in the holder,the right‑side telescope is fixed,therefore we will first adjust the optical axis of the left‑side telescope.

The individual adjustment of optical axis of the left‑side telescope will follow the usual procedure.

(1) Fasten the binoculars so that the right‑side telescope is directed to the collimator 3 ; then turn the left‑side telescope by hand around the hinge axis to a position being roughly directed to the collimator 1 ;further rotate the double rhomboid prism hinges until the operator sees the image of cross‑lines of collimator 1 through the measuring telescope.

(2) Rotate the knobs 8 and 9 to bring the left‑side telescope into the aiming state at position 1.

(3) Again , turn the left‑side telescope by hand around the hinge axis to the reading state at position 2 . As a result, we obtain two components γ_y and γ_z of the image shift , from which we can determine the thickness of the spacers and their location.

Since the right‑side telescope is fixed in the holder , the procedure of its adjustment of optical axis as well as the calculation of adjustment will be similar to and simpler than those of the left‑side telescope , and here be omitted.

Therefore , the calculation of adjustment in the next section will also be worked out only for the left‑side telescope.

8.4 Calculation of Adjustment

(1) Calculate the optical axis deviation

From Eq.(7.6),

238

$$\xi_y = -\frac{1}{2}\,\gamma_y + \frac{1}{2}\,\cot\frac{\omega}{2}\cdot\gamma_z \left.\vphantom{\begin{matrix}a\\b\end{matrix}}\right\}$$

$$\xi_z = -\frac{1}{2}\,\cot\frac{\omega}{2}\cdot\gamma_y - \frac{1}{2}\,\gamma_z \qquad (8.1)$$

Actually, the optical axis deviation $\vec{\xi}$ is orientated relative to a coordinate system $x_1 y_1 z_1$ which is rigidly attached to the binoculars and assumed to be (approximately) parallel to the coordinate system xyz of the measuring telescope, as shown in Fig. 8.6.

Thus, we have

$$\xi_{y_1} = \xi_y = -\frac{1}{2}\,\gamma_y + \frac{1}{2}\,\cot\frac{\omega}{2}\cdot\gamma_z \left.\vphantom{\begin{matrix}a\\b\end{matrix}}\right\}$$

$$\xi_{z_1} = \xi_z = -\frac{1}{2}\,\cot\frac{\omega}{2}\cdot\gamma_y - \frac{1}{2}\,\gamma_z \qquad (8.2)$$

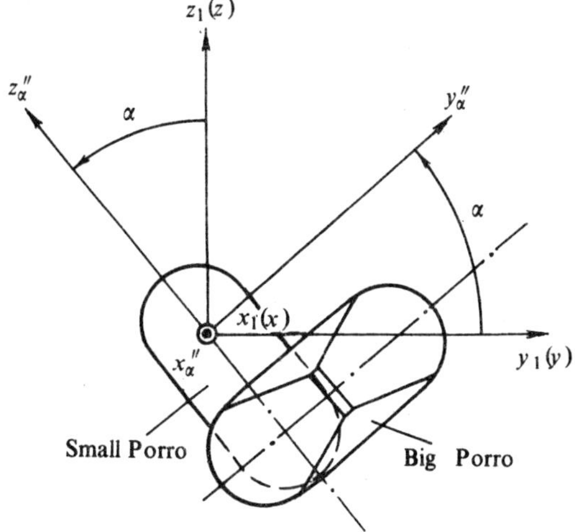

Fig. 8.6. The relative orientation of three coordinate systems

(2) Transform the components of the optical axis deviation ξ from the coordinate system $x_1 y_1 z_1$ to the coordinate system $x_\alpha'' y_\alpha'' z_\alpha''$, where the axis x_α'' is coincident with the axis x_1 while the axes y_α'' and z_α'' are parallel to the optical-axis sections of two prisms, respectively.

For the convenience of calculating the thickness of the required spacers, we will find the components $\xi_{y_\alpha''}$ and $\xi_{z_\alpha''}$ of the same optical axis deviation in the coordinate system $x_\alpha''\,y_\alpha''\,z_\alpha''$.

From Fig. 8.6,

$$\xi_{y_\alpha''} = \cos\alpha\cdot\xi_{y_1} + \sin\alpha\cdot\xi_{z_1} \left.\vphantom{\begin{matrix}a\\b\end{matrix}}\right\}$$

$$\xi_{z_\alpha''} = -\sin\alpha\cdot\xi_{y_1} + \cos\alpha\cdot\xi_{z_1} \qquad (8.3)$$

Putting Eq. (8.2) into Eq. (8.3) yields

$$\left.\begin{aligned}
\xi_{y_\alpha''} &= -\frac{1}{2}\left(\cos\alpha+\cot\frac{\omega}{2}\sin\alpha\right)\gamma_y \\[2mm]
&\quad -\frac{1}{2}\left(\sin\alpha-\cot\frac{\omega}{2}\cos\alpha\right)\gamma_z \\[2mm]
\xi_{z_\alpha''} &= \frac{1}{2}\left(\sin\alpha-\cot\frac{\omega}{2}\cos\alpha\right)\gamma_y \\[2mm]
&\quad -\frac{1}{2}\left(\cos\alpha+\cot\frac{\omega}{2}\sin\alpha\right)\gamma_z
\end{aligned}\right\}, \qquad (8.4)$$

where α denotes the angle between the axes y_α'' and y_1.

(3) Determine the components $\Delta S_{\xi',y_\alpha''}''$ and $\Delta S_{\xi,z_\alpha''}''$ of image point (F'') displacement corresponding to the existing optical axis deviation.

According to the definition and sign-convention of the optical axis deviation specially adopted in these two chapters (7,8), we obtain

$$\left.\begin{aligned}
\Delta S_{\xi',y_\alpha''}'' &= -\xi_{y_\alpha''}\cdot f_e' \\[2mm]
\Delta S_{\xi,z_\alpha''}'' &= -\xi_{z_\alpha''}\cdot f_e'
\end{aligned}\right\}, \qquad (8.5)$$

where f_e' represents the focal length of the eyepiece.

(4) Determine the components $\Delta S_{y_\alpha''}''$ and $\Delta S_{z_\alpha''}''$ of image point (F'') displacement caused by the adjustment of the prisms.

Obviously, these two terms should be equal to the negative of those respective terms found in (3), and thus we have

$$\left.\begin{aligned}
\Delta S_{y_\alpha''}'' &= -\Delta S_{\xi,y_\alpha''}'' = \xi_{y_\alpha''} f_e' \\[2mm]
\Delta S_{z_\alpha''}'' &= -\Delta S_{\xi,z_\alpha''}'' = \xi_{z_\alpha''} f_e'
\end{aligned}\right\}. \qquad (8.6)$$

(5) Construct all the coordinate systems as well as the related vectors and points which are needed in using the basic equation (4.2):

$$\Delta S' = \Delta\theta\, P\times\vec{\rho} + (-1)^{l-1}\Delta\theta\, P'\times\vec{\rho}'. \qquad (4.2)$$

Referring to Fig.8.7, we first calculate the terms of length $\left(\frac{L_s}{n}+b\right)$ and $\left(\frac{L_b}{n}+a+\frac{L_s}{n}+b\right)$

where a denotes the separation between two prisms while b the distance of the image point F'' from the exit face of the small Porro.

240

Note that all the previously used symbols will preserve their original meaning; only the subscripts s and b are used to distinguish the quantities referring to the small Porro from those referring to the big Porro.

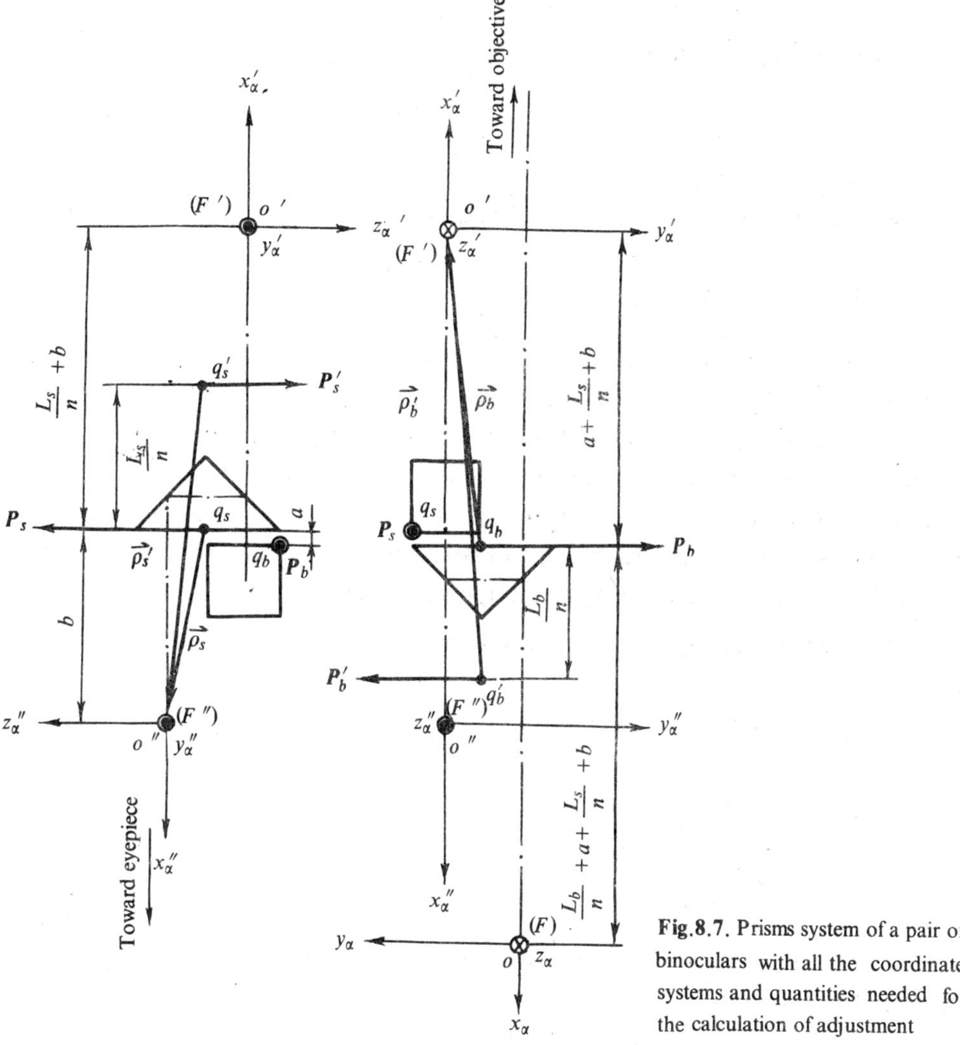

Fig.8.7. Prisms system of a pair of binoculars with all the coordinate systems and quantities needed for the calculation of adjustment

Construct two paris of completely conjugate coordinate systems for both prisms : one pair , $o'x_\alpha' y_\alpha' z_\alpha'$ and $o''x_\alpha'' y_\alpha'' z_\alpha''$ for the small Porro , and the other $o x_\alpha y_\alpha z_\alpha$ and $o'x_\alpha' y_\alpha' z_\alpha'$ for the big Porro. Of course , the origins o, o' and o'' are located at the object and image points F , F' and F'', respectively.

Let P_s and P_b represent the axes of rotation of two prisms which correspond to interposing spacers at the specified spots of the small Porro and big Porro , respectively. These two axes P_s and P_b and all the other related vectors and points P_s' , P_b' , $\vec{\rho}_s, \vec{\rho}_s'$, $\vec{\rho}_b$, $\vec{\rho}_b'$, q_s , q_s' , q_b and q_b' have been shown in Fig. 8.7 because of their simple geometrical relations in the present case.

(6) Derive the relationship between the image point (F'') displacement $\Delta S_{y_\alpha''}''$ and the

241

small angular displacement $\Delta\theta_s\, \boldsymbol{P}_s$.

From Eq. (4.2),

$$\Delta S_{y_\alpha''}'' = \Delta\theta_s\,(P_{sz_\alpha''}\,\rho_{sx_\alpha''} - P_{sx_\alpha''}\,\rho_{sz_\alpha''})$$

$$- \Delta\theta_s(P_{sz_\alpha''}'\,\rho_{sx_\alpha''}' - P_{sx_\alpha''}'\,\rho_{sz_\alpha''}')\ . \tag{8.7}$$

From Fig. 8.7, we have

$$P_{sz_\alpha''}=1,\ P_{sx_\alpha''}=0\ ,\ P_{sz_\alpha''}'=-1\ ,\ P_{sx_\alpha''}'=0,$$

$$\rho_{sx_\alpha''}'=b\ ,\qquad \rho_{sz_\alpha''}'=\frac{L_s}{n}+b.$$

Putting the above data into Eq. (8.7) yields

$$\Delta S_{y_\alpha''}''=\Delta\theta_s\left(\frac{L_s}{n}+2b\right). \tag{8.8}$$

As a matter of fact, the small angular displacement $\Delta\theta_s\, \boldsymbol{P}_s$ of the small Porro causes no component of image point (F'') displacement $\Delta S''$ along the z_α'' direction while the small angular displacement $\Delta\theta_b\boldsymbol{P}_b$ of the big Porro will cause no component of the image point (F'') displacement $\Delta S''$ along the y_α'' direction, that is to say, the optical axis deviation that has occurred in the plane $x_\alpha''\,y_\alpha''$, namely $\xi_{y_\alpha''}$, can be adjusted only by interposing a spacer under the small Porro while the optical axis deviation that has occurred in the plane $x_\alpha''\,z_\alpha''$, namely $\xi_{z_\alpha''}$, can be balanced only by interposing a spacer under the big Porro.

Thus, by combining Eq. (8.8) and the first formula of Eq. (8.6), we obtain

$$\Delta S_{y_\alpha''}''=\Delta\theta_s\left(\frac{L_s}{n}+2b\right)=\xi_{y_\alpha''}f_e'$$

or

$$\Delta\theta_s=\frac{\xi_{y_\alpha''}\,f_e'}{\dfrac{L_s}{n}+2b}\ . \tag{8.9}$$

Let l_s be the width of the hypotenuse face of the small Porro, and Δh_s be the thickness of the spacer under this prism. Then, we have

$$\Delta h_s=\frac{\xi_{y_\alpha''}\,f_e'\,l_s}{\dfrac{L_s}{n}+2b}$$

or

$$\Delta h_s=2.9\times10^{-4}\cdot\frac{\xi_{y_\alpha''}f_e'\,l_s}{\dfrac{L_s}{n}+2b}\ , \tag{8.10}$$

where the unit of optical axis deviation $\xi_{y_\alpha''}$ is expressed in min. of arc and the unit of all the length quantities are in mm.

(7) Derive the relationship between the image point (F') displacement $\Delta S_{z_\alpha}'$ and the small angular displacement $\Delta\theta_b\boldsymbol{P}_b$.

242

In this case , the small Porro is considered to be fixed , therefore the component $\Delta S'_{z_\alpha}$ of the image point $(F\,')$ displacement can readily be converted into the corresponding component $\Delta S''_{z_\alpha}$ of the image point $(F\,'')$ displacement by the following equation:

$$\Delta S''_{z_\alpha} = \Delta S'_{z_\alpha} \quad , \tag{8.11}$$

and according to the second formula of Eq. (8.6) we have

$$\Delta S'_{z_\alpha} = \xi_{z_\alpha''}\, f_e' \quad . \tag{8.12}$$

Now, from the same equation (4.2) ,

$$\Delta S'_{z_\alpha} = \Delta\theta_b (\, P_{bx_\alpha''}\, \rho_{b\,y_\alpha'} - P_{by_\alpha'}\, \rho_{bx_\alpha'}\,)$$
$$- \Delta\theta_b\, (P'_{bx_\alpha'}\, \rho'_{b\,y_\alpha'} - P'_{by_\alpha'}\, \rho'_{bx_\alpha'}\,). \tag{8.13 *}$$

Also from Fig. 8.7 ,

$$P_{bx_\alpha'} = 0, \qquad P_{by_\alpha'} = 1, \; P'_{bx_\alpha'} = 0 , \; P'_{by_\alpha'} = -1 ,$$

$$\rho_{b\,x_\alpha'} = a + \frac{L_s}{n} + b, \; \rho'_{bx_\alpha'} = a + \frac{L_s}{n} + b + \frac{L_b}{n} \quad .$$

Putting the above data into Eq. (8.13) and considering Eq. (8.12) gives

$$\Delta S'_{z_\alpha} = -\Delta\theta_b \left[\frac{L_b}{n} + 2\left(a + \frac{L_s}{n} + b\right) \right] = \xi_{z_\alpha''}\, f_e' \quad .$$

Thus,

$$\Delta\theta_b = - \frac{\xi_{z_\alpha''}\, f_e'}{\dfrac{L_b}{n} + 2\left(a + \dfrac{L_s}{n} + b\right)} \quad . \tag{8.14}$$

Let l_b be the width of the hypotenuse face of the big Porro , and Δh_b be the thickness of the spacer under this prism . Then , we obtain

$$\Delta h_b = - \frac{\xi_{z_\alpha''}\, f_e'\, l_b}{\left[\dfrac{L_b}{n} + 2\left(a + \dfrac{L_s}{n} + b\right) \right]}$$

or

$$\Delta h_b = -2.9 \times 10^{-4} \cdot \frac{\xi_{z_\alpha''} f_e'\, l_b}{\left[\dfrac{L_b}{n} + 2\left(a + \dfrac{L_s}{n} + b\right) \right]} \quad , \tag{8.15}$$

---

* In using this equation we should first check if the coordinate system $o'x_\alpha'\, y_\alpha'\, z_\alpha'$ is right-handed.

where the unit of optical axis deviation ξ_{z_α}'' is in minute of arc and the unit of all the length quantities are in mm.

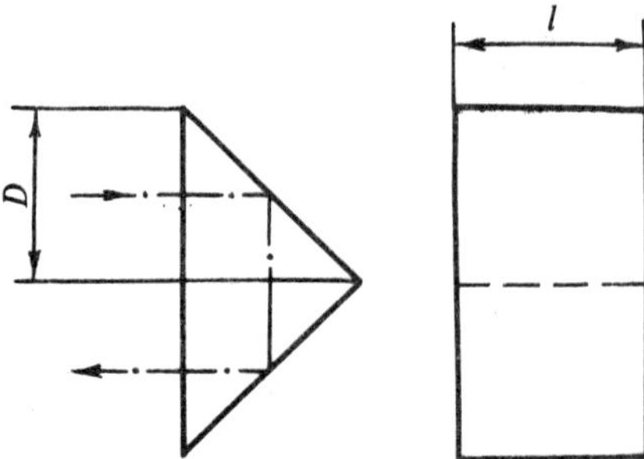

Fig.8.8. Dimensional parameters of a Porro prism

(8) Derive the relationships between the thickness of spacers Δh_s, Δh_b and the components of image shift γ_y, γ_z.

Putting Eq. (8.4) into Eqs. (8.10) and (8.15) to cancel the intermediate variables ξ_{y_α}'' and ξ_{z_α}'' yields

$$\Delta h_s = 2.9 \times 10^{-4} \cdot \frac{f_e' \, l_s}{\left(\dfrac{L_s}{n} + 2b\right)} \left[-\frac{1}{2}\left(\cos\alpha + \cot\frac{\omega}{2}\sin\alpha\right)\gamma_y \right.$$

$$\left. -\frac{1}{2}\left(\sin\alpha - \cot\frac{\omega}{2}\cos\alpha\right)\gamma_z \right] , \tag{8.16}$$

$$\Delta h_b = -2.9 \times 10^{-4} \cdot \frac{f_e' \, l_b}{\left[\dfrac{L_b}{n} + 2\left(\dfrac{L_s}{n} + a + b\right)\right]} \left[\frac{1}{2}\left(\sin\alpha \right.\right.$$

$$\left.\left. -\cot\frac{\omega}{2}\cos\alpha\right)\gamma_y - \frac{1}{2}\left(\cos\alpha + \cot\frac{\omega}{2}\sin\alpha\right)\gamma_z \right] . \tag{8.17}$$

According to the right-handed rule, the positive values of Δh indicate that the spacer should be put at the inner spots of the two prisms, i.e., B.I. or S.I. while the negative values of Δh at the outer spots of the two prisms, i.e., B.O. or S.O..

Now, let us take the real product as an example whose data are given below:

$$\omega = 60^\circ, \qquad \alpha = 45^\circ, \qquad f_e' = 16.7, \qquad a = 2, \qquad b = 20.3,$$
$$D_s = 16.5, \qquad D_b = 17.5, \qquad l_s = 16.5; \qquad l_b = 17.5,$$
$$n = 1.5163.$$

Thus,

$$\frac{L_s}{n} = \frac{2D_s}{n} = 21.7 \, ,$$

$$\frac{L_b}{n} = \frac{2D_b}{n} = 23.0 \, .$$

Putting the above data into Eqs. (8.16) and (8.17), we obtain

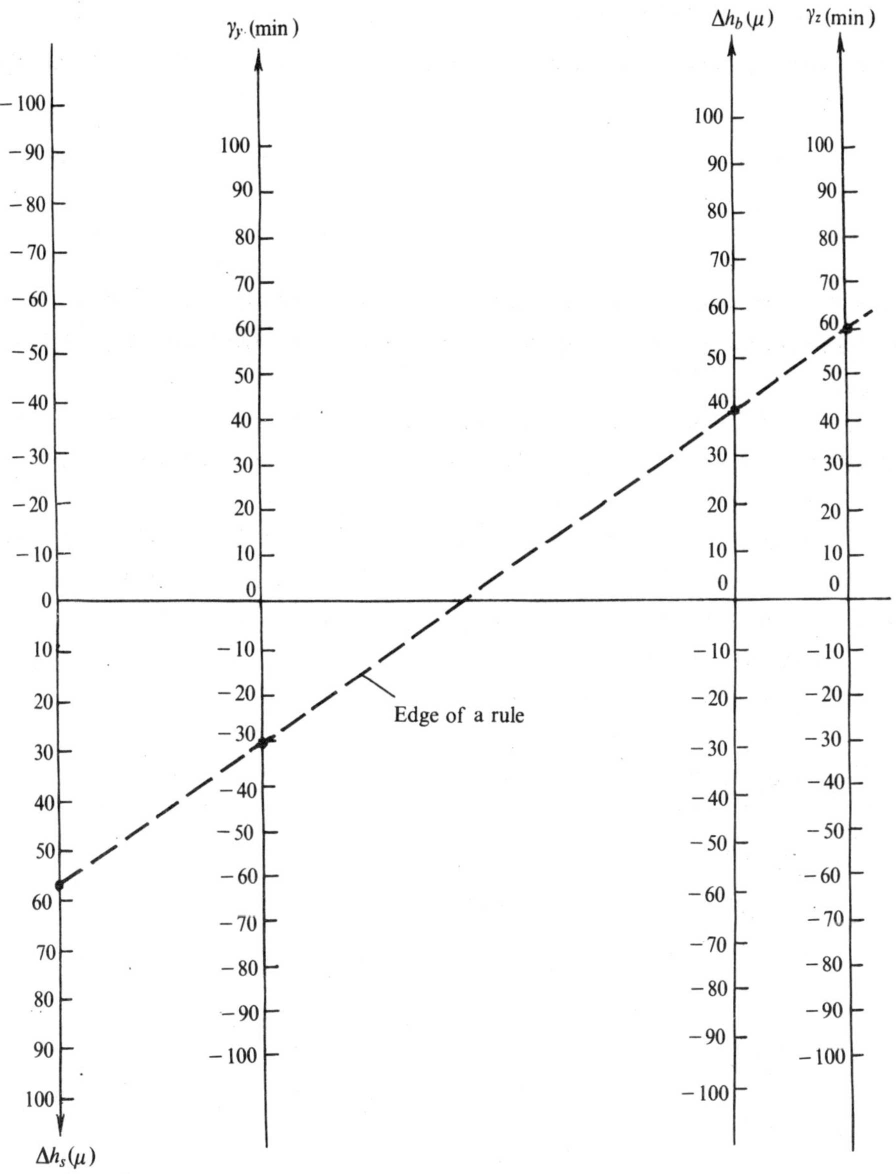

Fig.8.9. Plotting of $\Delta h - \gamma$ relationships in parallel coordinate axes

$$\Delta h_s = (-1.23\,\gamma_y + 0.33\,\gamma_z) \times 10^{-3}, \qquad (8.18)$$

$$\Delta h_b = (0.19\,\gamma_y + 0.74\,\gamma_z) \times 10^{-3}. \qquad (8.19)$$

For convenience of using these resulting formulas in assembly shops , the $\Delta h - \gamma$ relationships are plotted in parallel coordinate axes , as shown in Fig. 8.9.

For instance , if the detected image shift appears to be : $\gamma_y = -30'$ and $\gamma_z = 60'$, a plastic rule with its edge to intersect the γ_y and γ_z axes at the given values will then cut the Δh_s and Δh_b axes at the values you need . In this case , we have

$$\Delta h_s = 57\mu\,, \qquad \Delta h_b = 39\,\mu\,.$$

These spacers of positive value will thus be put under the inner spots of two prisms , namely S.I. and B.I.

Problems

8.1 Choose a rotation axis of either one of the two Porro prisms for adjusting the image lean of the binoculars, as shown in Fig. 8. 1, and give the reasoning that leads to your choice.

8.2 Work out a reasonable procedure of adjustment of such a pair of prism binoculars regarding mainly two major terms: the image lean and the parallelism of the optical axes .

8.3 When the measuring telescope is mounted on a simple hinged arm , as shown in Fig. 8. 10, instead of being equipped with a double rhomboid prism head, how could you deal with the problem of the individual adjustment of optical axis in this situation if you still want to use the equations (8. 16) and (8. 17)?

Hint: the problem is about how to handle the angle α .

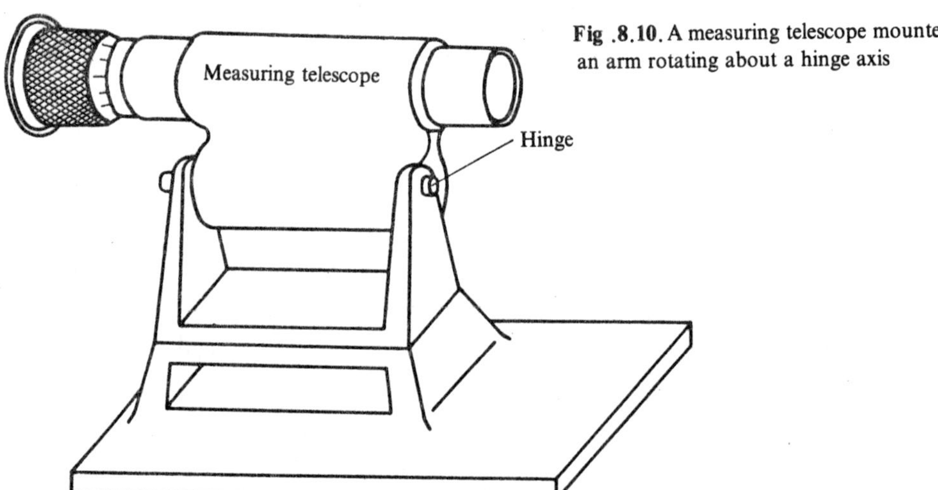

Fig .8.10. A measuring telescope mounted on an arm rotating about a hinge axis

Chapter 9
Calculation of Adjustment of a Stereoscopic Telescope

The adjustment of a stereoscopic telescope discussed in the first section of the chapter is another example which illustrates the individual adjustment of optical axis in those cases where the calculation of adjustment is not necessary after a reading operation.

In addition , the looped calculation of adjustment mentioned in the second section of the chapter is also a typical performance frequently encountered in practice.

9.1 Individual Adjustment of Optical axis

9.1.1 A Brief on the Structure of a Stereoscopic Telescope and the Equipment for Adjustment of Optical Axis

Referring to Fig. 9.1, a stereoscopic telescope includes two major parts: the upper part, a

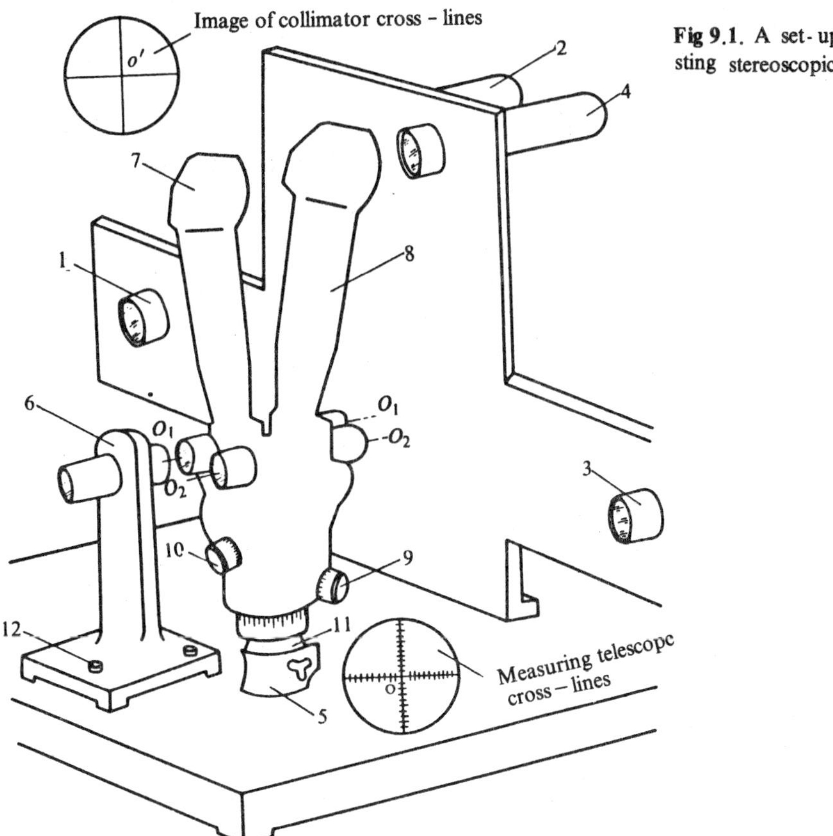

Fig 9.1. A set-up for adjusting stereoscopic telescopes

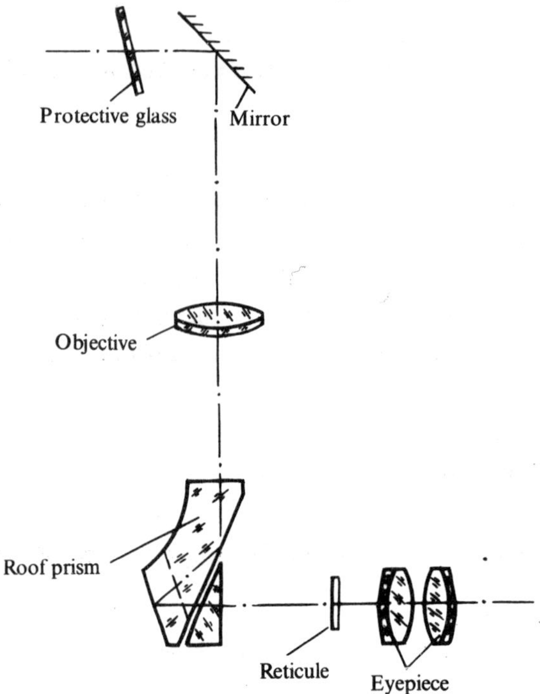

pair of viewing binoculars , and the lower part , an angle measuring device. The viewing binocu-lars contains two telescope-members : the left- side one 7, and the right-side one 8. Each telescope may rotate around its own axis O_1O_1 or O_2O_2. When two telescopes are folded, they act as a periscope , while when unfolded , they will provide a strong depth sensation . The angle measuring device consists of two worm gearing pairs with two knobs 9 and 10 rigidly attached to the worms , by the turning of which the viewing binoculars will be rotated in horizontal and vertical planes.

The whole stereoscopic telescope may be set on a tripod through the ball pivot at its bottom.

The optical systems of the two telescope- members are basically the same ; the only difference is that the right- side telescope , as shown in Fig.9.2, is possessed of an additional reticule.

During the individual adjustment of optical axis of each telescope- member , we need to change the direction of the parallel beam emerging from the product- telescope by adjusting a certain element of it , and this will be carried out by a slight rotation of the upper mirror for both left- side and right- side telescopes.

The structure of the mirror assembly is shown in Fig. 9. 3. The mirror 9 is fastened to a platform 1 with two pressure plates 10 and screws , and through a sphere pivot 5 the platform 1 is then attached to the cover 2. The screw ring 7 and the washer 6 are used to clamp the sphere pivot .The whole assembly is finally connected to the instrument by screwing the cover on the telescope tube .

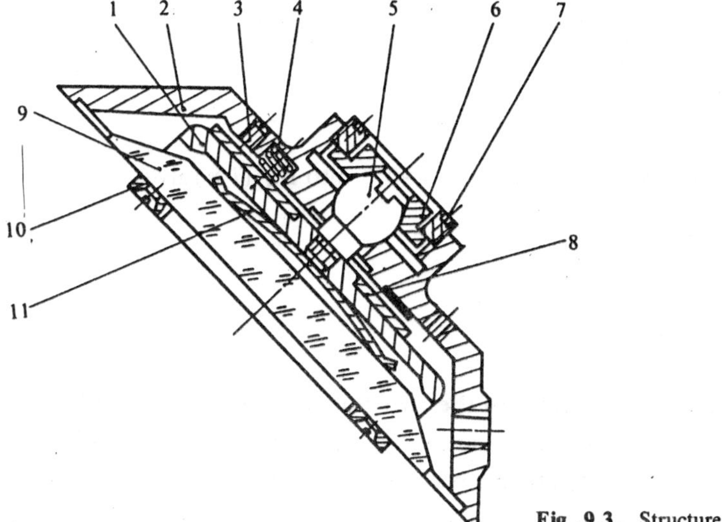

Fig .9.3. Structure of the mirror assembly

Before adjusting the optical axis of the produc t - telescope , the screw ring 7 should be loosened a little bit ; then by adjusting the three setting screws 3 , the mirror will turn around an arbitrary axis orientated parallel to the mirror plane.

Most parts of the set-up for adjusting the stereoscopic telescope , see Fig. 9. 1, have been mentioned in Chap.7. Collimators 1 and 2 are used to make the individual adjustment of optical axis for the left - side telescope while collimators 3 and 4 for the right- side telescope.

9.1.2 Individual Adjustment of Optical Axis

Referring again to Fig. 9. 1 , suppose that for both the left - side and right - side tele-scope -members the aiming operation is performed while the telescope tube is directed horizontal and the reading operation is completed while the telescope tube is directed ver-tical . Therefore, we have $\omega = -90°$ for the left- side telescope and $\omega = 90°$ for the right-side telescope.

In this case , the adjusting structure is accessible, so during performing the adjustment the operator may keep on viewing through the measuring telescope . According to the principle discussed in Sect.7.4 , the adjustment should be made in such a way so as to bring the image point o' of the center of the collimator cross-lines to point c which is the center of the circular locus of the moving point o' .

For this purpose , calculate the coordinates of point c in the coordinate system yoz representing the cross-line of the measuring telescope in reading state.It has been seen in Sect. 7.4 that the coordinates of point c are none other than the two components γ_y' and γ_z' of the reserved image shift.

For the left- side telescope , putting $\omega = -90°$ into Eq. (7.8) yields

249

$$\gamma_y{}' = \frac{\gamma_z + \gamma_y}{2} \left.\begin{array}{c} \\ \\ \\ \\ \end{array}\right\} .$$

$$\gamma_z{}' = \frac{\gamma_z - \gamma_y}{2}$$

(9.1)

For the right-side telescope , putting $\omega = 90$ ° into Eq. (7.8) gives

$$\gamma_y{}' = \frac{-\gamma_z + \gamma_y}{2} \left.\begin{array}{c} \\ \\ \\ \\ \end{array}\right\} .$$

$$\gamma_z{}' = \frac{\gamma_z + \gamma_y}{2}$$

(9.2)

In particular cases when either of the two components of the image shift happens to be equal to zero, or when the absolute values of two components of the image shift are equal to each other, then the rules for adjustment become much more simple , which are shown in Fig. 9.4.

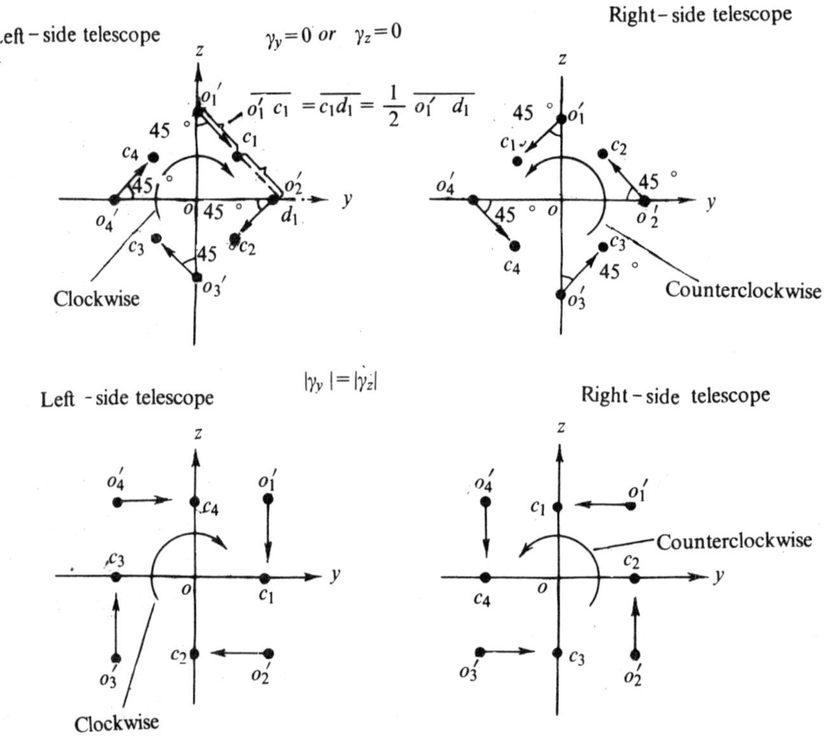

Fig.9.4. Rules for adjustment in particular cases

9.2 Adjustment of Image Lean —Looped Calculation of Adjustment

9.2.1 Concept of Looped Calculation of Adjustment

Very often , there are two reflecting prisms in an optical system , where one is used to adjust the optical axis deviation and the other will be used to adjust the image lean . For some reasons , these two steps of adjustment could not be independent of each other , that is to say , the first adjustment is upset by the adjustment that follows, and moreover redoing (or recovering) the previous adjustment might again disturb the presently adjusted term. Therefore , to get a better approximation the same procedures will be repeated for many cycles. For the example to be discussed, one cycle includes the following three procedures: adjustment of optical axis ; adjustment of image lean ; and again, adjustment of optical axis.

The so - called *looped calculation of adjustment* is intended to consider the influence of two successive procedures on each other so that the whole adjustment can be completed in one cycle.

To make this concrete , we will take the optical system of a stereoscopic telescope (see Fig. 9.2) as an example.

First of all, suppose that the adjustment of optical axes of two telescope- members has been completed. The principle and method in this regard are discussed in full detail in Chap. 7 as well as in Sect. 9.1.

At this moment , generally speaking there must be for the optical system still some image lean which remains to be adjusted by slightly rotating the roof prism through a small angle about a certain axis.

These are the starting conditions on which the following looped calculation of adjustment will be based .

Now, to calculate the small angular displacement $\Delta\theta P$ of the roof prism , we should consider two parts of image lean which are directly and indirectly caused by this small rotation $\Delta\theta P$.

We will explain how the indirect part of image lean is produced.

It is evident that a small rotation of the roof prism will cause not only an image lean (the direct part) but also an optical axis deviation . To eliminate the newly derived optical axis deviation , or in other words , to recover the optical axis*, we have to turn the mirror again. Similarly, the additional small rotation of the mirror will also cause an image lean . After this image lean is converted from the image space of the mirror into the image space of the roof prism , it will become that part of image lean indirectly caused by the small rotation of the prism.

We will then sum up these two parts of image lean caused directly and indirectly by the small angular displacement of the roof prism , and let this sum be equal to the existing image

---

* That means to set again the optical axis of the telescope parallel to the hinge axis.

lean of the optical system in magnitude but opposite in sign. Thus , we can avoid the need for one cycle more.

9.2.2 Derivation of Necessary Equations

As mentioned above , a looped calculation corresponds to a complete cycle of adjustment including a few procedures , and each physical process will be described by certain formulas . Therefore , prior to tackling a complete looped calculation of adjustment , we have to derive all the equations for each constituent part of the cycle as well as some formulas needed to transform the relevant quantities from one to another space.

(1) The optical axis deviation $\Delta\mu'_{(y')}$ and $\Delta\mu'_{(z')}$, and the image lean $\Delta\mu'_{(x')}$ caused by a small angular displacement $\Delta\theta_m P$ of the mirror

In Fig.9.5, is shown the adjustment diagram of a right - angle prism DI - 90 °which has the same adjustment behavior as that of a mirror intersecting the optical axis at 45°in a collimated beam.

Since a rotation of a mirror around its normal will cause nothing , the rotation axis P of the mirror is thus orientated in a plane parallel to the mirror plane , and the direction of P is represented by the angle γ measured from w , as shown in the view along the D direction.

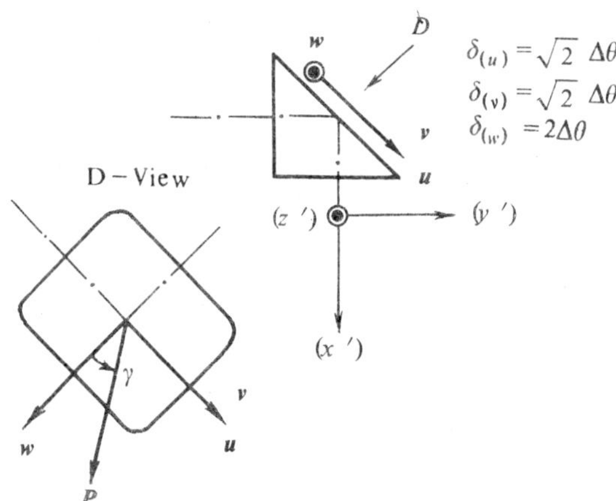

Fig.9.5. Adjustment diagram of a mirror at 45 °

From Fig . 9.5 , we obtain

$$\Delta\mu'_{(x')} = \sqrt{2}\ \sin\gamma \cdot \Delta\theta_m , \tag{9.3}$$

$$\Delta\mu'_{(y')} = \sqrt{2}\ \sin\gamma \cdot \Delta\theta_m , \tag{9.4}$$

$$\Delta\mu'_{(z')} = 2\cos\gamma \cdot \Delta\theta_m , \tag{9.5}$$

where (x') (y') (z') denotes the coordinate system in the image space of the mirror and has round brackets to avoid being confused with the coordinate system $x'y'z'$ in the image space of the roof prism.

(2) The image leans $\Delta\mu'_{x'P_1}$ and $\Delta\mu'_{x'P_2}$ caused by the respective small angular displace-

252

ments $\Delta\theta_{1p}\,\boldsymbol{P}_1$ and $\Delta\theta_{2p}\,\boldsymbol{P}_2$ of the roof prism .

As mentioned in Sect. 5. 4, while adjusting the image lean the roof prism may rotate slightly around either the axis $\boldsymbol{P}_1$ or $\boldsymbol{P}_2$ (see Figs. 5.11 and 5.12).

Similarly, we have the adjustment diagram of the roof prism $FX_J - 90°$, as shown in Fig. 9. 6.

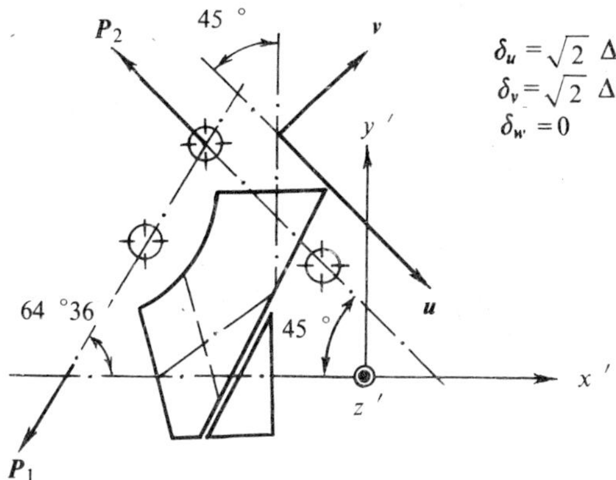

Fig .9.6. Adjustment diagram of a roof prism $FX_J - 90$ °

From Fig. 9. 6, we can readily obtain

$$\Delta\mu_{x'}{}'_{P1} = \sqrt{2}\ \Delta\theta_{1p} \cdot \cos 70°\,24\,' = 0.\,48\ \Delta\theta_{1p} , \tag{9.6}$$

$$\Delta\mu_{x'}{}'_{P2} = \sqrt{2}\ \Delta\theta_{2p} \cdot \cos 180° = -\sqrt{2}\ \Delta\theta_{2p} . \tag{9.7}$$

(3) The $y\,'$ and $z\,'$ image point displacements $\Delta S_{y'}{}'_{P1}$, $\Delta S_{z'}{}'_{P1}$, $\Delta S_{y'}{}'_{P2}$ and $\Delta S_{z'}{}'_{P2}$ caused by the respective small angular displacements $\Delta\theta_{1p}\,P_1$ and $\Delta\theta_{2p}\,P_2$ of the roof prism .

These relationship have been obtained from the solution of the example discussed in Sect. 5.4, therefore we may just rewrite the results:

$$\Delta S_{y'}{}'_{P1} = -11.\,1\Delta\theta_{1p} , \tag{5.13}$$

$$\Delta S_{z'}{}'_{P1} = 126.\,4\Delta\theta_{1p} , \tag{5.14}$$

$$\Delta S_{y'}{}'_{P2} = 0 , \tag{5.15}$$

$$\Delta S_{z'}{}'_{P2} = 39.\,2\ \Delta\theta_{2p} . \tag{5.16}$$

All the coordinate systems related to the above formulas are indicated in Fig. 9. 7, where $o\,x\,y\,z$ and $o\,'x\,'y\,'\,z\,'$ represent a pair of completely conjugate coordinate systems for the roof prism whose origins o and $o\,'$ are in coincidence with a conjugate pair of points F_0 and $F\,'$ for the same prism , respectively.

(4) Transformation of the optical axis deviation from the components of image point displacement $\Delta S_{y'}'$ and $\Delta S_{z'}'$ in the image space $o\,'\,x\,'\,y\,'z\,'$ of the roof prism into the components of image rotation $\Delta\mu_{(y\,')}'$ and $\Delta\mu_{(z\,')}'$ in the image space $(x\,')(y\,')(z\,')$ of the mirror

From the object and image relation for the roof prism, we have

253

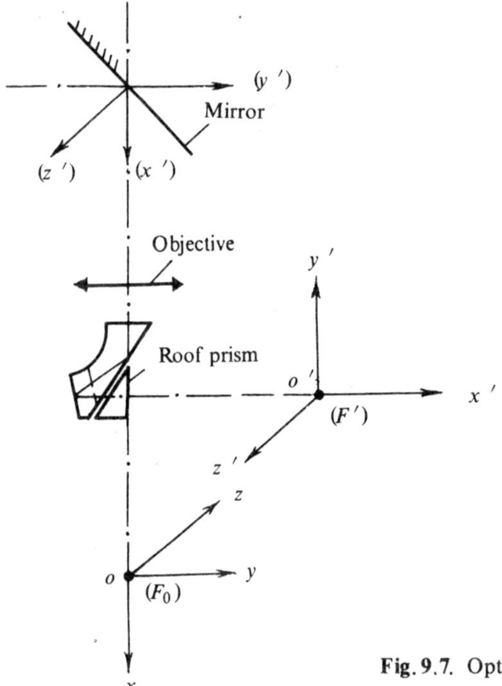

Fig. 9.7. Optical system with coordinate systems in different spaces

$$\Delta S_y = \Delta S_{y'}' \quad , \quad \Delta S_z = \Delta S_{z'}' \, ,$$

where ΔS_y and ΔS_z represent the displacement components of the object point F_0 in the yoz plane while $\Delta S_{y'}'$ and $\Delta S_{z'}'$ the corresponding displacement components of the image point F' in the $y'o'z'$ plane .

Then , according to the relative orientation of the coordinate system $oxyz$ · with respect to the coordinate system $(x')\ (y')\ (z')$, we have

$$\Delta\mu_{(y')}' = \frac{\Delta S_z}{f_o'} = \frac{\Delta S_{z'}'}{f_o'} \quad , \tag{9.8}$$

$$\Delta\mu_{(z')}' = \frac{\Delta S_y}{f_o'} = \frac{\Delta S_{y'}'}{f_o'} \quad , \tag{9.9}$$

where f_o' denotes the focal length of the objective.

(5) Transformation of the image lean from the component of image rotation $\Delta\mu_{(x')}'$ in the image space $(x')\ (y')\ (z')$ of the mirror into the component of image rotation $\Delta\mu_{x'}'$ in the image space $x'\ y'\ z'$ of the roof prism.

It should first be pointed out that neither the magnitude nor the sign of the image lean will be affected by the objective lens. And thanks to the same direction of the axes (x')

254

and x , we have

$$\Delta\mu_x = \Delta\mu'_{(x')}.$$

Then , according to the object and image relationship for the roof prism , considering $t = 3$ we obtain

$$\Delta\mu'_{x'} = (-1)^3 \Delta\mu_x = -\Delta\mu_x .$$

Thus, ultimately

$$\Delta\mu'_{x'} = -\Delta\mu'_{(x')}. \tag{9.10}$$

9.2.3 Looped Calculation of Adjustment

Consider first the case when the roof prism has rotated through a small angle $\Delta\theta_{1p}$ about the axis P_1 .

Let δ_1 be the sum of the two parts of image lean which are caused directly and indirectly by the small angular displacement $\Delta\theta_{1p} P_1$ of the roof prism.

Thus,

$$\delta_1 = \Delta\mu'_{x'P_1} + \Delta\mu'_{x'} . \tag{9.11}$$

Obviously , $\Delta\mu'_{x'P_1}$ representing the image lean of the part directly caused by a slight rotation of the root prism can readily be determined using Eq. (9.6)

$$\Delta\mu'_{x'P_1} = 0.48 \, \Delta\theta_{1p} . \tag{9.12}$$

Only, $\Delta\mu'_{x'}$ representing the image lean of that part indirectly caused by the slight rotation of the roof prism remains to be found in the following steps.

(1) Determine the y' and z' image point (F') displacements $\Delta S'_{y'P_1}$ and $\Delta S'_{z'P_1}$ due to $\Delta\theta_{1p} P_1$

From Eqs. (5.13) and (5.14),

$$\Delta S'_{y'P_1} = -11.1 \, \Delta\theta_{1p} , \tag{9.13}$$

$$\Delta S'_{z'P_1} = 126.4 \Delta\theta_{1p} . \tag{9.14}$$

(2) Convert the optical axis deviation from $\Delta S'_{y'P_1}$ and $\Delta S'_{z'P_1}$ into $\Delta\mu'_{(y')p}$ and $\Delta\mu'_{(z')p}$

From Eqs. (9.8) and (9.9),

$$\Delta\mu'_{(y')p} = \frac{126.4\Delta\theta_{1p}}{f'_o} , \tag{9.15}$$

$$\Delta\mu'_{(z')p} = -\frac{11.1\Delta\theta_{1p}}{f'_o} . \tag{9.16}$$

(3) Calculate the small rotation of the mirror $\Delta\theta_m$ and γ that are needed to recover

the parallelism of the optical axes

Changing the sign of the values of $\Delta\mu'_{(y')p}$ and $\Delta\mu'_{(z')p}$ obtained from Eqs. (9.15) and (9.16), and then putting them into Eqs. (9.4) and (9.5) yields

$$-\frac{126.4\Delta\theta_{1p}}{f'_o} = \sqrt{2}\,\Delta\theta_m \sin\gamma \, ,$$

$$\frac{11.1\Delta\theta_{1p}}{f'_o} = 2\Delta\theta_m \cos\gamma \, .$$

Thus, we have

$$\text{tg}\,\gamma = -16.1, \ \gamma = 93°\,33' ,$$

$$\Delta\theta_m = -\frac{89.7}{f'_o} \cdot \Delta\theta_{1p} \, .$$

(4) Determine the image lean $\Delta\mu'_{(x')}$ caused by the additional rotation of the mirror

Putting the obtained values of γ and $\Delta\theta_m$ into Eq. (9.3) leads to

$$\Delta\mu'_{(x')} = -\frac{126.4}{f'_o} \cdot \Delta\theta_{1p} \, .$$

(5) Determine $\Delta\mu'_{x'}$

From Eq. (9.10),

$$\Delta\mu'_{x'} = \frac{126.4}{f'_o} \cdot \Delta\theta_{1p} \, .$$

Putting $f'_o = 199.8$ mm into the above equation gives

$$\Delta\mu'_{x'} = 0.63\Delta\theta_{1p} \, . \tag{9.17}$$

Finally, from Eqs. (9.11), (9.12), and (9.17), we find

$$\delta_1 = \Delta\mu'_{x'P_1} + \Delta\mu'_{x'} = 1.11\Delta\theta_{1p} \, . \tag{9.18}$$

While considering the case when the roof prism has rotated through a small angle $\Delta\theta_{2p}$ about the axis P_2, let δ_2 be the total image lean caused by this small angular displacement $\Delta\theta_{2p}P_2$, and then in a similar way, we can find

$$\delta_2 = -1.22\Delta\theta_{2p} \, . \tag{9.19}$$

The opposite signs of the resulting equations (9.18) and (9.19) indicate that the fastening structure of the prism enables us to adjust image leans in different directions by padding at spot 1 or 2 (see Figs. 5.11 and 5.12).

Problems

9.1 Try to derive Eq. (9.19) in detail.

9.2 Try to choose such an axis of rotation of the roof prism for adjusting the image lean that the present step of adjustment will not upset the alignment of optical axis which has been previously completed.

Chapter 10

Layout of a Line of Sight (LOS) Steering and Stabilization Head*

The LOS steering and stabilization head discussed in this chapter is part of a navigational instrument. We will not attempt to involve ourselves in the specialized knowledge of the whole instrument , but we should at least be aware that the instrument is mounted on a platform of an aircraft autopilot course stabilizer . The instrument including the head is directly stabilized about the yaw axis by the stabilizer , and by use of the instrument , the aircraft is indirectly kept by the autopilot on a predetermined course and in level flight.

The material included in the present chapter is mainly intended to illustrate how to apply the previously developed theory to a typical configuration of stabilized prism assembly as in Fig. 10.1. However , a portion of this chapter is intended to remedy the deficiency encountered in Chap. 6 by including a treatment of a prism which acts simultaneously as a scanning and compensating element.

10.1 An Overview of the Layout

In Fig. 10.1 is depicted in diagrammatic outline a possible layout of a LOS steering and stabilization head . The whole head is mounted on the telescope housing through two bearings of the axle of the outer gimbal ring , and the telescope in turn is supported by a platform which will again be mounted on an aircraft.

The optical device is designed for measuring the position of a ground point with respect to the aircraft in terms of some angles measured relative to a certain reference frame.

For this purpose , the system is equipped with a viewing telescope which will provide a line of sight to track a ground point.

A scanning prism assembly which is disposed in front of the objective comprises two double-Dove prisms. The upper one is used to steer the LOS in the lateral direction and thus called the lateral prism while the lower one is used to steer the LOS in the longitudinal direction and thus called the longitudinal prism.

To ensure a certain stabilized reference frame for angular measurements , we make two scanning prisms serve simultaneously as compensating elements.

---

* As will be seen later in Eq. (10.13), in designing this head we have actually here a complete solution to the problem of image stabilization involving three degrees of freedom .

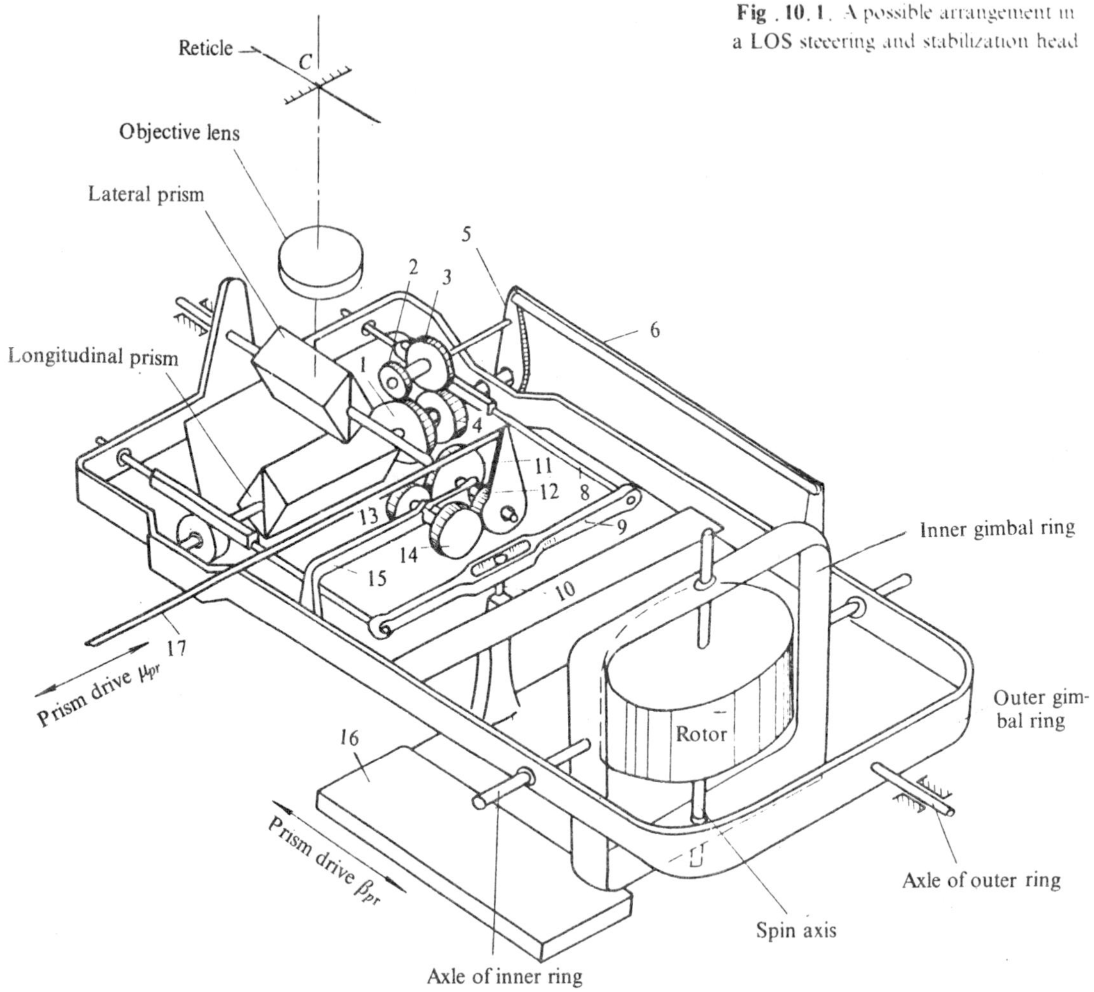

Fig . 10. 1. A possible arrangement in a LOS steering and stabilization head

Hence, the shafts of the two prisms are mounted on the outer gimbal ring of a gyroscope and receive both the scanning movements and the compensating motions from the prism drives 16, 17, and from the inner and outer gimbal rings through two gear differentials , respectively , as can be seen in Fig. 10. 1.

The gear differential belongs to a mechanism having two degrees of freedom which enables it to transmit two input-motions simultaneously and independently to one driven member ; in this case , the latter is the prism.

To make it clear , one gear differential driving the longitudinal prism is again shown in Fig. 10. 2, numbered to correspond with Fig. 10. 1. Here , the scanning movement is inputted through the sun gear 4 while one component of the compensating motion through the arm 5. The longitudinal prism is rotated together with the driven sun gear 1.

From the theory of mechanism, we know that the velocity ratio $i_{5,1}$ from the arm 5 to the sun gear 1 is equal to

260

$$i_{5,1} = \frac{n_5}{n_1} = \frac{Z_3 Z_1}{Z_3 Z_1 - Z_4 Z_2} , \tag{10.1}$$

where z_i represents the number of teeth of the i gear while n_i the turns of the i member.

It is apparent that stabilizing the longitudinal double-Dove prism in its mechanical rotation axis, as shown in Fig. 10.2 or 10.1, will cause overcompensation due to the angle doubling. To avoid this, the prism should be rotated by half the required angular compensation. This may be achieved by choosing the proper numbers of teeth for the gear differential to obtain an appropriate value of velocity ratio

$$i_{5,1} = 2 . \tag{10.2}$$

Another gear differential driving the lateral prism can be handled in the same way.

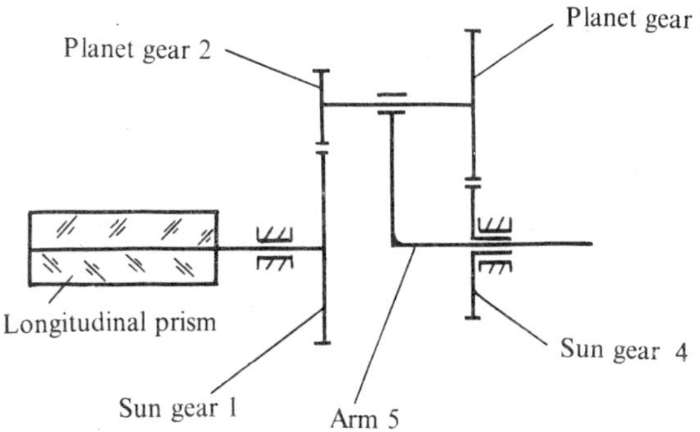

Fig. 10.2. The gear differential driving the longitudinal prism

10.2 LOS Steering System

Referring to Fig.10.3, an ordinary Keplerian telescope with a reticle at the focus of the objective lens and a scanning prism assembly in front of it is designed for a LOS steering system.

Assume that the light is travelling in the reverse direction. A divergent pencil of rays starting from the reticle center C will become a collimated beam A after its passage through the objective lens. Suppose the two prisms are momentarily in their home positions; then the entering parallel beam A just penetrates them, emerging along its original direction. This will also be the zero direction of the line of sight, and is said to be vertical if the aircraft would be kept absolutely horizontal.

Now, during the scanning process, suppose that the lateral and longitudinal prisms 1 and 2 have rotated through angles μ_{pr} and β_{pr}, respectively, at a certain instant. The collimated beam A is first deflected by the prism 1, emerging along the direction A', and will then be deflected again by the prism 2, projected finally into the measurement space along the direction A'' of the line of sight.

Assume $x'y'z'$ to be a reference coordinate system in the measurement space, where the x' axis represents the vertical while the $x'y'$ plane represents the vertical plane in the space. Also, the $x'y'$ plane will be perpendicular to the mechanical rotation axis of the longitudinal prism if the aircraft is strictly in level flight.

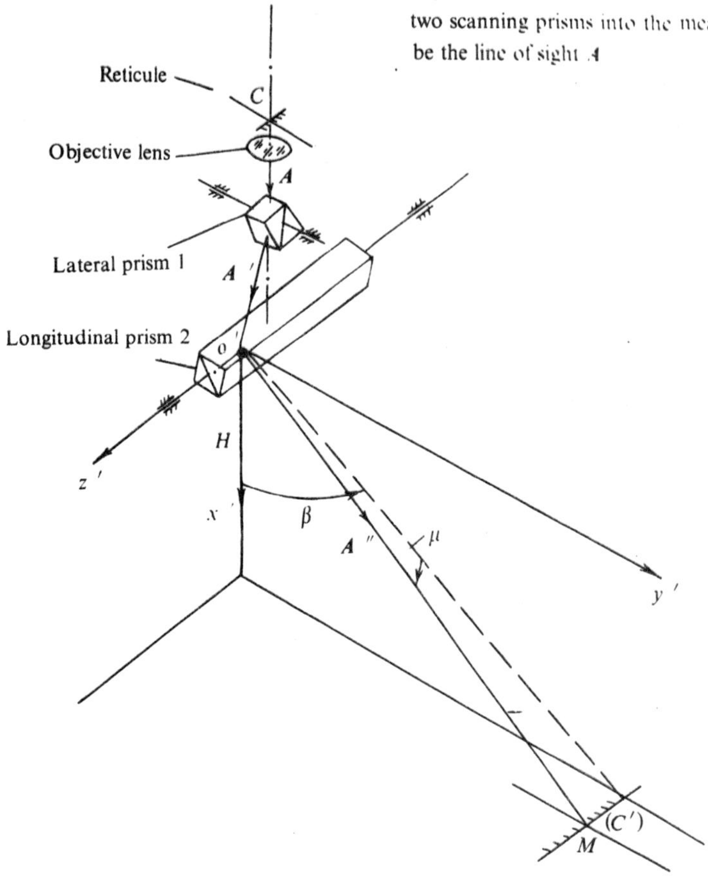

Fig. 10.3. The reticle vector A is projected through two scanning prisms into the measurement space to be the line of sight A

To measure the relative position of a ground point M with respect to the aircraft*, the LOS should be steered to track this ground point. Then ,the position of point M will be expressed in terms of the direction of the LOS represented by the following two angles β and μ , where the angle β is measured in the vertical plane $x'y'$ relative to the vertical axis x' and is called the longitudinal angle of LOS while the angle μ is measured relative to the vertical plane $x'y'$ and is called the lateral angle of LOS.

Let us derive the relationships between the angles β and μ of LOS and the angles of rotation β_{pr} and μ_{pr} of two scanning prisms.

To do this ,the normal N_1 to the interface of the lateral prism, the nomal N_2 to the interface of the longitudinal prism ,and the input vector A are all indicated in Fig. 10.4 according to their relative directions with respect to the reference frame $x'y'z'$. Also ,the intermediate output vector A' and the final output vector A'' are shown in the same figure.

From Fig. 10.4,

$$A_{x'} = 1, A_{y'} = 0, A_{z'} = 0 ,$$

---

* Strictly speaking, we should use 'the reference frame' instead of the aircraft .

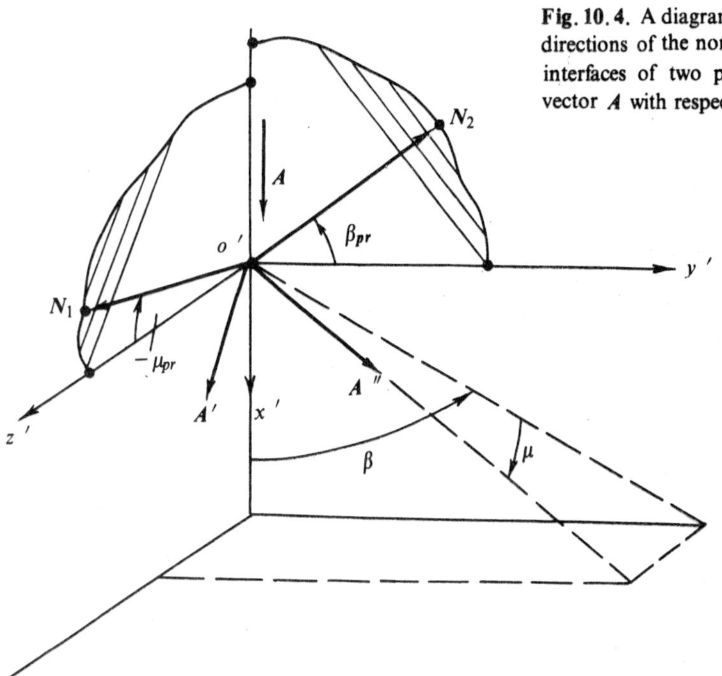

Fig. 10.4. A diagram to indicate the relative directions of the normals N_1 and N_2 to the interfaces of two prisms, and the input vector A with respect to the frame $x\,'y\,'z\,'$

$$N_{1x'} = -\sin(-\mu_{pr}) = \sin\mu_{pr} \, , \; N_{1y'} = 0 \, , \; N_{1z'} = \cos(-\mu_{pr}) = \cos\mu_{pr} \, .$$

Using the reflection vector formula (1.1) first for the lateral prism, we have

$$A' = A - 2(A \cdot N_1)N_1 \, .$$

Putting the known data into the above equation gives

$$-2(A \cdot N_1) = -2\sin\mu_{pr} \, ,$$

and

$$A_x' = A_{x'} - 2(A \cdot N_1)N_{1x'} = \cos2\mu_{pr} \, ,$$

$$A_y' = A_{y'} - 2(A \cdot N_1)N_{1y'} = 0,$$

$$A_z' = A_{z'} - 2(A \cdot N_1)N_{1z'} = -\sin2\mu_{pr} \, .$$

Similarly, from Fig. 10.4,

$$N_{2x'} = -\sin\beta_{pr} \, , \; N_{2y'} = \cos\beta_{pr} \, , \; N_{2z'} = 0 \, .$$

Again, using the reflection vector formula for the longitudinal prism yields

$$-2(A' \cdot N_2) = 2\cos2\mu_{pr}\sin\beta_{pr} \, ,$$

and

$$A_x'' = A_x' - 2(A' \cdot N_2)N_{2x'} = \cos2\mu_{pr}\cos2\beta_{pr} \, ,$$

$$A_y'' = A_y' - 2(A' \cdot N_2)N_{2y'} = \cos2\mu_{pr}\sin2\beta_{pr} \, ,$$

$$A_z'' = A_z' - 2(A' \cdot N_2)N_{2z'} = -\sin2\mu_{pr} \, .$$

Finally, from Fig. 10.5,

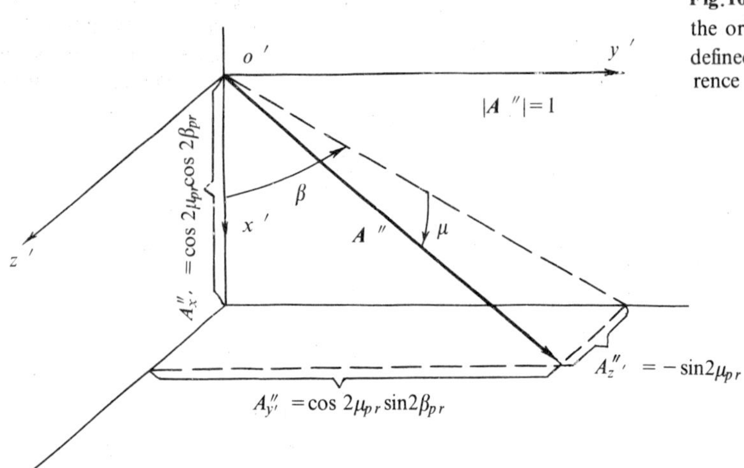

Fig. 10.5. A diagram to show the orientation of the LOS defined relative to the reference frame $x'y'z'$

$$\text{tg}\,\beta = \frac{A''_{y'}}{A''_{x'}} = \text{tg}\,2\beta_{pr}\,,$$

$$\sin\mu = \frac{A''_{z'}}{1} = -\sin 2\mu_{pr}\,.$$

Therefore,

$$\beta = 2\beta_{pr}\,,\tag{10.3}$$

$$\mu = -2\mu_{pr}\,.\tag{10.4}$$

The resulting relationships between the direction angles (β, μ) of LOS and the rotation angles (β_{pr}, μ_{pr}) of prisms appear to be very simple . The negative sign in Eq. (10.4) depends on the sign-convention as indicated in Fig. 10.4.

Hence , by continuously tracking a ground point we can obtain the data of its position information through the angular displacements of two scanning prisms.

All the relations mentioned in the present section including the deduced formulas as well as the base-plane and base-line for angular measurements should be retained regardless of the random angular vibration of the aircraft relying on a measure called the LOS stabilization which will be discussed in the next section.

10.3 LOS Stabilization

The derivation of relationships between a small angular perturbation of the whole system and small angular displacements of the two prisms for the present LOS stabilization structure can be regarded as an example of application of the principles and equations discussed in Chap. 6. However , it is felt that we still have to stress some points prior to the mathematical development.

a. Referring to Figs. 10.1 , 10.3 and 10.6 ,it may be remembered that we presume the direction of the ray path from the reticle vector (A) space to the scanning space of the LOS to be the reverse direction, so the opposite direction of the ray path from the scanning space to the reticle vector space will be regarded as the forward direction.

The LOS stabilization problem for the system in question can be handled either in the forward direction or in the reverse direction. The former way of treatment seems to belong to the type of relative image stabilization while the latter the type of absolute image stabilization.

In this case, the relative image stablilization · and the absolute image stabilization may be referred to as two aspects of the same problem . As a matter of fact, either of them will lead to the same result.

b. As a rule , assume a fixed coordinate system $x' y' z'$ attached to the fixed space , and a moving coordinate system $x_2'\ y_2'\ z_2'$ rigidly connected to the moving telescope housing . Of course, the initial orientation of the moving coordinate axes $x_2'\ y_2'\ z_2'$ are assumed to be parallel to the fixed coordinate axes $x'y'z'$, respectively, as shown in Fig. 10.6.

c. No matter whether the problem is treated by means of relative image stabilization or by means of absolute image stabilization ,the moving coordinate system $x_2'\ y_2'\ z_2'$ will convenietly be used throughout the LOS steering and stabilization assembly.

d. The conditions we have in the present case are different from those in Sect. 6.2.2. Here , two prisms 1 and 2 both serve simultaneously as the scanning and compensating elements . While the LOS is steered , the two prisms are driven by using certain scanning mechanisms so as to rotate about the axes P_1 and P_2 through the angles μ_{pr} and β_{pr} , respectively. During compensation for image oscillation caused by a small angular perturbation $\Delta\vec{\varepsilon}$, the same prisms 1 and 2 will in addition receive motions of small angular displacement $\Delta\vec{\theta_1}$ and $\Delta\vec{\theta_2}$, respectively, from the gimbal rings of a gyroscope.

In general , we have finite angles of LOS and thus finite angles of scan $\beta,\ \mu,\ \beta_{pr}$ and μ_{pr} but quite small angle of perturbation and thus small angles of compensation $\Delta\varepsilon,\ \Delta\theta_1$ and $\Delta\theta_2$.

Therefore ,it is convenient to deal with angles $\mu_{pr},\ \beta_{pr}$ and angles $\Delta\theta_1,\ \Delta\theta_2$ separately although these angles of rotation $\mu_{pr},\ \Delta\theta_1$ and $\beta_{pr},\ \Delta\theta_2$ are associated with thc same prisms 1 and 2 , respectively.

Now ,we can see that in dealing with the scanning angles μ_{pr} and β_{pr} of the two prisms we will apply the principle for finite rotation of mirrors (prisms) discussed in Chap. 3 while in dealing with the angular compensation $\Delta\vec{\theta_1}$ and $\Delta\vec{\theta_2}$ of the same prisms we will employ the principle for small rotation of prisms developed in Chap .4.

Actually we attemept to develop the theory in both Chaps. 3 and 4 further for solving some typical problems in which scanning and adjustment,i.e. , steering and stabilization , are simultaneously involved.

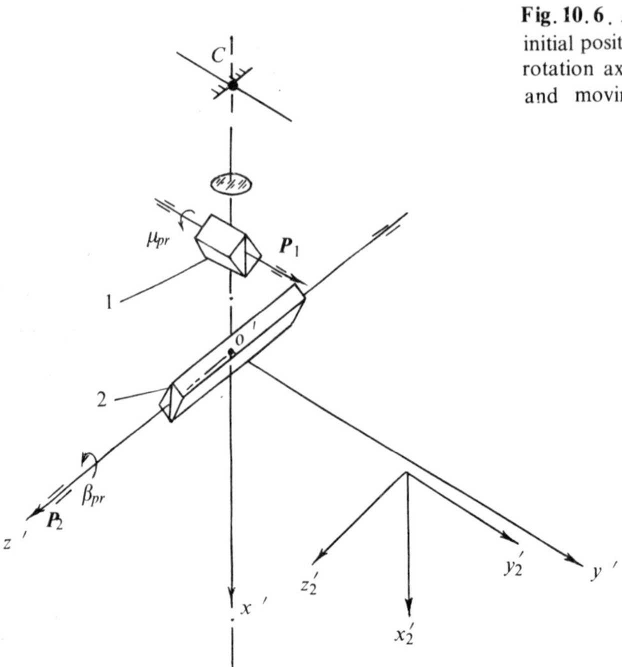

Fig. 10.6. A diagram to show the initial position of two prisms, their rotation axes, and a pair of fixed and moving coordinate systems

It should be noted that in this case the angular displacements $\Delta\vec{\theta}_1$ and $\Delta\vec{\theta}_2$ for LOS compensation are measured relative to the moving coordinate system $x_2'\ y_2'\ z_2'$,namely the telescope housing.

Let us derive the relationships for the present case, following the manner of absolute image stabilization.

According to symbols adopted in Chap.6, assume that the image rotation $\Delta\vec{\mu}_c''$ appearing in the image space of the last stabilized prism (in this situation , the longitudinal prisms 2) and measured relative to the moving coordinate system $x_2'\ y_2'\ z_2'$ is caused by the small angular compensation $\Delta\vec{\theta}_1$ and $\Delta\vec{\theta}_2$ of two prisms 1 and 2 during the scanning process.

Similarly $\Delta\vec{\mu}_c''$ consists of two parts:

$$\Delta\vec{\mu}_c'' = \Delta\vec{\mu}_{1,2}'' + \Delta\vec{\mu}_2'' \ , \tag{10.5}$$

where $\Delta\vec{\mu}_{1,2}''$ and $\Delta\vec{\mu}_2''$ are the contributions due to the respective angular displacements $\Delta\vec{\theta}_1$ and $\Delta\vec{\theta}_2$.

Relying on the previously developed theory in Chaps.3,4,and 6,we have

$$(\Delta\vec{\mu}_2'') = S_{P_2,\beta_{pr}}[E + (-1)^{l_2-1}R_2]\ S_{P_2,-\beta_{pr}}(\Delta\vec{\theta}_2) \tag{10.6}$$

or

$$(\Delta\vec{\mu}_2'') = [E + (-1)^{l_2-1}S_{P_2,\beta_{pr}}R_2\ S_{P_2,-\beta_{pr}}](\Delta\vec{\theta}_2), \tag{10.7}$$

and

266

$$(\Delta \vec{\mu}_{1,2}'') = K_{1,2} (\Delta \vec{\mu}_1')$$
$$= (-1)^{l_2} S_{P_2,\beta pr} R_2 S_{P_2,-\beta pr} [E + (-1)^{l_1-1} S_{P_1,\mu pr} R_1 S_{P_1,-\mu pr}] (\Delta \vec{\theta}_1). \qquad (10.8)$$

Thus, we obtain the total image rotation for compensation $\Delta \vec{\mu}_c''$:

$$(\Delta \vec{\mu}_c'') = [E + (-1)^{l_2-1} S_{P_2,\beta pr} R_2 S_{P_2,-\beta pr}] (\Delta \vec{\theta}_2) + (-1)^{l_2} S_{P_2,\beta pr} R_2 S_{P_2,-\beta pr} [E$$
$$+ (-1)^{l_1-1} S_{P_1,-\mu pr} R_1 S_{P_1,-\mu pr}] (\Delta \vec{\theta}_1). \qquad (10.9)$$

Two new symbols are defined as:

$$R_{2,\beta pr} = S_{P_2,\beta pr} R_2 S_{P_2,-\beta pr}, \qquad (10.10)$$

$$R_{1,\mu pr} = S_{P_1,\mu pr} R_1 S_{P_1,-\mu pr}, \qquad (10.11)$$

Then, Eq. (10.9) can be written

$$(\Delta \vec{\mu}_c'') = [E + (-1)^{l_2-1} R_{2,\beta pr}] (\Delta \vec{\theta}_2) + (-1)^{l_2} R_{2,\beta pr} [E + (-1)^{l_1-1} R_{1,\mu pr}] (\Delta \vec{\theta}_1). \quad (10.12)$$

Note that the subscripts '2' of all the column matrices in the above equations are omitted since the meaning is clear.

On the other hand, an angular perturbation $\Delta \vec{\varepsilon}$ of the telescope with respect to the fixed scanning space is regarded as if the latter inertial space experiences an angular displacement $-\Delta \vec{\varepsilon}$ relative to the telescope which is considered fixed in space for the moment.

Then, to achieve the LOS stabilization in the absolute coordinate system $x'y'z'$ — absolute stabilization, we have to meet the following condition:

$$(\Delta \vec{\mu}_c'') = (-\Delta \vec{\varepsilon}). \qquad (10.13)$$

Putting Eq. (10.12) into Eq. (10.13) ultimately gives

$$[E + (-1)^{l_2-1} R_{2,\beta pr}] (\Delta \vec{\theta}_2) + (-1)^{l_2} R_{2,\beta pr} [E + (-1)^{l_1-1} R_{1,\mu pr}] (\Delta \vec{\theta}_1) = (-\Delta \vec{\varepsilon}). \qquad (10.14)$$

Now, let us apply the resulting equation (10.14) to the arrangement as in Fig. 10.1. Referring to Fig. 10.6, we have

$$R_1 = \begin{pmatrix} 1 & 0 & 0 \\ 0 & 1 & 0 \\ 0 & 0 & -1 \end{pmatrix}, \qquad R_2 = \begin{pmatrix} 1 & 0 & 0 \\ 0 & -1 & 0 \\ 0 & 0 & 1 \end{pmatrix}.$$

While determining the rotation matrices $S_{P1,\mu pr}$ and $S_{P2,\beta pr}$, we can approximately assume that the rotation axes P_1 and P_2 are parallel to the coordinate axes y_2' and z_2', respectively. Thus, ws obtain

$$S_{P_1,\mu pr} = \begin{pmatrix} \cos \mu_{pr} & 0 & \sin \mu_{pr} \\ 0 & 1 & 0 \\ -\sin \mu_{pr} & 0 & \cos \mu_{pr} \end{pmatrix},$$

$$
S_{P_2,\,\beta_{pr}} = \begin{pmatrix} \cos\beta_{pr} & -\sin\beta_{pr} & 0 \\ \sin\beta_{pr} & \cos\beta_{pr} & 0 \\ 0 & 0 & 1 \end{pmatrix}.
$$

Considering that $S_{P_1,\,-\mu_{pr}} = S'_{P_1,\,\mu_{pr}}$ and $S_{P_2,\,-\beta_{pr}} = S'_{P_2,\,\beta_{pr}}$, and then substituting the relevant matrices into Eqs. (10.10) and (10.11) yields

$$
R_{1,\,\mu_{pr}} = \begin{pmatrix} \cos2\mu_{pr} & 0 & -\sin2\mu_{pr} \\ 0 & 1 & 0 \\ -\sin2\mu_{pr} & 0 & -\cos2\mu_{pr} \end{pmatrix}, \tag{10.15}
$$

$$
R_{2,\,\beta_{pr}} = \begin{pmatrix} \cos2\beta_{pr} & \sin2\beta_{pr} & 0 \\ \sin2\beta_{pr} & -\cos2\beta_{pr} & 0 \\ 0 & 0 & 1 \end{pmatrix}. \tag{10.16}
$$

Referring again to Fig. 10.1, since the gyroscope is insensitive to the rotation of the instrument around the x' axis, i.e., the spin axis, we may reasonably suppose that the instrument is mounted on an autopilot course stabilizer, so that one component $\Delta\varepsilon_{x'}$ of an angular perturbation $\Delta\vec{\varepsilon}$ of the aircraft will not be transmitted to the instrument. Thus, a merely two-dimensional angular perturbation of the telescope can be assumed as follows:

$$
(\Delta\vec{\varepsilon}) = \begin{pmatrix} 0 \\ \Delta\varepsilon_{y'} \\ \Delta\varepsilon_{z'} \end{pmatrix}. \tag{10.17}
$$

According to the mounting construction of the mechanical rotation axes of two prisms and the kinematical trains in the gimbal of the present LOS steering and stabilization head, it is apparent that after such a two-dimensional angular perturbation $\Delta\vec{\varepsilon}$ has happened, the axis P_2 of the longitudinal prism will be kept in the horizontal plane in the direction approximately parallel to its original position while the axis P_1 of the lateral prism will turn together with one component $\Delta\varepsilon_{z'}$ of the angular perturbation of the telescope but remain perpendicular to the axis p_2. Besides, the recovering motions of the gimbal rings will also be imparted to two prisms through gear differentials. Thus, the angular compensation $\Delta\vec{\theta}_1$ and $\Delta\vec{\theta}_2$ may be assumed to be

$$
(\Delta\vec{\theta}_1) = \begin{pmatrix} 0 \\ \Delta\theta_{1y_2'} \\ 0 \end{pmatrix}, \qquad (\Delta\vec{\theta}_2) = \begin{pmatrix} 0 \\ \Delta\theta_{2y_2'} \\ \Delta\theta_{2z_2'} \end{pmatrix}. \tag{10.18}
$$

Putting Eqs. (10. 15) through (10. 18) into Eq. (10. 14) and considering $t_1 = t_2 = 1$ gives

$$\begin{pmatrix} \Delta\mu''_{cx_2'} \\ \Delta\mu''_{cy_2'} \\ \Delta\mu''_{cz_2'} \end{pmatrix} = \begin{pmatrix} (\Delta\theta_{2y_2'} - 2\Delta\theta_{1y_2'})\sin 2\beta_{pr} \\ \Delta\theta_{2y_2'} + (2\Delta\theta_{1y_2'} - \Delta\theta_{2y_2'})\cos 2\beta_{pr} \\ 2\Delta\theta_{2z_2'} \end{pmatrix} = \begin{pmatrix} 0 \\ -\Delta\varepsilon_{y'} \\ -\Delta\varepsilon_{z'} \end{pmatrix} . \tag{10.19}$$

Then, we have

$$(\Delta\theta_{2y_2'} - 2\Delta\theta_{1y_2'})\sin 2\beta_{pr} = 0 \quad , \tag{10.20}$$

$$\Delta\theta_{2y_2'} + (2\Delta\theta_{1y_2'} - \Delta\theta_{2y_2'})\cos 2\beta_{pr} = -\Delta\varepsilon_{y'} \quad , \tag{10.21}$$

$$2\Delta\theta_{2z_2'} = -\Delta\varepsilon_{z'} \quad . \tag{10.22}$$

From the last two equations (10.21) and (10.22), we find

$$\left. \begin{aligned} \Delta\theta_{2y_2'} &= -\Delta\varepsilon_{y'} \\ \Delta\theta_{2z_2'} &= -\frac{1}{2}\Delta\varepsilon_{z'} \end{aligned} \right\} \quad , \tag{10.23}$$

$$\Delta\theta_1 = \Delta\theta_{1y_2'} = \frac{1}{2}\Delta\theta_{2y_2'} = -\frac{1}{2}\Delta\varepsilon_{y'} . \tag{10.24}$$

The results $\Delta\theta_1 = \Delta\theta_{1y_2'} = -\frac{1}{2}\Delta\varepsilon_{y'}$ and $\Delta\theta_{2z_2'} = -\frac{1}{2}\Delta\varepsilon_{z'}$ indicate that the two prisms rotate first together with the telescope as a unit through the angles $\Delta\varepsilon_{y'}$ and $\Delta\varepsilon_{z'}$, respectively , and then turn back alone relative to the telescope through half the above angles $-\frac{1}{2}\Delta\varepsilon_{y'}$ and $-\frac{1}{2}\Delta\varepsilon_{z'}$, respectively. Another result $\Delta\theta_{2y_2'} = -\Delta\varepsilon_{y'}$ means that the mechanical rotation axis P_2 of the longitudinal prism should be held horizontal all the time . These conditions can be ensured by the structure of the head as was mentioned before.

With the above condition and results , the equation (10. 20) will also be satisfied.

Moreover, the absence of the $2\mu_{pr}$- terms and the zero-valued coefficients of the terms $\sin 2\beta_{pr}$ and $\cos 2\beta_{pr}$ in the set of Eqs. (10.20) through (10.22) shows that the stabilization of the line of sight will constantly be in effect regardless of its direction in space.

In general , the LOS stabilization problem may not be solcvd exactly , and thus a vector $\Delta\vec{\mu}''_e$ defined as $\Delta\vec{\mu}''_e = \Delta\vec{\mu}''_c - (-\Delta\vec{\varepsilon})$ may be called the error of LOS stabilization.

Then , we have

$$\Delta\vec{\mu}''_e = \Delta\vec{\mu}''_c + \Delta\vec{\varepsilon} , \tag{10.25}$$

where

$$(\Delta\vec{\varepsilon}) = \begin{pmatrix} \Delta\varepsilon_{x'} \\ \Delta\varepsilon_{y'} \\ \Delta\varepsilon_{z'} \end{pmatrix} , \tag{10.26}$$

and the total image rotation for compensation $\Delta\vec{\mu}_c''$ is determined according to the actual construction of the LOS steering and stabilization head in question.

Now, let us calculate the direction error of the line of sight due to the error $\Delta\vec{\mu}_e''$ of LOS stabilization.

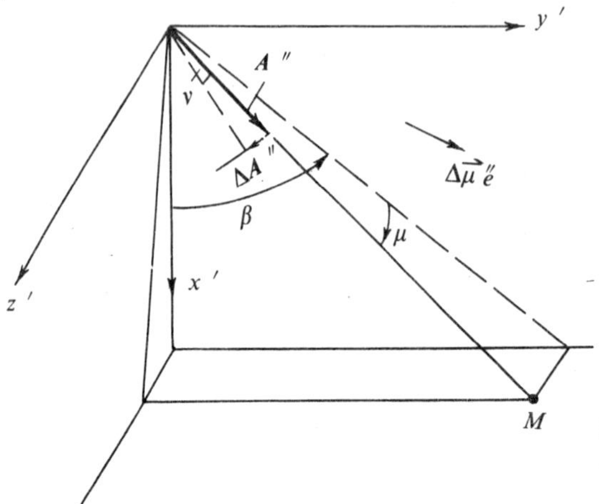

Fig.10.7. A diagram to show the direction error of a line of sight due to the error $\Delta\vec{\mu}_e''$ of LOS stabilization

In Fig.10.7, let A'' represent the direction of the line of sight and $\Delta A''$ denote the change of A'' due to $\Delta\vec{\mu}_e''$.

From Eq. (1.15),

$$\Delta A'' = \Delta\vec{\mu}_e'' \times A''.$$

$$(10.27)$$

Let v be the direction error of the LOS.

Then,

$$v = \frac{|\Delta A''|}{|A''|}.$$

$$(10.28)$$

Putting Eqs. (10.27) and (10.25) into Eq. (10.28) leads to

$$v = \frac{|(\Delta\vec{\mu}_c'' + \Delta\vec{\varepsilon}) \times A''|}{|A''|} = |(\Delta\vec{\mu}_c'' + \Delta\vec{\varepsilon}) \times A''|.$$

$$(10.29)$$

As mentioned before, we can also treat this problem, following the manner of relative image stabilization.

Utilizing the principle discussed in Secs. 6.2.2 and 10.3, the final equation can be written straight away

$$[E + (-1)^{i_1-1} R_{1,\mu_{pr}}^{-1}] (\Delta\vec{\theta_1}) + (-1)^{i_1} R_{1,\mu_{pr}}^{-1} [E + (-1)^{i_2-1} R_{2,\beta_{pr}}^{-1}] (\Delta\vec{\theta_2})$$

$$= -(-1)^{(i_2+i_1)} R_{1,\mu_{pr}}^{-1} R_{2,\beta_{pr}}^{-1} (-\Delta\vec{\varepsilon}),$$

$$(10.30)$$

where the meaning of the previously used symbols remains unchanged; $R_{1,\mu_{pr}}^{-1}$ and $R_{2,\beta_{pr}}^{-1}$ are the inverse of matrices $R_{1,\mu_{pr}}$ and $R_{2,\beta_{pr}}$, respectively:

270

$$R_{1,\,\mu pr}^{-1} = S_{P_1,\,\mu pr}\; R_1^{-1}\, S_{P_1,\,-\mu pr}\,, \qquad (10.31)$$

$$R_{2,\,\beta pr}^{-1} = S_{P_2,\,\beta pr}\; R_2^{-1}\, S_{P_2,\,-\beta pr}\,. \qquad (10.32)$$

It is readily seen that by multiplying each side of Eq. (10.30) on the left by a factor $(-1)^{(l_2+l_1)-1}\,R_{2,\,\beta pr}\,R_{1,\,\mu pr}$, the equation (10.30) will again be converted to the equation (10.14).

On some occasions the relationship for absolute image stabilization does differ from that for relative image stabilization provided the fixed reference frame is located in the image space of the optical system but the difference is on the whole immaterial.

Finally, it should again be noticed that the present stabilized prism assembly together with the course stabilizer actually provides a complete image stabilization which has three degrees of freedom.

Problems

10.1 Referring to the LOS steering and stabilized prism assembly as in Fig. 10.1, consider the scanning angles μ_{pr} and β_{pr} of the two prisms and the angular compensation $\Delta\vec{\theta}_1$ and $\Delta\vec{\theta}_2$ depending on the angular perturbation components $\Delta\varepsilon_{y'}$ and $\Delta\varepsilon_{z'}$ according to Eqs. (10.23) and (10.24) as the inputs to the prism assembly, and then trace the reticle vector A along the forward direction of the ray path to find the line of sight in the scanning space so as to verify that the direction of the LOS relative to the fixed coordinate system $x'y'z'$ is dependent on the two scanning angles only.

10.2 Explain how a problem of LOS stabilization differs from that of optical-axis stabilization.

10.3 Referring to Fig. 10.7, explain the meaning of the scalar product $\Delta\vec{\mu}_e'' \cdot A''$.

Postscript

Reflecting prisms are extensively used in most optical systems. The reason for this is that reflecting prisms perform various special functions, and in particular they can be used as moving parts in optical systems. Therefore, system designers and optical engineers usually deal with optical components of this type from time to time.

In designing an optical system involving reflecting prisms, the method of determining the prism sizes has been well resolved by unfolding it into a plane-parallel plate and then reducing the latter still further to its equivalent air thickness. As for a rotating reflecting system, the question of a finite rotation angle has also been discussed in considerable depth and some relevant laws have been revealed.

However, the significance of making a study of the relations between two conjugate spaces for reflecting prisms under various conditions in a systematic and comprehensive way as well as of the laws governing these relations for some reason had not been realized over a lengthy period, although research on these problems is really closely related to the fundamental principle, design and manufacture of optical instruments, and particularly to many practical engineering tasks.

In China, some scientific researchers in optics have all along been devoting great attention to the study of the theory of conjugation for reflecting prisms. The monograph "Adjustment of Reflecting Prisms" published in Chinese in 1978 by the National Defence Industry Press may be considered a truly significant volume, which covers almost all the research achievements previously made in this field. Since then the study of the theory of conjugation for reflecting prisms, entering a new stage of its development, has been made in an all-round and deepening way. Due to the contributions made by Chinese scientific researchers over several decades, the academic level of the theory of conjugation for reflecting prisms in China has to date attained such advanced standards that not only have we established a relatively complete system for the theory itself but have also founded a new school of research, i. e., the so-called *rigid body's kinematics* school, and the latter is distinguished as well as unique.

Professor Lian Tongshu is the chief founder of the theory of conjugation for reflecting prisms, and he is also the originator of the *rigid body's kinematics* school. The monograph *Theory of Conjugation for Reflecting Prisms* published in Chinese in 1988 by the Beijing Institute of Technology Press gathers together the research achievements over nearly four decades in China which are mainly his own.

Professor Lian has been one of my closest friends, and I am much impressed with his qualities in both teaching and research work — his scholarship, industry, and dedication.

The present book written in English is a revised and improved version of the Chinese edition. I believe that its publication will make a contribution to the spread of this material world-wide, and will lead to a further development of the theory and its application.

Hou Qian
Professor of Applied Optics

Appendix : Tabulation of Reflecting Prisms

Fifty-four commonly used reflecting prisms are tabulated in the present Appendix . Except four prisms which can only be used in collimated light ,for each prism a table is given with one adjustment diagram ,one matrix ,six formulas ,and twenty characteristic parameters .

There is considerable confusion in the names used for reflecting prisms ,which stems from the difficulty in tracing back to the original designer's description . All the prisms described in the Appendix have been given code names according to the standard WJ /Z2 - 65 adopted in China .

Table A.1 DI − 0°

Adjustment Diagram

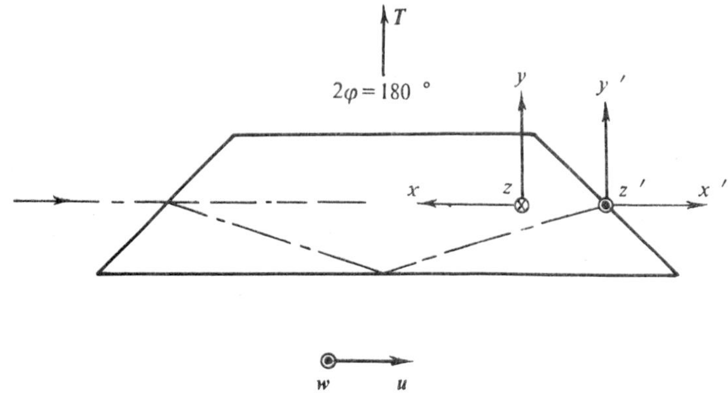

Fig. A.1.

Characteristic Parameters of Image Formation

Direction cosine		Unit vectors of image coordinate system			$2\varphi = 180°$
		$i'(x')$	$j'(y')$	$k'(z')$	T
Unit vectors of coordinate system in object space	$i(x)$	-1	0	0	0
	$j(y)$	0	1	0	1
	$k(z)$	0	0	-1	0

$$S_{T,2\varphi} = \begin{pmatrix} -1 & 0 & 0 \\ 0 & 1 & 0 \\ 0 & 0 & -1 \end{pmatrix}$$

$$t = 1 \qquad L = \frac{2nD}{\sqrt{2n^2-1}\ -1} \qquad\qquad 2d =$$

$$\overrightarrow{M_0 M'} =$$

$$r_0 =$$

$$r_i =$$

Characteristic Parameters of Adjustment

	Direction cosine	Image coordinate system		
		x'	y'	z'
Extreme-valued axis direction of r' image rotation	u	1	0	0
	v	/	/	/
	w	0	0	1
Extreme value of r' image rotation		$\delta_u = 2\Delta\theta$	$\delta_v = 0$	$\delta_w = 2\Delta\theta$
Extreme-valued shift direction of r' image displacement	a			
	b			
	c			
Extreme value of r' image displacement		$\delta_a =$	$\delta_b =$	$\delta_c =$

Planar three-dimensional zero-valued poles (Meaningless in collimated beams)	C_{Π_1}	$x_1' =$
		$y_1' =$
		$z_1' =$
		Associated plane Π_1 :
	C_{Π_2}	$x_2' =$
		$y_2' =$
		$z_2' =$
		Associated plane Π_1 :

Equations of Image Motion

Parallel beam (Image rotation $\Delta\vec{\mu'}$)

Image lean

$$\Delta\mu_{x'}' = 2\Delta\theta\, P_{x'}$$

Optical axis deviation

$$\Delta\mu_{y'}' = 0$$

$$\Delta\mu_{z'}' = 2\Delta\theta\, P_{z'}$$

Adjustment Diagram

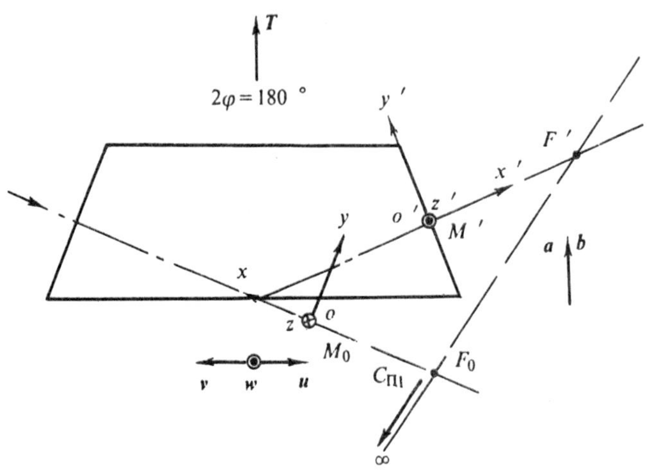

Fig .A .2 .

Characteristic Parameters of Image Formation

Direction cosine		Unit vectors of image coordinate system			$2\varphi = 180°$
		$i'(x')$	$j'(y')$	$k'(z')$	T
Unit vectors of coordinate system in object space	$i(x)$		-0.707	0.707	0	0.383
	$j(y)$	$S_{T,2\varphi} =$	0.707	0.707	0	0.924
	$k(z)$		0	0	-1	0

$$t = 1 \qquad L = 2.414\,D \qquad\qquad 2d =$$

$$\overrightarrow{M_0 M'} = \left(2.060D - 1.707\,\frac{D}{n}\right) i' + \left(-0.853\,D + 1.707\,\frac{D}{n}\right) j'$$

$$r_0 = \infty$$

$$r_i = \infty$$

Characteristic Parameters of Adjustment

Direction cosine		Image coordinate system		
		x'	y'	z'
Extreme-valued axis direction of r' image rotation	u	0.924	-0.383	0
	v	-0.924	0.383	0
	w	0	0	1
Extreme value of r' image rotation		$\delta_u = 1.848\Delta\theta$	$\delta_v = 0.765\Delta\theta$	$\delta_w = 2\Delta\theta$
Extreme-valued shift direction of r' image displacement	a	0.383	0.924	0
	b	0.383	0.924	0
	c	/	/	/
Extreme value of r' image displacement		$\delta_a = 0.765\Delta g$	$\delta_b = 1.848\Delta g$	$\delta_c = 0$

Planar three-dimensional zero-valued poles	C_{Π_1} (Approaches to infinity along $\overrightarrow{F_0 F'}$)	$x_1' = \infty$
		$y_1' = \infty$
		$z_1' = 0$
		Associated plane Π_1: Conjugate optical-axis section $x'o'y'$
	C_{Π_2} (Does not exist)	$x_2' =$
		$y_2' =$
		$z_2' =$
		Associated plane Π_2:

Equations of Image Motion

Parallel beam (Image rotation $\Delta\vec{\mu}'$)

Image lean
$$\Delta\mu_x' = \Delta\theta\,(1.707 P_{x'} - 0.707 P_{y'})$$

Optical axis deviation
$$\Delta\mu_y' = \Delta\theta\,(-0.707 P_{x'} + 0.293 P_{y'})$$
$$\Delta\mu_z' = 2\Delta\theta\, P_{z'}$$

Convergent beam (Image point displacement $\Delta S_F'$)

Parallax

$$\Delta S_{F'x'}' = \Delta\theta\,[\,0.707\,P_{x'}z_q' - 0.293 P_{y'}z_q' + P_{z'}(-0.707 x_q' + 0.293 y_q' - 0.853D)\,]$$

$$+ \Delta g\,(0.293\,D_{x'} + 0.707\,D_{y'})$$

Optical axis deviation

$$\Delta S_{F'y'}' = \Delta\theta\left[1.707\,P_{x'}z_q' - 0.707\,P_{y'}z_q' + P_{z'}\left(-1.707\,x_q' + 0.707\,y_q' + 2b' \right.\right.$$

$$\left.\left. - 2.060\,D + 2.414\frac{D}{n}\right)\right] + \Delta g\,(0.707\,D_{x'} + 1.707\,D_{y'})$$

$$\Delta S_{F'z'}' = \Delta\theta\left[P_{x'}\left(0.707\,b' - 0.853D + 1.707\,\frac{D}{n}\right)\right.$$

$$\left. + P_{y'}\left(-0.293b' - 2.060D + 1.707\,\frac{D}{n}\right)\right]$$

Table A .3 DI – 60 °

Adjustment diagram

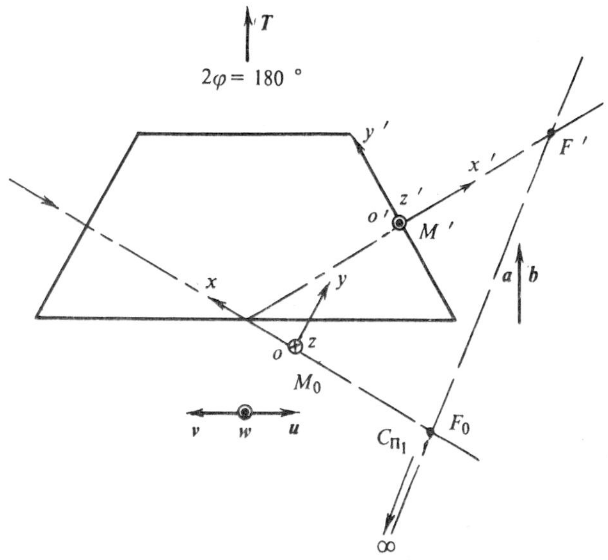

Fig . A . 3 .

Characteristic Parameters of Image Formation

Direction cosine		Unit vectors of image coordinate system			2φ = 180 °
		$i\,'(x\,')$	$j\,'(y\,')$	$k\,'(z\,')$	T
Unit vectors of coordinate system in object space	$i\,(x)$	$S_{T,2\varphi}=\begin{pmatrix}$ -0.5	0.866	0	0.5
	$j\,(y)$	0.866	0.5	0	0.866
	$k\,(z)$	0	0	$-1\end{pmatrix}$	0

$t = 1 \qquad L = 1.732\,D$ $2d =$

$$\overrightarrow{M_0 M\,'} = \left(1.299\,D - 0.866\,\frac{D}{n}\right) i\,' + \left(-0.75D + 1.5\,\frac{D}{n}\right) j\,'$$

$r_0 = \infty$

$r_i = \infty$

Characteristic Parameters of Adjustment

Direction cosine		Image coordinate system		
		x'	y'	z'
Extreme - valued axis direction of r' image rotation	u	0.866	-0.5	0
	v	-0.866	0.5	0
	w	0	0	1
Extreme value of r' image rotation		$\delta_u = 1.732\Delta\theta$	$\delta_v = \Delta\theta$	$\delta_w = 2\Delta\theta$
Extreme- valued shift direction of r' image displacement	a	0.5	0.866	0
	b	0.5	0.866	0
	c	/	/	/
Extreme value of r' image displacement		$\delta_a = \Delta g$	$\delta_b = 1.732\Delta g$	$\delta_c = 0$

Planar three-dimensional zero- valued poles	C_{Π_1} (Approaches to infinity along $\overrightarrow{F_0 F'}$)	$x_1' = \infty$
		$y_1' = \infty$
		$z_1' = 0$
		Associated plane Π_1 : Conjugate optical- axis section $x'o'y'$
	C_{Π_2} (Does not exist)	$x_2' =$
		$y_2' =$
		$z_2' =$
		Associated plane Π_2:

Equations of Image Motion

Parallel beam (Image rotation $\Delta\overrightarrow{\mu'}$)

Image lean

$$\Delta\mu_{x'}' = \Delta\theta\,(1.5\,P_{x'} - 0.866\,P_{y'})$$

Optical axis deviation

$$\Delta\mu_{y'}' = \Delta\theta\,(-0.866\,P_{x'} + 0.5\,P_{y'})$$

$$\Delta\mu_{z'}' = 2\Delta\theta\,P_{z'}$$

Convergent beam (Image point displacement $\Delta S_{F'}'$)

Parallax

$$\Delta S_{F'x}^{'} = \Delta\theta\,[\,0.866\,P_{x'}z_{q}^{'} - 0.5\,P_{y'}z_{q}^{'} + P_{z'}(-0.866\,x_{q}^{'} + 0.5y_{q}^{'} - 0.75D\,)]$$
$$+\,\Delta g\,(0.5\,D_{x'} + 0.866D_{y'})$$

Optical axis deviation

$$\Delta S_{F'y}^{'} = \Delta\theta\left[\,1.5\,P_{x'}z_{q}^{'} - 0.866\,P_{y'}z_{q}^{'} + P_{z'}\left(-1.5x_{q}^{'} + 0.866y_{q}^{'} + 2b' - 1.299D\right.\right.$$
$$\left.\left.+\,1.732\frac{D}{n}\right)\right] + \Delta g\left(0.866\,D_{x'} + 1.5D_{y'}\right)$$
$$\Delta S_{F'z}^{'} = \Delta\theta\left[\,P_{x'}\left(0.866\,b' - 0.75\,D + 1.5\frac{D}{n}\right) + P_{y'}\left(-0.5b' - 1.299\,D + 0.866\frac{D}{n}\right)\right]$$

Table A.4   DI – 80 °

Adjustment Diagram

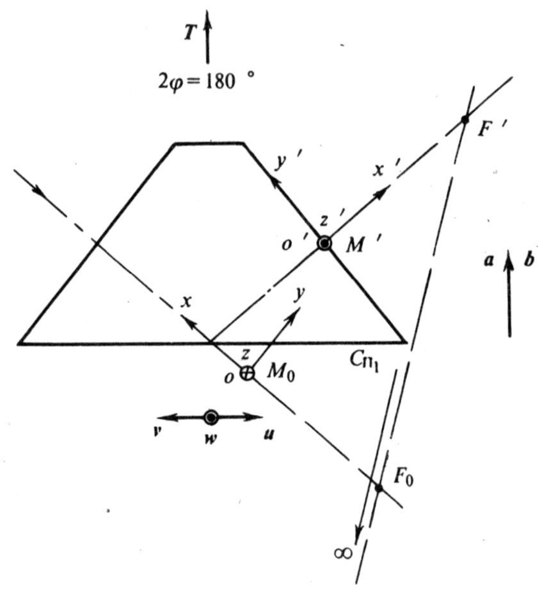

Fig. A.4.

Characteristic Parameters of Image Formation

Direction cosine		Unit vectors of image coordinate system			$2\varphi = 180$ °
		$i\,'(x\,')$	$j\,'(y\,')$	$k\,'(z\,')$	T
Unit vectors of coordinate system in object space	$i\,(x)$	$S_{T,2\varphi}=\begin{pmatrix} -0.174 & 0.985 & 0 \\ 0.985 & 0.174 & 0 \\ 0 & 0 & -1 \end{pmatrix}$			0.643
	$j\,(y)$				0.766
	$k\,(z)$				0

$t = 1$	$L = 1.192\,D$	$2d =$

$$\overrightarrow{M_0 M\,'} = \left(0.7D - 0.207\,\frac{D}{n}\right) i\,' + \left(-0.587D + 1.174\,\frac{D}{n}\right) j\,'$$

$r_0 = \infty$

$r_i = \infty$

Characteristic Parameters of Adjustment

Direction cosine		Image coordinate system		
		x'	y'	z'
Extreme-valued axis direction of r' image rotation	u	0.766	-0.643	0
	v	-0.766	0.643	0
	w	0	0	1
Extreme value of r' image rotation		$\delta_u = 1.532\Delta\theta$	$\delta_v = 1.286\Delta\theta$	$\delta_w = 2\Delta\theta$
Extreme-valued shift direction of r' image displacement	a	0.643	0.766	0
	b	0.643	0.766	0
	c	/	/	/
Extreme value of r' image displacement		$\delta_a = 1.286\Delta g$	$\delta_b = 1.532\Delta g$	$\delta_c = 0$

Planar three-dimensional zero-valued poles	C_{Π_1} (Approaches to infinity along $\overrightarrow{F_0F'}$)	$x_1' = \infty$
		$y_1' = \infty$
		$z_1' = 0$
		Associated plane Π_1: Conjugate optical-axis section $x'o'y'$
	C_{Π_2} (Does not exist)	$x_2' =$
		$y_2' =$
		$z_2' =$
		Associated plane Π_2:

Equations of Image Motion

Parallel beam (Image rotation $\Delta\overrightarrow{\mu'}$)

Image lean

$$\Delta\mu_{x'}' = \Delta\theta\,(1.174\,P_{x'} - 0.985P_{y'})$$

Optical axis deviation

$$\Delta\mu_{y'}' = \Delta\theta\,(-0.985\,P_{x'} + 0.826\,P_{y'})$$

$$\Delta\mu_{z'}' = 2\,\Delta\theta\,P_{z'}$$

Convergent beam (Image point displacement $\Delta S_{F'}'$)

Parallax

$$\Delta S'_{F'_{x'}} = \Delta\theta\,[\,0.985\,P_{x'}z'_q - 0.826 P_{y'}z'_q + P_{z'}(-0.985\,x'_q + 0.826 y'_q - 0.587 D\,)]$$
$$+\Delta g\,(0.826 D_{x'} + 0.985 D_{y'})$$

Optical axis deviation

$$\Delta S'_{F'_{y'}} = \Delta\theta\left[1.174 P_{x'}z'_q - 0.985\,P_{y'}z'_q + P_{z'}\left(-1.174 x'_q + 0.985 y'_q + 2b' - 0.7 D\right.\right.$$
$$\left.\left.+1.192\frac{D}{n}\right)\right] + \Delta g\,(0.985\,D_{x'} + 1.174 D_{y'})$$
$$\Delta S'_{F'_{z'}} = \Delta\theta\left[P_{x'}\left(0.985\,b' - 0.587 D + 1.174\frac{D}{n}\right) + P_{y'}\left(-0.826 b' - 0.7 D + 0.207\frac{D}{n}\right)\right].$$

Adjustment Diagram

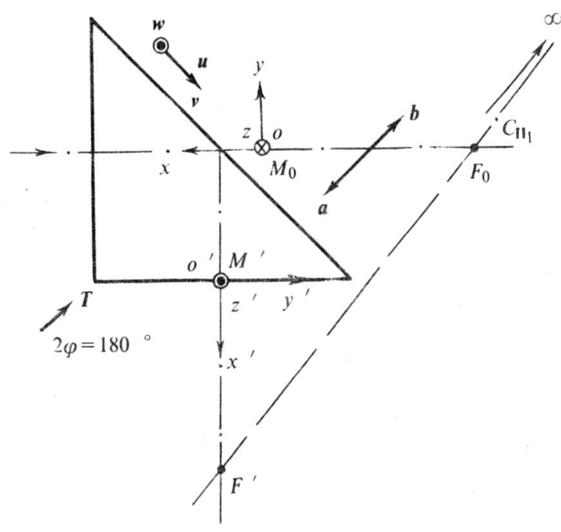

Fig. A .5 .

Characteristic Parameters of Image Formation

Direction cosine		Unit vectors of image coordinate system			$2\varphi = 180°$
		$i'(x')$	$j'(y')$	$k'(z')$	**T**
Unit vectors of coordinate system in object space	$i(x)$		0 $\qquad$ -1 $\qquad$ 0		-0.707
	$j(y)$	$S_{T,2\varphi} =$ $\begin{pmatrix} 0 & -1 & 0 \\ -1 & 0 & 0 \\ 0 & 0 & -1 \end{pmatrix}$			0.707
	$k(z)$				0

$t = 1 \qquad L = D \qquad\qquad\qquad\qquad\qquad 2d =$

$$\overrightarrow{M_0 M'} = 0.5\,D\,i' + \left(0.5\mathrm{D} - \frac{D}{n}\right) j'$$

$r_0 = \infty$

$r_i = \infty$

Characteristic Parameters of Adjustment

Direction cosine		Image coordinate system		
		x'	y'	z'
Extreme-valued axis direction of r' image rotation	u	0.707	0.707	0
	v	0.707	0.707	0
	w	0	0	1
Extreme value of r' image rotation		$\delta_u = 1.414\Delta\theta$	$\delta_v = 1.414\Delta\theta$	$\delta_w = 2\Delta\theta$
Extreme-valued shift direction of r' image displacement	a	0.707	-0.707	0
	b	-0.707	0.707	0
	c	/	/	/
Extreme value of r' image displacement		$\delta_a = 1.414\Delta g$	$\delta_b = 1.414\Delta g$	$\delta_c = 0$

Planar three-dimensional zero-valued poles	C_{Π_1} (Approaches to infinity along $\overrightarrow{F_0 F'}$)	$x_1' = \infty$
		$y_1' = \infty$
		$z_1' = 0$
		Associated plane Π_1: Conjugate optical-axis section $x'o'y'$
	C_{Π_2} (Does not exist)	$x_2' =$
		$y_2' =$
		$z_2' =$
		Associated plane Π_2:

Equations of image Motion

Parallel beam   (Image rotation $\Delta\overrightarrow{\mu'}$)
Image lean

$$\Delta\mu_{x'}' = \Delta\theta\,(P_{x'} + P_{y'})$$

Optical axis deviation

$$\Delta\mu_{y'}' = \Delta\theta\,(P_{x'} + P_{y'})$$

$$\Delta\mu_{z'}' = 2\Delta\theta\,P_{z'}$$

Convergent beam (Image point displacement $\Delta S_{F'}'$)
Parallax

288

$$\Delta S'_{F\,'x\,'} = \Delta\theta\,[\,-P_x\,'z\,'_q - P_y\,'z\,'_q + P_z\,'\,(x'_q + y'_q + 0.5\,D\,)\,] + \Delta g\,(D_x\,' - D_y\,')$$

Optical axis deviation

$$\Delta S'_{F\,'y\,'} = \Delta\theta\left[\,P_x\,'z\,'_q + P_y\,'z\,'_q + P_z\,'\left(-x_q\,' - y_q\,' + 2\,b\,' - 0.5D + \frac{D}{n}\right)\right] + \Delta g\,(-D_x\,' + D_y\,')$$

$$\Delta S'_{F\,'z\,'} = \Delta\theta\left[\,P_x\,'\left(-b\,' + 0.5\,D - \frac{D}{n}\right) + P_y\,'\left(-b - 0.5\,D\right)\right]$$

Adjustment Diagram

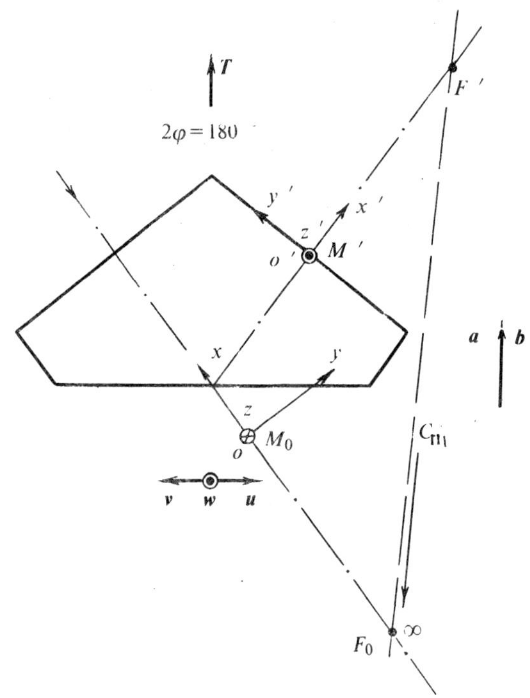

Fig . A .6 .

Characteristic Parameters of Image Formation

Direction cosine		Unit vectors of image coordinate system			2φ = 180 °
		$i\,'(x\,')$	$j\,'(y\,')$	$k\,'(z\,')$	T
Unit vectors of coordinate system in object space	$i\,(x)$	$S_{T,2\varphi}=\big($ 0.259	0.966	0	0.793
	$j\,(y)$	0.966	− 0.259	0	0.609
	$k\,(z)$	0	0	− 1 $\big)$	0

$t = 1$	$L = 1.303D$		$2d =$

$$\overrightarrow{M_0 M\,'} = \left(0.483\,D - 0.337\,\frac{D}{n}\right) i\,' + \left(-0.629\,D + 1.258\,\frac{D}{n}\right) j\,'$$

$r_0 = \infty$

$r_i = \infty$

Characteristic Parameters of Adjustment

Direction cosine		Image coordinate system		
		x'	y'	z'
Extreme - valued axis direction of r' image rotation	u	0.609	-0.793	0
	v	-0.609	0.793	0
	w	0	0	1
Extreme value of r' image rotation		$\delta_u = 1.218\Delta\theta$	$\delta_v = 1.587\Delta\theta$	$\delta_w = 2\Delta\theta$
Extreme - valued shift direction of r' image displacement	a	0.793	0.609	0
	b	0.793	0.609	0
	c	/	/	/
Extreme value of r' image displacement		$\delta_a = 1.587\Delta g$	$\delta_b = 1.218\Delta g$	$\delta_c = 0$

Planar three-dimensional zero- valued poles	C_{Π_1} (Approaches to infinity along $\overrightarrow{F_0 F'}$)	$x_1' = \infty$
		$y_1' = \infty$
		$z_1' = 0$
		Associated plane Π_1: Conjugate optical- axis section $x'o'y'$
	C_{Π_2} (Does not exist)	$x_2' = $
		$y_2' = $
		$z_2' = $
		Associated plane Π_2:

Equations of Image Motion

Parallel beam (Image rotation $\Delta\overrightarrow{\mu'}$)
Image lean

$$\Delta\mu_{x'}' = \Delta\theta\,(0.741\,P_{x'} - 0.966\,P_{y'})$$

Optical axis deviation

$$\Delta\mu_{y'}' = \Delta\theta\,(-0.966\,P_{x'} + 1.259\,P_{y'})$$

$$\Delta\mu_{z'}' = 2\,\Delta\theta\,P_{z'}$$

Convergent beam   (Image point displacement $\Delta S_{F'}'$)

Parallax

$$\Delta S'_{F\,x'} = \Delta\theta\left[0.966\,P_{x'}z'_q - 1.259 P_{y'}z'_q + P_{z'}\left(-0.966\,x'_q + 1.259 y'_q - 0.629 D\right.\right.$$
$$\left.\left. + 0.651\,\frac{D}{n}\right)\right] + \Delta g\,(1.259\,D_{x'} + 0.966 D_{y'})$$

Optical axis deviation

$$\Delta S'_{F\,y'} = \Delta\theta\left[0.741\,P_{x'}z'_q - 0.966\,P_{y'}z_q' + P_{z'}\left(-0.741\,x'_q + 0.966\,y'_q + 2\,b' - 0.483 D\right.\right.$$
$$\left.\left. + 1.128\,\frac{D}{n}\right)\right] + \Delta g\,(0.966 D_{x'} + 0.741 D_{y'})$$

$$\Delta S'_{F\,z'} = \Delta\theta\left[P_{x'}\left(0.966 b' - 0.629 D + 1.258\,\frac{D}{n}\right) + P_{y'}\left(-1.259 b' - 0.483 D + 0.337\,\frac{D}{n}\right)\right]$$

Adjustment Diagram

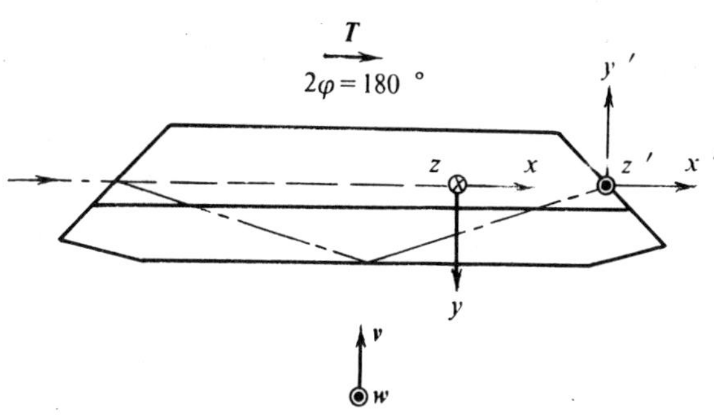

Fig.A.7.

Characteristic Parameters of Image Formation

Direction cosine		Unit vectors of image coordinate system			$2\varphi = 180°$
		$i'(x')$	$j'(y')$	$k'(z')$	T
Unit vectors of coordinate system in object space	$i(x)$		1	0	0	1
	$j(y)$	$S_{T,2\varphi} =$	0	-1	0	0
	$k(z)$		0	0	-1	0

$t = 2$	$L = \dfrac{2.824\,n\,D}{\sqrt{2n^2-1}\ -1}$	$2d =$

$\overrightarrow{M_0 M'} =$

$r_0 =$

$r_i =$

Characteristic Parameters of Adjustment

Direction cosine		Image coordinate system		
		x'	y'	z'
Extreme-valued axis direction of r' image rotation	u	/	/	/
	v	0	1	0
	w	0	0	1
Extreme value of r' image rotation		$\delta_u = 0$	$\delta_v = 2\,\Delta\theta$	$\delta_w = 2\,\Delta\theta$
Extreme-valued shift direction of r' image displacement	a			
	b			
	c			
Extreme value of r' image displacement				

Planar three-dimensional zero-valued poles (Meaningless in collimated beams)	C_{Π_1}	$x_1' =$
		$y_1' =$
		$z_1' =$
		Associated plane Π_1 :
	C_{Π_2}	$x_2' =$
		$y_2' =$
		$z_2' =$
		Associated plane Π_2:

Equations of Image Motion

Parallel beam (Image rotation $\Delta\vec{\mu}'$)

Image lean

$$\Delta\mu_{x'}' = 0$$

Optical axis deviation

$$\Delta\mu_{y'}' = 2\,\Delta\theta\,P_{y'}$$

$$\Delta\mu_{z'}' = 2\,\Delta\theta\,P_{z'}$$

Adjustment Diagram

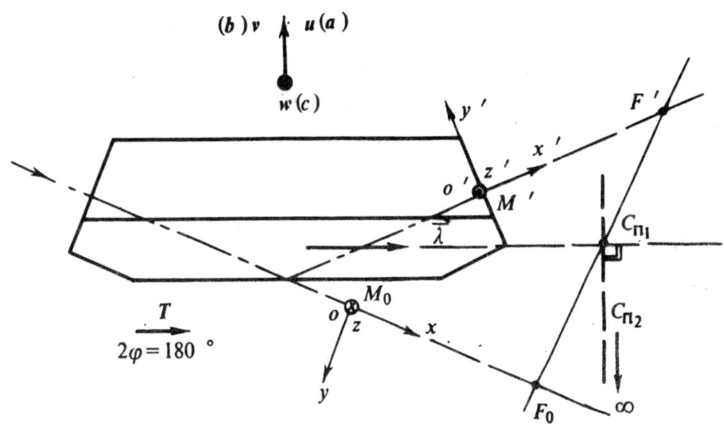

F ig.A .8 .

Characteristic Parameters of Image Formation

Direction cosine		Unit vectors of image coordinate system			2φ = 180 °
		$i\,'(x\,')$	$j'(y\,')$	$k\,'(z\,')$	T
Unit vectors of coordinate system in object space	$i\,(x)$	$S_{T,2\varphi}=\begin{pmatrix} 0.707 \\ -0.707 \\ 0 \end{pmatrix}$	−0.707	0	0.924
	$j\,(y)$		−0.707	0	−0.383
	$k\,(z)$		0	−1	0

$t = 2 \qquad L = 3.558\,D$

$2d = 3.288\,D - 3.288\,\dfrac{D}{n}$

$$\overrightarrow{M_0 M\,'} = \left(3.037D - 2.516\,\frac{D}{n}\right)i\,' + \left(-1.258D + 2.516\,\frac{D}{n}\right)j\,'$$

$$r_0 = -0.261\,\frac{D}{n}\,i\,' - 0.628\,\frac{D}{n}\,j\,'$$

$$r_i =$$

Characteristic Parameters of Adjustment

Direction cosine		Image coordinate system		
		x'	y'	z'
Extreme-valued axis direction of r' image rotation	u	0.383	0.924	0
	v	0.383	0.924	0
	w	0	0	1
Extreme-value of r' image rotation		$\delta_u = 0.765\Delta\theta$	$\delta_v = 1.848\Delta\theta$	$\delta_w = 2\Delta\theta$
Extreme-valued shift direction of r' image displacement	a	0.383	0.924	0
	b	0.383	0.924	0
	c	0	0	1
Extreme value of r' image displacement		$\delta_a = 0.765\Delta g$	$\delta_b = 1.848\Delta g$	$\delta_c = 2\Delta g$

Planar three-dimensional zero-valued poles	C_{Π_1} (Mid-point of segment $\overline{F_0 F'}$)	$x_1' = \frac{1}{2}\left(1.707b' - 3.037D + 2.516\dfrac{D}{n}\right)$
		$y_1' = \frac{1}{2}\left(-0.707b' + 1.258D - 2.516\dfrac{D}{n}\right)$
		$z_1' = 0$
		Associated plane Π_1: Conjugate optical-axis section $x'o'y'$
	C_{Π_2} (Approaches to infinity along the intersecting line of planes Π_1 and Π_2)	$x_2' = \infty$
		$y_2' = \infty$
		$z_2' = 0$
		Associated plane Π_2: Plane passing through C_{Π_1} and perpendicular to $\vec{\lambda}$

Equations of Image Motion

Parallel beam (Image rotation $\Delta\vec{\mu}'$)

Image lean

$$\Delta\mu_{x'}' = \Delta\theta\,(0.293\,P_{x'} + 0.707\,P_{y'})$$

Optical axis deviation

$$\Delta\mu_{y'}' = \Delta\theta\,(0.707\,P_{x'} + 1.707\,P_{y'})$$

$$\Delta\mu_{z'}' = 2\Delta\theta\,P_{z'}$$

Convergent beam (Image point displacement $\Delta S_{F'}'$)

Parallax

$$\Delta S_{F'x'}' = \Delta\theta\,[\,0.707\,P_{x'}z_q' - 0.293\,P_{y'}z_q' + P_{z'}(-0.707x_q' + 0.293y_q' - 1.258D)\,]$$
$$+ \Delta g\,(0.293\,D_{x'} + 0.707\,D_{y'})$$

Optical axis deviation

$$\Delta S'_{F'_{y'}} = \Delta\theta\left[1.707\,P_{x'}z'_q - 0.707P_{y'}z'_q + P_{z'}\left(-1.707x'_q + 0.707\,y'_q + 2b'\right.\right.$$
$$\left.\left. - 3.037D + 3.558\,\frac{D}{n}\right)\right] + \Delta g\,(0.707D_{x'} + 1.707D_{y'})\quad .$$

$$\Delta S'_{F'_{z'}} = \Delta\theta\left[P_{x'}\left(-2y'_q - 0.707b' + 1.258D - 2.516\,\frac{D}{n}\right) + P_{y'}\left(2x'_q\right.\right.$$
$$\left.\left. - 1.707b' + 3.037\,D - 2.516\,\frac{D}{n}\right)\right] + 2\,\Delta g\,D_{z'}$$

Adjustment Diagram

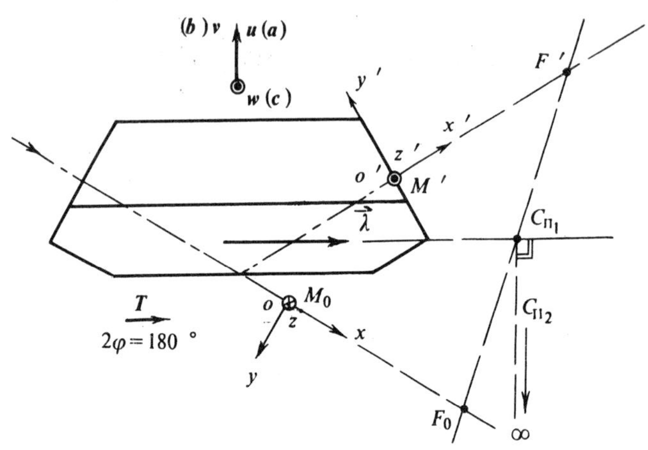

Fig . A .9 .

Characteristic Parameters of Image Formation

Direction cosine		Unit vectors of image coordinate system			2φ = 180 °
		$i'(x')$	$j'(y')$	$k'(z')$	T
Unit vectors of coordinate system in object space	$i(x)$	0.5	−0.866	0	0.866
	$j(y)$	−0.866	−0.5	0	−0.5
	$k(z)$	0	0	−1	0

$$S_{T,2\varphi} = \begin{pmatrix} 0.5 & -0.866 & 0 \\ -0.866 & -0.5 & 0 \\ 0 & 0 & -1 \end{pmatrix}$$

$$t = 2 \qquad L = 2.646\,D \qquad\qquad 2d = 2.292\,D - 2.291\,\frac{D}{n}$$

$$\overrightarrow{M_0 M'} = \left(1.985D - 1.323\,\frac{D}{n}\right) i' + \left(-1.146D + 2.291\,\frac{D}{n}\right) j'$$

$$r_0 = -0.331\,\frac{D}{n}\,i' - 0.573\,\frac{D}{n}\,j'$$

$$r_i =$$

Characteristic Parameters of Adjustment

Direction cosine		Image coordinate system		
		x'	y'	z'
Extreme- valued axis direction of r' image rotation	u	0.5	0.866	0
	v	0.5	0.866	0
	w	0	0	1
Extreme value of r' image rotation		$\delta_u = \Delta\theta$	$\delta_v = 1.732\Delta\theta$	$\delta_w = 2\Delta\theta$
Extreme- valued shift direction of r' image displacement	a	0.5	0.866	0
	b	0.5	0.866	0
	c	0	0	1
Extreme value of r' image displacement		$\delta_a = \Delta g$	$\delta_b = 1.732\Delta g$	$\delta_c \doteq 2\Delta g$

Planar three-dimensional zero- valued poles	C_{Π_1} (Mid-point of segment $\overline{F_0 F'}$)	$x_1' = \dfrac{1}{2}\left(1.5\, b' - 1.984 D + 1.323\,\dfrac{D}{n}\right)$
		$y_1' = \dfrac{1}{2}\left(-0.866\, b' + 1.146\, D - 2.291\,\dfrac{D}{n}\right)$
		$z_1' = 0$
		Associated plane Π_1 : Conjugate optical- axis section $x'o'y'$
	C_{Π_2} (Approaches to infinity along the intersecting line of planes Π_1 and Π_2)	$x_2' = \infty$
		$y_2' = \infty$
		$z_2' = 0$
		Associated plane Π_2: Plane passing through C_{Π_1} and perpendicular to $\vec{\lambda}$

Equations of Image Motion

Parallel beam (Image rotation $\Delta\vec{\mu}'$)

Image lean

$$\Delta\mu'_{x'} = \Delta\theta\,(0.5\, P_{x'} + 0.866 P_{y'})$$

Optical axis deviation

$$\Delta\mu'_{y'} = \Delta\theta\,(0.866\, P_{x'} + 1.5 P_{y'})$$

$$\Delta\mu'_{z'} = 2\Delta\theta\, P_{z'}$$

Convergent beam (Image point displacement $\Delta S_{F'}'$)

Parallax

$$\Delta S'_{F'x'} = \Delta\theta[0.866 P_{x'}\, z'_q - 0.5\, P_{y'}\, z'_q + P_{z'}\,(-0.866 x'_q + 0.5 y'_q - 1.146 D)]$$

$$+ \Delta g\, (0.5 D_{x'} + 0.866 D_{y'})$$

Optical axis deviation

$$\Delta S'_{F'y'} = \Delta\theta \left[1.5 P_{x'} z'_q - 0.866 P_{y'} z'_q + P_{z'}\left(-1.5 x'_g + 0.866 y'_q + 2 b' - 1.984 D \right.\right.$$

$$\left.\left. + 2.646\,\frac{D}{n} \right)\right] + \Delta g\, (0.866 D_{x'} + 1.5 D_{y'})$$

$$\Delta S'_{F'z'} = \Delta\theta \left[P_{x'}\left(-2 y'_q - 0.866 b' + 1.146 D - 2.291\,\frac{D}{n} \right) + P_{y'}\left(2 x'_q - 1.5 b' \right.\right.$$

$$\left.\left. + 1.984 D - 1.323\,\frac{D}{n} \right)\right] + 2\Delta g\, D_{z'}$$

Adjustment Diagram

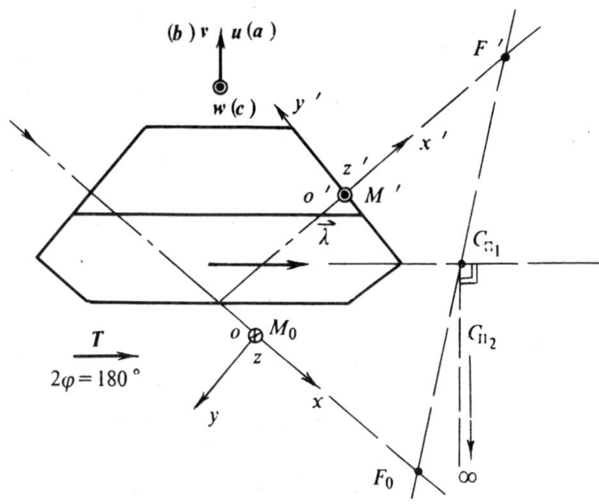

Fig . A . 10 .

Characteristic Parameters of Image Formation

Direction cosine		Unit vectors of image coordinate system			$2\varphi = 180°$
		$i'(x')$	$j'(y')$	$k'(z')$	T
Unit vectors of coordinate system in object space	$i(x)$	$S_{T.2\varphi} = \begin{pmatrix} 0.174 & -0.985 & 0 \\ -0.985 & -0.174 & 0 \\ 0 & 0 & -1 \end{pmatrix}$			0.766
	$j(y)$				-0.643
	$k(z)$				0

$t = 2$	$L = 1.960D$		$2d = 1.501D - 1.502\dfrac{D}{n}$

$$\overrightarrow{M_0 M'} = \left(1.150D - 0.340\frac{D}{n}\right)i' + \left(-0.965D + 1.931\frac{D}{n}\right)j'$$

$$r_0 = -0.405\frac{D}{n}i' - 0.483\frac{D}{n}j'$$

$$r_i =$$

Characteristic Parameters of Adjustment

Direction cosine		Image coordinate system		
		x'	y'	z'
Extreme-valued axis direction of r' image rotation	u	0.643	0.766	0
	v	0.643	0.766	0
	w	0	0	1
Extreme value of r' image rotation		$\delta_u = 1.286\Delta\theta$	$\delta_v = 1.532\Delta\theta$	$\delta_w = 2\Delta\theta$
Extreme-valued shift direction of r' image displacement	a	0.643	0.766	0
	b	0.643	0.766	0
	c	0	0	1
Extreme value of r' image displacement		$\delta_a = 1.286\Delta g$	$\delta_b = 1.532\Delta g$	$\delta_c = 2\Delta g$

Planar three-dimensional zero-valued poles	C_{Π_1} (Mid-point of segment $\overline{F_0F'}$)	$x_1' = \dfrac{1}{2}\left(1.174b' - 1.150D + 0.340\dfrac{D}{n}\right)$
		$y_1' = \dfrac{1}{2}\left(-0.985b' + 0.965D - 1.931\dfrac{D}{n}\right)$
		$z_1' = 0$
		Associated plane Π_1: Conjugate optical-axis section $x'o'y'$
	C_{Π_2} (Approaches to infinity along the intersecting line of planes Π_1 and Π_2.)	$x_2' = \infty$
		$y_2' = \infty$
		$z_2' = 0$
		Associated plane Π_2: Plane passing through C_{Π_1} and perpendicular to $\vec{\lambda}$

Equations of Image Motion

Parallel beam (Image rotation $\Delta\vec{\mu'}$)

Image lean

$$\Delta\mu_{x'}' = \Delta\theta\,(0.826P_{x'} + 0.985P_{y'})$$

Optical axis deviation

$$\Delta\mu_{y'}' = \Delta\theta\,(0.985P_{x'} + 1.174P_{y'})$$

$$\Delta\mu_{z'}' = 2\Delta\theta\,P_{z'}$$

Convergent beam (Image point displacement $\Delta S_{F'}'$)

302

Parallax

$$\Delta S_{F'x'}' = \Delta\theta\,[\,0.985 P_{x'}\,z_q' - 0.826 P_{y'}\,z_q' + P_{z'}\,(-0.985 x_q' + 0.826 y_q' - 0.965 D)\,]$$

$$+\,\Delta g\,(0.826\,D_{x'} + 0.985 D_{y'})$$

Optical axis deviation

$$\Delta S_{F'y'}' = \Delta\theta\left[1.174\,P_{x'}\,z_q' - 0.985 P_{y'}\,z_q' + P_{z'}\left(-1.174 x_q' + 0.985 y_q' + 2b'\right.\right.$$

$$\left.\left.-\,1.150 D + 1.960\,\frac{D}{n}\right)\right] + \Delta g\,(0.985 D_{x'} + 1.174 D_{y'})$$

$$\Delta S_{F'z'}' = \Delta\theta\left[P_x'\left(-2 y_q' - 0.985 b' + 0.965 D - 1.931\,\frac{D}{n}\right) + P_{y'}\,(2 x_q'\right.$$

$$\left.-\,1.174 b' + 1.150 D - 0.340\,\frac{D}{n}\right)\bigg] + 2\Delta g\,D_{z'}$$

Adjustment Diagram

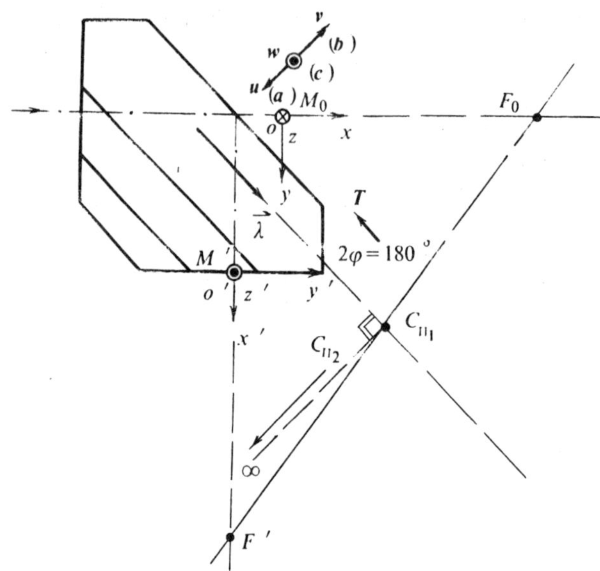

Fig . A . 11 .

Characteristic Parameters of Image Formation

Direction cosine		Unit vectors of image coordinate system			$2\varphi = 180$ °
		$i'(x')$	$j'(y')$	$k'(z')$	T
Unit vectors of coordinate system in object space	$i(x)$		0	1	0	-0.707
	$j(y)$	$S_{T,2\varphi}=$	1	0	0	-0.707
	$k(z)$		0	0	-1	0

$$t = 2 \qquad L = 1.732D \qquad\qquad 2d = -1.225D + 1.225\frac{D}{n}$$

$$\overrightarrow{M_0 M'} = 0.866\,D\,i' + \left(0.866D - 1.732\frac{D}{n}\right)j'$$

$$r_0 = -0.433\frac{D}{n}\,i' + 0.433\frac{D}{n}\,j'$$

$$r_i =$$

Characteristic Parameters of Adjustment

Direction cosine		Image coordinate system		
		x'	y'	z'
Extreme-valued axis direction of r' image rotation	u	0.707	-0.707	0
	v	-0.707	0.707	0
	w	0	0	1
Extreme value of r' image rotation		$\delta_u = 1.414\Delta\theta$	$\delta_v = 1.414\Delta\theta$	$\delta_w = 2\Delta\theta$
Extreme-valued shift direction of r' image displacement	a	0.707	-0.707	0
	b	-0.707	0.707	0
	c	0	0	1
Extreme value of r' image displacement		$\delta_a = 1.414\Delta g$	$\delta_b = 1.414\Delta g$	$\delta_c = 2\Delta g$

Planar three-dimensional zero-valued poles	C_{Π_1} (Mid-point of segment $\overline{F_0F'}$)	$x_1' = \dfrac{1}{2}(b' - 0.866D)$
		$y_1' = \dfrac{1}{2}\left(b' - 0.866D + 1.732\dfrac{D}{n}\right)$
		$z_1' = 0$
		Associated plane Π_1: Conjugate optical-axis section $x'o'y'$
	C_{Π_2} (Approaches to infinity along the intersecting line of planes Π_1 and Π_2)	$x_2' = \infty$
		$y_2' = \infty$
		$z_2' = 0$
		Associated plane Π_2: Plane passing through C_{Π_1} and perpendicular to $\vec{\lambda}$

Equations of Image Motion

Parallel beam (Image rotation $\Delta\vec{\mu}'$)
Image lean

$$\Delta\mu_{x'}' = \Delta\theta(P_{x'} - P_{y'})$$

Optical axis deviation

$$\Delta\mu_{y'}' = \Delta\theta(-P_{x'} + P_{y'})$$

$$\Delta\mu_{z'}' = 2\Delta\theta P_{z'}$$

Convergent beam (Image point displacement $\Delta S_{F'}'$)

Parallax

$$\Delta S'_{F'_{x'}} = \Delta \theta [-P_{x'} z'_q - P_{y'} z'_q + P_{z'} (x'_q + y'_q + 0.866D)] + \Delta g (D_{x'} - D_{y'})$$

Optical axis deviation

$$\Delta S'_{F'_{y'}} = \Delta \theta \left[P_{x'} z'_q + P_{y'} z'_q + P_{z'} \left(-x'_q - y'_q + 2b' - 0.866 D + 1.732 \frac{D}{n} \right) \right]$$

$$+ \Delta g (-D_{x'} + D_{y'})$$

$$\Delta S'_{F'_{z'}} = \Delta \theta \left[P_x \left(-2 y'_q + b' - 0.866D + 1.732 \frac{D}{n} \right) + P_{y'} (2 x'_q - b' + 0.866D) \right]$$

$$+ 2\Delta g D_{z'}$$

Adjustment Diagram

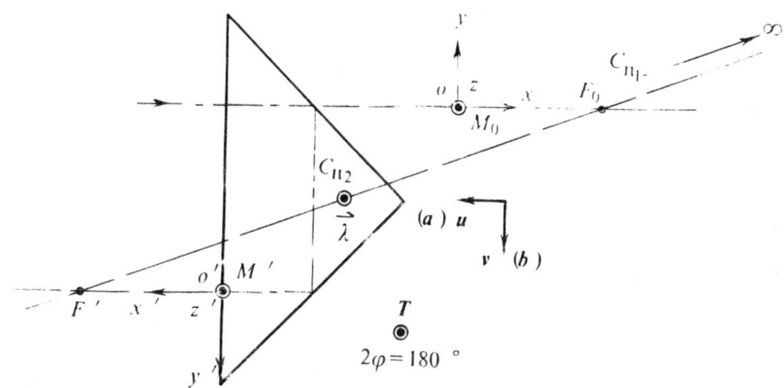

Fig. A .12 .

Characteristic Parameters of Image Formation

Direction cosine		Unit vectors of image coordinate system			$2\varphi = 180°$
		$i'(x')$	$j'(y')$	$k'(z')$	T
Unit vectors of coordinate system in object space	$i(x)$		-1	0	0	0
	$j(y)$	$S_{T.2\varphi} =$	0	-1	0	0
	$k(z)$		0	0	1	1

$t = 2 \qquad L = 2D$ $\qquad\qquad 2d = 0$

$$\overrightarrow{M_0 M'} = 2\frac{D}{n}\, i' + D j'$$

$$r_0 = -\frac{D}{n}\, i + 0.5\, D\, j'$$

$$r_i =$$

Characteristic Parameters of Adjustment

Direction cosine		Image coordinate system		
		x'	y'	z'
Extreme-valued axis direction of r' image rotation	u	1	0	0
	$\dot{v}$	0	1	0
	w	/	/	/
Extreme value of r' image rotation		$\delta_u = 2\Delta\theta$	$\delta_v = 2\Delta\theta$	$\delta_w = 0$
Extreme-valued shift direction of r' image displacement	a	1	0	0
	b	0	1	0
	c	/	/	/
Extreme value of r' image displacement		$\delta_a = 2\Delta g$	$\delta_b = 2\Delta g$	$\delta_c = 0$

Planar three-dimensional zero-valued poles	C_{Π_1} (Approaches to infinity along $\overrightarrow{F_0 F'}$)	$x_1' = \infty$
		$y_1' = \infty$
		$z_1' = 0$
		Associated plane Π_1: Conjugate optical-axis section $x'o'y'$
	C_{Π_2}	$x_2' = -\dfrac{D}{n}$
		$y_2' = -\dfrac{D}{2}$
		$z_2' = 0$
		Associated plane Π_2: Plane passing through $\overrightarrow{\lambda}$ and parallel to $\overrightarrow{F_0 F'}$

Equations of Image Motions

Parallel beam (Image rotation $\Delta\overrightarrow{\mu'}$)

Image lean

$$\Delta\mu'_{x'} = 2\Delta\theta\, P_{x'}$$

Optical axis deviation

$$\Delta\mu'_{y'} = 2\Delta\theta\, P_{y'}$$

$$\Delta\mu'_{z'} = 0$$

Convergent beam (Image point displacement $\Delta S'_{F'}$)

Parallax

308

$$\Delta S'_{F'_x} = \Delta\theta\,[\,-2P_{y'}z'_q + P_{z'}\,(2y'_q + D)\,] + 2\,\Delta g\,D_{x'}$$

Optical axis deviation

$$\Delta S'_{F'_y} = \Delta\theta\left[2P_{x'}z'_q + P_{z'}\left(-2x_{q'} - 2\frac{D}{n}\right)\right] + 2\Delta g\,D_{y'}$$

$$\Delta S'_{F'_z} = \Delta\theta\left[DP_{x'} + P_{y'}\left(-2b' - 2\frac{D}{n}\right)\right]$$

Adjustment Diagram

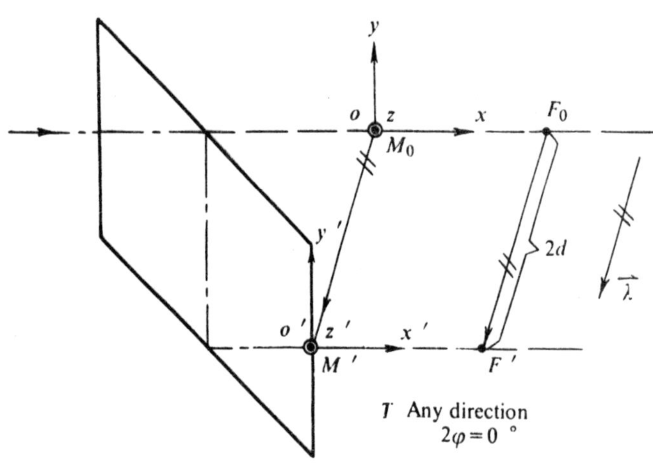

Fig . A . 13 .

Characteristic Parameters of Image Formation

Direction cosine		Unit vectors of image coordinate system			$2\varphi = 0°$ T (Any direction)
		$i'(x')$	$j'(y')$	$k'(z')$	
Unit vectors of coordinate system in object space	$i(x)$	$S_{T.2\varphi} = \begin{pmatrix} 1 & & \\ & & \end{pmatrix}$	0	0	
	$j(y)$	0	1	0	
	$k(z)$	0	0	1	

$t = 2$ $L = 2D$ $2d = \sqrt{(D - 2\dfrac{D}{n})^2 + D^2}$

$$\overrightarrow{M_0 M'} = \left(D - 2\frac{D}{n}\right)i' - Dj'$$

$r_0 =$ A vector of any length , and directed perpendicularly to $\overrightarrow{M_0 M'}$

$r_i =$

Characteristic Parameters of Adjustment

Direction cosine		Image coordinate system		
		x'	y'	z'
Extreme-valued axis direction of r' image rotation	u			
	v			
	w			
Extreme value of r' image rotation		$\delta_u = 0$	$\delta_v = 0$	$\delta_w = 0$
Extreme-valued shift direction of r' image displacement	a			
	b			
	c			
Extreme value of r' image displacement		$\delta_a = 0$	$\delta_b = 0$	$\delta_c = 0$

Spatial three-dimensional zero-valued pole (Approaches to infinity along $\overrightarrow{M_0M}'$)	C	$x' = \infty$
		$y' = \infty$
		$z' = 0$

Note : C is located at infinity along $\overrightarrow{M_0M}'$, that is ,all lines parallel to $\overrightarrow{M_0M}'$ are three-dimensional zero-valued axes .

Equations of Image Motion

Parallel beam (Image rotation $\overrightarrow{\Delta\mu}'$)
Image lean
$$\Delta\mu'_{x'} = 0$$
Optical axis deviation
$$\Delta\mu'_{y'} = 0$$
$$\Delta\mu'_{z'} = 0$$

Convergent beam (Image point displacement $\Delta S'_F$)
Parallax
$$\Delta S'_{F'x'} = D\Delta\theta\, P_{z'}$$

Optical axis deviation

$$\Delta S'_{F'_{y'}} = \Delta\theta\, P_{z'}\left(D - 2\,\frac{D}{n}\right)$$

$$\Delta S'_{F'_{z'}} = \Delta\theta\left[-DP_{x'} + P_{y'}\left(-D + 2\,\frac{D}{n}\right)\right]$$

Table A. 14   BII– 40 °

Adjustment Diagram

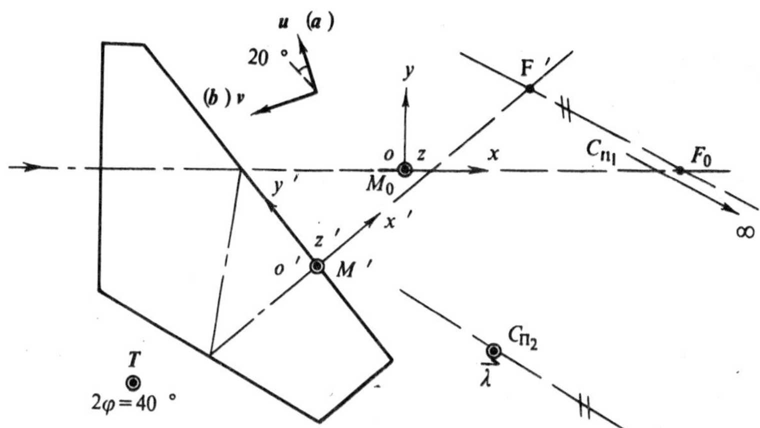

Fig . A . 14 .

Characteristic Parameters of Image Formation

Direction cosine		Unit vectors of image coordinate system			$2\varphi = 40$ °
		$i'(x')$	$j'(y')$	$k'(z')$	**T**
Unit vectors of coordinate system in object space	$i(x)$	0.766	−0.643	0	0
	$j(y)$	$S_{T,2\varphi} =$ 0.643	0.766	0	0
	$k(z)$	0	0	1	1

$t = 2$ $L = 1.969 D$ $2d = 0$

$$\overrightarrow{M_0 M'} = \left(0.456D - 1.509\frac{D}{n}\right) i' + \left(-0.883D + 1.266\frac{D}{n}\right) j'$$

$$r_0 = \left(0.985D - 0.985\frac{D}{n}\right) i' + \left(1.068D - 2.706\frac{D}{n}\right) j'$$

$$r_i =$$

Characteristic Parameters of Adjustment

Direction cosine		Image coordinate system		
		x'	y'	z'
Extreme - valued axis direction of r' image rotation	u	0.342	0.940	0
	v	-0.940	0.342	0
	w			
Extreme value of r' image rotation		$\delta_u = 0.684\Delta\theta$	$\delta_v = 0.684\,\Delta\theta$	$\delta_w = 0$
Extreme - valued shift direction of r' image displacement	a	0.342	0.940	0
	b	-0.940	0.342	0
	c			
Extreme value of r' image displacement		$\delta_a = 0.684\Delta g$	$\delta_b = 0.684\Delta g$	$\delta_c = 0$

Planar three-dimensional zero-valued poles	C_{Π_1} (Approaches to infinity along $\overrightarrow{F_0 F}'$)	$x_1' = \infty$
		$y_1' = \infty$
		$z_1' = 0$
		Associated plane Π_1: Conjugate optical- axis section $x'o'y'$
	C_{Π_2}	$x_2' = 0.975D - 0.975\,\dfrac{D}{n}$
		$y_2' = 1.051D - 2.679\,\dfrac{D}{n}$
		$z_2' = 0$
		Associated plane Π_2: Plane passing through $\vec{\lambda}$ and parallel to $\overrightarrow{F_0 F}'$

Equations of Image Motion

Parallel beam   (Image rotation $\Delta\vec{\mu}'$)

Image lean

$$\Delta\mu_{x'}' = \Delta\theta\,(0.234 P_{x'} + 0.643\,P_{y'})$$

Optical axis deviation

$$\Delta\mu_{y'}' = \Delta\theta\,(-0.643 P_{x'} + 0.234\,P_{y'})$$

$$\Delta\mu_{z'}' = 0$$

Convergent beam   (Image point displacement $\Delta S_{F'}'$)

Parallax

$$\Delta S'_{F'_{x'}} = \Delta\theta\,[\,0.643 P_{x'} z'_q - 0.234 P_{y'} z'_q + P_{z'}(-0.643 x'_q + 0.234 y'_q + 0.383 D)\,]$$
$$+\Delta g\,(0.234 D_{x'} + 0.643 D_{y'})$$

Optical axis deviation

$$\Delta S'_{F'_{y'}} = \Delta\theta\left[\,0.234 P_{x'} z'_q + 0.643 P_{y'} z'_q + P_{z'}\left(-0.243 x'_q - 0.643 y'_q + 0.917 D\right.\right.$$
$$\left.\left.-1.969\frac{D}{n}\right)\right] + \Delta g\,(-0.643 D_{x'} + 0.243 D_{y'})$$

$$\Delta S'_{F'_{z'}} = \Delta\theta\left[P_{x'}\left(0.643 b' - 0.883 D + 1.266\frac{D}{n}\right) + P_{y'}\left(-0.234 b' - 0.456 D\right.\right.$$
$$\left.\left. + 1.509\frac{D}{n}\right)\right]$$

Table A.15   BII − 45 °

Adjustment Diagram

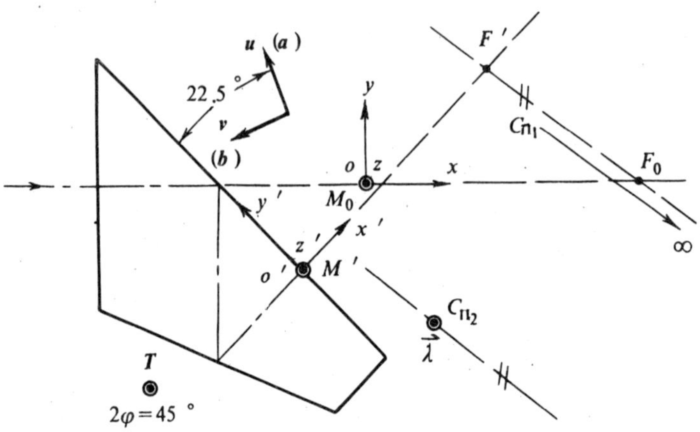

Fig. A . 15 .

Characteristic Parameters of Image Formation

Direction cosine		Unit vectors of image coordinate system			$2\varphi = 45°$
		$i\,'(x\,')$	$j\,'(y\,')$	$k\,'(z\,')$	**T**
Unit vectors of coordinate system in object space	$i\,(x)$	$S_{T.2\varphi}= \begin{pmatrix} 0.707$	-0.707	0	0
	$j\,(y)$	0.707	0.707	0	0
	$k\,(z)$	0	0	$1 \end{pmatrix}$	1

$t = 2$ $L = 1.707\,D$ $2d = 0$

$$\overrightarrow{M_0 M\,'} = \left(0.354D - 1.207\frac{D}{n}\right)i\,' + \left(-0.854D + 1.207\frac{D}{n}\right)j\,'$$

$$r_0 = \left(0.854D - 0.854\frac{D}{n}\right)i\,' + \left(0.854D - 2.060\frac{D}{n}\right)j\,'$$

$$r_i =$$

316

Characteristic Parameters of Adjustment

Direction cosine		Image coordinate system		
		x'	y'	z'
Extreme-valued axis direction of r' image rotation	u	0.383	0.924	0
	v	-0.924	0.383	0
	w	/	/	/
Extreme value of r' image rotation		$\delta_u = 0.765\,\Delta\theta$	$\delta_v = 0.765\,\Delta\theta$	$\delta_w = 0$
Extreme-valued shift direction of r' image displacement	a	0.384	0.924	0
	b	-0.924	0.383	0
	c	/	/	/
Extreme value of r' image displacement		$\delta_a = 0.765\Delta g$	$\delta_b = 0.765\Delta g$	$\delta_c = 0$

Planar three-dimensional zero-valued poles	C_{Π_1} (Approaches to infinity along $\overrightarrow{F_0 F'}$)	$x_1' = \infty$
		$y_1' = \infty$
		$z_1' = 0$
		Associated plane Π_1: Conjugate optical-axis section $x'o'y'$
	C_{Π_2}	$x_2' = 0.855D - 0.854\dfrac{D}{n}$
		$y_2' = -0.854D - 2.061\dfrac{D}{n}$
		$z_2' = 0$
		Associated plane Π_2: Plane passing through $\vec{\lambda}$ and parallel to $\overrightarrow{F_0 F'}$

Equations of Image Motion

Parallel beam   (Image rotation $\Delta\vec{\mu}'$)
Image lean

$$\Delta\mu_{x'}' = \Delta\theta\,(0.293P_{x'} + 0.707P_{y'})$$

Optical axis deviation

$$\Delta\mu_{y'}' = \Delta\theta\,(-0.707P_{x'} + 0.293P_{y'})$$

$$\Delta\mu_{z'}' = 0$$

Convergent beam   (Image point displacement $\Delta S_{F'}'$)
Parallax

$$\Delta S'_{F'x'} = \Delta\theta\,[\,0.707\,P_{x'}z'_q - 0.293P_{y'}z'_q + P_{z'}(-0.707x'_q + 0.293y'_q + 0.354D)\,]$$
$$+\Delta g\,(0.293D_{x'} + 0.707D_{y'})$$

Optical axis deviation

$$\Delta S'_{F'y'} = \Delta\theta\left[0.293P_{x'}z'_q + 0.707P_{y'}z'_q + P_{z'}\left(-0.293x'_q - 0.707y'_q + 0.854D\right.\right.$$
$$\left.\left. - 1.707\frac{D}{n}\right)\right] + \Delta g\,(-0.707D_{x'} + 0.293D_{y'})$$

$$\Delta S'_{F'z'} = \Delta\theta\left[P_{x'}\left(0.707b' - 0.854D + 1.207\frac{D}{n}\right)\right.$$
$$\left. + P_{y'}\left(-0.293b' - 0.354D + 1.207\frac{D}{n}\right)\right]$$

Adjustment Diagram

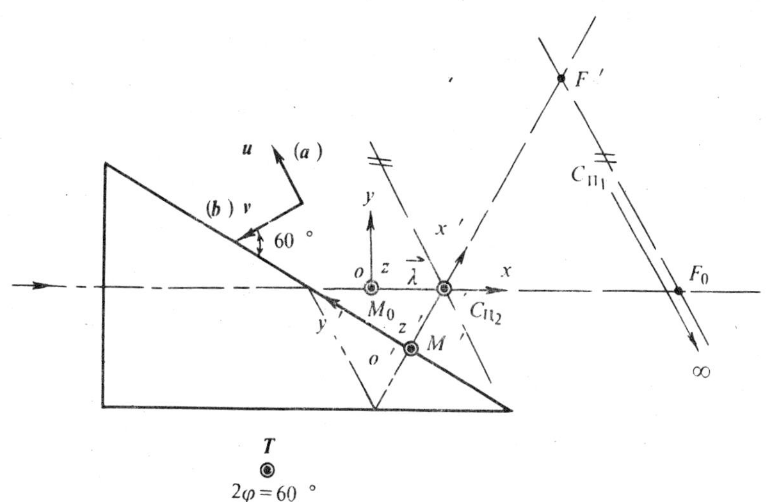

Fig. A .16 .

Characteristic Parameters of Image Formation

Direction cosine		Unit vectors of image coordinate system			$2\varphi = 60°$
		$i'(x')$	$j'(y')$	$k'(z')$	T
Unit vectors of coordinate system in object space	$i(x)$	$S_{T,2\varphi} = \big($ 0.5	-0.866	0 $\big)$	0
	$j(y)$	0.866	0.5	0	0
	$k(z)$	0	0	1	1

$t = 2$ $\qquad L = 1.732D$		$2d = 0$

$$\overrightarrow{M_0 M'} = \left(0.433D - 0.866\,\frac{D}{n}\right)i' + \left(-1.25D + 1.5\,\frac{D}{n}\right)j'$$

$$r_0 = \left(0.866D - 0.866\,\frac{D}{n}\right)i' + \left(D - 1.5\,\frac{D}{n}\right)j'$$

$$r_i =$$

Characteristic Parameters of Adjustment

Direction cosine		Image coordinate system		
		x'	y'	z'
Extreme - valued axis direction of r' image rotation	u	0.5	0.866	0
	v	-0.866	0.5	0
	w			
Extreme value of r' image rotation		$\delta_u = \Delta\theta$	$\delta_v = \Delta\theta$	$\delta_w = 0$
Extreme - valued shift direction of r' image displacement	a	0.5	0.866	0
	b	-0.866	0.5	0
	c			
Extreme value of r' image displacement		$\delta_a = \Delta g$	$\delta_b = \Delta g$	$\delta_c = 0$

Planar three-dimensional zero-valued poles	C_{Π_1} (Approaches to infinity along $\overrightarrow{F_0 F'}$)	$x_1' = \infty$
		$y_1' = \infty$
		$z_1' = 0$
		Associated plane Π_1: Conjugate optical-axis section $x'o'y'$
	C_{Π_2}	$x_2' = 0.866D - 0.866\dfrac{D}{n}$
		$y_2' = D - 1.5\dfrac{D}{n}$
		$z_2' = 0$
		Associated plane Π_2: Plane passing through $\overrightarrow{\lambda}$ and parallel to $\overrightarrow{F_0 F'}$

Equations of Image Motion

Parallel beam   (Image rotation $\overrightarrow{\Delta\mu}'$)
Image lean

$$\Delta\mu_{x'}' = \Delta\theta\,(0.5P_{x'} + 0.866P_{y'})$$

Optical axis deviation

$$\Delta\mu_{y'}' = \Delta\theta\,(-0.866P_{x'} + 0.5\,P_{y'})$$

$$\Delta\mu_{z'}' = 0$$

Convergent beam   (Image point displacement $\Delta S_{F'}'$)
Parallax

320

$$\Delta S'_{F'x'} = \Delta\theta\,[\,0.866\,P_{x'}z'_q - 0.5\,P_{y'}z'_q + P_{z'}(-0.866x'_q + 0.5\,y'_q + 0.25\,D)]$$
$$+\,\Delta g\,(0.5\,D_{x'} + 0.866\,D_{y'})$$

Optical axis deviation

$$\Delta S'_{F'y'} = \Delta\theta\left[0.5\,P_{x'}z'_q + 0.866\,P_{y'}z'_q + P_{z'}\left(-0.5\,x'_q - 0.866y'_q + 1.299D\right.\right.$$
$$\left.\left. -\,1.732\frac{D}{n}\right)\right] + \Delta g\,(-0.866\,D_{x'} + 0.5\,D_{y'})$$

$$\Delta S'_{F'z'} = \Delta\theta\left[P_{x'}\left(0.866\,b' - 1.25\,D + 1.5\frac{D}{n}\right)\right.$$
$$\left. +\,P_{y'}\left(-0.5\,b' - 0.433D + 0.866\frac{D}{n}\right)\right]$$

Adjustment   Diagram

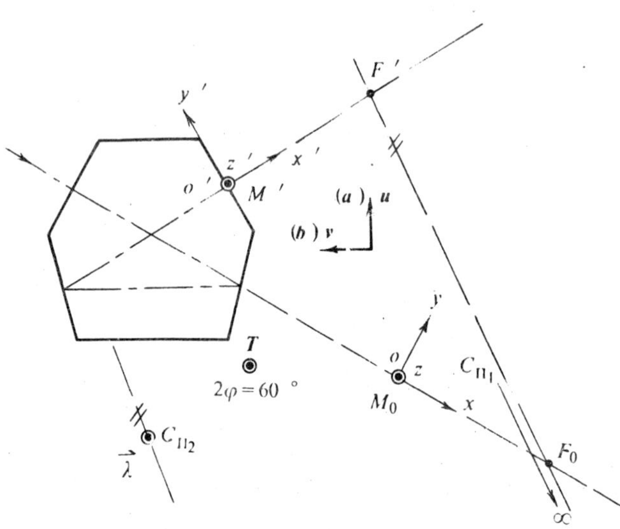

Fig . A. 17.

Characteristic Parameters of Image Formation

Direction cosine		Unit vectors of image coordinate system			$2\varphi = 60$ °
		$i'(x')$	$j'(y')$	$k'(z')$	T
Unit vectors	$i(x)$	0.5	−0.866	0	0
of coordinate					
system in	$j(y)$	0.866	0.5	0	0
object space	$k(z)$	0	0	1	1

$$S_{T.2\varphi} = \begin{pmatrix} 0.5 & -0.866 & 0 \\ 0.866 & 0.5 & 0 \\ 0 & 0 & 1 \end{pmatrix}$$

$t = 2$	$L = 5.464D$	$2d = 0$

$$\overrightarrow{M_0 M'} = \left(1.299D - 2.732\frac{D}{n}\right)i' + \left(-0.75D + 4.732\frac{D}{n}\right)j'$$

$$r_0 = -2.732\frac{D}{n}i' + \left(1.5D - 4.732\frac{D}{n}\right)j'$$

$$r_i =$$

Characteristic Parameters of Adjustment

Direction cosine		Image coordinate system		
		x'	y'	z'
Extreme-valued axis direction of r' image rotation	u	0.5	0.866	0
	v	-0.866	0.5	0
	w			
Extreme value of r' image rotation		$\delta_u = \Delta\theta$	$\delta_v = \Delta\theta$	$\delta_w = 0$
Extreme-valued shift direction of r' image displacement	a	0.5	0.866	0
	b	-0.866	0.5	0
	c			
Extreme value of r' image displacement		$\delta_a = \Delta g$	$\delta_b = \Delta g$	$\delta_c = 0$

Planar three-dimensional zero-valued poles	C_{Π_1} (Approaches to infinity along $\overrightarrow{F_0F'}$)	$x_1' = \infty$
		$y_1' = \infty$
		$z_1' = 0$
		Associated plane Π_1: Conjugate optical-axis section $x'o'y'$
	C_{Π_2}	$x_2' = -2.732\,\dfrac{D}{n}$
		$y_2' = 1.5\,D - 4.732\,\dfrac{D}{n}$
		$z_2' = 0$
		Associated plane Π_2: Plane passing through $\overrightarrow{\lambda}$ and parallel to $\overrightarrow{F_0F'}$

Equations of Image Motion

Parallel beam   (Image rotation $\Delta\overrightarrow{\mu}'$)
Image lean

$$\Delta\mu_{x'}' = \Delta\theta\,(0.5\,P_{x'} + 0.866 P_{y'})$$

Optical axis deviation

$$\Delta\mu_{y'}' = \Delta\theta\,(-0.866 P_{x'} + 0.5\,P_{y'})$$

$$\Delta\mu_{z'}' = 0$$

Convergent beam   (Image point displacement $\Delta S_{F'}'$)
Parallax

$$\Delta S_{F'x'}' = \Delta\theta\,[0.866\,P_{x'}z_q' - 0.5\,P_{y'}z_q' + P_{z'}(-0.866x_q' + 0.5y_q' - 0.75D)]$$
$$+ \Delta g\,(0.5\,D_{x'} + 0.866D_{y'})$$

Optical axis deviation

$$\Delta S_{F'y'}' = \Delta\theta\left[0.5\,P_{x'}z_q' + 0.866\,P_{y'}z_q' + P_{z'}\left(-0.5x_q' - 0.866y_q' + 1.299D\right.\right.$$
$$\left.\left. - 5.464\frac{D}{n}\right)\right] + \Delta g\,(-0.866D_{x'} + 0.5\,D_{y'})$$

$$\Delta S_{F'z'}' = \Delta\theta\left[P_{x'}\left(0.866b' - 0.75D + 4.732\frac{D}{n}\right)\right.$$
$$\left. + P_{y'}\left(-0.5b' - 1.299D + 2.732\frac{D}{n}\right)\right]$$

Adjustment Diagram

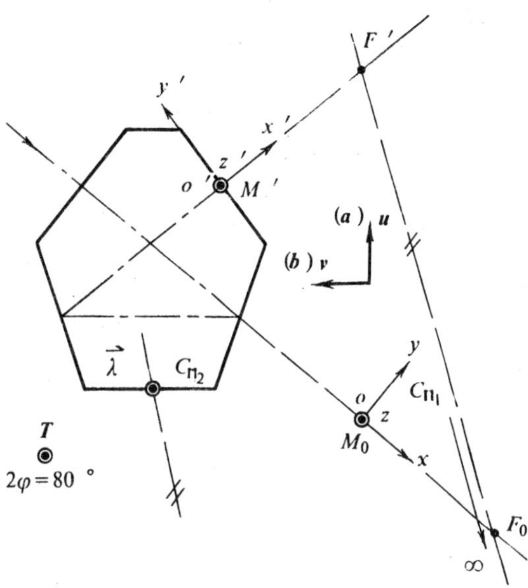

Fig . A .18 .

Characteristic Parameters of Image Formation

Direction cosine		Unit vectors of image coordinate system			$2\varphi = 80°$
		$i'(x')$	$j'(y')$	$k'(z')$	T
Unit vectors of coordinate system in object space	$i(x)$	$S_{T.2\varphi} =$ 0.174	−0.985	0	0
	$j(y)$	0.985	0.174	0	0
	$k(z)$	0	0	1	1

$t = 2$	$L = 3.940\,D$	$2d = 0$

$$\overrightarrow{M_0 M'} = \left(0.7D - 0.684\frac{D}{n}\right)i' + \left(-0.587D + 3.880\frac{D}{n}\right)j'$$

$$r_0 = -1.970\frac{D}{n}\,i' + \left(0.711D - 2.348\frac{D}{n}\right)j'$$

$$r_i =$$

Characteristic Parameters of Adjustment

Direction cosine		Image coordinate system		
		x'	y'	z'
Extreme-valued axis direction of r' image rotation	u	0.643	0.766	0
	v	−0.766	0.643	0
	w			
Extreme value of r' image rotation		$\delta_u = 1.286\Delta\theta$	$\delta_v = 1.286\Delta\theta$	$\delta_w = 0$
Extreme-valued shift direction of r' image displacement	a	0.643	0.766	0
	b	−0.766	0.643	0
	c			
Extreme value of r' image displacement		$\delta_a = 1.286\Delta g$	$\delta_b = 1.286\Delta g$	$\delta_c = 0$

Planar three-dimensional zero-valued poles	C_{Π_1} (Approaches to infinity along $\overrightarrow{F_0F'}$)	$x_1' = \infty$
		$y_1' = \infty$
		$z_1' = 0$
		Associated plane Π_1: Conjugate optical-axis section $x'o'y'$
	C_{Π_2}	$x_2' = -1.970\dfrac{D}{n}$
		$y_2' = 0.711D - 2.349\dfrac{D}{n}$
		$z_2' = 0$
		Associated plane Π_2: Plane passing through $\overrightarrow{\lambda}$ and parallel to $\overrightarrow{F_0F'}$

Equations of Image Motion

Parallel beam   (Image rotation $\Delta\overrightarrow{\mu}'$)
Image lean

$$\Delta\mu_{x'}' = \Delta\theta\,(0.826P_{x'} + 0.985P_{y'})$$

Optical axis deviation

$$\Delta\mu_{y'}' = \Delta\theta\,(-0.985P_{x'} + 0.826P_{y'})$$

$$\Delta\mu_{z'}' = 0$$

Convergent beam   (Image point displacement $\Delta S_F'$)
Parallax

326

$$\Delta S_{F'x'}' = \Delta\theta [0.985 P_{x'} z_q' - 0.826 P_{y'} z_q' + P_{z'} (-0.985 x_q' + 0.826 y_q' - 0.587D)]$$

$$+ \Delta g\,(0.866 D_{x'} + 0.985 D_{y'})$$

Optical axis deviation

$$\Delta S_{F'y'}' = \Delta\theta \left[0.826\,P_{x'} z_q' + 0.985 P_{y'} z_q' + P_{z'} \left(-0.826 x_q' - 0.985 y_q' + 0.7\,D \right.\right.$$

$$\left.\left. -3.940\,\frac{D}{n} \right) \right] + \Delta g\,(-0.985 D_{x'} + 0.826\,D_{y'})$$

$$\Delta S_{F'y'}' = \Delta\theta \left[P_{x'} \left(0.985b' - 0.587D + 3.880\,\frac{D}{n} \right) + P_{y'} \left(-0.826b' - 0.7D \right.\right.$$

$$\left.\left. + 0.684\,\frac{D}{n} \right) \right]$$

Adjustment   Diagram

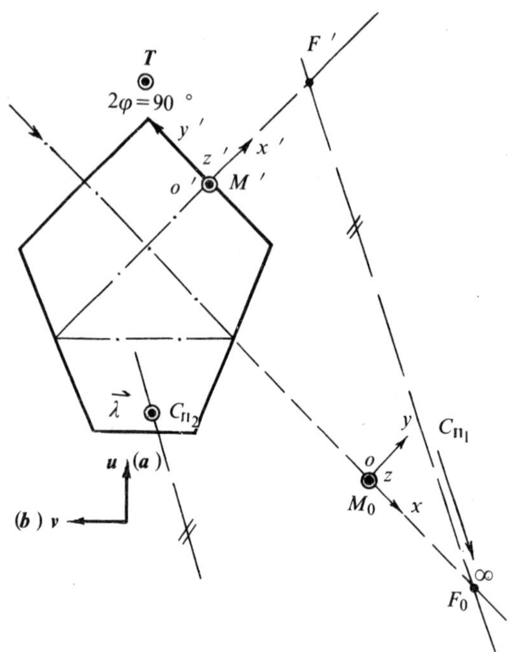

Fig . A .19 .

Characteristic Parameters of Image Formation

Direction cosine		Unit vectors of image coordinate system			$2\varphi = 90°$
		$i\,'(x\,')$	$j\,'(y\,')$	$k\,'(z\,')$	T
Unit vectors of coordinate system in object space	$i\,(x)$	0	-1	0	0
	$j\,(y)$ $S_{T,2\varphi} =$	1	0	0	0
	$k\,(z)$	0	0	1	1

$t = 2$ $L = 3.414D$		$2d = 0$

$$\overrightarrow{M_0 M}\,' = 0.5\,D\,i\,' + \left(-0.5\,D + 3.414\frac{D}{n}\right)j\,'$$

$$r_0 = -1.707\frac{D}{n}\,i\,' + \left(0.5D - 1.707\frac{D}{n}\right)j\,'$$

$$r_i =$$

Characteristic Parameters of Adjustment

Direction cosine		Image coordinate system		
		x'	y'	z'
Extreme-valued axis direction of r' image rotation	u	0.707	0.707	0
	v	-0.707	0.707	0
	w	/	/	/
Extreme value of r' image rotation		$\delta_u = 1.414\,\Delta\theta$	$\delta_v = 1.414\,\Delta\theta$	$\delta_w = 0$
Extreme-valued shift direction of r' image displacement	a	0.707	0.707	0
	b	-0.707	0.707	0
	c	/	/	/
Extreme value of r' image displacement		$\delta_a = 1.414\Delta g$	$\delta_b = 1.414\Delta g$	$\delta_c = 0$

Planar three-dimensional zero-valued poles	C_{Π_1} (Approaches to infinity along $\overrightarrow{F_0 F}\,'$)	$x_1' = \infty$
		$y_1' = \infty$
		$z_1' = 0$
		Associated plane Π_1: Conjugate optical-axis section $x'o'y'$
	C_{Π_2}	$x_2' = -1.707\dfrac{D}{n}$
		$y_2' = 0.5D - 1.707\dfrac{D}{n}$
		$z_2' = 0$
		Associated plane Π_2: Plane passing through $\overrightarrow{\lambda}$ and parallel to $\overrightarrow{F_0 F}\,'$

Equations of Image Motion

Parallel beam (Image rotation $\Delta\overrightarrow{\mu}\,'$)

Image lean

$$\Delta\mu_{x'}' = \Delta\theta\,(P_{x'} + P_{y'})$$

Optical axis deviation

$$\Delta\mu_{y'}' = \Delta\theta\,(-P_{x'} + P_{y'})$$

$$\Delta\mu_{z'}' = 0$$

Convergent beam (Image point displacement $\Delta S_F'$)

Parallax

$$\Delta S_{F'x'}' = \Delta\theta\left[\,P_{x'}z_q' - P_{y'}z_q' + P_{z'}(-x_q' + y_q' - 0.5\,D)\right] + \Delta g\,(D_{x'} + D_{y'})$$

Optical axis deviation

$$\Delta S_{F'y'}' = \Delta\theta\left[P_{x'}z_q' + P_{y'}z_q' + P_z\left(-x_q' - y_q' + 0.5D - 3.414\frac{D}{n}\right)\right]$$

$$+ \Delta g\,(-D_{x'} + D_{y'})$$

$$\Delta S_{F'z'}' = \Delta\theta\left[P_{x'}\left(b' - 0.5\,D + 3.414\frac{D}{n}\right) + P_{y'}(-b' - 0.5\,D)\right]$$

Adjustment Diagram

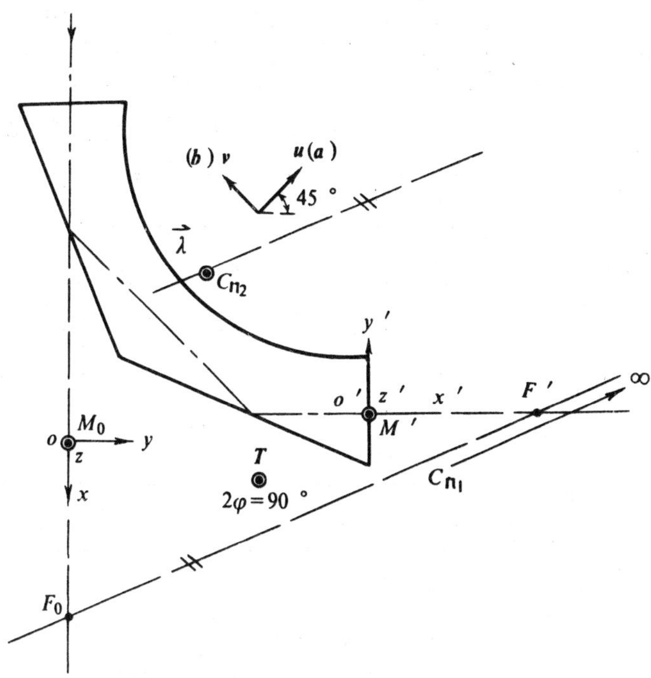

Fig .A .20 .

Characteristic Parameters of Image Formation

Direction cosine		Unit vectors of image coordinate system			$2\varphi = 90$ °
		$i'(x')$	$j'(y')$	$k'(z')$	T
Unit vectors of coordinate system in object space	$i(x)$	$S_{T,2\varphi} = \begin{pmatrix} 0 \\ 1 \\ 0 \end{pmatrix}$	$\begin{pmatrix} -1 \\ 0 \\ 0 \end{pmatrix}$	$\begin{pmatrix} 0 \\ 0 \\ 1 \end{pmatrix}$	0
	$j(y)$				0
	$k(z)$				1

$t=2$	$L=4.828 D$	$2d=0$

$$\overrightarrow{M_0 M'} = 2.914 D\, i' + \left(-2.914D + 4.828\frac{D}{n}\right) j'$$

$$r_0 = -2.414\frac{D}{n}\, i' + \left(2.914D - 2.414\frac{D}{n}\right) j'$$

$$r_i =$$

Characteristic Parameters of Adjustment

Direction cosine		Image coordinate system		
		x'	y'	z'
Extreme-valued axis direction of r' image rotation	u	0.707	0.707	0
	v	-0.707	0.707	0
	w			
Extreme value of r' image rotation		$\delta_u = 1.414\,\Delta\theta$	$\delta_v = 1.414\,\Delta\theta$	$\delta_w = 0$
Extreme-valued shift direction of r' image displacement	a	0.707	0.707	0
	b	-0.707	0.707	0
	c			
Extreme value of r' image displacement		$\delta_a = 1.414\,\Delta g$	$\delta_b = 1.414\,\Delta g$	$\delta_c = 0$

Planar three-dimensional zero-valued poles	C_{Π_1} (Approaches to infinity along $\overrightarrow{F_0 F'}$)	$x_1' = \infty$
		$y_1' = \infty$
		$z_1' = 0$
		Associated plane Π_1: Conjugate optical-axis section $x'o'y'$
	C_{Π_2}	$x_2' = -2.414\,\dfrac{D}{n}$
		$y_2' = 2.914D - 2.414\,\dfrac{D}{n}$
		$z_2' = 0$
		Associated plane Π_2: Plane passing through $\overrightarrow{\lambda}$ and parallel to $\overrightarrow{F_0 F'}$

Equations of Image Motion

Parallel beam (Image rotation $\Delta\overrightarrow{\mu}'$)
Image lean

$$\Delta\mu_{x'}' = \Delta\theta\,(P_{x'} + P_{y'})$$

Optical axis deviation

$$\Delta\mu_{y'}' = \Delta\theta\,(-P_{x'} + P_{y'})$$

$$\Delta\mu_{z'}' = 0$$

Convergent beam (Image point displacement $\Delta S_F'$)
Parallax

$$\Delta S_{F'x'}' = \Delta \theta [\, P_{x'} z_q' - P_{y'} z_q' + P_{z'}(-x_q' + y_q' - 2.914D)] + \Delta g (D_{x'} + D_{y'})$$

Optical axis deviation

$$\Delta S_{F'y'}' = \Delta \theta \left[P_{x'} z_q' + P_{y'} z_q' + P_{z'} \left(-x_q' - y_q' + 2.914D - 4.828 \frac{D}{n} \right) \right]$$

$$+ \Delta g (-D_{x'} + D_{y'})$$

$$\Delta S_{F'z'}' = \Delta \theta \left[P_{x'} \left(b' - 2.914D + 4.828 \frac{D}{n} \right) + P_{y'} (-b' - 2.914D) \right]$$

Adjustment Diagram

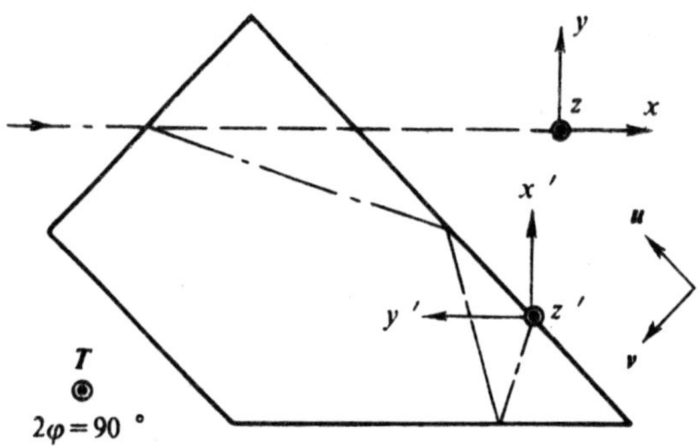

Fig . A .21 .

Characteristic Parameters of Image Formation

Direction cosine		Unit vectors of image coordinate system			$2\varphi = 90$ °
		$i\,'(x\,')$	$j\,'(y\,')$	$k\,'(z\,')$	T
Unit vectors of coordinate system in object space	$i\,(x)$	0	-1	0	0
	$j\,(y)$	1	0	0	0
	$k\,(z)$	0	0	1	1

$$S_{T,2\varphi} = \begin{pmatrix} 0 & -1 & 0 \\ 1 & 0 & 0 \\ 0 & 0 & 1 \end{pmatrix}$$

$t = 2 \qquad L = 2nD \qquad\qquad 2d =$

$$\overrightarrow{M_0 M\,'} =$$

$$r_0 =$$

$$r_i =$$

Characteristic Parameters of Adjustment

Direction cosine		Image coordinate system		
		x'	y'	z'
Extreme - valued axis direction of r' image rotation	u	0.707	0.707	0
	v	-0.707	0.707	0
	w	/	/	/
Extreme value of r' image rotation		$\delta_u = 1.414\Delta\theta$	$\delta_v = 1.414\Delta\theta$	$\delta_w = 0$
Extreme - valued shift direction of r' image displacement	a			
	b			
	c			
Extreme value of r' image displacement				

Planar three-dimensional zero-valued poles (Meaningless in collimated beams)	C_{Π_1}	$x_1' =$	
		$y_1' =$	
		$z_1' =$	
		Associated plane Π_1 :	
	C_{Π_2}	$x_2' =$	
		$y_2' =$	
		$z_2' =$	
		Associated plane Π_2 :	

Equations of Image Motion

Parallel beam (Image rotation $\overrightarrow{\Delta\mu}'$)
Image lean

$$\Delta\mu_{x'}' = \Delta\theta\,(P_{x'} + P_{y'})$$

Optical axis deviation

$$\Delta\mu_{y'}' = \Delta\theta\,(-P_{x'} + P_{y'})$$

$$\Delta\mu_{z'}' = 0$$

Adjustment Diagram

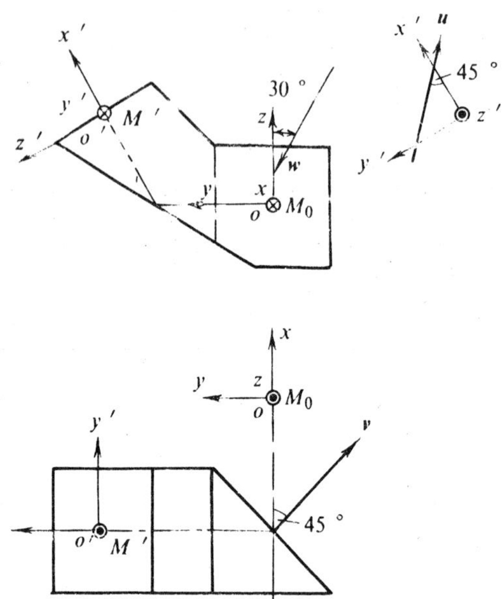

Fig. A .22 .

Characteristic Parameters of Image Formation

Direction cosine		Unit vectors of image coordinate system			$2\varphi = 221\,°24\,'$
		$i\,'(x\,')$	$j\,'(y\,')$	$k\,'(z\,')$	T
Unit vectors	$i(x)$	0	1	0	0.655
of coordinate	$j(y)$	$S_{T.2\varphi} =$ 0.5	0	0.866	0.655
system in object space	$k(z)$	0.866	0	−0.5	0.378

$t = 2 \qquad L = 2.366\,D$	$2d = 1.550D - 1.550\dfrac{D}{n}$

$$\overrightarrow{M_0 M\,'} = 1.366D\,i\,' + \left(0.5D - 2.366\frac{D}{n}\right)j\,' + 0.866D\,k\,'$$

$$r_0 = \left(-0.247D - 0.677\frac{D}{n}\right)i\,' + \left(0.267D + 0.676\frac{D}{n}\right)j\,' - 0.033D\,k\,'$$

$$r_i =$$

Characteristic Parameters of Adjustment

Direction cosine		Image coordinate system		
		x'	y'	z'
Extreme-valued axis direction of r' image rotation	u	0.707	-0.707	0
	v	-0.354	0.707	-0.612
	w	-0.5	0	0.866
Extreme value of r' image rotation		$\delta_u = 1.414\Delta\theta$	$\delta_v = 1.414\Delta\theta$	$\delta_w = 1.732\Delta\theta$
Extreme-valued shift direction of r' image displacement	a	0.707	-0.707	0
	b	-0.354	0.707	-0.612
	c	-0.5	0	0.866
Extreme value of r' image displacement		$\delta_a = 1.414\Delta g$	$\delta_b = 1.414\Delta g$	$\delta_c = 1.732\Delta g$
Planar three-dimensional zero-valued poles (Do not exist)	C_{Π_1}	$x_1' =$		
		$y_1' =$		
		$z_1' =$		
		Associated plane Π_1 :		
	C_{Π_2}	$x_2' =$		
		$y_2' =$		
		$z_2' =$		
		Associated plane Π_2 :		

Equations of Image Motion

Parallel beam (Image rotation $\Delta\vec{\mu}'$)
Image lean

$$\Lambda\mu_{x'}' = \Delta\theta\,(P_{x'} - P_{y'})$$

Optical axis deviation

$$\Delta\mu_{y'}' = \Delta\theta\,(-0.5\,P_{x'} + P_{y'} - 0.866\,P_{z'})$$

$$\Delta\mu_{z'}' = \Delta\theta\,(-0.866\,P_{x'} + 1.5\,P_{z'})$$

Convergent beam (Image point displacement $\Delta S_{F'}'$)

Parallax

$$\Delta S_{F'x'}' = \Delta\theta\left[P_{x'}(-z_q'-0.866D)-P_{y'}z_q'+P_{z'}(x_q'+y_q'+1.366D)\right]+\Delta g(D_{x'}-D_{y'})$$

Optical axis deviation

$$\Delta S_{F'y'}' = \Delta\theta\left[P_{x'}\left(0.866y_q'+z_q'-0.866b'+0.433D-2.049\frac{D}{n}\right)+P_{y'}(-0.866x_q'\right.$$

$$\left.+0.5z_q'-0.75D)+P_{z'}\left(-x_q'-0.5y_q'+1.5b'-0.25D+1.183\frac{D}{n}\right)\right]$$

$$+\Delta g(-0.5D_{x'}+D_{y'}-0.866D_{z'})$$

$$\Delta S_{F'z'}' = \Delta\theta\left[P_{x'}\left(-1.5y_q'+0.5b'-0.25D+1.183\frac{D}{n}\right)+P_{y'}(1.5x_q'+0.866z_q'-b'\right.$$

$$\left.+1.433D)+P_{z'}\left(-0.866y_q'+0.866b'-0.433D+2.049\frac{D}{n}\right)\right]$$

$$+\Delta g(-0.866D_{x'}+1.5D_{z'})$$

Adjustment Diagram

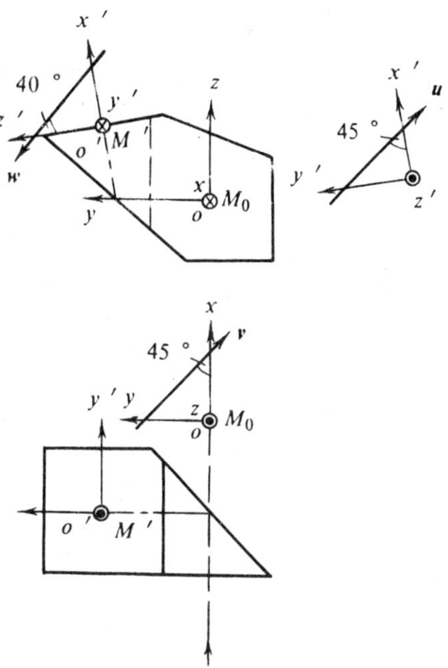

Fig.A.23.

Characteristic Parameters of Image Formation

Direction cosine		Unit vectors of image coordinate system			$2\varphi = 234°4'$
		$i'(x')$	$j'(y')$	$k'(z')$	T
Unit vectors of coordinate system in object space	$i(x)$	$S_{T.2\varphi} = \begin{pmatrix} 0 & 1 & 0 \\ 0.174 & 0 & 0.985 \\ 0.985 & 0 & -0.174 \end{pmatrix}$			0.608
	$j(y)$				0.608
	$k(z)$				0.510

$t = 2$	$L = 1.874D$	$2d = 1.139D - 1.139\dfrac{D}{n}$	

$$\overrightarrow{M_0 M'} = 0.731D\,i' + \left(0.5D - 1.874\frac{D}{n}\right)j' + 0.766D\,k'$$

$$r_0 = \left(-0.073D - 0.59\frac{D}{n}\right)i' + \left(0.12D + 0.59\frac{D}{n}\right)j' - 0.057D\,k'$$

$$r_i =$$

Characteristic Parameters of Adjustment

Direction cosine		Image coordinate system		
		x'	y'	z'
Extreme-valued axis direction of r' image rotation	u	0.707	-0.707	0
	v	-0.123	0.707	-0.696
	w	-0.643	0	0.766
Extreme value of r' image rotation		$\delta_u = 1.414\Delta\theta$	$\delta_v = 1.414\Delta\theta$	$\delta_w = 1.532\Delta\theta$
Extreme-valued shift direction of r' image displacement	a	0.707	-0.707	0
	b	-0.123	0.707	-0.696
	c	-0.643	0	0.766
Extreme value of r' image displacement		$\delta_a = 1.414\Delta g$	$\delta_b = 1.414\Delta g$	$\delta_c = 1.532\Delta g$

Planar three-dimensional zero-valued poles (Do not exist)	C_{Π_1}	$x_1' =$
		$y_1' =$
		$z_1' =$
		Associated plane Π_1 :
	C_{Π_2}	$x_2' =$
		$y_2' =$
		$z_2' =$
		Associated plane Π_2 :

Equations of Image Motion

Parallel beam (Image rotation $\Delta\vec{\mu}'$)

Image lean

$$\Delta\mu_{x'}' = \Delta\theta\,(P_{x'} - P_{y'})$$

Optical axis deviation

$$\Delta\mu_{y'}' = \Delta\theta\,(-0.174P_{x'} + P_{y'} - 0.985\,P_{z'})$$

$$\Delta\mu_{z'}' = \Delta\theta\,(-0.985P_{x'} + 1.174P_{z'})$$

Convergent beam (Image point displacement $\Delta S_F'$)

Parallax

$$\Delta S_{F'x'}' = \Delta\theta\,[\,P_{x'}(-z_q'-0.766D)-P_{y'}z_q'+P_{z'}(x_q'+y_q'+0.731D)]+\Delta g\,(D_{x'}-D_{y'})$$

Optical axis deviation

$$\Delta S_{F'y'}' = \Delta\theta\left[P_{x'}\left(0.985y_q'+z_q'-0.985b'+0.493D-1.845\frac{D}{n}\right)+P_{y'}(-0.985x_q'\right.$$

$$+0.174z_q'-0.586D)+P_{z'}\left(-x_q'-0.174y_q'+1.174b'-0.087D+0.325\frac{D}{n}\right)\Big]$$

$$+\Delta g\,(-0.174\,D_{x'}+D_{y'}-0.985D_{z'})$$

$$\Delta S_{F'z'}' = \Delta\theta\left[P_{x'}\left(-1.174y_q'+0.174b'-0.087D+0.325\frac{D}{n}\right)+P_{y'}(1.174x_q'+0.985z_q'\right.$$

$$-b'+0.881D)+P_{z'}\left(-0.985y_q'+0.985b'-0.493D+1.846\frac{D}{n}\right)\Big]$$

$$+\Delta g\,(-0.985D_{x'}+1.174D_{z'})$$

Adjustment Diagram

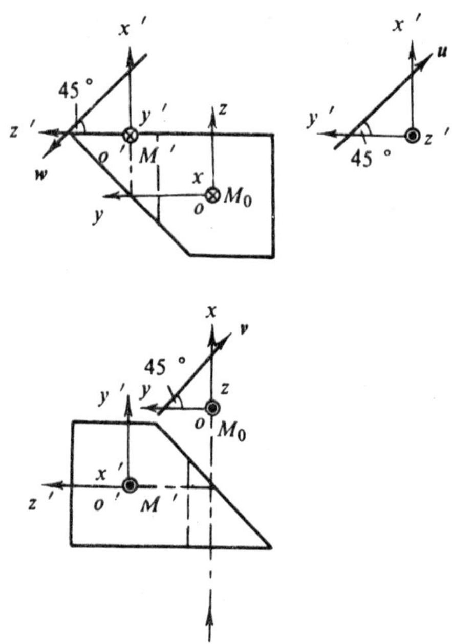

Fig . A .24 .

Characteristic Parameters of Image Formation

Direction cosine		Unit vectors of image coordinate system			$2\varphi = 240$ °
		$i\,'(x\,')$	$j\,'(y\,')$	$k\,'(z\,')$	T
Unit vectors of coordinate system in object space	$i\,(x)$		0	1	0	0.577
	$j\,(y)$	$S_{T,2\varphi} = \begin{pmatrix} & & \\ & & \\ & & \end{pmatrix}$ 0	0	1	0.577
	$k\,(z)$	1	0	0	0.577

$t = 2$	$L = 1.707D$	$2d = 0.985D - 0.985\dfrac{D}{n}$

$$\overrightarrow{M_0M\,'} = 0.5\,D\,i\,' + \left(0.5D - 1.707\frac{D}{n}\right)j\,' + 0.707D\,k\,'$$

$$r_0 = -0.569\,\frac{D}{n}\,i\,' + \left(0.069D + 0.569\frac{D}{n}\right)j\,' - 0.069D\,k\,'$$

$$r_i =$$

Characteristic Parameters of Adjustment

Direction cosine		Image coordinate system		
		x'	y'	z'
Extreme-valued axis direction of r' image rotation	u	0.707	-0.707	0
	v	0	0.707	-0.707
	w	-0.707	0	0.707
Extreme value of r' image rotation		$\delta_u = 1.414\Delta\theta$	$\delta_v = 1.414\Delta\theta$	$\delta_w = 1.414\Delta\theta$
Extreme-valued shift direction of r' image displacement	a	0.707	-0.707	0
	b	0	0.707	-0.707
	c	-0.707	0	0.707
Extreme value of r' image displacement		$\delta_a = 1.414\Delta g$	$\delta_b = 1.414\Delta g$	$\delta_c = 1.414\Delta g$

Planar three-dimensional zero-valued poles (Do not exist)	C_{Π_1}	$x_1' =$	
		$y_1' =$	
		$z_1' =$	
		Associated plane Π_1 :	
	C_{Π_2}	$x_2' =$	
		$y_2' =$	
		$z_2' =$	
		Associated plane Π_2 :	

Equations of Image Motion

Parallel beam (Image rotation $\Delta\vec{\mu}'$)

Image lean

$$\Delta\mu_{x'}' = \Delta\theta\,(P_{x'} - P_{y'})$$

Optical axis deviation

$$\Delta\mu_{y'}' = \Delta\theta\,(P_{y'} - P_{z'})$$

$$\Delta\mu_{z'}' = \Delta\theta\,(-P_{x'} + P_{z'})$$

Convergent beam (Image point displacement $\Delta S_{F'}'$)

Parallax

$$\Delta S_{F'x'} = \Delta\theta\,[\,P_x\cdot(-z_q' - 0.707D) - P_y\cdot z_q' + P_z\cdot(x_q' + y_q' + 0.5D\,)\,] + \Delta g\,(D_{x'} - D_{y'})$$

Optical axis deviation

$$\Delta S_{F'y'} = \Delta\theta\left[\,P_x\cdot\left(y_q' + z_q' - b' + 0.5D - 1.707\,\frac{D}{n}\,\right) + P_y\cdot(-x_q' - 0.5D)\right.$$

$$\left. + P_z\cdot(-x_q' + b')\right] + \Delta g\,(D_{y'} - D_{z'})$$

$$\Delta S_{F'z'} = \Delta\theta\left[\,-P_x\cdot y_q' + P_y\cdot(x_q' + z_q' - b' + 0.707D\,) + P_z\cdot\left(-y_q' + b' - 0.5D\right.\right.$$

$$\left.\left. + 1.707\,\frac{D}{n}\,\right)\right] + \Delta g\,(-D_{x'} + D_{z'})$$

Table A.25   KII − 100°−90°

Adjustment Diagram

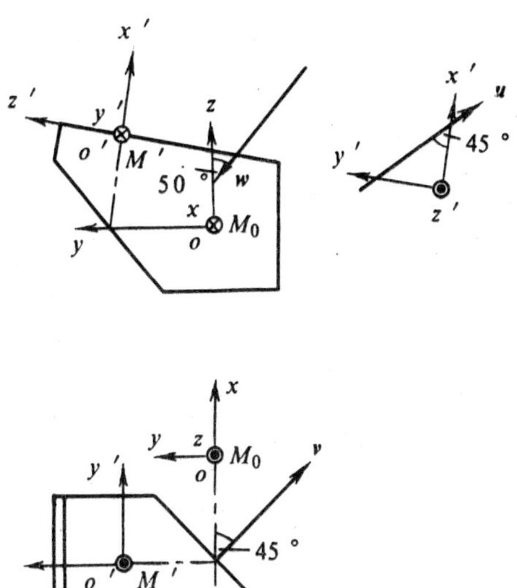

Fig. A.25.

Characteristic Parameters of Image Formation

Direction cosine		Unit vectors of image coordinate system			$2\varphi = 245°\,36'$
		$i'(x')$	$j'(y')$	$k'(z')$	T
Unit vectors of coordinate system in object space	$i(x)$	$S_{T,2\varphi} = \begin{pmatrix} 0 \\ -0.174 \\ 0.985 \end{pmatrix}$	$\begin{pmatrix} 1 \\ 0 \\ 0 \end{pmatrix}$	$\begin{pmatrix} 0 \\ 0.985 \\ 0.174 \end{pmatrix}$	0.541
	$j(y)$				0.541
	$k(z)$				0.645

$t = 2$ $\qquad L = 1.954D$ $\qquad\qquad 2d = 1.057D - 1.057\dfrac{D}{n}$

$$\overrightarrow{M_0 M'} = 0.5D\,i' + \left(0.5D - 1.954\frac{D}{n}\right) j' + 0.800D\,k'$$

$$r_0 = -0.692\frac{D}{n}\,i' + \left(0.071D + 0.691\frac{D}{n}\right) j' - 0.059D\,k'$$

$$r_i =$$

Characteristic Parameters of Adjustment

Direction cosine		Image coordinate system		
		x'	y'	z'
Extreme-valued axis direction of r' image rotation	u	0.707	-0.707	0
	v	0.123	0.707	-0.696
	w	-0.766	0	0.643
Extreme value of r' image rotation		$\delta_u = 1.414\Delta\theta$	$\delta_v = 1.414\Delta\theta$	$\delta_w = 1.285\Delta\theta$
Extreme-valued shift direction of r' image displacement	a	0.707	-0.707	0
	b	0.123	0.707	-0.696
	c	-0.766	0	0.643
Extreme value of r' image displacement		$\delta_a = 1.414\Delta g$	$\delta_b = 1.414\Delta g$	$\delta_c = 1.285\Delta g$

Planar three-dimensional zero-valued poles (Do not exist)	C_{Π_1}	$x_1' =$
		$y_1' =$
		$z_1' =$
		Associated plane Π_1 :
	C_{Π_2}	$x_2' =$
		$y_2' =$
		$z_2' =$
		Associated plane Π_2 :

Equations of Image Motion

Parallel beam (Image rotation $\Delta\vec{\mu}'$)

Image lean

$$\Delta\mu_{x'}' = \Delta\theta (P_{x'} - P_{y'})$$

Optical axis deviation

$$\Delta\mu_{y'}' = \Delta\theta (0.174 P_{x'} + P_{y'} - 0.985 P_{z'})$$

$$\Delta\mu_{z'}' = \Delta\theta (-0.985 P_{x'} + 0.826 P_{z'})$$

Convergent beam (Image point displacement $\Delta S_F'$)

Parallax

$$\Delta S_{F'x'}' = \Delta\theta \left[P_{x'}(-z_q' - 0.800\,D) - P_{y'}z_q' + P_{z'}(x_q' + y_q' + 0.5\,D) \right] + \Delta g\,(D_{x'} - D_{y'})$$

Optical axis deviation

$$\Delta S_{F'y'}' = \Delta\theta \left[P_{x'}\left(0.985\,y_q' + z_q' - 0.985\,b' + 0.493\,D - 1.924\,\frac{D}{n}\right) + P_{y'}(-0.985\,x_q' \right.$$

$$\left. - 0.174\,z_q' - 0.632\,D) + P_{z'}\left(-x_q' + 0.174\,y_q' + 0.826\,b' + 0.087\,D - 0.340\,\frac{D}{n}\right) \right]$$

$$+ \Delta g\,(0.174\,D_{x'} + D_{y'} - 0.985\,D_{z'})$$

$$\Delta S_{F'z'}' = \Delta\theta \left[P_{x'}\left(-0.826\,y_q' - 0.174\,b' + 0.087\,D - 0.340\,\frac{D}{n}\right) + P_{y'}(0.826\,x_q' + 0.985\,z_q' \right.$$

$$\left. - b' + 0.701\,D) + P_{z'}\left(-0.985\,y_q' + 0.985\,b' - 0.492\,D + 1.924\,\frac{D}{n}\right) \right]$$

$$+ \Delta g\,(-0.985\,D_{x'} + 0.826\,D_{z'})$$

Adjustment Diagram

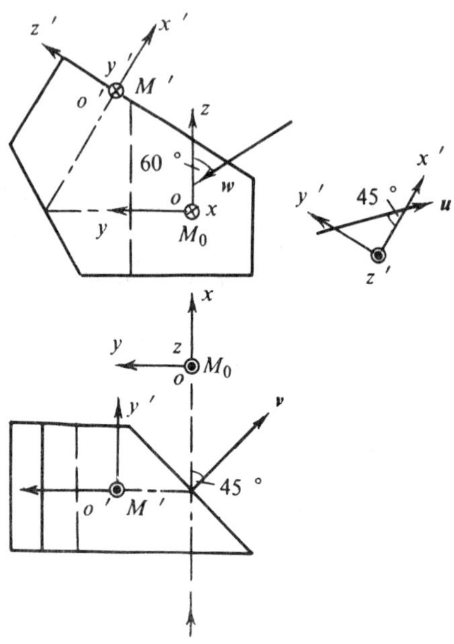

Fig . A .26 .

Characteristic Parameters of Image Formation

Direction cosine		Unit vectors of image coordinate system			$2\varphi = 255°31'$
		$i'(x')$	$j'(y')$	$k'(z')$	T
Unit vectors of coordinate system in object space	$i(x)$	$S_{T,2\varphi}=\begin{pmatrix} 0 \\ -0.5 \\ 0.866 \end{pmatrix}$	$\begin{pmatrix} 1 \\ 0 \\ 0 \end{pmatrix}$	$\begin{pmatrix} 0 \\ 0.866 \\ 0.5 \end{pmatrix}$	0.447
	$j(y)$				0.447
	$k(z)$				0.775

$$t = 2 \qquad L = 2.823D \qquad \qquad 2d = 1.263D - 1.262\frac{D}{n}$$

$$\overrightarrow{M_0 M'} = 0.5\,Di' + \left(0.5D - 2.823\frac{D}{n}\right)j' + 1.053D\,k'$$

$$r_0 = -1.129\frac{D}{n}\,i' + \left(0.065D + 1.129\frac{D}{n}\right)j' - 0.037D\,k'$$

$$r_i =$$

Characteristic Parameters of Adjustment

Direction cosine		Image coordinate system		
		x'	y'	z'
Extreme - valued axis direction of r' image rotation	u	0.707	-0.707	0
	v	0.353	0.707	-0.612
	w	-0.866	0	0.5
Extreme value of r' image rotation		$\delta_u = 1.414\Delta\theta$	$\delta_v = 1.414\Delta\theta$	$\delta_w = \Delta\theta$
Extreme - valued shift direction of r' image displacement	a	0.707	-0.707	0
	b	0.353	0.707	-0.612
	c	-0.866	0	0.5
Extreme value of r' image displacement		$\delta_a = 1.414\Delta g$	$\delta_b = 1.414\Delta g$	$\delta_c = \Delta g$

Planar three-dimensional zero-valued poles (Do not exist)	C_{Π_1}	$x_1' =$
		$y_1' =$
		$z_1' =$
		Associated plane Π_1 :
	C_{Π_2}	$x_2' =$
		$y_2' =$
		$z_2' =$
		Associated plane Π_2 :

Equations of Image Motion

Parallel beam (Image rotation $\Delta\vec{\mu}'$)
Image lean

$$\Delta\mu_{x'}' = \Delta\theta\,(P_{x'} - P_{y'})$$

Optical axis deviation

$$\Delta\mu_{y'}' = \Delta\theta\,(0.5\,P_{x'} + P_{y'} - 0.866\,P_{z'})$$

$$\Delta\mu_{z'}' = \Delta\theta\,(-0.866\,P_{x'} + 0.5\,P_{z'})$$

Convergent beam (Image point displacement $\Delta S_F'$)

Parallax

$$\Delta S_{F'x'}' = \Delta\theta\,[\,P_{x'}(-z_q'-1.053D)-P_{y'}z_q'+P_{z'}(x_q'+y_q'+0.5D)] + \Delta g\,(D_{x'}-D_{y'})$$

Optical axis deviation

$$\Delta S_{F'y'}' = \Delta\theta\left[P_{x'}\left(0.866y_q'+z_q'-0.866b'+0.433D-2.445\frac{D}{n}\right)+P_{y'}(-0.866x_q'\right.$$

$$\left.-0.5z_q'-0.959D)+P_{z'}\left(-x_q'+0.5y_q'+0.5b'+0.25D-1.411\frac{D}{n}\right)\right]$$

$$+\Delta g\,(0.5D_{x'}+D_{y'}-0.866D_{z'})$$

$$\Delta S_{F'z'}' = \Delta\theta\left[P_{x'}\left(-0.5y_q'-0.5b'+0.25D-1.411\frac{D}{n}\right)+P_{y'}(0.5x_q'+0.866z_q'\right.$$

$$\left.-b'+0.662D)+P_{z'}\left(-0.866y_q'+0.866b'-0.433D+2.445\frac{D}{n}\right)\right]$$

$$+\Delta g\,(-0.866D_{x'}+0.5D_{z'})$$

Adjustment Diagram

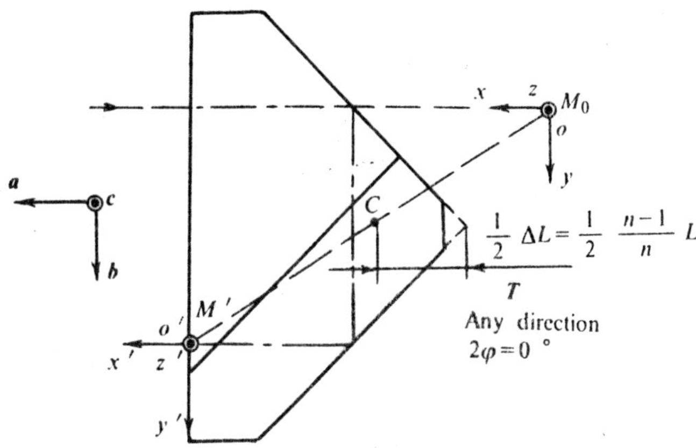

Fig . A . 27 .

Characteristic Parameters of Image Formation

Direction cosine		Unit vectors of image coordinate system			$2\varphi = 0\,^\circ$ T (Any direction)
		$i\,'(x\,')$	$j\,'(y')$	$k\,'(z\,')$	
Unit vectors of coordinate system in object space	$i\,(x)$	$S_{T,2\varphi} = \begin{pmatrix} 1 & 0 & 0 \\ & & \\ 0 & 1 & 0 \\ & & \\ 0 & 0 & 1 \end{pmatrix}$			
	$j\,(y)$				
	$k\,(z)$				

| $t=3 \qquad L = 2.957D$ | | | $2d = 0$ | | |

$$\overrightarrow{M_0 M}\,' = 2.957\,\frac{D}{n}\,i' + 1.225 D j'$$

$$r_0 = \text{Any value}$$

$$r_i = -1.479\,\frac{D}{n}\,i\,' - 0.613 D j\,'$$

Characteristic Parameters of Adjustment

Direction cosine		Image coordinate system		
		x'	y'	z'
Extreme-valued	u	/	/	/
axis direction of	v	/	/	/
r' image rotation	w	/	/	/
Extreme value of r' image rotation		$\delta_u = 0$	$\delta_v = 0$	$\delta_w = 0$
Extreme-valued	a	1	0	0
shift direction of	b	0	1	0
r' image displacement	c	0	0	1
Extreme value of r' image displacement		$\delta_a = 2\Delta g$	$\delta_d = 2\Delta g$	$\delta_c = 2\Delta g$

Spatial three-dimensional zero-valued pole	C	$x' = -1.479\,\dfrac{D}{n}$
		$y' = -0.613\,D$
		$z' = 0$

Note: Two planar three-dimensional zero-valued poles C_{Π_1} and C_{Π_2} appear to be coincident at C so that they form a spatial three-dimensional zero-valued pole at C.

Equations of Image Motion

Parallel beam (Image rotation $\Delta\vec{\mu}'$)

Image lean

$$\Delta\mu'_{x'} = 0$$

Optical axis deviation

$$\Delta\mu'_{y'} = 0$$

$$\Delta\mu'_{z'} = 0$$

Convergent beam (Image point displacement $\Delta S_{F'}'$)

Parallax

$$\Delta S'_{F'x'} = \Delta\theta\,[\,-2P_{y'}z'_q + P_{z'}(2y'_q + 1.225\,D)] + 2\Delta g\,D_{x'}$$

Optical axis deviation

$$\Delta S'_{F'y'} = \Delta\theta\left[\,2P_{x'}z'_q + P_{z'}\left(-2x'_q - 2.957\,\frac{D}{n}\right)\right] + 2\Delta g\,D_{y'}$$

$$\Delta S'_{F'z'} = \Delta\theta\left[\,P_{x'}\left(-2y'_q - 1.225\,D\right) + P_{y'}\left(2x'_q + 2.957\,\frac{D}{n}\right)\right] + 2\Delta g\,D_{z'}$$

352

Table A.28 B II$_J$ -45 °

Adjustment Diagram

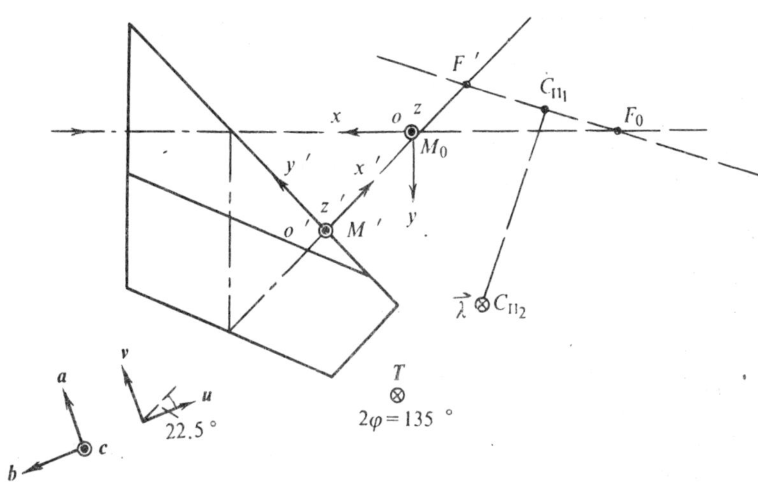

Fig.A.28.

Characteristic Parameters of Image Formation

Direction cosine		Unit vectors of image coordinate system			$2\varphi = 135$ °
		$i'(x')$	$j'(y')$	$k'(z')$	T
Unit vectors of coordinate system in object space	$i(x)$	-0.707	0.707	0	0
	$j(y)$	-0.707	-0.707	0	0
	$k(z)$	0	0	1	-1

$$S_{T,2\varphi} = \begin{pmatrix} -0.707 & 0.707 & 0 \\ -0.707 & -0.707 & 0 \\ 0 & 0 & 1 \end{pmatrix}$$

$t = 3$	$L = 2.111D$	$2d = 0$

$$\overrightarrow{M_0M'} = \left(0.354\,D - 1.493\,\frac{D}{n}\right) i' + \left(-1.021\,D + 1.493\,\frac{D}{n}\right) j'$$

$$r_0 = \left(1.056\,D - 1.056\,\frac{D}{n}\right) i' + \left(0.938\,D - 2.549\,\frac{D}{n}\right) j'$$

$$r_i = \left(1.056\,D - 1.056\,\frac{D}{n}\right) i' + \left(0.938\,D - 2.549\,\frac{D}{n}\right) j'$$

Characteristic Parameters of Adjustment

Direction cosine		Image coordinate system		
		x'	y'	z'
Extreme - valued axis direction of r' image rotation	u	0.924	-0.383	0
	v	0.383	0.924	0
	w	/	/	/
Extreme value of r' image rotation		$\delta_u = 1.848\Delta\theta$	$\delta_v = 1.848\Delta\theta$	$\delta_w = 0$
Extreme- valued shift direction of r' image displacement	a	0.383	0.924	0
	b	-0.924	0.383	0
	c	0	0	1
Extreme value of r' image displacement		$\delta_a = 0.765\Delta g$	$\delta_b = 0.765\Delta g$	$\delta_c = 2\Delta g$
planar three-dimensional zero- valued poles	C_{Π_1}	$x_1' = \dfrac{1}{2}\left(1.707b' - 0.354D + 1.493\dfrac{D}{n}\right)$		
		$y_1' = \dfrac{1}{2}\left(-0.707b' + 1.021D - 1.493\dfrac{D}{n}\right)$		
		$z_1' = 0$		
		Associated plane Π_1 : Conjugate optical- axis section $x'o'y'$		
	C_{Π_2}	$x_2' = 1.056D - 1.056\dfrac{D}{n}$		
		$y_2' = 0.937D - 2.548\dfrac{D}{n}$		
		$z_2' = 0$		
		Associated plane Π_2 : Plane passing through $\vec{\lambda}$ and C_{Π_1}		

Equations of Image Motion

Parallel beam (Image rotation $\Delta\vec{\mu}'$)

Image lean

$$\Delta\mu_x' = \Delta\theta\,(1.707P_{x'} - 0.707P_{y'})$$

Optical axis deviation

$$\Delta\mu_y' = \Delta\theta\,(0.707P_{x'} + 1.707P_{y'})$$
$$\Delta\mu_z' = 0$$

Convergent beam (Image point displacement $\Delta S_F'$)

Parallax

$$\Delta S_{F'x'}' = \Delta\theta\,[\,0.707 P_{x'}\,z_q' - 0.293 P_{y'}\,z_q' + P_{z'}\,(-0.707 x_q' + 0.293 y_q' + 0.472\,D\,)] \\ + \Delta g\,(0.293 D_{x'} + 0.707 D_{y'})$$

Optical axis deviation

$$\Delta S_{F'y'}' = \Delta\theta\left[\,0.293 P_{x'}\,z_q' + 0.707 P_{y'}\,z_q' + P_{z'}\left(-0.293 x_q' - 0.707 y_q' + 0.972\,D - 2.111\frac{D}{n}\right)\right] \\ + \Delta g\,(-0.707 D_{x'} + 0.293 D_{y'})$$

$$\Delta S_{F'z'}' = \Delta\theta\left[P_{x'}\left(-2 y_q' - 0.707 b' + 1.021 D - 1.493\frac{D}{n}\right) + P_{y'}\left(2 x_q' - 1.707 b' + 0.354\,D\right. \\ \left.\left. - 1.493\frac{D}{n}\right)\right] + 2\Delta g\,D_{z'}$$

Adjustment Diagram

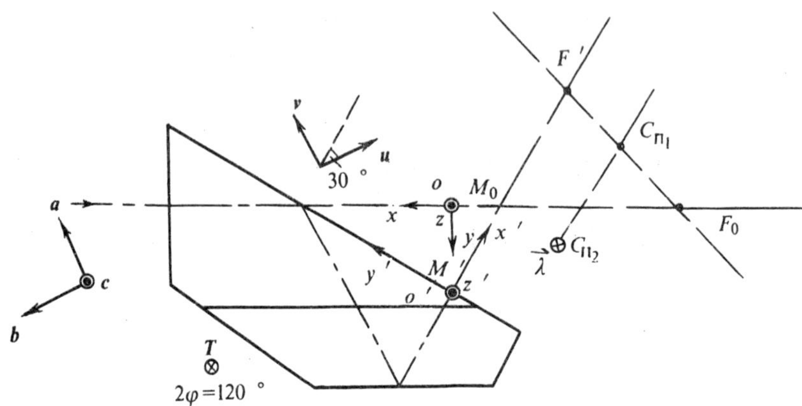

Fig. A.29.

Characteristic Parameters of Image Formation

Direction cosine		Unit vectors of image coordinate system			$2\varphi = 120°$
		$i'(x')$	$j'(y')$	$k'(z')$	T
Unit vectors of coordinate system in object space	$i(x)$		-0.5	0.866	0	0
	$j(y)$	$S_{T,2\varphi} =$	-0.866	-0.5	0	0
	$k(z)$		0	0	1	-1

$t=3$	$L=2.801D$	$2d=0$

$$\overrightarrow{M_0 M}' = \left(0.433\,D - 1.401\,\frac{D}{n}\right) i' + \left(-1.922\,D + 2.426\,\frac{D}{n}\right) j'$$

$$r_0 = \left(1.448\,D - 1.4\,\frac{D}{n}\right) i' + \left(1.336\,D - 2.426\,\frac{D}{n}\right) j'$$

$$r_i = \left(1.448\,D - 1.4\,\frac{D}{n}\right) i' + \left(1.336\,D - 2.426\,\frac{D}{n}\right) j'$$

Characteristic Parameters of Adjustment

	Direction cosine	Image coordinate system		
		x'	y'	z'
Extreme-valued axis direction of r' image rotation	u	0.866	-0.5	0
	v	0.5	0.866	0
	w	/	/	/
Extreme value of r' image rotation		$\delta_u = 1.732\Delta\theta$	$\delta_v = 1.732\Delta\theta$	$\delta_w = 0$
Extreme-valued shift direction of r' image displactment	a	0.5	0.866	0
	b	-0.866	0.5	0
	c	0	0	1
Extreme value of r' image displacement		$\delta_a = \Delta g$	$\delta_b = \Delta g$	$\delta_c = 2\Delta g$
Planar three-dimensional zero-valued poles	C_{Π_1}	$x_1' = \dfrac{1}{2}\left(1.5b' - 0.433\,D + 1.401\dfrac{D}{n}\right)$		
		$y_1' = \dfrac{1}{2}\left(-0.866\,b' + 1.922\,D - 2.426\dfrac{D}{n}\right)$		
		$z_1' = 0$		
		Associated plane Π_1: Conjugate optical-axis section $x'o'y'$		
	C_{Π_2}	$x_2' = 1.448\,D - 1.401\dfrac{D}{n}$		
		$y_2' = 1.336\,D - 2.426\dfrac{D}{n}$		
		$z_2' = 0$		
		Associated plane Π_2: Plane passing through $\vec{\lambda}$ and C_{Π_1}		

Equations of Image Motion

Parallel beam (Image rotation $\Delta\vec{\mu}'$)

Image lean

$$\Delta\mu_{x'}' = \Delta\theta\,(1.5P_{x'} - 0.866P_{y'})$$

Optical axis deviation

$$\Delta\mu_{y'}' = \Delta\theta\,(0.866P_{x'} + 1.5P_{y'})$$

$$\Delta\mu_{z'}' = 0$$

Convergent beam (Image point displacement $\Delta S_F'$)

Parallax

$$\Delta S_{F'x'}' = \Delta\theta\,[\,0.866\,P_{x'}\,z_{q}' - 0.5\,P_{y'}\,z_{q}' + P_{z'}\,(-0.866x_{q}' + 0.5y_{q}' + 0.586D\,)]$$
$$+ \Delta g\,(0.5D_{x'} + 0.866D_{y'}\,)$$

Optical axis deviation

$$\Delta S_{F'x'}' = \Delta\theta\left[0.5P_{x'}\,z_{q}' + 0.866P_{y'}\,z_{q}' + P_{z'}\left(-0.5x_{q}' - 0.866y_{q}' + 1.881D - 2.801\frac{D}{n}\right)\right]$$
$$+ \Delta g\,(-0.866D_{x'} + 0.5D_{y'}\,)$$

$$\Delta S_{F'z'}' = \Delta\theta\left[P_{x'}\left(-2y_{q}' - 0.866b' + 1.922D - 2.426\frac{D}{n}\right) + P_{y'}\left(2x_{q}' - 1.5b' + 0.433D\right.\right.$$
$$\left.\left. - 1.401\frac{D}{n}\right)\right] + 2\Delta g\,D_{z'}$$

Adjustment Diagram

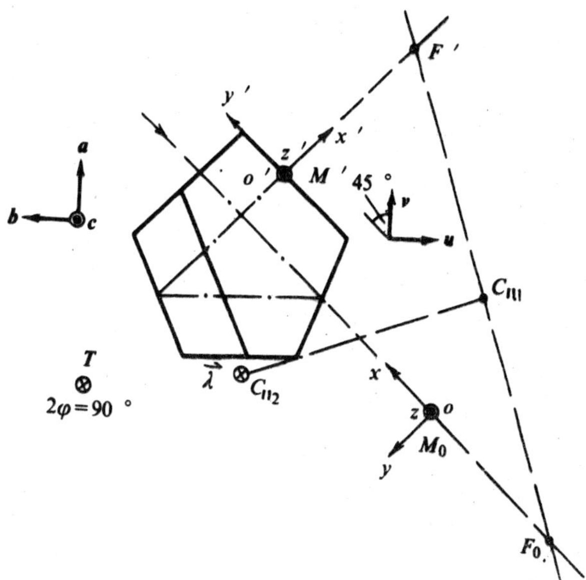

Fig . A . 30 .

Characteristic Parameters of Image Formation

Direction cosine		Unit vectors of image coordinate system			$2\varphi=90$ °
		$i'(x')$	$j'(y')$	$k'(z')$	T
Unit vectors of coordinate system in object space	$i(x)$	$S_{T,2\varphi}=$ ⎛ 0	1	0 ⎞	0
	$j(y)$	−1	0	0	0
	$k(z)$	⎝ 0	0	1 ⎠	−1
$t=3$ $\qquad$ $L=4.223D$			$2d=0$		

$$\overrightarrow{M_0M}\,'=0.5D\ i'+\left(-0.5D+4.223\frac{D}{n}\right)j'$$

$$r_0=-2.112\frac{D}{n}\ i'+\left(0.5D-2.112\frac{D}{n}\right)j'$$

$$r_i=-2.112\frac{D}{n}\ i'+\left(0.5D-2.112\frac{D}{n}\right)j'$$

Characteristic Parameters of Adjustment

Direction cosine		Image coordinate system		
		x'	y'	z'
Extreme-valued axis direction of r' image rotation	u	0.707	-0.707	0
	v	0.707	0.707	0
	w	/	/	/
Extreme value of r' image rotation		$\delta_u = 1.414\Delta\theta$	$\delta_v = 1.414\Delta\theta$	$\delta_w = 0$
Extreme-valued shift direction of r' image displacement	a	0.707	0.707	0
	b	-0.707	0.707	0
	c	0	0	1
Extreme value of r' image displacement		$\delta_a = 1.414\Delta g$	$\delta_b = 1.414\Delta g$	$\delta_c = 2\Delta g$

Planar three-dimensional zero-valued poles	C_{Π_1}	$x_1' = \dfrac{1}{2}\left(b' - 0.5D\right)$
		$y_1' = \dfrac{1}{2}\left(-b' + 0.5D - 4.223\dfrac{D}{n}\right)$
		$z_1' = 0$
		Associated plane Π_1: Conjugate optical-axis section $x'o'y'$
	C_{Π_2}	$x_2' = -2.112\dfrac{D}{n}$
		$y_2' = 0.5D - 2.112\dfrac{D}{n}$
		$z_2' = 0$
		Associated plane Π_2: Plane passing through $\vec{\lambda}$ and C_{Π_1}

Equations of Image Motion

Parallel beam (Image rotation $\Delta\vec{\mu}'$)

Image lean

$$\Delta\mu_{x'}' = \Delta\theta\,(P_{x'} - P_{y'})$$

Optical axis deviation

$$\Delta\mu_{y'}' = \Delta\theta\,(P_{x'} + P_{y'})$$
$$\Delta\mu_{z'}' = 0$$

Convergent beam (Image point displacement $\Delta S_F'$)

Parallax

$$\Delta S_{F'x'}' = \Delta\theta\,[\,P_{x'}z_q' - P_{y'}z_q' + P_{z'}(-x_q' + y_q' - 0.5\,D)\,] + \Delta g\,(D_{x'} + D_{y'})$$

Optical axis deviation

$$\Delta S_{F'y'}' = \Delta\theta\left[P_{x'}z_q' + P_{y'}z_q' + P_{z'}\left(-x_q' - y_q' + 0.5D - 4.223\,\frac{D}{n}\right)\right] + \Delta g\,(-D_{x'} + D_{y'})$$

$$\Delta S_{F'z'}' = \Delta\theta\left[P_{x'}\left(-2y_q' - b' + 0.5D - 4.223\,\frac{D}{n}\right) + P_{y'}\left(2x_q' - b' + 0.5D\right)\right] + 2\Delta g\,D_{z'}$$

Adjustment Diagram

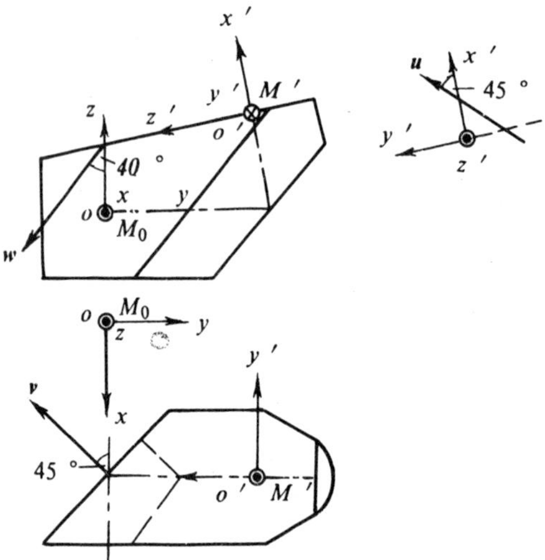

Fig . A . 31 .

Characteristic Parameters of Image Formation

Direction cosine		Unit vectors of image coordinate system			$2\varphi = 234°\,4\,'$
		$i\,'(x\,')$	$j\,'(y\,')$	$k\,'(z\,')$	T
Unit vectors	$i\;(x)$	0	− 1	0	− 0 .608
of coordinate system in	$j\;(y)$	$S_{T.2\varphi} =$ − 0.174	0	− 0 .985	0 .608
object space	$k\;(z)$	0 .985	0	− 0 .174	− 0 .510

$t = 3$	$L = 2 .532\,D$	$2d = 0$

$$\overrightarrow{M_0 M\,'} = 0.5\,D\;i\,' + \left(0.5\,D - 2 .532\,\frac{D}{n}\right) j\,' - 1 .286\,D\;k\,'$$

$$r_0 = \left(-0 .966\,D - 0 .797\,\frac{D}{n}\right) i\,' + \left(-1 .067\,D + 0 .798\,\frac{D}{n}\right) j\,' + \left(-0 .12\,D + 1 .901\,\frac{D}{n}\right) k\,'$$

$$r_i = \left(-0 .766\,D - 1 .265\,\frac{D}{n}\right) i\,' + \left(-1 .266\,D + 1 .266\,\frac{D}{n}\right) j\,' + \left(0 .047D + 1 .508\,\frac{D}{n}\right) k\,'$$

Characteristic Parameters of Adjustment

Direction cosine		Image coordinate system		
		x'	y'	z'
Extreme- valued	u	0.707	0.707	0
axis direction of	v	0.123	0.707	0.697
r' image rotation	w	-0.643	0	0.766
Extreme value of r' image rotation		$\delta_u = 1.414\Delta\theta$	$\delta_v = 1.414\Delta\theta$	$\delta_w = 1.532\Delta\theta$
Extreme- valued	a	0.707	-0.707	0
shift direction of	b	-0.123	0.707	-0.697
r' image displacement	c	0.766	0	0.643
Extreme value of r' image displacement		$\delta_a = 1.414\Delta g$	$\delta_b = 1.414\Delta g$	$\delta_c = 1.285\Delta g$

Planar three- dimensional zero- valued poles (Do not exist)	C_{Π_1}	$x_1' =$
		$y_1' =$
		$z_1' =$
		Associated plane Π_1 :
	C_{Π_2}	$x_2' =$
		$y_2' =$
		$z_2' =$
		Associated plane Π_2 :

Equations of Image Motion

Parallel beam (Image rotation $\Delta\vec{\mu}'$)

Image lean

$$\Delta\mu_{x'}' = \Delta\theta\,(P_{x'} + P_{y'})$$

Optical axis deviation

$$\Delta\mu_{y'}' = \Delta\theta\,(0.174P_{x'} + P_{y'} + 0.985P_{z'})$$

$$\Delta\mu_{z'}' = \Delta\theta\,(-0.985P_{x'} + 1.174P_{z'})$$

Convergent beam (Image point displacement $\Delta S_{F'}'$)

Parallax

$$\Delta S_{F'x'}' = \Delta\theta\,[\,P_{x'}\,(-z_q' + 1.287D) - P_{y'}z_q' + P_{z'}\,(x_q' + y_q' + 0.500D)] + \Delta g\,(D_{x'} - D_{y'})$$

Optical axis deviation

$$\Delta S_{F'x'}' = \Delta\theta\left[P_{x'}\left(0.986y_q' + z_q' - 0.985b' + 0.473D - 2.496\frac{D}{n}\right) + P_{y'}\,(-0.986x_q' + 0.174z_q' - 0.716D)\right.$$

$$\left. + P_{z'}\left(-x_q' - 0.174y_q' + 1.174b' - 0.087D + 0.440\frac{D}{n}\right)\right] + \Delta g\,(-0.174D_{x'} + D_{y'} - 0.985D_{z'})$$

$$\Delta S_{F'z'}' = \Delta\theta\left[P_{x'}\left(-0.826y_q' - 0.174b' + 0.087D - 0.440\frac{D}{n}\right) + P_{y'}\,(0.826x_q' - 0.986z_q' - b' + 1.180D)\right.$$

$$\left. + P_{z'}\left(0.986y_q' - 0.985b' + 0.493D - 2.496\frac{D}{n}\right)\right] + \Delta g\,(0.985D_{x'} + 0.826D_{z'})$$

Adjustment Diagram

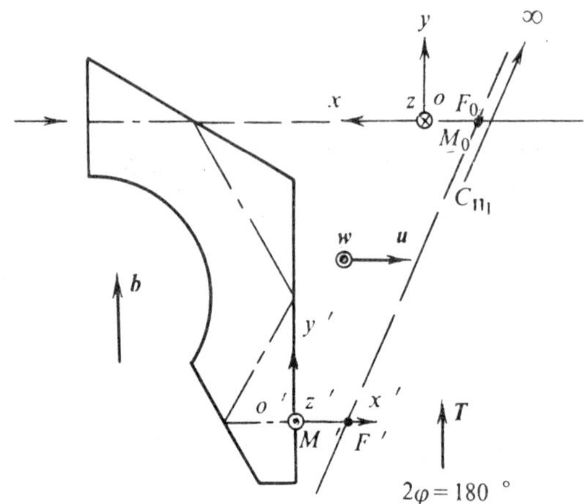

Fig . A . 32 .

Characteristic Parameters of Image Formation

Direction cosine		Unit vectors of image coordinate system			$2\varphi = 180$
		$i'(x')$	$j'(y')$	$k'(z')$	T
Unit vectors of coordinate system in object space	$i(x)$	$S_{T,2\varphi} = \begin{pmatrix} -1 \\ 0 \\ 0 \end{pmatrix}$	$\begin{pmatrix} 0 \\ 1 \\ 0 \end{pmatrix}$	$\begin{pmatrix} 0 \\ 0 \\ -1 \end{pmatrix}$	0
	$j(y)$				1
	$k(z)$				0

$t = 3$ $L = 4.330\,D$ $\qquad\qquad\qquad$ $2d =$

$$\overrightarrow{M_0 M}' = \left(1.732\,D - 4.330\,\frac{D}{n} \right)\, i' - 2.5D\; j'$$

$r_0 = \infty$

$r_i = \infty$

Characteristic Parameters of Adjustment

Direction cosine		Image coordinate system		
		x'	y'	z'
Extreme-valued axis direction of r' image rotation	u	1	0	0
	v	/	/	/
	w	0	0	1
Extreme value of r' image rotation		$\delta_u = 2\Delta\theta$	$\delta_v = 0$	$\delta_w = 2\Delta\theta$
Extreme-valued shift direction of r' image displacement	a	/	/	/
	b	0	1	0
	c	/	/	/
Extreme value of r' image displacement		$\delta_a = 0$	$\delta_b = 2\Delta g$	$\delta_c = 0$

Planar three-dimensional zero-valued poles	C_{Π_1} (Approaches to infinity along $\overrightarrow{F_o F'}$)	$x'_1 = \infty$
		$y'_1 = \infty$
		$z'_1 = 0$
		Associated plane Π_1: Conjugate optical-axis section $x'o'y'$
	C_{Π_2} (Does not exist)	$x'_2 =$
		$y'_2 =$
		$z'_2 =$
		Associated plane Π_2:

Equations of Image Motion

Parallel beam (Image rotation $\Delta\vec{\mu}'$)

Image lean

$$\Delta\mu'_{x'} = 2\Delta\theta\, P_x$$

Optical axis deviation

$$\Delta\mu'_{y'} = 0$$

$$\Delta\mu'_{z'} = 2\Delta\theta\, P_{z'}$$

Convergent beam (Image point displacement $\Delta S'_{F'}$)

Parallax

$$\Delta S'_{F'x'} = 2.5 D\Delta\theta\, P_{z'}$$

Optical axis deviation

$$\Delta S'_{F'y'} = \Delta\theta \left[2P_{x'}z'_{q} + P_{z'} \left(-2x'_{q} + 2b' - 1.732\,D + 4.330\,\frac{D}{n} \right) \right] + 2\Delta g\, D_{y'}$$

$$\Delta S'_{F'z'} = \Delta\theta \left[-2.5D\,P_{x'} + P_{y'} \left(-1.732\,D + 4.330\,\frac{D}{n} \right) \right]$$

Table A .33 D III − 45 °

Adjustment Diagram

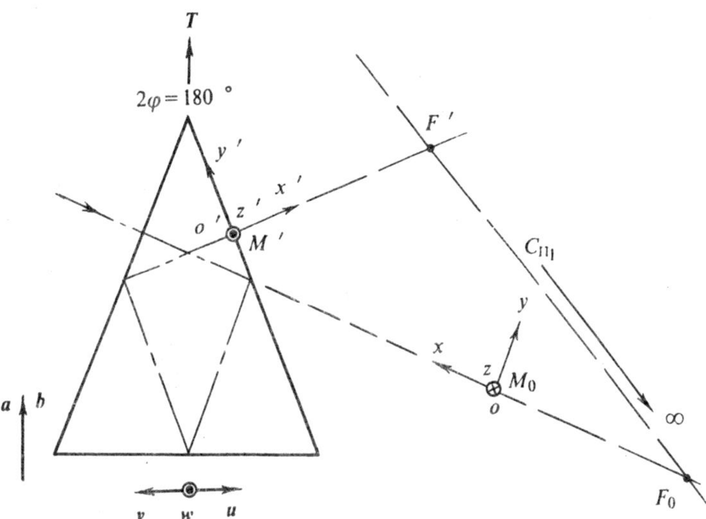

Fig . A . 33 .

Characteristic Parameters of Image Formation

Direction cosine		Unit vectors of image coordinate system			$2\varphi = 180$ °
		$i'(x')$	$j'(y')$	$k'(z')$	T
Unit vectors of coordinate system in object space	$i(x)$	$S_{T,2\varphi} =$ ⎛ −0.707	0.707	0	0.383
	$j(y)$	0.707	0.707	0	0.924
	$k(z)$	0	0	−1 ⎞	0

$t = 3$	$L = 2.414D$	$2d =$

$$\overrightarrow{M_0M}' = \left(0.354D - 1.707\,\frac{D}{n}\right)i' + \left(-0.146D + 1.707\,\frac{D}{n}\right)j'$$

$r_0 = \infty$

$r_i = \infty$

Characteristic Parameters of Adjustment

Direction cosine		Image coordinate system		
		x'	y'	z'
Extreme- valued axis direction of r' image rotation	u	0.924	-0.383	0
	v	-0.924	0.383	0
	w	0	0	1
Extreme value of r' image rotation		$\delta_u = 1.848\Delta\theta$	$\delta_v = 0.765\Delta\theta$	$\delta_w = 2\Delta\theta$
Extreme- valued shift direction of r' image displacement	a	0.383	0.924	0
	b	0.383	0.924	0
	c	/	/	/
Extreme value of r' image displacement		$\delta_a = 0.765\Delta g$	$\delta_b = 1.848\Delta g$	$\delta_c = 0$

Planar three-dimensional zero- valued poles	C_{Π_1} (Approaches to infinity along $\overrightarrow{F_o F'}$)	$x'_1 = \infty$
		$y'_1 = \infty$
		$z'_1 = 0$
		Associated plane Π_1 : Conjugate optical-axis section $x' o' y'$
	C_{Π_2} (Does not exist)	$x'_2 =$
		$y'_2 =$
		$z'_2 =$
		Associated plane Π_2 :

Equations of Image Motion

Parallel beam (Imge rotation $\Delta\vec{\mu}'$)

Image lean

$$\Delta\mu'_{x'} = \Delta\theta (1.707P_{x'} - 0.707P_{y'})$$

Optical axis deviation

$$\Delta\mu'_{y'} = \Delta\theta (-0.707P_{x'} + 0.293P_{y'})$$

$$\Delta\mu'_{z'} = 2\Delta\theta P_{z'}$$

Convergent beam (Image point displacement $\Delta S'_F$)

Parallax

$$\Delta S'_{F'x} = \Delta\theta[0.707P_{x'}z'_q - 0.293P_{y'}z'_q + P_{z'}(-0.707x'_q + 0.293y'_q - 0.146D)]$$
$$+ \Delta g (0.293D_{x'} + 0.707D_{y'})$$

Optical axis deviation

$$\Delta S'_{F'y} = \Delta\theta\left[1.707P_{x'}z'_q - 0.707P_{y'}z'_q + P_{z'}\left(-1.707x'_q + 0.707y'_q + 2b' - 0.354D + 2.414\frac{D}{n}\right)\right]$$
$$+ \Delta g (0.707 D_{x'} + 1.707D_{y'})$$

$$\Delta S'_{F'z} = \Delta\theta\left[P_{x'}\left(0.707b' - 0.146D + 1.707\frac{D}{n}\right) + P_{y'}\left(-0.293b' - 0.354D + 1.707\frac{D}{n}\right)\right]$$

Adjustment Diagram

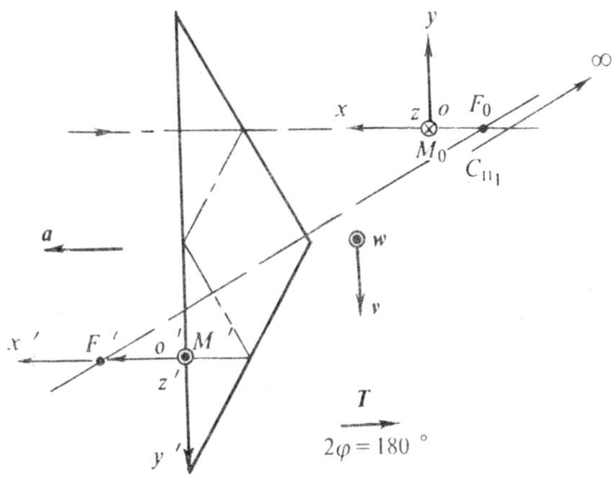

Fig . A .34 .

Characteristic Parameters of Image Formation

Direction cosine		Unit vectors of image coordinate system			$2\varphi = 180°$ T
		$i\,'(x\,')$	$j\,'(y\,')$	$k\,'(z\,')$	
Unit vectors of coordinate system in object space	$i\,(x)$	$S_{T.2\varphi}=\begin{pmatrix} 1$	0	0	-1
	$j\,(y)$	0	-1	0	0
	$k\,(z)$	0	0	$-1 \end{pmatrix}$	0

$t = 3$	$L = 1.732\,D$	$2d =$

$$\overrightarrow{M_0 M}\,' = 1.732\,\frac{D}{n}\,i\,' + Dj\,'$$

$$r_0 = \infty$$

$$r_i = \infty$$

Characteristic Parameters of Adjustment

Direction cosine		Image coordinate system		
		x'	y'	z'
Extreme-valued axis direction of r' image rotation	u	/	/	/
	v	0	1	0
	w	0	0	1
Extreme value of r' image rotation		$\delta_u = 0$	$\delta_v = 2\Delta\theta$	$\delta_w = 2\Delta\theta$
Extreme-valued shift direction of r' image displacement	a	1	0	0
	b	/	/	/
	c	/	/	/
Extreme value of r' image displacement		$\delta_a = 2\Delta g$	$\delta_b = 0$	$\delta_c = 0$
Planar three-dimensional zero-valued poles	C_{Π_1} (Approaches to infinity along $\overrightarrow{F_0 F'}$)	$x_1' = \infty$ $y_1' = \infty$ $z_1' = 0$ Associated plane Π_1 : Conjugate optical-axis section $x'o'y'$		
	C_{Π_2} (Does not exist)	$x_2' =$ $y_2' =$ $z_2' =$ Associated plane Π_2 :		

Equations of Image Motion

Parallel beam (Image rotation $\overrightarrow{\Delta\mu}'$)
Image lean

$$\Delta\mu_{x'}' = 0$$

Optical axis deviation

$$\Delta\mu_{y'}' = 2\Delta\theta\, P_{y'}$$

$$\Delta\mu_{z'}' = 2\Delta\theta P_{z'}$$

Convergent beam (Image point displacement $\Delta S_{F'}'$)
Parallax

$$\Delta S_{F'x}' = \Delta\theta\,[-2P_{y'}z_q' + P_{z'}(2y_q' + D)] + 2\Delta g\, D_{x'}$$

Optical axis deviation

$$\Delta S_{F'y}' = \Delta\theta\, P_{z'}\left(2b' + 1.732\frac{D}{n}\right)$$

$$\Delta S_{F'z}' = \Delta\theta\left[D P_{x'} + P_{y'}\left(-2b' - 1.732\frac{D}{n}\right)\right]$$

Adjustment Diagram

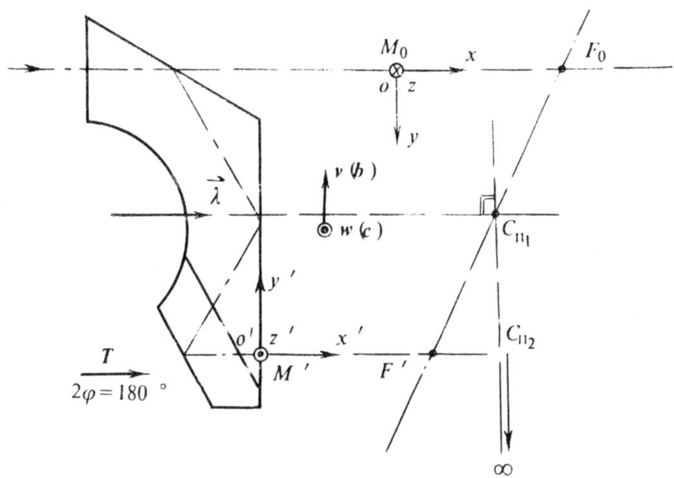

Fig . A .35 .

Characteristic Parameters of Image Formation

Direction cosine		Unit vectors of image coordinate system			$2\varphi = 180^\circ$
		$i'(x')$	$j'(y')$	$k'(z')$	T
Unit vectors of coordinate system in object space	i (x)	1	0	0	1
	j (y)	$S_{T.2\varphi} = $ 0	-1	0	0
	k (z)	0	0	-1	0

$t = 4$	$L = 4.619D$	$2d = 1.732D - 4.619 \dfrac{D}{n}$

$$\overrightarrow{M_0 M}\,' = \left(1.732D - 4.619 \frac{D}{n}\right) i\,' - 2.667D\, j\,'$$

$$r_0 = 1.334D\, j\,'$$

$$r_i =$$

Characteristic Parameters of Adjustment

Direction cosine		Image coordinate system		
		x'	y'	z'
Extreme-valued axis direction of r' image rotation	u	/	/	/
	v	0	1	0
	w	0	0	1
Extreme value of r' image rotation		$\delta_u = 0$	$\delta_v = 2\Delta\theta$	$\delta_w = 2\Delta\theta$
Extreme-valued shift direction of r' image displacement	a	/	/	/
	b	0	1	0
	c	0	0	1
Extreme value of r' image displacement		$\delta_a = 0$	$\delta_b = 2\Delta g$	$\delta_c = 2\Delta g$
Planar three-dimensional zero-valued poles	C_{Π_1}	$x_1' = \dfrac{1}{2}\left(2b' - 1.732\,D + 4.619\dfrac{D}{n}\right)$ $y_1' = 1.334\,D$ $z_1' = 0$ Associated plane Π_1 : Conjugate optical-axis section $x'o'y'$		
	C_{Π_2} (Approaches to infinity along the intersecting line of planes Π_1 and Π_2)	$x_2' = x_1'$ $y_2' = \infty$ $z_2' = 0$ Associated plane Π_2 : Plane passing through C_{Π_1} and perpendicular to $\vec{\lambda}$		

Equations of Image Motion

Parallel beam (Image rotation $\Delta\vec{\mu}'$)

Image lean

$$\Delta\mu_x' = 0$$

Optical axis deviation

$$\Delta\mu_y' = 2\Delta\theta\,P_{y'}$$

$$\Delta\mu_z' = 2\Delta\theta\,P_{z'}$$

Convergent beam (Image point displacement $\Delta S_F'$)

Parallax

$$\Delta S_{F'x'}' = 2.667 D\Delta\theta P_{z'}$$

Optical axis deviation

$$\Delta S_{F'y'}' = \Delta\theta\left[2P_{x'}z_q' + P_{z'}\left(-2x_q' + 2b' - 1.732D + 4.619\frac{D}{n}\right)\right] + 2\Delta g\,D_{y'}$$

$$\Delta S_{F'z'}' = \Delta\theta\left[P_{x'}(-2y_q' + 2.667D) + P_{y'}\left(2x_q' - 2b' + 1.732D - 4.619\frac{D}{n}\right)\right] + 2\Delta g\,D_{z'}$$

Adjustment Diagram

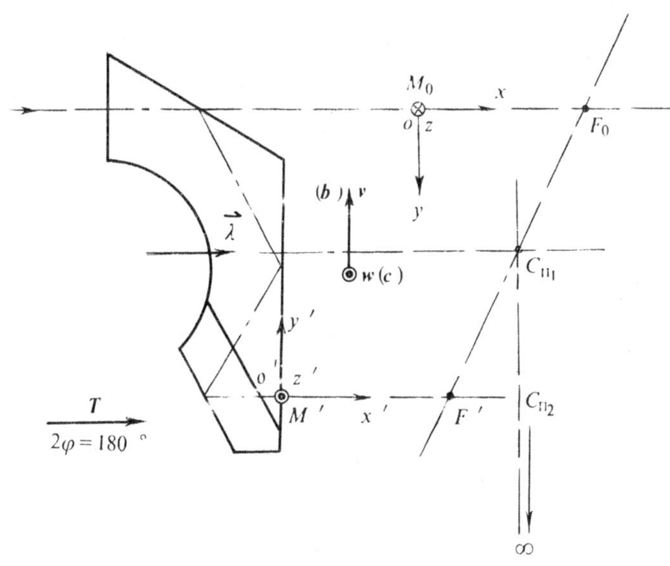

Fig . A . 36 .

Characteristic Parameters of Image Formation

Direction cosine		Unit vectors of image coordinate system			$2\varphi = 180$ $^{\circ}$ T
		$i'(x')$	$j'(y')$	$k'(z')$	
Unit vectors of coordinate system in object space	$i(x)$	1	0	0	1
	$j(y)$	$S_{T,2\varphi} = $ $\begin{pmatrix} 0 \end{pmatrix}$	-1	0	0
	$k(z)$	0	0	-1	0

$t=4$	$L=5.196D$	$2d=1.732D-5.196\dfrac{D}{n}$

$$\overrightarrow{M_0M'} = \left(1.732D - 5.196\frac{D}{n}\right) i' - 3Dj'$$

$$r_0 = 1.5\, Dj'$$

$$r_i =$$

Characteristic Parameters of Adjustment

Direction cosine		x'	y'	z'
Extreme-valued axis direction of r' image rotation	u	/	/	/
	v	0	1	0
	w	0	0	1
Extreme value of r' image rotation		$\delta_u = 0$	$\delta_v = 2\Delta\theta$	$\delta_w = 2\Delta\theta$
Extreme-valued shift direction of r' image displacement	a	/	/	/
	b	0	1	0
	c	0	0	1
Extreme value of r' image displacement		$\delta_a = 0$	$\delta_b = 2\Delta g$	$\delta_c = 2\Delta g$
Planar three-dimensional zero-valued poles	C_{Π_1}	$x'_1 = \dfrac{1}{2}\left(2b' - 1.732D + 5.196\dfrac{D}{n}\right)$ $y'_1 = 1.5D$ $z'_1 = 0$ Associated plane Π_1 : Conjugate optical-axis section $x'o'y'$		
	C_{Π_2} (Approaches to infinity along the intersecting line of planes Π_1 and Π_2)	$x'_2 = x'_1$ $y'_2 = \infty$ $z'_2 = 0$ Associated plane Π_2 : Plane passing through C_{Π_1} and perpendicular to $\vec{\lambda}$		

Equations of Image Motion

Parallel beam (Image rotation $\Delta\vec{\mu}'$)

Image lean

$$\Delta\mu'_{x'} = 0$$

Optical axis deviation

$$\Delta\mu'_{y'} = 2\Delta\theta\, P_{y'}$$

$$\Delta\mu'_{z'} = 2\Delta\theta\, P_{z'}$$

Convergent beam (Image point displacement $\Delta S'_F$)

Parallax

$$\Delta S'_{F\,x'} = 3D\,\Delta\theta\, P_{z'}$$

Optical axis deviation

$$\Delta S'_{F\,y'} = \Delta\theta\left[2P_{x'}z'_q + P_{z'}\left(-2x'_q + 2b' - 1.732D + 5.196\frac{D}{n}\right)\right] + 2\Delta g\, D_{y'}$$

$$\Delta S'_{F\,z'} = \Delta\theta\left[P_{x'}\left(-2y'_q + 3D\right) + P_{y'}\left(2x'_q - 2b' + 1.732D - 5.196\frac{D}{n}\right)\right] + 2\Delta g\, D_{z'}$$

Adjustment Diagram

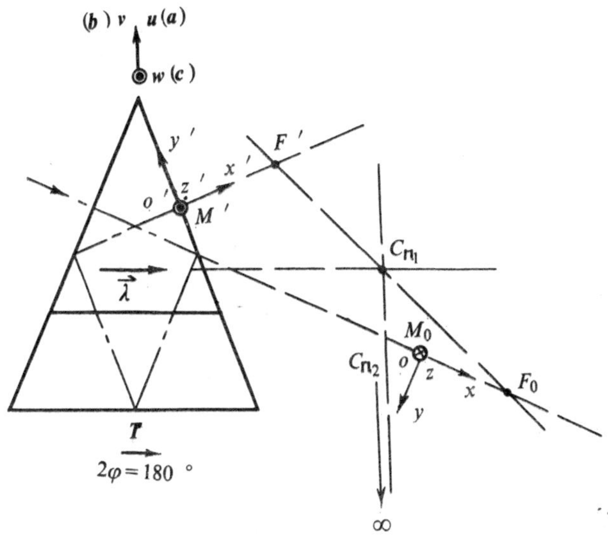

Fig . A . 37 .

Characteristic Parameters of Image Formation

Direction cosine		Unit vectors of image coordinate system			2φ = 180 ° T
		$i'(x')$	$j'(y')$	$k'(z')$	
Unit vectors of coordinate system in object space	$i(x)$	$S_{T,2\varphi} = \begin{pmatrix} 0.707 \\ -0.707 \\ 0 \end{pmatrix}$	-0.707	0	0.924
	$j(y)$	-0.707	-0.707	0	-0.383
	$k(z)$	0	0	-1	0

$t=4$	$L=3.04\ D$	$2d = 0.482D - 2.810\,\dfrac{D}{n}$

$$\overrightarrow{M_0M}\,' = \left(0.445D - 2.150\,\frac{D}{n}\right)i\,' + \left(-0.185D + 2.150\,\frac{D}{n}\right)j\,'$$

$$r_0 = -0.223\,\frac{D}{n}\ i\,' - 0.537\,\frac{D}{n}\ j\,'$$

$$r_i =$$

Characteristic Parameters of Adjustment

Direction cosine		Image coordinate system		
		x'	y'	z'
Extreme-valued	u	0.383	0.924	0
axis direction of	v	0.383	0.924	0
r' image rotation	w	0	0	1
Extreme value of r' image rotation		$\delta_u = 0.765\Delta\theta$	$\delta_v = 1.848\Delta\theta$	$\delta_w = 2\Delta\theta$
Extreme-valued	a	0.383	0.924	0
shift direction of	b	0.383	0.924	0
r' image displacement	c	0	0	1
Extreme value of r' image displacement		$\delta_a = 0.765\Delta g$	$\delta_b = 1.848\Delta g$	$\delta_c = 2\Delta g$
Planar three-dimensional zero-valued poles	C_{Π_1}	$x_1' = \dfrac{1}{2}\left(1.707b' - 0.445D + 2.150\dfrac{D}{n}\right)$ $y_1' = \dfrac{1}{2}\left(-0.707b' + 0.185D - 2.150\dfrac{D}{n}\right)$ $z_1' = 0$ Associated plane Π_1 : Conjugate optical-axis section $\quad x'o'y'$		
	C_{Π_2} (Approaches to infinity along the intersecting line of planes Π_1 and Π_2)	$x_2' = \infty$ $y_2' = \infty$ $z_2' = 0$ Associated plane Π_2 : Plane passing through C_{Π_1} and perpendicular to $\vec{\lambda}$		

Equations of Image Motion

Parallel beam (Image rotation $\Delta\vec{\mu}'$)

Image lean

$$\Delta\mu_{x'}' = \Delta\theta\,(0.293 P_{x'} + 0.707 P_{y'})$$

Optical axis deviation

$$\Delta\mu_{y'}' = \Delta\theta\,(0.707 P_{x'} + 1.707 P_{y'})$$

$$\Delta\mu_{z'}' = 2\Delta\theta\,P_{z'}$$

Convergent beam (Image point displacement $\Delta S_F'$)

Parallax

$$\Delta S_{F\,x}' = \Delta\theta[\,0.707 P_{x'}z_q' - 0.293 P_{y'}z_q' + P_{z'}(-0.707 x_q' + 0.293 y_q' - 0.185D)\,]$$
$$+ \Delta g\,(0.293 D_{x'} + 0.707 D_{y'})$$

Optical axis deviation

$$\Delta S_{F\,y}' = \Delta\theta\left[\,1.707 P_{x'}z_q' - 0.707 P_{y'}z_q' + P_{z'}\left(-1.707 x_q' + 0.707 y_q' + 2b' - 0.445D + \right.\right.$$

$$3.04\frac{D}{n}\bigg)\bigg] + \Delta g\,(0.707D_{x'} + 1.707D_{y'})$$

$$\Delta S_{F'z'} = \Delta\theta\bigg[P_{x'}\bigg(-2y'_q - 0.707b' + 0.185D - 2.150\frac{D}{n}\bigg) + P_{y'}\bigg(2x'_q - 1.707b' + 0.445D$$

$$- 2.150\frac{D}{n}\bigg)\bigg] + 2\Delta g\,D_{z'}$$

Adjustment Diagram

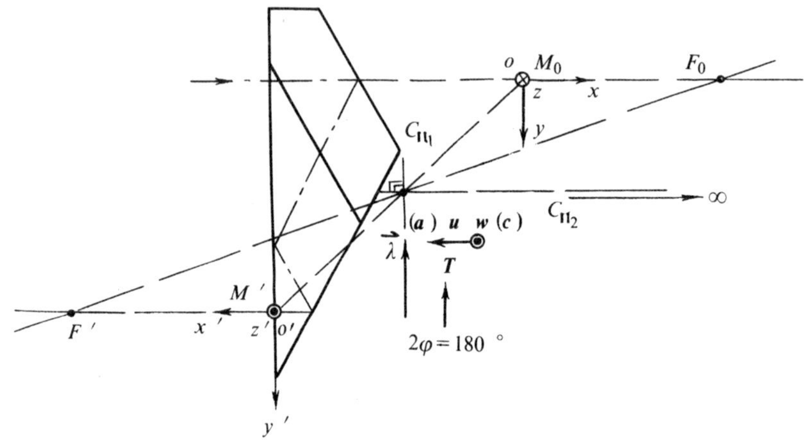

Fig . A .38 .

Characteristic Parameters of Image Formation

Direction cosine		Unit vectors of image coordinate system			$2\varphi = 180$ $^\circ$
		$i'(x')$	$j'(y')$	$k'(z')$	T
Unit vectors of coordinate system in object space	$i(x)$	$S_{T,2\varphi} = \begin{pmatrix} -1$	0	$0 \end{pmatrix}$	0
	$j(y)$	0	1	0	-1
	$k(z)$	0	0	-1	0

$t = 4$ $\quad$ $L = 2 .802 D$	$2d = -1 .681 D$

$$\overrightarrow{M_0 M}\,' = 2 .802 \frac{D}{n} i' + 1 .618\ D j'$$

$$r_0 = -1 .401 \frac{D}{n}\ i'$$

$$r_i =$$

Characteristic Parameters of Adjustment

Direction cosine		Image coordinate system		
		x'	y'	z'
Extreme-valued axis direction of r' image rotation	u	1	0	0
	v	/	/	/
	w	0	0	1
Extreme value of r' image rotation		$\delta_u = 2\Delta\theta$	$\delta_v = 0$	$\delta_w = 2\Delta\theta$
Extreme-valued shift direction of r' image displacement	a	1	0	0
	b	/	/	/
	c	0	0	1
Extreme value of r' image displacement		$\delta_a = 2\Delta g$	$\delta_b = 0$	$\delta_c = 2\Delta g$
Planar three-dimensional zero-valued poles	C_{Π_1}	$x_1' = -1.401\dfrac{D}{n}$		
		$y_1' = -0.809D$		
		$z_1' = 0$		
		Associated plane Π_1 : Conjugate optical-axis section $x'o'y'$		
	C_{Π_2} (Approaches to infinity along the intersecting line of planes Π_1 and Π_2)	$x_2' = \infty$		
		$y_2' = y_1'$		
		$z_2' = 0$		
		Associated plane Π_2 : Plane passing through C_{Π_1} and perpendicular to $\vec{\lambda}$		

Equations of Image Motion

Parallel beam (Image rotation $\Delta\vec{\mu}'$)

Image lean

$$\Delta\mu_{x'}' = 2\Delta\theta\, P_{x'}$$

Optical axis deviation

$$\Delta\mu_{y'}' = 0$$

$$\Delta\mu_{z'}' = 2\Delta\theta\, P_{z'}$$

Convergent beam (Image point displacement $\Delta S_F'$)

Parallax

$$\Delta S_{F'x'}' = \Delta\theta\left[-2P_{y'}z_q' + P_{z'}(2y_q' + 1.618D)\right] + 2\Delta g\, D_{x'}$$

Optical axis deviation

$$\Delta S_{F'y'}' = \Delta\theta\, P_{z'}\left(2b' + 2.802\frac{D}{n}\right)$$

$$\Delta S_{F'z'}' = \Delta\theta\left[P_{x'}(-2y_q' - 1.618D) + P_{y'}\left(2x_q' + 2.802\frac{D}{n}\right)\right] + 2\Delta g\, D_{z'}$$

Table A.39 FL– 0 °

Adjustment Diagram

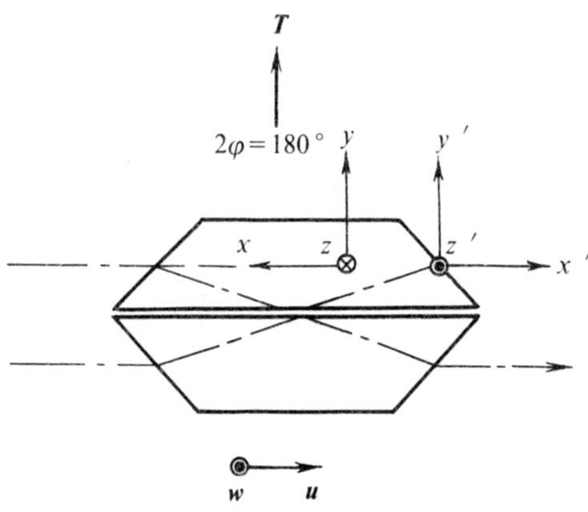

Fig. A.39.

Characteristic Parameters of Image Formation

Direction cosine		Unit vectors of image coordinate system			$2\varphi = 180°$ T
		$i'(x')$	$j'(y')$	$k'(z')$	
Unit vectors of coordinate system in object space	$i(x)$	$S_{T.2\varphi} = \begin{pmatrix} -1 \\ \\ 0 \\ \\ 0 \end{pmatrix}$	$\begin{pmatrix} 0 \\ \\ 1 \\ \\ 0 \end{pmatrix}$	$\begin{pmatrix} 0 \\ \\ 0 \\ \\ -1 \end{pmatrix}$	0
	$j(y)$				1
	$k(z)$				0

$t = 1$	$L = \dfrac{n}{\sqrt{2n^2 - 1} - 1} D$	$2d =$

$\overrightarrow{M_0 M'} =$

$r_0 =$

$r_i =$

Characteristic Parameters of Adjustment

Direction cosine		Image coordinate system		
		x'	y'	z'
Extreme-valued axis direction of r' image rotation	u	1	0	0
	v	/	/	/
	w	0	0	1
Extreme value of r' image rotation		$\delta_u = 2\Delta\theta$	$\delta_v = 0$	$\delta_w = 2\Delta\theta$
Extreme-valued shift direction of r' image displacement	a			
	b			
	c			
Extreme value of r' image displacement		$\delta_a =$	$\delta_b =$	$\delta_c =$
Planar three-dimensional zero-valued poles (Meaningless in collimated beams)	C_{Π_1}	$x'_1 =$		
		$y'_1 =$		
		$z'_1 =$		
		Associated plane Π_1 :		
	C_{Π_2}	$x'_2 =$		
		$y'_2 =$		
		$z'_2 =$		
		Associated plane Π_2 :		

Equations of Image Motion

Parallel beam (Image rotation $\Delta\vec{\mu}'$)

Image lean

$$\Delta\mu'_{x'} = 2\Delta\theta P_{x'}$$

Optical axis deviation

$$\Delta\mu'_{y'} = 0$$

$$\Delta\mu'_{z'} = 2\Delta\theta P_{z'}$$

Adjustment Diagram

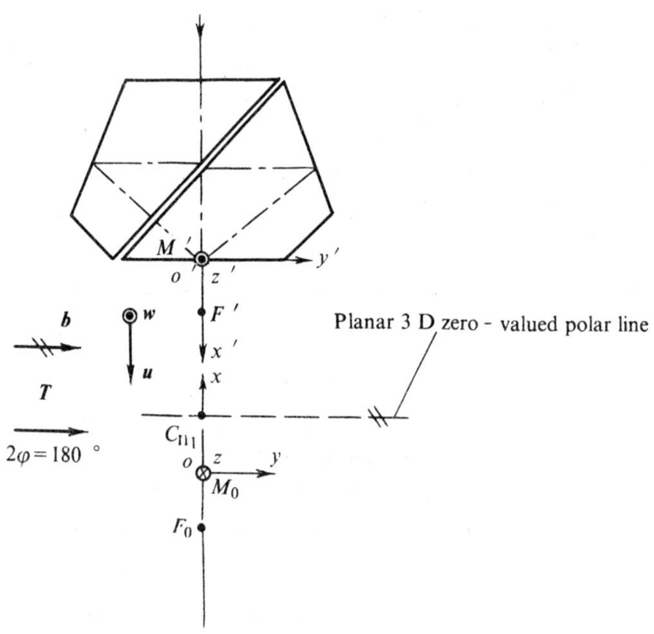

Fig . A . 40 .

Characteristic Parameters of Image Formation

Direction cosine		Unit vectors of image coordinate system			$2\varphi = 180°$
		$i'(x')$	$j'(y')$	$k'(z')$	T
Unit vectors of coordinate system in object space	$i(x)$	$S_{T.2\varphi} = \begin{pmatrix} -1$	0	0	0
	$j(y)$	0	1	0	1
	$k(z)$	0	0	$-1 \end{pmatrix}$	0

$t = 5$	$L = 4.621D$	$2d =$

$$\overrightarrow{M_0M}' = \left(1.207D - 4.621\frac{D}{n}\right) i'$$

$$r_0 = \infty$$

$$r_i = \infty$$

Characteristic Parameters of Adjustment

Direction cosine		Image coordinate system		
		x'	y'	z'
Extreme-valued axis direction of r' image rotation	u	1	0	0
	v	/	/	/
	w	0	0	1
Extreme value of r' image rotation		$\delta_u = 2\Delta\theta$	$\delta_v = 0$	$\delta_w = 2\Delta\theta$
Extreme-valued shift direction of r' image displacement	a	/	/	/
	b	0	1	0
	c	/	/	/
Extreme value of r' image displacement		$\delta_a = 0$	$\delta_b = 2\Delta g$	$\delta_c = 0$

Planar three-dimensional zero-valued polar line	C_{Π_1}	$x'_1 = \dfrac{1}{2}\left(2b' - 1.207D + 4.621\dfrac{D}{n}\right)$
		$y'_1 = $ Any value
		$z'_1 = 0$
		Associated plane Π_1 : Planes passing through C_{Π_1} and perpendicular to b
	Direction	Parallel to b

Equations of Image Motion

Parallel beam (Image rotation $\Delta\vec{\mu}'$)

Image lean

$$\Delta\mu'_{x'} = 2\Delta\theta\, P_{y'}$$

Optical axis deviation

$$\Delta\mu'_{y'} = 0$$

$$\Delta\mu'_{z'} = 2\Delta\theta\, P_{z'}$$

Convergent beam (Image point displacement $\Delta S'_F$)

Parallax

$$\Delta S'_{F'x'} = 0$$

Optical axis deviation

$$\Delta S'_{F'y'} = \Delta\theta\left[2P_{x'}z'_q + P_{z'}\left(-2x'_q + 2b' - 1.207D + 4.612\dfrac{D}{n}\right)\right] + 2\Delta g\, D_{y'}$$

$$\Delta S'_{F'z'} = \Delta\theta\, P_{y'}\left(-1.207D + 4.621\dfrac{D}{n}\right)$$

Table A . 41 FA − 0 °

Adjustment Diagram

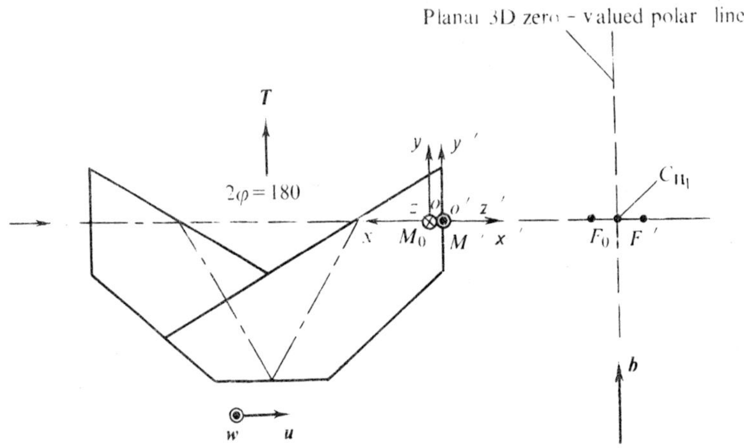

Fig . A . 41 .

Characteristic Parameters of Image Formation

Direction cosine		Unit vectors of image coordinate system			$2\varphi = 180\,°$ T
		$i\,'(x\,')$	$j\,'(y\,')$	$k\,'(z\,')$	
Unit vectors of coordinate system in object space	$i\,(x)$	$S_{T.2\varphi}=$ $\Big($ -1	0	0 $\Big)$	0
	$j\,(y)$	0	1	0	1
	$k\,(z)$	0	0	-1	0

$t=3$ $L=5.196D$	$2d=$

$$\overrightarrow{M_0M}\,' = \left(3.464D - 5.196\,\frac{D}{n}\right) i\,'$$

$$r_0 = \infty$$

$$r_i = \infty$$

Characteristic Parameters of Adjustment

Direction cosine		Image coordinate system		
		x'	y'	z'
Extreme-valued axis direction of r' image rotation	u	1	0	0
	v	/	/	/
	w	0	0	1
Extreme value of r' image rotation		$\delta_u = 2\Delta\theta$	$\delta_v = 0$	$\delta_w = 2\Delta\theta$
Extreme-valued shift direction of r' image displacement	a	/	/	/
	b	0	1	0
	c	/	/	/
Extreme value of r' image displacement		$\delta_a = 0$	$\delta_b = 2\Delta g$	$\delta_c = 0$

Planar three-dimensional zero-valued polar line	C_{Π_1}	$x_1' = \dfrac{1}{2}\left(2b' - 3.464\,D + 5.196\,\dfrac{D}{n}\right)$
		$y_1' = $ Any value
		$z_1' = 0$
		Associated planes Π_1 : Planes passing through C_{Π_1} and perpendicular to b
	Direction	Parallel to b

Equations of Image Motion

Parallel beam (Image rotation $\Delta\vec{\mu}'$)

Image lean

$$\Delta\mu_{x'}' = 2\Delta\theta\,P_{x'}$$

Optical axis deviation

$$\Delta\mu_{y'}' = 0$$

$$\Delta\mu_{z'}' = 2\Delta\theta\,P_{z'}$$

Convergent beam (Image point displacement $\Delta S_F'$)

Parallax

$$\Delta S_{F'x'}' = 0$$

Optical axis deviation

$$\Delta S_{F'y'}' = \Delta\theta\left[2P_{x'}z_q' + P_{z'}\left(-2x_q' + 2b' - 3.464D + 5.196\,\frac{D}{n}\right)\right] + 2\Delta g\,D_{y'}$$

$$\Delta S_{F'z'}' = \Delta\theta\,P_{y'}\left(-3.464D + 5.196\,\frac{D}{n}\right)$$

Adjustment Diagram

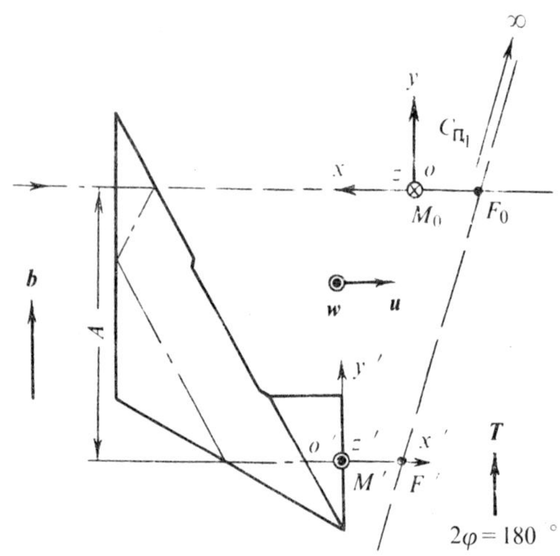

Fig. A.42.

Characteristic Parameters of Image Formation

Direction cosine		Unit vectors of image coordinate system			$2\varphi = 180$ °
		$i'(x')$	$j'(y')$	$k'(z')$	T
Unit vectors of coordinate system in object space	$i(x)$	$S_{T.2\varphi}=\begin{pmatrix} -1 & & \\ & & \\ & & \\ \end{pmatrix}$ -1	0	0	0
	$j(y)$	0	1	0	1
	$k(z)$	0	0	-1	0

$t=3$	$L=1.155\,(A+D)$	$2D=$

$$\overrightarrow{M_0 M'}=\left[\,0.577\,(A+D)-1.155\,\frac{(A+D)}{n}\,\right]i'-A\,j'$$

$r_0 = \infty$

$r_i = \infty$

Characteristic Parameters of Adjustment

Direction cosine		Image coordinate system		
		x'	y'	z'
Extreme-valued axis direction of r' image rotation	u	1	0	0
	v	/	/	/
	w	0	0	1
Extreme value of r' image rotation		$\delta_u = 2\Delta\theta$	$\delta_v = 0$	$\delta_w = 2\Delta\theta$
Extreme-valued shift direction of r' image displacement	a	/	/	/
	b	0	1	0
	c	/	/	/
Extreme value of r' image displacement		$\delta_a = 0$	$\delta_b = 2\Delta g$	$\delta_c = 0$

Planar three-dimensional zero-valued poles	C_{Π_1} (Approaches to infinity along $\overrightarrow{F_0 F'}$)	$x'_1 = \infty$
		$y'_1 = \infty$
		$z'_1 = 0$
		Associated plane Π_1: Conjugate optical-axis section $x'o'y'$
	C_{Π_2} (Does not exist)	$x'_2 =$
		$y'_2 =$
		$z'_2 =$
		Associated plane Π_2:

Equations of Image Motion

Parallel beam (Image rotation $\Delta\overrightarrow{\mu'}$)

Image lean

$$\Delta\mu'_{x'} = 2\Delta\theta P_{x'}$$

Optical axis deviation

$$\Delta\mu'_{y'} = 0$$
$$\Delta\mu'_{z'} = 2\Delta\theta P_{z'}$$

Convergent beam (Image point displacement $\Delta S'_{F'}$)

Parallax

$$\Delta S'_{F'x'} = A\Delta\theta P_{z'}$$

Optical axis deviation

$$\Delta S'_{F'y'} = \Delta\theta\left\{2P_{x'}z'_q + P_{z'}\left[-2x'_q + 2b' - 0.577(A+D) + 1.155\frac{(A+D)}{n}\right]\right\} + 2\Delta gD_y$$

$$\Delta S'_{F'z'} = \Delta\theta\left\{-AP_{x'} + P_{y'}\left[-0.577(A+D) + 1.155\frac{(A+D)}{n}\right]\right\}$$

Table A .43 FY – 60 °

Adjustment Diagram

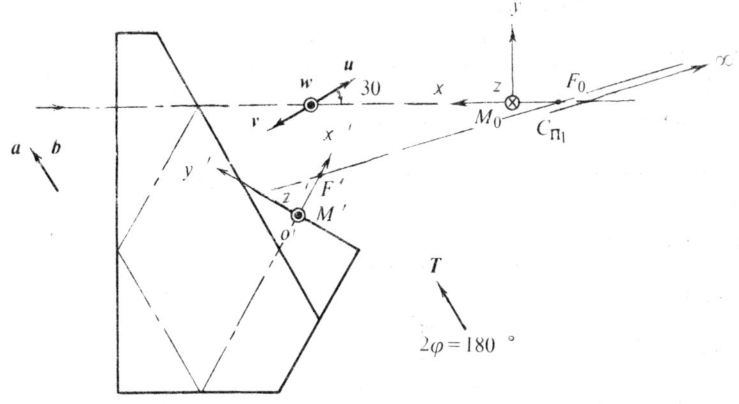

Fig . A .43 .

Characteristic Parameters of Image Formation

Direction cosine		Unit vectors of image coordinate system			$2\varphi = 180°$
		$i'(x')$	$j'(y')$	$k'(z')$	T
Unit vectors of coordinate system in object space	$i(x)$	-0.5	0.866	0	0.5
	$j(y)$	0.866	0.5	0	0.866
	$k(z)$	0	0	-1	0

$S_{T.2\varphi}=$ (matrix above)

$t=3$	$L=4.330D$	$2d=$

$$\overrightarrow{M_0M'} = -2.165\,\frac{D}{n}\,i' + \left(-1.5D + 3.750\,\frac{D}{n}\right)j'$$

$r_0 = \infty$

$r_i = \infty$

Characteristic Parameters of Adjustment

Direction cosine		Image coordinate system		
		x'	y'	z'
Extreme-valued axis direction of r' image rotation	u	0.866	-0.5	0
	v	-0.866	0.5	0
	w	0	0	1
Extreme value of r' image rotation		$\delta_u = 1.732\Delta\theta$	$\delta_v = \Delta\theta$	$\delta_w = 2\Delta\theta$
Extreme-valued shift direction of r' image displacement	a	0.5	0.866	0
	b	0.5	0.866	0
	c	/	/	/
Extreme value of r' image displacement		$\delta_a = \Delta g$	$\delta_b = 1.732\Delta g$	$\delta_c = 0$

Planar three-dimensional zero-valued poles	C_{Π_1} (Approaches to infinity along $\overrightarrow{F_0 F'}$)	$x'_1 = \infty$
		$y'_1 = \infty$
		$z'_1 = 0$
		Associated plane Π_1: Conjugate optical-axis section $x' o' y'$
	C_{Π_2} (Does not exist)	$x'_2 =$
		$y'_2 =$
		$z'_2 =$
		Associated plane Π_2:

Equations of Image Motion

Parallel beam (Image rotation $\Delta\vec{\mu}'$)

Image lean

$$\Delta\mu'_{x'} = \Delta\theta\,(1.5P_{x'} - 0.866P_{y'})$$

Optical axis deviation

$$\Delta\mu'_{y'} = \Delta\theta\,(-0.866P_{x'} + 0.5P_{y'})$$

$$\Delta\mu'_{z'} = 2\Delta\theta P_{z'}$$

Convergent beam (Image point displacement $\Delta S'_{F'}$)

Parallax

$$\Delta S'_{F'_{x'}} = \Delta\theta[0.866P_{x'}z'_q - 0.5P_{y'}z'_q + P_{z'}(-0.866x'_q + 0.5y'_q + 0.75D)]$$
$$+ \Delta g\,(0.5D_{x'} + 0.866D_{y'})$$

Optical axis deviation

$$\Delta S'_{F'_{y'}} = \Delta\theta\left[1.5P_{x'}z'_q - 0.866P_{y'}z'_q + P_{z'}\left(-1.5x'_q + 0.866y'_q + 2b' - 1.299D + 4.330\frac{D}{n}\right)\right]$$
$$+ \Delta g\,(0.866D_{x'} + 1.5D_{y'})$$

$$\Delta S'_{F'_{z'}} = \Delta\theta\left[P_{x'}\left(0.866b' - 1.5D + 3.750\frac{D}{n}\right) + P_{y'}\left(-0.5b' + 2.165\frac{D}{n}\right)\right]$$

Adjustment Diagram

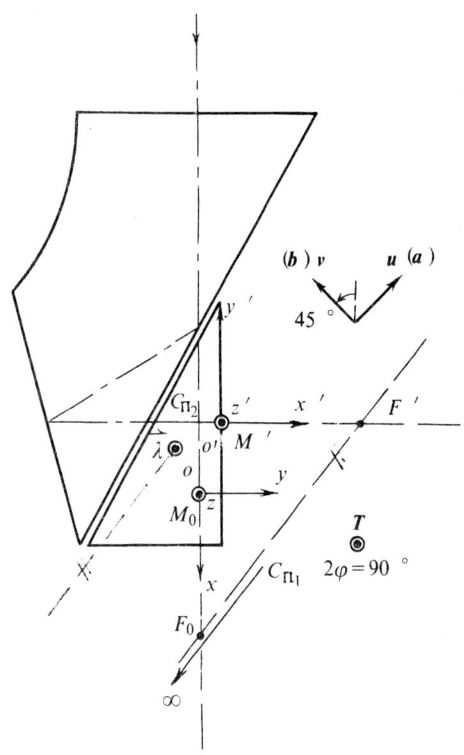

Fig . A . 44 .

Characteristic Parameters of Image Formation

Direction cosine		Unit vectors of image coordinate system			2φ = 90 °
		$i'(x')$	$j'(y')$	$k'(z')$	T
Unit vectors of coordinate system in object space	$i(x)$	0	-1	0	0
	$j(y)$ $S_{T,2\varphi}=$	1	0	0	0
	$k(z)$	0	0	1	1

$t=2$	$L = 2.309D$		$2d=0$

$$\overrightarrow{M_0 M} = 0.077D\,i' + \left(-1.232D + 2.309\,\frac{D}{n}\right)j'$$

$$r_0 = \left(0.577D - 1.155\,\frac{D}{n}\right)i' + \left(0.655D - 1.155\,\frac{D}{n}\right)j'$$

$$r_i =$$

Characteristic Parameters of Adjustment

Direction cosine		$x\,'$	$y\,'$	$z\,'$
		Image coordinate system		
Extreme-valued axis direction of $r\,'$ image rotation	u	0.707	0.707	0
	v	-0.707	0.707	0
	w	/	/	/
Extreme value of $r\,'$ image rotation		$\delta_u = 1.414\Delta\theta$	$\delta_v = 1.414\Delta\theta$	$\delta_w = 0$
Extreme-valued shift direction of $r\,'$ image displacement	a	0.707	0.707	0
	b	-0.707	0.707	0
	c	/	/	/
Extreme value of $r\,'$ image displacement		$\delta_a = 1.414\Delta g$	$\delta_b = 1.414\Delta g$	$\delta_c = 0$
Planar three-dimensional zero-valued poles	C_{Π_1} (Approaches to infinity along $\overrightarrow{F_0 F\,'}$)	$x\,'_1 = \infty$		
		$y\,'_1 = \infty$		
		$z\,'_1 = 0$		
		Associated plane Π_1: Conjugate optical-axis section $x\,'o\,'y\,'$		
	C_{Π_2}	$x\,'_2 = 0.578D - 1.155\dfrac{D}{n}$		
		$y\,'_2 = 0.655D - 1.155\dfrac{D}{n}$		
		$z\,'_2 = 0$		
		Associated plane Π_2: plane passing through $\overrightarrow{\lambda}$ and parallel to $\overrightarrow{F_0 F\,'}$		

Equations of Image Motion

Parallel beam (Image rotation $\Delta\vec{\mu}\,'$)

Image lean

$$\Delta\mu\,'_{x\,'} = \Delta\theta\,(P_{x\,'} + P_{y\,'})$$

Optical axis deviation

$$\Delta\mu\,'_{y\,'} = \Delta\theta\,(-P_{x\,'} + P_{y\,'})$$

$$\Delta\mu\,'_{z\,'} = 0$$

Convergent beam (Image point displacement $\Delta S_{F'}'$)

Parallax

$$\Delta S_{F'x'}' = \Delta\theta\, [\, P_{x'}\, z_q' - P_{y'}\, z_q' + P_{z'}\, (-x_q' + y_q' - 0.077D)] + \Delta g\, (D_{x'} + D_{y'})$$

Optical axis deviation

$$\Delta S_{F'y'}' = \Delta\theta\left[P_{x'}\, z_q' + P_{y'}\, z_q' + P_{z'}\left(-x_q' - y_q' + 1.232D - 2.309\,\frac{D}{n}\right)\right] + \Delta g\,(-D_{x'} + D_{y'})$$

$$\Delta S_{F'z'}' = \Delta\theta\left[P_{x'}\left(b' - 1.232D + 2.309\,\frac{D}{n}\right) + P_{y'}\,(-b' - 0.077D)\right]$$

Adjustment Diagram

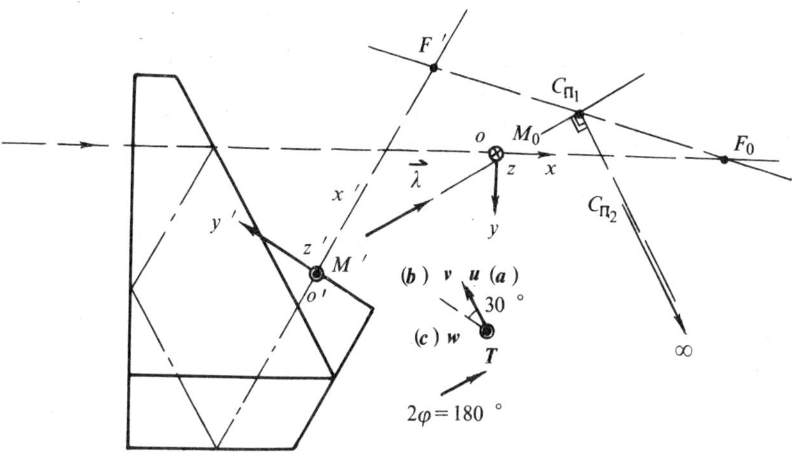

Fig . A .45 .

Characteristic Parameters of Image Formation

Direction cosine		Unit vectors of image coordinate system			$2\varphi = 180°$
		$i'(x')$	$j'(y')$	$k'(z')$	T
Unit vectors of coordinate system in object space	$i(x)$	$S_{T,2\varphi} = \Big($ 0.5	-0.866	0	0.866
	$j(y)$	-0.866	-0.5	0	-0.5
	$k(z)$	0	0	-1 $\Big)$	0

$t=4$	$L=4.468D$	$2d=0.750D-3.869\dfrac{D}{n}$

$$\overrightarrow{M_0 M'}=\left(-0.069D-2.234\frac{D}{n}\right)i'+\left(-1.620D+3.869\frac{D}{n}\right)j'$$

$$r_0=\left(0.359D-0.558\frac{D}{n}\right)i'+\left(0.622D-0.967\frac{D}{n}\right)j'$$

$$r_i=$$

Characteristic Parameters of Adjustment

<table>
<tr><td colspan="3" rowspan="2">Direction
cosine</td><td colspan="3">Image coordinate system</td></tr>
<tr><td>x'</td><td>y'</td><td>z'</td></tr>
<tr><td rowspan="3">Extreme-valued
axis direction of
r' image rotation</td><td>u</td><td>0.5</td><td>0.866</td><td>0</td></tr>
<tr><td>v</td><td>0.5</td><td>0.866</td><td>0</td></tr>
<tr><td>w</td><td>0</td><td>0</td><td>1</td></tr>
<tr><td colspan="2">Extreme value of
r' image rotation</td><td>$\delta_u = \Delta\theta$</td><td>$\delta_v = 1.732\Delta\theta$</td><td>$\delta_w = 2\Delta\theta$</td></tr>
<tr><td rowspan="3">Extreme-valued
shift direction of
r' image displacement</td><td>a</td><td>0.5</td><td>0.866</td><td>0</td></tr>
<tr><td>b</td><td>0.5</td><td>0.866</td><td>0</td></tr>
<tr><td>c</td><td>0</td><td>0</td><td>1</td></tr>
<tr><td colspan="2">Extreme value of
r' image displacement</td><td>$\delta_a = \Delta g$</td><td>$\delta_b = 1.732\Delta g$</td><td>$\delta_c = 2\Delta g$</td></tr>
<tr><td rowspan="6">Planar three-
dimensional
zero-valued
poles</td><td rowspan="3">C_{Π_1}
(Mid-point
of segment
$\overline{F_0F'}$)</td><td colspan="4">$x_1' = \dfrac{1}{2}\left(1.5b' + 0.069D + 2.234\dfrac{D}{n}\right)$</td></tr>
<tr><td colspan="4">$y_1' = \dfrac{1}{2}\left(-0.866b' + 1.620D - 3.869\dfrac{D}{n}\right)$</td></tr>
<tr><td colspan="4">$z_1' = 0$</td></tr>
<tr><td colspan="4">Associated plane Π_1:
    Conjugate optical-axis section $x'o'y'$</td></tr>
<tr><td rowspan="2">C_{Π_2}
(Approaches
to infinity along
the intersecting
line of planes
Π_1 and Π_2)</td><td colspan="4">$x_2' = \infty$
$y_2' = \infty$
$z_2' = 0$</td></tr>
<tr><td colspan="4">Associated plane Π_2: Plane passing
    through C_{Π_1} and perpendicular to $\vec{\lambda}$</td></tr>
</table>

Equations of Image Motion

Parallel beam (Image rotation $\Delta\vec{\mu}'$)

Image lean

$$\Delta\mu_{x'}' = \Delta\theta\,(0.5P_{x'} + 0.866P_{y'})$$

Optical axis deviation

$$\Delta\mu_{y'}' = \Delta\theta\,(0.866P_{x'} + 1.5P_{y'})$$

$$\Delta\mu_z' = 2\Delta\theta P_z'$$

Convergent beam (Image point displacement $\Delta S_F'$)

Parallax

$$\Delta S_{F'x'}' = \Delta\theta\,[\,0.866P_{x'}z_q' - 0.5P_{y'}z_q' + P_{z'}(-0.866x_q' + 0.5y_q' + 0.870D)\,]$$
$$+ \Delta g\,(0.5D_{x'} + 0.866D_{y'})$$

Optical axis deviation

$$\Delta S_{F'y'}' = \Delta\theta\left[\,1.5P_{x'}z_q' - 0.866P_{y'}z_q' + P_{z'}\left(-1.5x_q' + 0.866y_q' + 2b' - 1.368D + 4.468\frac{D}{n}\right)\right]$$
$$+ \Delta g\,(0.866D_{x'} + 1.5D_{y'})$$
$$\Delta S_{F'z'}' = \Delta\theta\left[P_{x'}\left(-2y_q' - 0.866b' + 1.620D - 3.869\frac{D}{n}\right) + P_{y'}\left(2x_q' - 1.5b' - 0.069D\right.\right.$$
$$\left.\left. - 2.234\frac{D}{n}\right)\right] + 2\Delta g D_{z'}$$

Adjustment Diagram

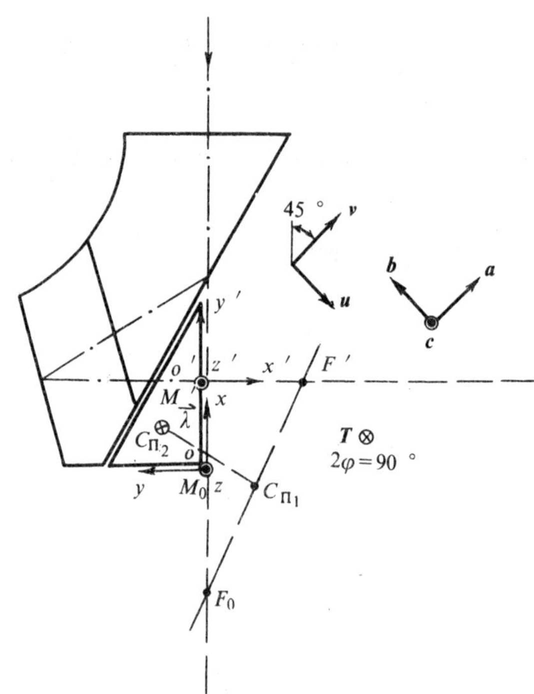

Fig . A .46 .

Characteristic Parameters of Image Formation

Direction cosine		Unit vectors of image coordinate system			$2\varphi = 90^\circ$
		$i'(x')$	$j'(y')$	$k'(z')$	T
Unit vectors of coordinate system in object space	$i(x)$	$S_{T,2\varphi} = \begin{pmatrix} 0 \\ -1 \\ 0 \end{pmatrix}$ ...	1	0	0
	$j(y)$	-1	0	0	0
	$k(z)$	0	0	1	-1

$t = 3$	$L = 2.981D$	$2d = 0$

$$\overrightarrow{M_0 M'} = -0.046D\,i' + \left(-1.445D + 2.981\,\frac{D}{n}\right)j'$$

$$r_0 = \left(0.746D - 1.491\,\frac{D}{n}\right)i' + \left(0.699D - 1.491\,\frac{D}{n}\right)j'$$

$$r_i = \left(0.746D - 1.491\,\frac{D}{n}\right)i' + \left(0.699D - 1.491\,\frac{D}{n}\right)j'$$

Characteristic Parameters of Adjustment

Direction cosine		Image coordinate system		
		x'	y'	z'
Extreme-valued axis direction of r' image rotation	u	0.707	-0.707	0
	v	0.707	0.707	0
	w	/	/	/
Extreme value of r' image rotation		$\delta_u = 1.414\Delta\theta$	$\delta_v = 1.414\Delta\theta$	$\delta_w = 0$
Extreme-valued shift direction of r' image displacement	a	0.707	0.707	0
	b	-0.707	0.707	0
	c	0	0	1
Extreme value of r' image displacement		$\delta_a = 1.414\Delta g$	$\delta_b = 1.414\Delta g$	$\delta_c = 2\Delta g$

Planar three dimensional zero-valued poles	C_{Π_1} (Mid-point of segment $\overline{F_0F'}$)	$x_1' = \dfrac{1}{2}(b' + 0.046D)$
		$y_1' = \dfrac{1}{2}\left(-b' + 1.445D - 2.981\dfrac{D}{n}\right)$
		$z_1' = 0$
		Associated plane Π_1: Conjugate optical-axis section $x'\hat{o}'y'$
	C_{Π_2}	$x_2' = 0.746D - 1.491\dfrac{D}{n}$
		$y_2' = 0.700D - 1.491\dfrac{D}{n}$
		$z_2' = 0$
		Associated Plane Π_2: Plane passing through $\vec{\lambda}$ and C_{Π_1}

Equations of Image Motion

Parallel beam (Image rotation $\Delta\vec{\mu}'$)

Image lean

$$\Delta\mu_{x'}' = \Delta\theta\,(P_{x'} - P_{y'})$$

Optical axis deviation

$$\Delta\mu_{y'}' = \Delta\theta \ (P_{x'} + P_{y'})$$

$$\Delta\mu_{z'}' = 0$$

Convergent beam (Image point displacement $\Delta S_F'$)

Parallax

$$\Delta S_{F'x'}' = \Delta\theta[P_{x'} z_q' - P_{y'} z_q' + P_{z'}(-x_q' + y_q' + 0.046D)] + \Delta g \ (D_{x'} + D_{y'})$$

Optical axis deviation

$$\Delta S_{F'y'}' = \Delta\theta \left[P_{x'} z_q' + P_{y'} z_q' + P_{z'} \left(-x_q' - y_q' + 1.445D - 2.981 \frac{D}{n} \right) \right] + \Delta g \ (-D_{x'} + D_{y'})$$

$$\Delta S_{F'z'}' = \Delta\theta \left[P_{x'} \left(-2y_q' - b' + 1.445D - 2.981 \frac{D}{n} \right) + P_{y'} \ (2x_q' - b' - 0.046D) \right] + 2\Delta g D_{z'}$$

Adjustment Diagram

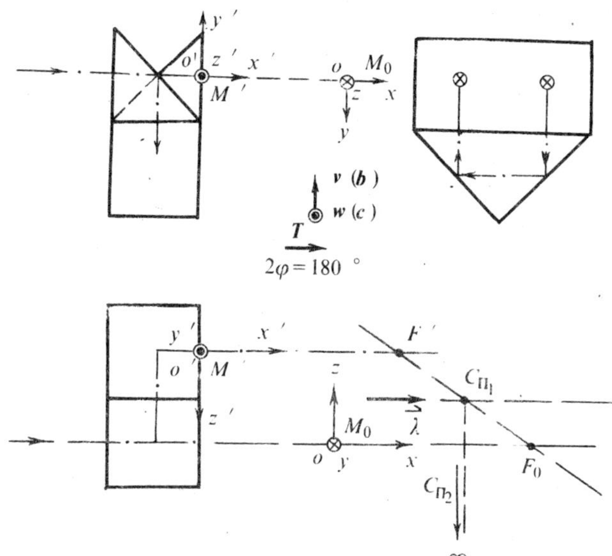

Fig . A . 47.

Characteristic Parameters of Image Formation

Direction cosine		Unit vectors of image coordinate system			$2\varphi = 180°$
		$i'(x')$	$j'(y')$	$k'(z')$	T
Unit vectors of coordinate system in object space	$i(x)$	$S_{T,2\varphi} = \begin{pmatrix} 1 \\ 0 \\ 0 \end{pmatrix}$	$0 \\ -1 \\ 0$	$0 \\ 0 \\ -1 \end{pmatrix}$	1
	$j(y)$				0
	$k(z)$				0

$t = 4$	$L = 4D$	$2d = D - 4\dfrac{D}{n}$

$$\overrightarrow{M_0 M'} = \left(D - 4\frac{D}{n} \right) i' - D k'$$

$$r_0 = 0.5 D k'$$

$$r_i =$$

Characteristic Parameters of Adjustment

	Direction cosine	Image coordinate system		
		x'	y'	z'
Extreme-valued axis direction of r' image rotation	u	/	/	/
	v	0	1	0
	w	0	0	1
Extreme value of r' image rotation		$\delta_u = 0$	$\delta_v = 2\Delta\theta$	$\delta_w = 2\Delta\theta$
Extreme-valued shift direction of r' image displacement	a	/	/	/
	b	0	1	0
	c	0	0	1
Extreme value of r' image displacement		$\delta_a = 0$	$\delta_b = 2\Delta g$	$\delta_c = 2\Delta g$

Planar three-dimensional zero-valued poles	C_{Π_1} (Mid-point of segment $\overline{F_0 F'}$)	$x_1' = \dfrac{1}{2}\left(2b' - D + 4\dfrac{D}{n}\right)$
		$y_1' = 0$
		$z_1' = \dfrac{1}{2} D$
		Associated plane Π_1: Conjugate optical-axis section $x'o'z'$
	C_{Π_2} (Approaches to infinity along the intersecting line of planes Π_1 and Π_2)	$x_2' = x_1'$
		$y_2' = 0$
		$z_2' = \infty$
		Associated plane Π_2: Plane passing through C_{Π_1} and perpendicular to $\vec{z}$

Equations of Image Motion

Parallel beam (Image rotation $\Delta\vec{\mu}'$)

Image lean

$$\Delta\mu_{x'}' = 0$$

Optical axis deviation

$$\Delta\mu_{y'}' = 2\Delta\theta\, P_{y'}$$

$$\Delta\mu_{z'}' = 2\Delta\theta\, P_{z'}$$

Convergent beam (Image point displacement $\Delta S'_{F'}$)

Parallax

$$\Delta S'_{F'_{x'}} = -\Delta\theta P_{y'} D$$

Optical axis deviation

$$\Delta S'_{F'_{y'}} = \Delta\theta\left[P_{x'}(2z'_q - D) + P_{z'}\left(-2x'_q + 2b' - D + 4\frac{D}{n}\right)\right] + 2\Delta g D_{y'}$$

$$\Delta S'_{F'_{z}} = \Delta\theta\left[-2P_{x'}y'_q + P_{y'}\left(2x'_q - 2b' + D - 4\frac{D}{n}\right)\right] + 2\Delta g D_{z'}$$

Adjustment Diagram

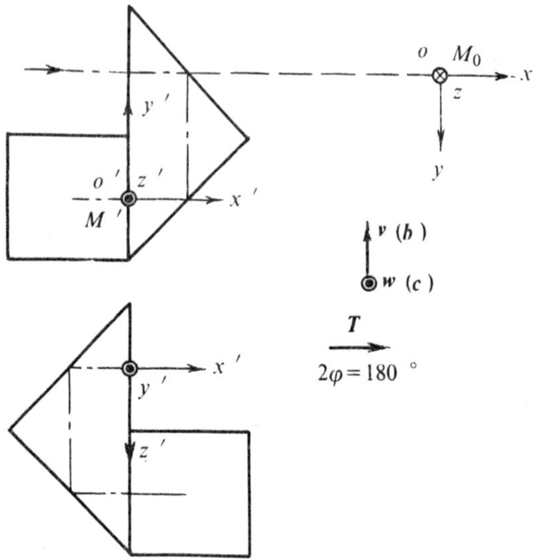

Fig . A. 48.

Characteristic Parameters of Image Formation

Direction cosine		Unit vectors of image coordinate system			$2\varphi = 180$ °
		$i'(x')$	$j'(y')$	$k'(z')$	T
Unit vectors of coordinate system in object space	$i(x)$	$S_{T.2\varphi} =$ $\begin{pmatrix} 1 \\ 0 \\ 0 \end{pmatrix}$	0 -1 0	0 0 -1 $\end{pmatrix}$	1
	$j(y)$				0
	$k(z)$				0

$t = 4$ $L = 4D$ $\qquad\qquad 2d = -4\dfrac{D}{n}$

$$\overrightarrow{M_0M'} = -4\frac{D}{n}\,i' - Dj' - Dk'$$

$$r_0 = 0.5Dj' + 0.5Dk'$$

$$r_i =$$

Characteristic Parameters of Adjustment

Direction cosine		Image coordinate system		
		x'	y'	z'
Extreme-valued axis direction of r' image rotation	u	/	/	/
	v	0	1	0
	w	0	0	1
Extreme value of r' image rotation		$\delta_u = 0$	$\delta_v = 2\Delta\theta$	$\delta_w = 2\Delta\theta$
Extreme-valued shift direction of r' image displacement	a	/	/	/
	b	0	1	0
	c	0	0	1
Extreme value of r' image displacement		$\delta_a = 0$	$\delta_b = 2\Delta g$	$\delta_c = 2\Delta g$
Planar three-dimensional zero-valued poles	C_{Π_1}	$x_1' = b' + 2\dfrac{D}{n}$		
		$y_1' = \dfrac{D}{2}$		
		$z_1' = \dfrac{D}{2}$		
		Associated plane Π_1 : Plane passing through C_{Π_1} and meeting the requirement : $P_{y'} = P_{z'}$		
	C_{Π_2} (Approaches to infinity along the intersecting line of planes Π_1 and Π_2)	$x_2' = b' + 2\dfrac{D}{n}$		
		$y_2' = \infty$		
		$z_2' = \infty$		
		Associated plane Π_2 : Plane passing through C_{Π_1} and perpendicular to $\vec{\lambda}$		

Equations of Image Motion

Parallel beam (Image rotation $\Delta\vec{\mu}'$)

Image lean

$$\Delta\mu_{x'}' = 0$$

Optical axis deviation

$$\Delta\mu_{y'}' = 2\Delta\theta P_{y'}$$

$$\Delta\mu_{z'}' = 2\Delta\theta\, P_{z'}$$

Convergent beam (Image point displacement $\Delta S_F'$)

Parallax

$$\Delta S_{F'x'}' = \Delta\theta\ (-DP_{y'} + DP_{z'})$$

Optical axis deviation

$$\Delta S_{F'y'}' = \Delta\theta\left[P_{x'}\,(2z_q' - D) + P_{z'}\left(-2x_q' + 2b' + 4\frac{D}{n}\right)\right] + 2\Delta g D_{y'}$$

$$\Delta S_{F'z'}' = \Delta\theta\left[P_{x'}\,(-2y_q'' + D) + P_{y'}\left(2x_q' - 2b' - 4\frac{D}{n}\right)\right] + 2\Delta g\, D_{z'}$$

Adjustment Diagram

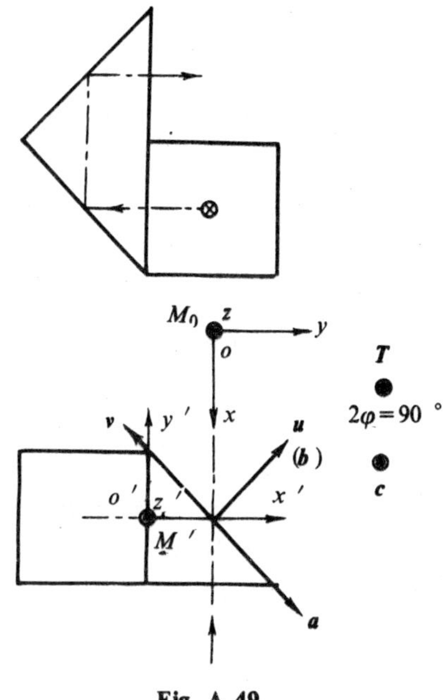

Fig . A. 49.

Characteristic Parameters of Image Formation

Direction cosine		Unit vectors of image coordinate system			$2\varphi = 90\,°$
		$i\,'(x\,')$	$j\,'(y\,')$	$k\,'(z\,')$	T
Unit vectors of coordinate system in object space	$i\,(x)$		0	-1	0	0
	$j\,(y)$	$S_{T,2\varphi} =$	1	0	0	0
	$k\,(z)$		0	0	1	1

$t = 3$	$L = 3D$	$2d = 0$

$$\overrightarrow{M_0 M\,'} = -0.5Di\,' + \left(0.5D - 3\frac{D}{n}\right)j\,' + Dk\,'$$

$$r_0 = \left(0.5D - 1.5\frac{D}{n}\right)i\,' + 1.5\frac{D}{n}j\,'$$

$$r_i = \left(0.5D - 1.5\frac{D}{n}\right)i\,' + 1.5\frac{D}{n}j\,' - 0.5Dk\,'$$

Characteristic Parameters of Adjustment

Direction cosine		Image coordinate system		
		x'	y'	z'
Extreme-valued axis direction of r' image rotation	u	0.707	0.707	0
	v	-0.707	0.707	0
	w	/	/	/
Extreme value of r' image rotation		$\delta_u = 1.414\Delta\theta$	$\delta_v = 1.414\Delta\theta$	$\delta_w = 0$
Extreme-valued shift direction of r' image displacement	a	0.707	-0.707	0
	b	0.707	0.707	0
	c	0	0	1
Extreme value of r' image displacement		$\delta_a = 1.414\Delta g$	$\delta_b = 1.414\Delta g$	$\delta_c = 2\Delta g$

Planar three-dimensional zero-valued poles (Do not exist)	C_{Π_1}	$x_1' =$
		$y_1' =$
		$z_1' =$
		Associated plane Π_1 :
	C_{Π_2}	$x_2' =$
		$y_2' =$
		$z_2' =$
		Associated plane Π_2 :

Equations of Image Motion

Parallel beam (Image rotation $\Delta\vec{\mu}'$)

Image lean

$$\Delta\mu_{x'}' = \Delta\theta\,(P_{x'} + P_{y'})$$

Optical axis deviation

$$\Delta\mu_{y'}' = \Delta\theta\,(-P_{x'} + P_{y'})$$
$$\Delta\mu_{z'}' = 0$$

Convergent beam (Image point displacement $\Delta S_F'$)

Parallax

$$\Delta S_{F'x'}' = \Delta\theta\,[P_{x'}(-z_q' - D) - P_{y'}z_q' + P_{z'}(x_q' + y_q' - 0.5D)] + \Delta g\,(D_{x'} - D_{y'})$$

Optical axis deviation

$$\Delta S'_{F'_{y'}} = \Delta\theta \left[P_{x'} z'_q + P_{y'} (-z'_q - D) + P_{z'} \left(-x'_q + y'_q + 0.5D - 3\frac{D}{n} \right) \right] + \Delta g \, (D_{x'} + D_{y'})$$

$$\Delta S'_{F'_z} = \Delta\theta \left[P_{x'} \left(-2y'_q + b' - 0.5D + 3\frac{D}{n} \right) + P_{y'} (2x'_q - b' - 0.5D) \right] + 2\Delta g \, D_{z'}$$

Adjustment Diagram

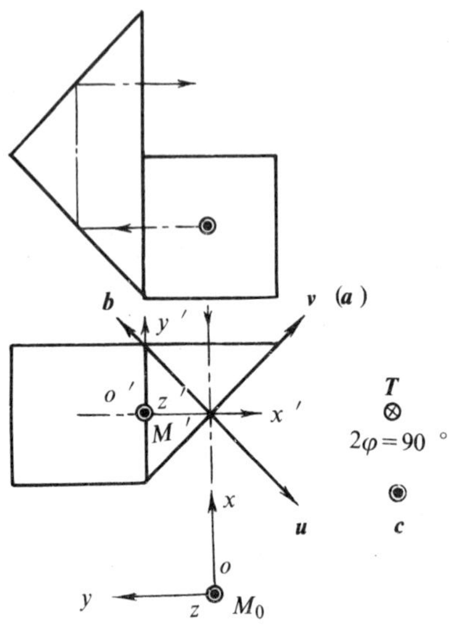

Fig. A.50.

Characteristic Parameters of Image Formation

Direction cosine		Unit vectors of image coordinate system			$2\varphi = 90\,°$
		$i\,'(x\,')$	$j\,'(y\,')$	$k\,'(z\,')$	T
Unit vectors of coordinate system in object space	$i\,(x)$	$S_{T,2\varphi} = \begin{pmatrix} 0 & 1 & 0 \\ -1 & 0 & 0 \\ 0 & 0 & 1 \end{pmatrix}$	1	0	0
	$j\,(y)$		0	0	0
	$k\,(z)$		0	1	-1

$t = 3$	$L = 3D$	$2d = 0$

$$\overrightarrow{M_0 M'} = -0.5Di\,' + \left(-0.5D + 3\frac{D}{n}\right)j\,' + Dk\,'$$

$$r_0 = \left(0.5D - 1.5\frac{D}{n}\right)i\,' - 1.5\frac{D}{n}\,j\,'$$

$$r_i = \left(0.5D - 1.5\frac{D}{n}\right)i\,' - 1.5\frac{D}{n}\,j\,' - 0.5Dk\,'$$

Characteristic Parameters of Adjustment

	Direction cosine	Image coordinate system		
		x'	y'	z'
Extreme-valued axis direction of r' image rotation	u	0.707	-0.707	0
	v	0.707	0.707	0
	w	/	/	/
Extreme value of r' image rotation		$\delta_u = 1.414\Delta\theta$	$\delta_v = 1.414\Delta\theta$	$\delta_w = 0$
Extreme-valued shift direction of r' image displacement	a	0.707	0.707	0
	b	-0.707	0.707	0
	c	0	0	1
Extreme value of r' image displacement		$\delta_a = 1.414\Delta g$	$\delta_b = 1.414\Delta g$	$\delta_c = 2\Delta g$

Planar three-dimensional zero-valued poles (Do not exist)	C_{Π_1}	$x_1' =$
		$y_1' =$
		$z_1' =$
		Associated plane Π_1 :
	C_{Π_2}	$x_2' =$
		$y_2' =$
		$z_2' =$
		Associated plane Π_2 :

Equations of Image motion

Parallel beam (Image rotation $\Delta\vec{\mu}'$)

Image lean

$$\Delta\mu_{x'}' = \Delta\theta\,(P_{x'} - P_{y'})$$

Optical axis deviation

$$\Delta\mu_{y'}' = \Delta\theta\,(P_{x'} + P_{y'})$$

$$\Delta\mu_{z'}' = 0$$

Convergent beam (Image point displacement $\Delta S_F'$)

Parallax

$$\Delta S_{F'x'}' = \Delta\theta\,[\,P_{x'}\,(z_q' + D) - P_{y'}\,z_q' + P_{z'}\,(-x_q' + y_q' + 0.5D)\,] + \Delta g\,(D_{x'} + D_{y'})$$

Optical axis deviation

$$\Delta S'_{F'_{y'}} = \Delta\theta\left[P_{x'}z'_q + P_{y'}\,(z'_q + D) + P_{z'}\left(-x'_q - y'_q + 0.5D - 3\frac{D}{n}\right)\right] + \Delta g\,(-D_{x'} + D_{y'})$$

$$\Delta S'_{F'_{z'}} = \Delta\theta\left[P_{x'}\left(-2y'_q - b' + 0.5D - 3\frac{D}{n}\right) + P_{y'}\,(2x'_q - b' - 0.5D)\right] + 2\Delta g D_{z'}$$

Table A.51 $FB_J - 0\,°$

Adjustment Diagram

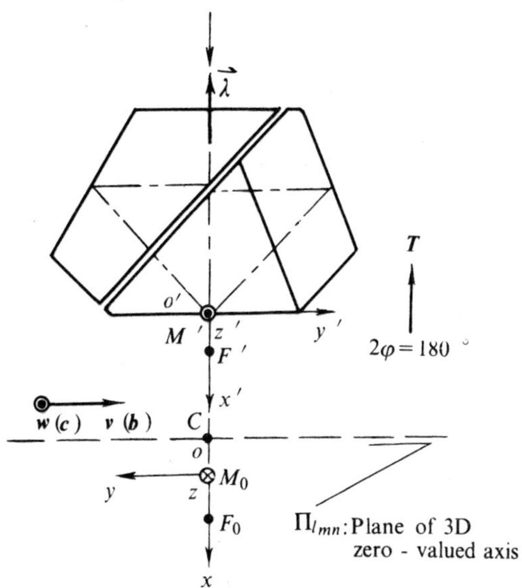

Fig.A.51.

Characteristic Parameters of Image Formation

Direction cosine		Unit vectors of image coordinate system			$2\varphi = 180\,°$
		$i'(x')$	$j'(y')$	$k'(z')$	T
Unit vectors of coordinate system in object space	$i(x)$	$S_{T,2\varphi} = \begin{pmatrix} 1 & 0 & 0 \\ 0 & -1 & 0 \\ 0 & 0 & -1 \end{pmatrix}$			-1
	$j(y)$				0
	$k(z)$				0

$t-6$	$L = 5.156D$	$2d = -1.299D + 5.156\dfrac{D}{n}$

$$\overrightarrow{M_0 M'} = \left(1.299D - 5.156\frac{D}{n}\right)i'$$

$$r_0 = 0$$

$$r_i =$$

Characteristic Parameters of Adjustment

Direction cosine		Image coordinate system		
		x'	y'	z'
Extreme-valued axis direction of r' image rotation	u	/	/	/
	v	0	1	0
	w	0	0	1
Extreme value of r' image rotation		$\delta_u = 0$	$\delta_v = 2\Delta\theta$	$\delta_w = 2\Delta\theta$
Extreme-valued shift direction of r' image displacement	a	/	/	/
	b	0	1	0
	c	0	0	1
Extreme value of r' image displacement		$\delta_a = 0$	$\delta_b = 2\Delta g$	$\delta_c = 2\Delta g$

Spatial three-dimensional zero-valued pole and plane of three-dimensional zero-valued axis	C (Mid-point of segment $\overline{F_0F'}$)	$x' = \dfrac{1}{2}\left(2b' - 1.299D + 5.156\dfrac{D}{n}\right)$
		$y' = 0$
		$z' = 0$
	Π_{lmn}	Passing through C and perpendicular to $\vec{\lambda}$

Equations of Image motion

Parallel beam (Image rotation $\Delta\vec{\mu}'$)

Image lean

$$\Delta\mu_{x'}' = 0$$

Optical axis deviation

$$\Delta\mu_{y'}' = 2\Delta\theta P_{y'}$$

$$\Delta\mu_{z'}' = 2\Delta\theta P_{z'}$$

Convergent beam (Image point displacement $\Delta S_{F'}'$)

Parallax

$$\Delta S_{F'x'}' = 0$$

Optical axis deviation

$$\Delta S'_{F'_{y'}} = \Delta\theta \left[2P_{x'} z'_q + P_{z'} \left(-2x'_q + 2b' - 1.299D + 5.156\frac{D}{n} \right) \right] + 2\Delta g\, D_{y'}$$

$$\Delta S'_{F'_{z'}} = \Delta\theta \left[-2P_{x'} y'_q + P_{y'} \left(2x'_q - 2b' + 1.299D - 5.156\frac{D}{n} \right) \right] + 2\Delta g D_{z'}$$

Adjustment Diagram

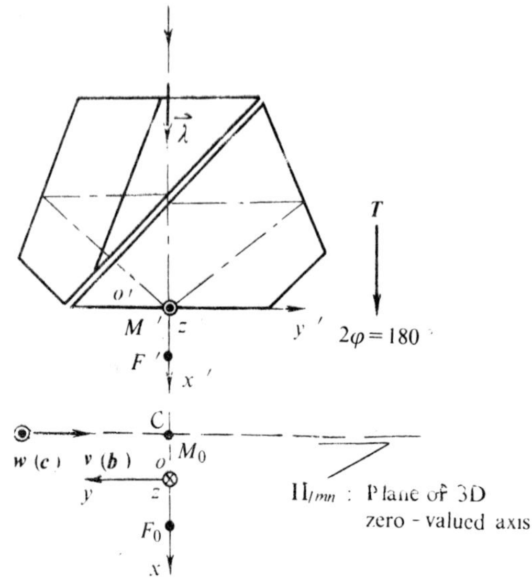

Fig . A .52 .

Characteristic Parameters of Image Formation

Direction cosine		Unit vectors of image coordinate system			$2\varphi = 180$ °
		$i'(x')$	$j'(y')$	$k'(z')$	T
Unit vectors of coordinate system in object space	$i(x)$	$S_{T.2\varphi} = \begin{pmatrix} 1 \\ \\ 0 \\ \\ 0 \end{pmatrix}$	$\begin{matrix} 0 \\ \\ -1 \\ \\ 0 \end{matrix}$	$\begin{pmatrix} 0 \\ \\ 0 \\ \\ -1 \end{pmatrix}$	1
	$j(y)$				0
	$k(z)$				0

$t = 6$	$L = 5.714D$	$2d = 1.492D - 5.714 \dfrac{D}{n}$

$$\overrightarrow{M_0 M'} = \left(1.492D - 5.714 \frac{D}{n} \right) i'$$

$$r_0 = 0$$

$$r_i =$$

Characteristic Parameters of Adjustment

Direction cosine		Image coordinate system		
		x'	y'	z'
Extreme-valued axis direction of r' image rotation	u	/	/	/
	v	0	1	0
	w	0	0	1
Extreme value of r' image rotation		$\delta_u = 0$	$\delta_v = 2\Delta\theta$	$\delta_w = 2\Delta\theta$
Extreme-valued shift direction of r' image displacement	a	/	/	/
	b	0	1	0
	c	0	0	1
Extreme value of r' image displacement		$\delta_a = 0$	$\delta_b = 2\Delta g$	$\delta_c = 2\Delta g$
Spatial three-dimensional zero-valued pole and plane of three-dimensional zero-valued axis	C (Mid-point of segment $\overline{F_0 F'}$)	$x' = \dfrac{1}{2}\left(2b' - 1.492D + 5.174\dfrac{D}{n}\right)$ $y' = 0$ $z' = 0$		
	Π_{lmn}	Passing through C and perpendicular to $\vec{\lambda}$		

Equations of Image Motion

Parallel beam (Image rotation $\Delta\vec{\mu}'$)

Image lean

$$\Delta\mu'_{x'} = 0$$

Optical axis deviation

$$\Delta\mu'_{y'} = 2\Delta\theta P_{y'}$$

$$\Delta\mu'_{z'} = 2\Delta\theta P_{z'}$$

Convergent beam (Image point displacement $\Delta S'_{F'}$)

Parallax

$$\Delta S'_{F'x'} = 0$$

Optical axis deviation

$$\Delta S'_{F'_{y'}} = \Delta\theta\left[2P_{x'}z'_q + P_{z'}\left(-2x'_q + 2b' - 1.492D + 5.714\frac{D}{n}\right)\right] + 2\Delta g\,D_{y'}$$

$$\Delta S'_{F'_{z'}} = \Delta\theta\left[-2P_{x'}y'_q + P_{y'}\left(2x'_q - 2b' + 1.492D - 5.714\frac{D}{n}\right)\right] + 2\Delta g D_{z'}$$

Adjustment Diagram

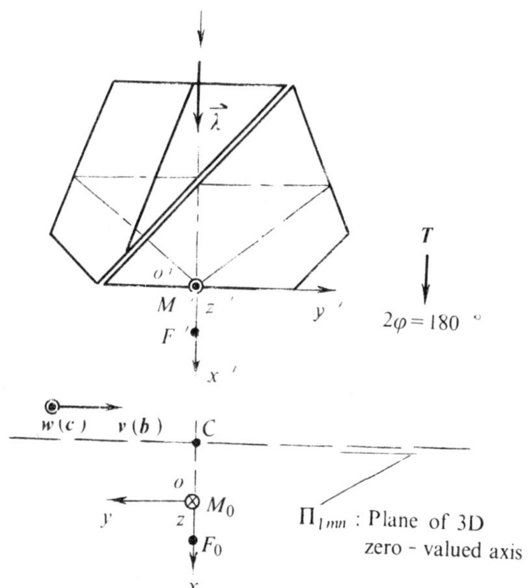

Fig . A . 53 .

Characteristic Parameters of Image Formation

Direction cosine		Unit vectors of image coordinate system			$2\varphi = 180\ ^\circ$
		$i'(x')$	$j'(y')$	$k'(z')$	T
Unit vectors of coordinate system in object space	$i(x)$	$S_{T,2\varphi} =$ $\begin{pmatrix} 1 \\\\ 0 \\\\ 0 \end{pmatrix}$	$\begin{pmatrix} 0 \\\\ -1 \\\\ 0 \end{pmatrix}$	$\begin{pmatrix} 0 \\\\ 0 \\\\ -1 \end{pmatrix}$	1
	$j(y)$				0
	$k(z)$				0

$t = 6$	$L = 5.312D$	$2d = 1.326D - 5.312\dfrac{D}{n}$

$$\overrightarrow{M_0 M'} = \left(1.326D - 5.312\frac{D}{n}\right) i'$$

$$r_0 = 0$$

$$r_i =$$

Characteristic Parameters of Adjustment

Direction cosine		Image coordinate system		
		x'	y'	z'
Extreme-valued axis direction of r' image rotation	u	/	/	/
	v	0	1	0
	w	0	0	1
Extreme value of r' image rotation		$\delta_u = 0$	$\delta_v = 2\Delta\theta$	$\delta_w = 2\Delta\theta$
Extreme-valued shift direction of r' image displacement	a	/	/	/
	b	0	1	0
	c	0	0	1
Extreme value of r' image displacement		$\delta_a = 0$	$\delta_b = 2\Delta g$	$\delta_c = 2\Delta g$
Spatial three-dimensional zero-valued pole and plane of three-dimensional zero-valued axis	C (Mid-point of segment $\overline{F_0 F'}$)	$x' = \dfrac{1}{2}\left(2b' - 1.326D + 5.312\dfrac{D}{n}\right)$ $y' = 0$ $z' = 0$		
	Π_{lmn}	Passing through C and perpendicular to $\vec{\lambda}$		

Equations of Image Motion

Parallel beam (Image rotation $\Delta\vec{\mu}'$)

Image lean

$$\Delta\mu'_{x'} = 0$$

Optical axis deviation

$$\Delta\mu'_{y'} = 2\Delta\theta P_{y'}$$

$$\Delta\mu'_{z'} = 2\Delta\theta P_{z'}$$

Convergent beam (Image point displacement $\Delta S'_{F'}$)

Parallax

$$\Delta S'_{F'x'} = 0$$

Optical axis deviation

$$\Delta S'_{F'y'} = \Delta\theta\left[2P_{x'}z'_q + P_{z'}\left(-2x'_q + 2b' - 1.326D + 5.312\frac{D}{n}\right)\right] + 2\Delta g\, D_{y'}$$

$$\Delta S'_{F'z'} = \Delta\theta\left[-2P_{x'}y'_q + P_{y'}\left(2x'_q - 2b' + 1.326D - 5.312\frac{D}{n}\right)\right] + 2\Delta g\, D_{z'}$$

Adjustment Diagram

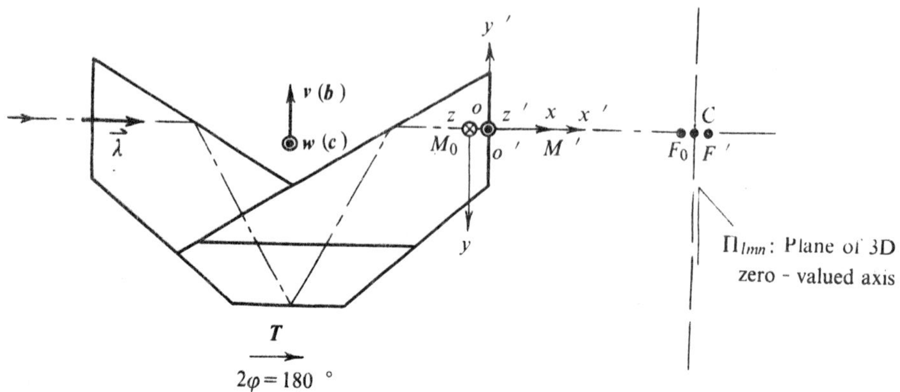

Fig . A . 54.

Characteristic Parameters of Image Formation

Direction cosine		Unit vectors of image coordinate system			$2\varphi = 180\,^\circ$
		$i\,'(x\,')$	$j\,'(y\,')$	$k\,'(z\,')$	T
Unit vectors of coordinate system in object space	$i\,(x)$	$S_{T,2\varphi} = \begin{pmatrix} 1 \\ 0 \\ 0 \end{pmatrix}$	$\begin{pmatrix} 0 \\ -1 \\ 0 \end{pmatrix}$	$\begin{pmatrix} 0 \\ 0 \\ -1 \end{pmatrix}$	1
	$j\,(y)$				0
	$k\,(z)$				0

$t = 4$	$L = 5.196D$	$2d = 3.464D - 5.196\dfrac{D}{n}$

$$\overrightarrow{M_0 M}\,' = \left(3.464D - 5.196\frac{D}{n}\right)i\,'$$

$$r_0 = 0$$

$$r_i =$$

Characteristic Parameters of Adjustment

<table>
<tr><td colspan="2" rowspan="2">Direction cosine</td><td colspan="3">Image coordinate system</td></tr>
<tr><td>x'</td><td>y'</td><td>z'</td></tr>
<tr><td rowspan="3">Extreme-valued axis direction of r' image rotation</td><td>u</td><td>/</td><td>/</td><td>/</td></tr>
<tr><td>v</td><td>0</td><td>1</td><td>0</td></tr>
<tr><td>w</td><td>0</td><td>0</td><td>1</td></tr>
<tr><td colspan="2">Extreme value of r' image rotation</td><td>$\delta_u = 0$</td><td>$\delta_v = 2\Delta\theta$</td><td>$\delta_w = 2\Delta\theta$</td></tr>
<tr><td rowspan="3">Extreme-valued shift direction of r' image displacement</td><td>a</td><td>/</td><td>/</td><td>/</td></tr>
<tr><td>b</td><td>0</td><td>1</td><td>0</td></tr>
<tr><td>c</td><td>0</td><td>0</td><td>1</td></tr>
<tr><td colspan="2">Extreme value of r' image displacement</td><td>$\delta_a = 0$</td><td>$\delta_b = 2\Delta g$</td><td>$\delta_c = 2\Delta g$</td></tr>
<tr><td rowspan="4">Spatial three-dimensional zero-valued pole and plane of three-dimensional zero-valued axis</td><td rowspan="3">C (Mid-point of segment $\overline{F_0 F'}$)</td><td colspan="3">$x' = \dfrac{1}{2}\left(2b' - 3.464D + 5.196\dfrac{D}{n}\right)$</td></tr>
<tr><td colspan="3">$y' = 0$</td></tr>
<tr><td colspan="3">$z' = 0$</td></tr>
<tr><td>Π_{lmn}</td><td colspan="3">Passing through C and perpendicular to $\vec{\lambda}$</td></tr>
</table>

Equations of Image Motion

Parallel beam (Image rotation $\Delta\vec{\mu}'$)

Image lean

$$\Delta\mu'_{x'} = 0$$

Optical axis deviation

$$\Delta\mu'_{y'} = 2\Delta\theta P_{y'}$$

$$\Delta\mu'_{z'} = 2\Delta\theta P_{z'}$$

Convergent beam (Image point displacement $\Delta S_{F'}'$)

Parallax

$$\Delta S_{F'\,x'}' = 0$$

Optical axis deviation

$$\Delta S'_{F'y'} = \Delta\theta\left[2P_{x'}z'_q + P_{z'}\left(-2x'_q + 2b' - 3.464D + 5.196\frac{D}{n}\right)\right] + 2\Delta g D_{y'}$$

$$\Delta S'_{F'z'} = \Delta\theta\left[-2P_{x'}y'_q + P_{y'}\left(2x'_q - 2b' + 3.464D - 5.196\frac{D}{n}\right)\right] + 2\Delta g\, D_{z'}$$

Program

The program enclosed here is used to compute most of the data needed for tabulation of reflecting prisms*. This program has been written with the so-called DIJ-8 FORTRAN language, which is not appreciably different from the commonly used FORTRAN, and therefore, is easy to understand.

The input data are the values of each basic characteristic parameter of image formation for a reflecting prism, i.e., t, T_x, T_y, T_z, 2φ, A_{01}, A_{02}, A_{03}, B_{01}, B_{02}, and B_{03}, of which the correspoding specified variables in the program are T, TX, TY, TZ, TWOFAI, A(1), A(2), A(3), B(1), B(2), and B(3), respectively.

It should be noticed that when writing this program the vector connecting two basic conjugate points was expressed in a slightly different way:

$$\overrightarrow{M\,'M_0} = \left(A_{01} \cdot D + B_{01} \cdot \frac{D}{n}\right) i\,' + \left(A_{02} \cdot D + B_{02} \cdot \frac{D}{n}\right) j\,'$$

$$+ \left(A_{03} \cdot D + B_{03} \cdot \frac{D}{n}\right) k\,' . \tag{P.1}$$

Then, considering $\overrightarrow{M\,'M_0} = -\overrightarrow{M_0M\,'}$ and comparing Eqs. (P.1) and (2.19), we have

$$A_{01} = -A_1 , \qquad A_{02} = -A_2 , \qquad A_{03} = -A_3 ,$$

$$B_{01} = -B_1 , \qquad B_{02} = -B_2 , \qquad B_{03} = -B_3 .$$

The program will be listed below without additional explanations, those wishing to know more about it may consult the reference [2].

```
       PAGE 1
       SUBROUTINE FRP (X, Y, Z)
       DIMENSION X (3, 3), Y (3), Z (3)
       DO 101 I = 1, 3
       Z (I) = 0. 0
       DO 101 J = 1, 3
101    Z (I) = Z (I) + X (I, J) * Y (J)
       RETURN
       END
       SUBROUTINE FXYZ (X, Y, Z)
       DIMENSION X (3), Y (3), Z (3)
       Z (1) = X (2) * Y (3) − X (3) * Y (2)
```

---

* To obtain all the necessary data, a little extension of this program is enough.

```
      Z(2) = X(3)*Y(1) - X(1)*Y(3)
      Z(3) = X(1)*Y(2) - X(2)*Y(1)
      RETURN
      END
      SUBROUTINE SF(X, P, A , B, G, OSF)
      INTEGER T
      DIMENSION X(6), P(3), A(3), B(3), G(3), OSF(3), AMO(3), D(3), RP(3),
     RAM(3), ERP(3), ETRG(3), H(3), RPRA(3), Q(3), E(3,3), R(3,3),
     C(3), ER(3,3), ETR(3 ,3)
      PUBLIC E, R, T, DELTST, ER, ETR
      C = 0. 0
      C(1) = 1. 0
      DO 111  I = 1,3
      AMO(I) = A(I) * X(5) + B(I) * X(6) - X(I)
111   D(I) = C(I) * X(4)
      CALL FRP(R, P, RP)
      CALL FRP(R, AMO, RAM)
      CALL FRP(ER, P, ERP)
      CALL FRP(ETR, G, ETRG)
      CALL FXYZ(P, X, H)
      CALL FXYZ(RP, RAM , RPRA)
      CALL FXYZ(ERP, D ,Q)
      DO 133  I = 1,3
133   OSF(I) = DELTST * (-H(I) + (-1)**(T-1) * RPRA(I) + Q(I) + ETRG(I)
      RETURN
      END
      PAGE  2
      MASTER MUISFK
      INTEGER MA, T
      REAL LEN, AS, DELTST, FTF, BS, BC
      DIMENSION A(3), B(3), OSF1(3), SFG1(3), SFG(3, 3), UVW(3, 3), ER
     (3, 3), ERP(3, 3), OSF(3,3,6), P(3), G(3), DL(3), DELT(3), X(6), E(3,3),
     R(3,3), ETR(3,3)
      PUBLIC  MA, E, R, T , DELTST, ER, ETR
      N = 0
      E = 0.0
      DO 2  I = 1,3
2     E (I,I) = 1, 0
99    READ  (1) T, LEN, TX, TY, TZ, TWOFAI, A, B, PRISM
      FTF = TWOFAI * 3. 14159/180. 0
      AS = (SIN(FTF/2. 0)) ** 2 * 2. 0
```

```
      BS = SIN (FTF )
      BC = COS (FTF )
      R (1,1 ) = BC + TX*TX*AS
      R (1,2 ) = - TZ*BS + TX*TY*AS
      R (1,3 ) = TY*BS + TX*TZ*AS
      R (2,1 ) = TZ*BS + TX*TY*AS
      R (2,2 ) = BC + TY*TY*AS
      R (2,3 ) = - TX*BS + TY*TZ*AS
      R (3,1 ) = - TY*BS + TX*TZ*AS
      R (3,2 ) = TX * BS + TY*TZ*AS
      R (3,3 ) = BC + TZ*TZ*AS
      DO  16  I = 1 , 3
      DL   (I ) = 0 . 0
      DO  1 6  J = 1, 3
      ER (I, J ) = E (I, J ) - R (I, J )
      ETR (I, J ) = E (I, J ) + ( - 1 ) * * (T - 1 ) * R (I, J )
   16 DL (I ) = DL (I ) + ER (I, J ) ** 2
      DO  19  I = 1, 3
      DO  19  J = 1, 3
      DELT (I ) = SQRT (DL (I ) )
      IF (.00004 - DELT (I ) )18, 18, 17
   17 UVW (I, J ) = 0.0
      DELT (I ) = 0.0
      GOTO  19
   18 UVW (I,J ) = ER (I, J )/DELT (I )
   19 CONTINUE
      DELTST = 1.0
      G = 0.0
      DO  31  J = 1, 3
      DO  31  K = 1, 6
      X = 0.0
      P = 0.0
      P   (J ) = 1.0
      X   (K ) = 1.0
      CALL SF (X, P, A, B, G, OSF1 )
      DO  31  I = 1, 3
   31 OSF (I, J, K ) = OSF1 (I )
      DELTST  = 0.0
      X = 0.0
      P = 0.0
      DO  32  J = 1, 3
```

```
      G = 0.0
      G (J) = 1.0
      CALL SF (X, P, A, B, G, SFG1)
      DO 32  I = 1,3
32    SFG (I, J) = SFG1 (I)
      PAGE  3
      WRITE   (2,5) PRISM
5     FORMAT  (///50X, 7HPRISM △   (,A8, 1H )/3  OX, 20  (5H ★ ★ ★ ★ ★ ) )
      WRITE (2, 10 ) T, LEN, TWOFAI, TX, TY, T Z, (A (I ), B (I ), I = 1, 3 )
10    FORMAT (/5X, 7H △ GIVEN:, 6X, 2HT = , I2/ 18X, 2HL = , F9.5/17X,
      3H2O = , F8.2/17X,  3HTX = , F9.5,6X, 3HTY = ,F9.5, 6X, 3HTZ = , F9.5/10X,
      7HM, MO = D (,F9.5,3H △ + △ , F9 . 5, 9H /N )I △ + △ D (,F9.5, 3H △ +
      △ ,  F9.5, 9H /N ) J △ + △ D (, F9.5, 3H △ + △ , F9.5,4H /N ) K )
      WRITE (2, 15 ) ( (R (I,J), J = 1, 3 ), I = 1, 3 )
15    FORMAT (/5X,  20HTRANSFOR △ - △ MATRIX △ R:, 3(/20X, 3F15.5 ) )
      WRITE (2, 20 )UVW
20    FORMAT (/5X, 23HCHARACTERI △ - △ DATA △ △ UVW:, 10X, 1HU,
      11X, 1HV, 11X, 1HW,/30 X, 1HI, 3F12.5/30X, 1HJ, 3F12.5/30X, 1 HK, 3F
      12.5/)
      WRITE (2, 25 )DELT
25    FORMAT (25X, 3HOX = ,F10.5, 7X, 3HOY = , F 10.5, 7X, 3HOZ = , F10.5/20X,
      50HPROBL EM △ OF △ PARALLEL △ RAYS △ (DEVIATION △ OF △
      IMAGE △ MU △ △ ) )
      WRITE (2,30 ) ((ER(I,J), J = 1, 3 ), I = 1 ,3 )
30    FORMAT(/5X, 10HIMAGE △ LEAN, 8H △ MX △ = △ O (, F10.5, 2HPX,
      3H △ + △ , F10.5, 2HPY, 3 H △ + △ ,F10.5,3HPZ )/5X, 9HDEVIATION ,
      1 X, 8H △ MY △ = △ O (, F10.5, 2HPX, 3H △ + △ , F10 . 5,2HPY, 3H △
      + △ , F10.5, 3HPZ )/15X,8H △ M Z △ = △ O (, F10.5, 2HPX, 3H △ +
      △ , F10.5, 2HP Y, 3H △ + △ ,F10.5, 3HPZ )/20X, 60HPROBLE M △ OF
      △ CONVERGENT △ RAYS (DISPLACEMENT △ OF △ IMAGE △POINT
      △ △ SF △ ) )
      WRITE (2,35 ) ((OSF(1, J, K), K = 1,6), J = 1,3 ), (SFG(1,J), J = 1,3 )
35    FORMAT (/5X,8HPARALLAX, 3X,12H .  SFX △ = △ O (PX (,F9.5, 5HXQ
      △ + △ ,F9.5, 5HYQ △ + △ , F9.5, 5HZQ △ + △ ,F9.5, 5HB △ △ + △ ,
      F9.5, 5HD △ △ + △ , F9.5, 6HD/N △ △ )/ 25X, 3HPY (, 6 (F9.5, 5X), 1H ),
      /25X, 3HPZ (, 6(F9.5, 5 X), 2H )),/28X, F9.5, 5H △ GX+ △ , F9.5, 5H △
      GY+ △ ,F9.5, 3H △ GZ )
      WRITE (2,40 ) ((OSF (2,J,K), K = 1,6), J = 1,3 ), (SFG(2,J), J = 1,3 )
40    FORMAT (/5X, 9HDEVIATION, 2X, 12H △ SFY △ = △ O (PX (, F9.5,
      5HXQ △ + △ , F9.5, 5HYQ △ + △ , F9.5, 5HZQ △ + △ , F9.5, 5HB △
      △ + △ ,F9.5, 5HD △ △ + △ ,F9.5, 6HD/N △ △ )/ 25X, 3HPY (, 6 (F9.5,
```

```
      5X),1H),/25X,3HPZ (,6 (F9.5, 5X) , 2H)),/28X,F9.5,5H△ GX+ △ ,F9.5,
      5H△ GY+ △ , F9.5,3H△ GZ)
      WRITE (2,45) ((OSF(3,J,K),K = 1,6),J = 1,3),  (SFG(3,J), J = 1,3)
   45 FORMAT (/16X, 12H△ SFZ△ = △O  (PX (,F9.5,  5HXQ△ + △ ,F9.5,
      5HYQ△ + △ ,F9.5,5HZQ△ + △ , F9.5, 5HB△ △ + △ , F9.5,5HD△
      △ + △ ,F9.5, 6HD/N△△ )/25X,3HPY (,6(F9.5, 5X), 1H), /25X, 3HPZ
      (,6(F9.5,5X),2H)),/28X, F9.5, 5H△ GX+ △ ,F9.5, 5H△ GY+ △ , F9.5,
      3H△ GZ, )
      N = N + 1
      IF (N − 2) 50, 60, 60
   50 PAUSE
   60 CONTINUE
      IF (N − MA) 99, 100, 100
  100 STOP
      END
      FINISH
```

Solutions to Selected Problems

Chapter 1

1.2 Using the result obtained from Prob.1.1

$$(A_p) = A \cdot P \; (P) = (P)(A \cdot P)$$
$$= (P)(P)'(A) .$$

On the other hand, by definition

$$(A_p) = M_P (A) .$$

Comparing the above two equations gives

$$M_P = (P)(P)'$$

$$= \begin{pmatrix} P_x \\ P_y \\ P_z \end{pmatrix} (P_x \quad P_y \quad P_z) ,$$

therefore

$$M_P = \begin{pmatrix} P_x^2 & P_x P_y & P_x P_z \\ P_y P_x & P_y^2 & P_y P_z \\ P_z P_x & P_z P_y & P_z^2 \end{pmatrix} . \tag{S1.1}$$

1.3 The vector product $D = A \times B$ can first be written in its expanded form

$$D_x i + D_y j + D_z k$$
$$= (A_y B_z - A_z B_y) i + (A_z B_x - A_x B_z) j + (A_x B_y - A_y B_x) k ,$$

and thus

$$D_x = 0 \cdot B_x - A_z B_y + A_y B_z ,$$
$$D_y = A_z B_x + 0 \cdot B_y - A_x B_z ,$$
$$D_z = - A_y B_x + A_x B_y + 0 \cdot B_z .$$

Then, by the definition of matrix multiplication the above set of linear equations can also be written

$$(D) = C_A (B) ,$$

where

$$C_A = \begin{pmatrix} 0 & -A_z & A_y \\ A_z & 0 & -A_x \\ -A_y & A_x & 0 \end{pmatrix}.$$

(S1.2)

1.5 Assume that the Hooke's joint 3, as shown in Fig. S1.1, is represented by two unit vectors A and B rigidly connected at 90° to each other.

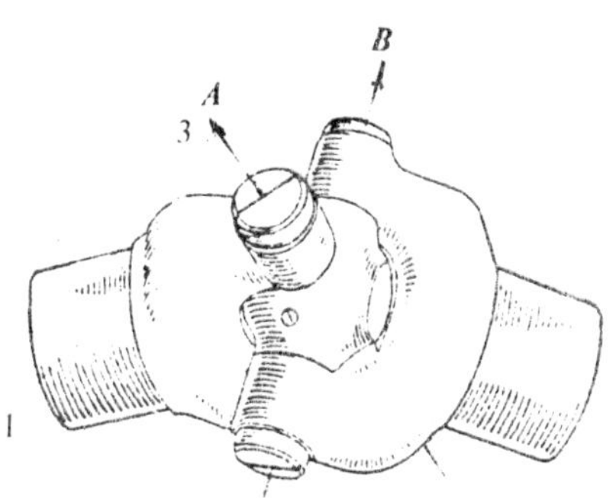

Fig.S1.1. A Hooke's joint connecting two shafts intersecting at angle

The plane view with a part of side view of the mechanism, as shown in Fig.S1.2, indicates

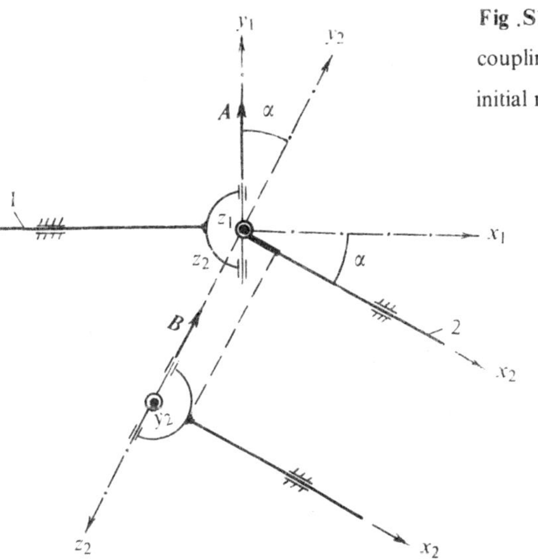

Fig.S1.2. Relative position of the coupling member and two shafts at the initial moment when $\theta = 0$ and $\varphi = 0$

the relative positions of the coupling member and two shafts at the initial moment when $\theta = 0$ and $\varphi = 0$. Assume two fixed coordinate systems $x_1 y_1 z_1$ and $x_2 y_2 z_2$. After the driving shaft 1 and the driven shaft 2 have rotated through respective angles θ and φ, the two unit vectors A

431

and B will be expressed in terms of their components relative to the coordinate systems $x_1 y_1 z_1$ and $x_2 y_2 z_2$, respectively

$$(A)_1 = \begin{pmatrix} A_{x_1} \\ A_{y_1} \\ A_{z_1} \end{pmatrix} = \begin{pmatrix} 0 \\ \cos\theta \\ \sin\theta \end{pmatrix} .$$

$$(B)_2 = \begin{pmatrix} B_{x_2} \\ B_{y_2} \\ B_{z_2} \end{pmatrix} = \begin{pmatrix} 0 \\ \sin\varphi \\ -\cos\varphi \end{pmatrix} .$$

By the transformation of coordinates of the vector B,

$$(B)_1 = \begin{pmatrix} B_{x_1} \\ B_{y_1} \\ B_{z_1} \end{pmatrix} = \begin{pmatrix} \cos\alpha & \sin\alpha & 0 \\ -\sin\alpha & \cos\alpha & 0 \\ 0 & 0 & 1 \end{pmatrix} \begin{pmatrix} 0 \\ \sin\varphi \\ -\cos\varphi \end{pmatrix} .$$

Thus,

$$(B)_1 = \begin{pmatrix} \sin\alpha \sin\varphi \\ \cos\alpha \sin\varphi \\ -\cos\varphi \end{pmatrix} .$$

Since A and B are orthogonal to each other, we have

$$(A)_1' (B)_1 = 0 .$$

Thus,

$$\cos\theta \cos\alpha \sin\varphi - \sin\theta \cos\varphi = 0 ,$$

and finally we obtain

$$\text{tg}\,\varphi = \frac{1}{\cos\alpha} \text{tg}\,\theta .$$

Chapter 2

2.1 The observer will see the optical tunnel for a Penta prism in its reduced form, i.e. the picture resulting from the whole process including both the unfolding and the reducing.

2.2 The observer will see the exit face $g''h''$ of a Penta prism at the exit face $g_0 h_0$ of its equivalent air plate.

2.3 We can find out the object and image relationship for a prism by making use of the unfolding and reducing processes simply because such a relationship is implied in these processes.

2.4 The concept of the equivalent air thickness is derived traditionally in a straight way for the purpose of determining prism sizes without showing explicitly the object and image relationship for the prism while the equivalent air plate is so treated in this text that the exit face of the equivalent air plate and the exit face of the prism are found to be a pair of object and image planes with respect to the prism. Since such a conjugate relationship can also be applied to calculating prism sizes, then why not try to kill two birds with one stone? And actually, only the object and image relationship for a prism can make the calculation of its sizes well founded.

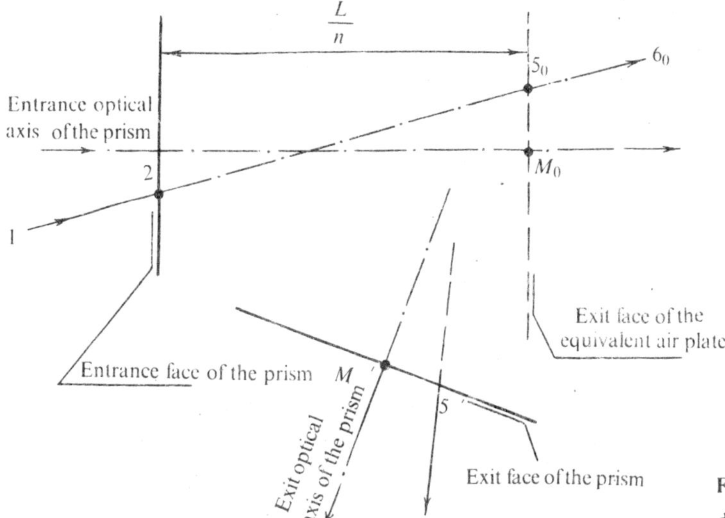

Fig. S2.1. A diagram for determining prism sizes

2.5 In Fig. S2.1, let 12 ···5′6′ be the path of a marginal ray passing through the prism and cutting its exit face at the extremity 5′. Thus, the incident ray 12 will determine the limiting aperture of the prism. In fact, we hardly ascertain which one of the bounding rays will ultimately determine the prism size by examining the real ray paths in the prism since refractions and reflections of rays happening during the passage of them through the prism must be taken into account all together. Of course, it will be much simplified if we can do these in the object space of the prism. For this purpose, we have defined an object plane conjugate to the exit face of the prism, the latter being taken as an image plane with respect to the prism ; and this object plane has been found to be the exit face of the so-called equivalent air plate. Now, since the equivalent air plate and the incident ray 12 are located in the same space, i.e., the object space of the prism, the ray 12 will travel in straight line to cut the exit face of the equivalent air plate at a height $\overline{M_0 5_0}$ by its extension. According to the relationship between two conjugate spaces, we have

$$\overline{M_0 5_0} = \overline{M'5'} .$$

Hence, the pattern of light occurring in the exit face of the prism will be completely reproduced in the exit face of the equivalent air plate except that the handedness of the pattern may be changed in some cases (t = odd), but the latter will be of no effect on the prism size.

2.6 The answer is substantially the same as that to Prob. 2.1, but in this case the prism is first unfolded to the image space of the reflecting part and then reduced to the image space of the prism.

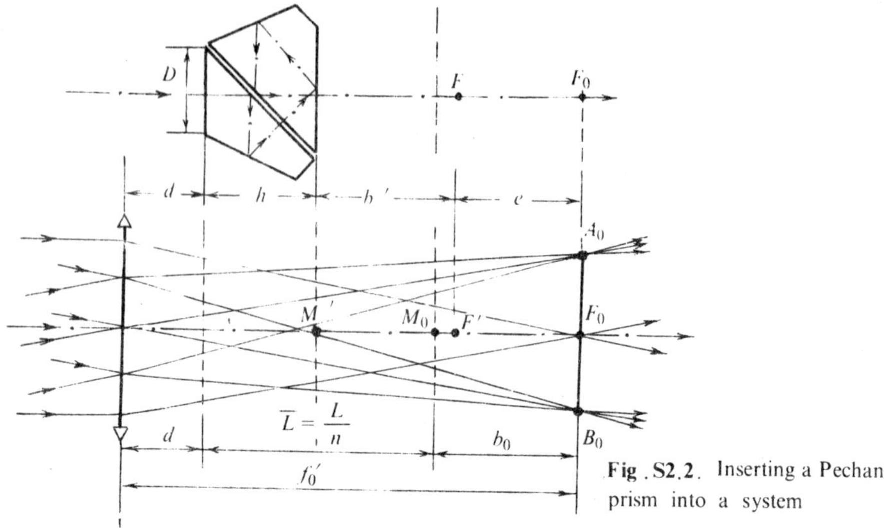

Fig. S2.2. Inserting a Pechan prism into a system

2.7 Referring to Fig. S2.2, after the Pechan prism has been inserted into the ray paths, you should first note that the point F_0, which we considered originally as being the image point relative to the objective, has now become the object point with respect to the prism. Then the image of the object point F_0 is generated at the point F' by the prism. Unfolding and subsequently reducing the prism we obtain its equivalent air plate of thickness $\dfrac{L}{n}$, and a pair of two basic conjugate points M_0, M'. From the diagram, we can readily write

$$d + \frac{L}{n} + b_0 = f_0', \tag{S2.1}$$

$$d + h + b' + e = f_0', \tag{S2.2}$$

$$b' = b_0, \tag{S2.3}$$

where e denotes the distance between the object point F_0 and the image point F'. Solving these three equations for e and b' yields

$$e = \frac{L}{n} - h, \tag{S2.4}$$

$$b' = b_0 = f_0' - d - \frac{L}{n}. \tag{S2.5}$$

From the Appendix of the this text, we get the data for a Pechan prism as follows:

$$L = kD = 4.6213D, \quad h = 1.207\,D. \tag{S2.6}$$

Finally, putting $f_0' = 120\,\text{mm}, d = 20\,\text{mm}, D = 20\,\text{mm}$, and $n = 1.5163$ into Eqs. (S2.6),(S2.4), and (S2.5)

$$e = 36.815 \text{ mm}; \quad b' = b_0 = 39.045 \text{ mm}.$$

434

The result shows that the insertion of a Pechan prism into the ray paths will cause the image of the objective shifted to the left .

Chapter 3

3.3   " With the object fixed , any image motion caused by a rotation of a mirror system through a finite angle θ about an object-axis p may be accomplished by a rotation of the image through an angle θ about the object-axis p , followed by another rotation of the intermediate image through an angle $(-1)^{t-1}\theta$ about the image-axis p_0' ,where t denotes the number of reflections ; p_0' represents the image of p for the mirror system in its final position ." It should be noticed that the image-axis p' as a rotation axis in the original statement of the theorem has been replaced by another image-axis p_0' in this alternative statement . The difference between these two statements implies that a combination of two finite rotations successively imparted to a rigid body can no longer be performed by simply using the principle of superposition by reason of what will further be explained by answering the next problem 3.4.

3.6   To determine the image point of an object point through a rotating mirror systerm , we have to carry out two kinds of shifts : one , the shift from the object space to the image space, and the other the shift from the fixed space to the moving space or in reverse. As for the former shift, it has been clear that the pair of conjugate points M and M' (or M_0 and M' for a reflecting prism) is utilized . As for the latter shift , we have the following two different cases: first, when this shift happens in the object space, an arbitrary point q lying on the rotation axis p ,as a border line,is utilized to complete the transfer from an object point F in the fixed space to another point M in the moving space; second ,when this shift happens in the image space , the conjugate point q' lying on the image-axis p' , as a border line , is utilized to complete the transfer from the conjugate point M' in the relatively fixed space to the image point F_0' in the relatively moving space .

Chapter 4

4.3   (1) Calculate $\overline{L}=L/n$, b_0 , and then locate a pair of conjugate points F_0 and F', where F' is the image point for a distant object through the objective-prism group.

   (2) Construct a pair of conjugate systems $o\,x\,y\,z$ and $o'\,x'\,y'\,z'$ with their originals o and o' located at F_0 and F' , respectively , as shown in Fig .4.4.

   (3) Determine the components of p along the axes x, y, z and x', y', z'.

$$P_x = -1 , \qquad P_y = 0 , \qquad P_z = 0 ,$$

$$P_{x'} = 0 , \qquad P_{y'} = 1 , \qquad P_{z'} = 0 .$$

(4) Determine the coordinates of q (see Fig. 4.4) in the coordinate systems $o\,x\,y\,z$ and $o'x'y'z'$.

$$x_q = -91.1\,, \qquad y_q = 20\,, \qquad z_q = 0\,,$$

$$x_q' = -60.1\,, \qquad y_q' = 0\,, \qquad z_q' = 0\,.$$

(5) Determine the components of p' and the coordinates of q' relative to the image coordinate system $o'x'y'z'$. According to the object and image relationship, we have

$$P_{x'}' = P_x = -1\,, \qquad P_{y'}' = P_y = 0\,, \qquad P_{z'}' = P_z = 0\,,$$

$$x_{q'}' = x_q = -91.1\,, \qquad y_{q'}' = y_q = 20\,, \qquad z_{q'}' = z_q = 0\,.$$

(6) Determine the components of the vectors $\vec{\rho}_F$ and $\vec{\rho}_F'$ along the axes x', y', z'. From the geometry in Fig. 4.1, we obtain

$$\rho_{F'x} = -x_q' = 60.7\,, \qquad \rho_{F'y} = -y_q' = 0\,, \qquad \rho_{F'z} = -z_q' = 0\,,$$

$$\rho_{F'x}' = -x_{q'}' = 91.1\,, \qquad \rho_{F'y}' = -y_{q'}' = -20\,, \qquad \rho_{F'z}' = -z_{q'}' = 0\,.$$

(7) Determine the y', z' components of the image point displacement $\Delta S_F'$. Substituting the relevant data into Eq. (4.2), we find

$$\Delta S_{F'y'}' = \Delta\theta\,[\,(P_z{}'\rho_{F'x} - P_x{}'\rho_{F'z}) + (P_z'\rho_{F'x}' - P_x'\rho_{F'z}')\,]$$
$$= 0\,,$$

$$\Delta S_{F'z'}' = \Delta\theta\,[\,(P_x{}'\rho_{F'y} - P_y{}'\rho_{F'x}) + (P_x'\rho_{F'y}' - P_y'\rho_{F'x}')\,]$$
$$= -40.7\Delta\theta\,.$$

(8) Determine the optical axis deviation. Let the focal length of the eyepiece be $f_e' = 30$ mm. Then, we can find two components $\xi_{z'}$ and $\xi_{y'}$ of the optical axis deviation as:

$$\xi_{z'} = -\frac{\Delta S_{F'y'}'}{f_e'} = 0\,,$$

$$\xi_{y'} = \frac{\Delta S_{F'z'}'}{f_e'} = \frac{-40.7\Delta\theta}{f_e'} = -67.8'\,.$$

(9) Determine the image lean. According to Eq. (4.1), we have

$$\Delta\mu_{x'}' = \Delta\theta\,P_{x'} + \Delta\theta\,P_{x'}' = \Delta\theta\,(P_{x'} + P_{x'}') = -50'\,.$$

Chapter 5

5.2 (1) Determine $S_{T.2\varphi}$ and $J_{\mu P}$. Considering $t = 2$, we have

$$S_{T.2\varphi} = G_{\mu\,\varphi} = \begin{pmatrix} \cos 90^\circ & \cos 0^\circ & \cos 90^\circ \\ \sin 10 & \cos 90^\circ & \cos 10^\circ \\ \cos 10 & \cos 90^\circ & -\sin 10^\circ \end{pmatrix}$$

$$= \begin{pmatrix} 0 & 1 & 0 \\ 0.174 & 0 & 0.985 \\ 0.985 & 0 & -0.174 \end{pmatrix},$$

$$J_{\mu P} = E - S_{T.2\varphi} = \begin{pmatrix} 1 & -1 & 0 \\ -0.174 & 1 & -0.985 \\ -0.985 & 0 & 1.174 \end{pmatrix}.$$

(2) Represent p in terms of its components relative to the coordinate system $x'y'z'$.

$$p = k'.$$

(3) Determine $\Delta\vec{\mu}'$.

$$(\Delta\vec{\mu}') = \Delta\theta J_{\mu P}(p) = \Delta\theta \begin{pmatrix} 1 & -1 & 0 \\ -0.174 & 1 & -0.985 \\ -0.985 & 0 & 1.174 \end{pmatrix} \begin{pmatrix} 0 \\ 0 \\ 1 \end{pmatrix}$$

$$= \begin{pmatrix} 0 \\ -0.99 \\ 1.17 \end{pmatrix} \Delta\theta.$$

Thus,

$$\Delta\mu'_x = 0, \qquad \Delta\mu'_y = -0.99\,\Delta\theta, \qquad \Delta\mu'_z = 1.17\,\Delta\theta.$$

Chapter 6

6.2 Some other circular beam deflectors are shown in Fig. S6.1.

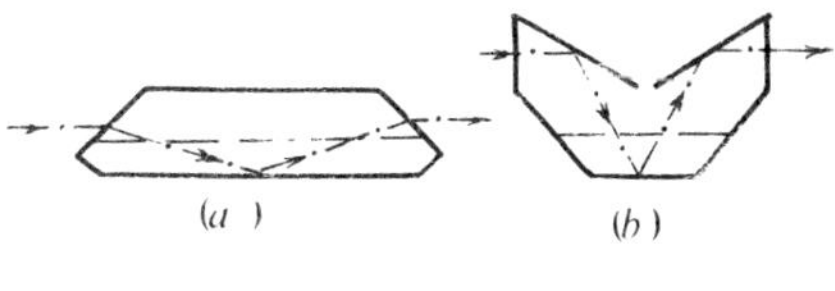

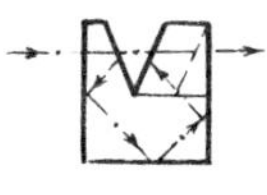

Fig. S6.1. Rotator prisms equipped with roof for circular beam deflection : (a) Wirth ; (b) roof Abbe; (c) roof Uppendahl

Chapter 7

7.2 To make the answer clear, a relevant part of the drawing in Fig. 7.4 has been singled out and reproduced in Fig. S7.1. If the adjustment is made to pull the image point o' back to

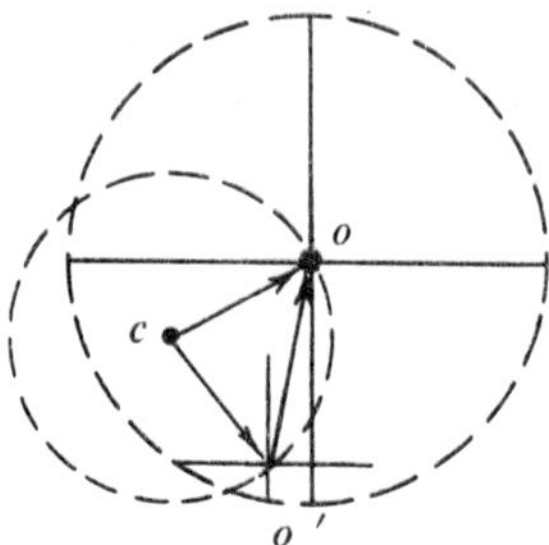

Fig. S7.1. Field of view of the measuring telescops during the reading state

the origin o, then the residual optical axis deviation $\vec{\xi}_r$ of the adjusted telescope will be

$$\vec{\xi}_r = \overrightarrow{co'} + \overrightarrow{o'o} = \overrightarrow{co} \; ,$$

which differs from the original optical axis deviation $\overrightarrow{co'}$ only in direction but its magnitude does not change any.

Chapter 8

8.1 & 8.2 The rotation axes P_s and P_b of two Porro prisms in accordance with interposing spacers at S.I.(S.O.) and B.I.(B.O.), respectively, as well as the extreme-valued axis directions u_s and u_b of the image lean for each prism are all shown in Fig. S8.1. In this case, since $P_s \perp u_s$ and $P_b \perp u_b$, the adjustment of the parallelism of optical axes will cause no additional image lean. Thus, it would be reasonable to arrange the adjustment procedure of the binoculars in such an order that the adjustment of the image lean is made first, and then the adjustment of the binocular axes will follow. As the image lean adjustment is made previous to all the other procedures, whether this first step will introduce additional errors to the system seems not to be a very critical factor, and thus more attention may be given to raising the sensitivity of the adjustment. For this purpose, a rotation axis P_1 parallel to the extreme-valued axis direction u_b for the big Porro, or a rotation axis P_2 parallel to the extreme-valued axis direction u_s for the small Porro is suggested. Mechanical aspects must be considered in any case. It is obvious that small rotations of the prisms around such axes will be accepted structurally without much difficulty.

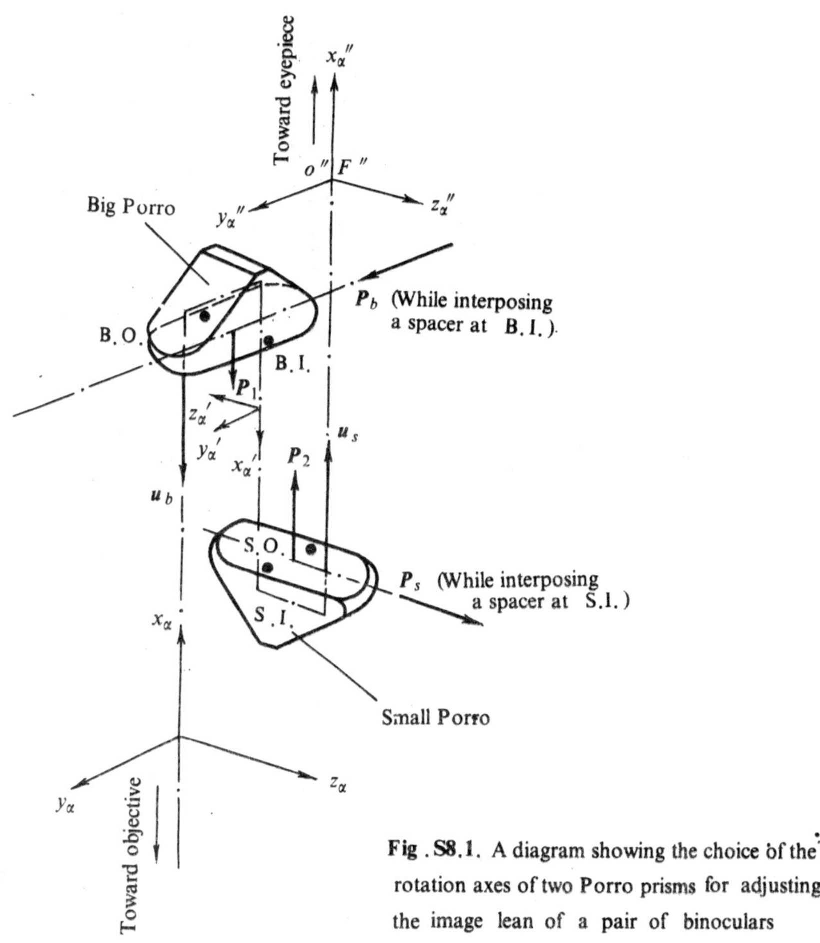

Fig. S8.1. A diagram showing the choice of the rotation axes of two Porro prisms for adjusting the image lean of a pair of binoculars

Chapter 9

9.2 We have the adjustment diagram of the roof prism $FX_J - 90°$, as shown in Fig. S9.1. Using the three-dimensional zero-valued pole C_{Π_1}, in its associated plane Π_1, i.e., the conjugate optical-axis section of the prism, draw a line ll passing through the pole C_{Π_1} and parallel to the gradient axis direction u of x' image rotation. Now, let the rotation axis P of the prism for adjusting the image lean be coincident with the line ll. Then, on one hand, because $P // u$, a small rotation angle $\Delta\theta$ around such an axis P will offer high sensitivity of adjustment of image lean; on the other hand, because $\delta_w = 0$, $P \perp v$, and P itself is a three-dimensional zero-valued axis, this small angular displacement $\Delta\theta P$ will not introduce any other errors at all. The choice of such a rotation axis can not be decided yet. Its feasibility should further be examined through the mechanical design of the prism mounting and even the experiment of the prototype.

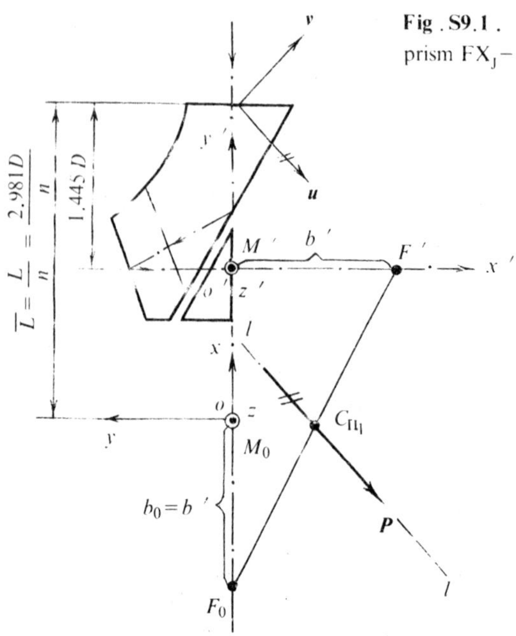

Fig . S9.1 . Adjustment diagram of a roof prism FX$_J$– 90 taken from the Appendix

Chapter 10

10.3 If the scalar product $\Delta\,\vec{\mu}''_e \cdot A''$ is not equal to zero, then it will be a measure of that part of the error of LOS stabilization which causes a vibration of the image around the line of sight.

References

[1] Lian Tongshu , *Adjustment of Reflecting Prisms*, National Defence Industry Press, 1978,Beijing.

[2] Lian Tongshu, *Adjustment of Reflecting Prisms (Principle and Tabulation)*, National Defence Industry Press,1979,Beijing.

[3] Tang Jiafen,Image Lean and Optical Axis Deviation Caused by Position Errors of Reflecting Prisms, *Yunguang Jishu*,1972,No.3,Kunming.

[4] He Shoyu and Zheng Changying,The Change of the Ray Path Caused by Linear and Angular Displacements of Reflecting Prisms, *Proceedings on Optical Design* ,Institute of Information,First Ministry of Machine Building,1973,Beijing.

[5] Lian Tongshu, *Adjustment of Reflecting Prisms (Calculation of Optical Axis Deviation and Image Lean)*,Beijing Institute of Technology,1974,Beijing.

[6] Lian Tongshu,Adjustment of Reflecting Prisms, *Yunguang Jishu*, 1976, No. 2, Kunming.

[7] Lian Tongshu, Calculation of Adjustment of Optical Axes of Binoculars, *Yunguang Jishu*,1977,No.1,Kunming.

[8] Tang Zhiyi, Xu Yaohui and Wang Zhijian , *Reflecting Prisms*, National Defence Industry Press,1981,Beijing.

[9] Deng Bixin, Mathematical Analysis of Rotation of Reflecting Prisms, *Guangxue Jishu*, 1982,No.3,Beijing.

[10] Ai Bihui and Yin Baiyun , Stabilization of Hand-Held Optical Instruments, *Yunguang Jishu,1983*,No.4,Kunming.

[11] Jinggong Shier, *Applied Optics*,National Defence Industry Press, 1973, Beijing.

[12] Chi Zeying and Xu Jinyong , *Applied Optics* , East-China Engineering Institute, 1984, Nanjing.

[13] G. B. Pogarev, *Problems of Optical Adjustment*, Machine Building Publishing House, 1974,Leningrad.

[14] R. E. Hopkins,Mirror and Prism Systems,in"*Applied Optics and Optical Engineering*" (Rudolf Kingslake,ed.),Vol.3,P.269,Academic Press,1965,New York.

[15] E. A. Greim, *Mirror and Reflecting Prism Systems*, Machine Building Publishing House,1981,USSR.

[16] John C. Polasek, Matrix Analysis of Gimbaled Mirror and Prism Systems, *JOSA*, Vol. 57,No.10,1967,USA .

[17] G. B. Pogarev, *Adjustment of Optical Instruments*, Machine Building Publishing House,1982,Leningrad .

[18] Rudolf Kingslake, *Optical System Design*,Academic Press,1983,New York.

[19] Mao Wenwei, Matrix Analysis of Adjustment of Prisms, *Yunguang Jishu*, 1981, No. 5, Kunming.

[20] Teaching and Research Group of Optical Instrumentation of BIT, *Assembly and Ad-*

justment of Optical Instruments, National Defence Industry Press, 1980, Beijing.

[21] Lian Tongshu, A New System of Theory of Adjustment for Reflecting Prisms Developed in China on the Basis of Kinematics of Rigid Bodies, *Yunguang Jishu*, 1984, No. 5, 6; 1985, No. 3, Kunming.

[22] Lian Tongshu, General Solution of Planar Three-Dimensional Zero-Valued Pole for Reflecting Prisms, *Instrumentation-Analysis-Monitoring*, 1986, No. 2, 3, Beijing.

[23] Lian Tongshu, The Theory of Image Formation and Adjustment of Reflecting Prisms, *Acta Polytechnica Scandinavica, Applied Physics Series*, No. 149, 1985, Helsinki.

[24] Lian Tongshu, Progress of Research on the Theory of Conjugation for Reflecting Prisms, *Yunguang Jishu*, 1987, No. 6, Kunming.

[25] Lian Tongshu, Theory of Image Stabilizing Reflecting Prisms and Its Application, *Yunguang Jishu*, 1988, No. 2, 3, Kunming.

[26] He Zhihua and Kang Songgao, Principle of Image Formation for Mirror Systems, *Guangxue Jishu*, 1983, No. 4, Beijing.

[27] Wang Wenhua, A Brief on Image Stabilized Apparatus of FUJI PHOTO OPTICS CO. LTD, *Yunguang Jishu*, 1987, No. 6, Kunming.

[28] Mao Yingtai, *Theory and Calculation of Errors*, National Defence Industry Press 1982, Beijing.

[29] Wang Yingming, *Design of Optical Measurement Instruments*, China Machine Press, 1982, Beijing.

[30] Zhang Yimo et al., *Applied Optics*, China Machine Press, 1982, Beijing.

[31] Zhejiang University, *Applied Optics*, China Industry Press, Beijing.

[32] Richard N. Shagam and William C. Sweatt, ed., Optical Alignment, *Proceedings of SPIE*, Vol. **251**, 1980, USA.

[33] Zhang Qingsheng, Xu Luping, Huang Bangrong, Zhang Xinli and Lian Tongshu, *Report on Experiment of a Image Stabilized Apparatus By Using a Roof Pechan Prism in a Convergent Beam*, 1975, Beijing Institute of Technology, Beijing.

[34] Gu Sumei, Matrix Equation of Optical Image Stabilization, *Engineering Optics*, 1980, No. 2, BIT, Beijing.

[35] Qiu Songfa and Ren Zhiwen, Calculation of Image Lean for Prisms, *Guangxue Jige Yu Qingbao*, Great Wall Optical Instrument Factory, 1973, No. 1, Beijing.

[36] Xie Yaoting, Calculation of the Influence of Position Errors of Reflecting Prisms, *Yunguang Jishu*, 1975, No. 6, Kunming.

[37] Jack D. Gaskill, *Linear Systems, Fourier Transforms, and Optics*, John Wiley & Sons 1978, New York.

[38] Lian Tongshu, *The Theory of Conjugation for Reflecting Prisms*, Beijing Institute of Technology Press, 1988, Beijing.

[39] Robert R. Shannon-James C. Wyant, ed., *Applied Optics and Optical Engineering*, Academic Press, 1965-1983, New York.

[40] Francis T. S. Yu, *Optical Information Processing*, John Wiley & Sons, 1982, New York.

[41] Eugene Hecht and Alferd Zajac, *Optics*, Addison-Wesley Publishing Company, 1974, Menlo Park, California.

[42] Walter G. Driscoll and William Vaughan, ed., *Handbook of Optics*, McGraw-Hill Book Company, 1978, New York.

[43] Bela I. Sandor, *Engineering Mechanics: Dynamics*, Prentice-Hall, 1983, Englewood Cliffs, New Jersey.

[44] Ovid W. Eshbach and Mott Souders, ed., *Handbook of Engineering Fundamentals*, John Wiley & Sons, 1975, New York.

[45] M. J. Beesley, *Lasers and Their Applications*, Taylor & Francis LTD, 1976, London.

[46] David Casasent, ed., *Optical Data Processing*, Springer-Verlag, 1978, New York.

[47] P. G. Harper and D. L. Weaire, *Introduction to Physical Mathematics*, Cambridge University Press, 1985, Cambridge.

[48] A. Davison, ed., *Handbook of Precision Engineering*, Vol. 6, Mechanical Design Applications, Whitefriar Press, 1972, London and Tonbridge.

[49] R. L. Weber, M. W. White and K. V. Manning, General Physics, Selections from *Physics for Science and Engineering, Elements of Physics* and *Modern University Physics*.

[50] W. A. Granville, P. F. Smith and W. R. Longley, *Elements of Calculus*, 1946.

[51] D. C. Murdoch, *Linear Algebra for Undergraduates*, John Wiley & Sons, 1957, New York.

[52] P. Schwamb, A. L. Merrill and W. H. James, *Elements of Mechanism*, Sixth Edition Revised by V. L. Doughtie, 1946.

[53] Virgil M. Faires, *Kinematics*, McGraw-Hill Book Company, 1959, New York.

[54] Gudmunn A. Slettemoen, Derivation of Phase Differences of Nonsymmetrical Interferometers Using Partitioned Transfer Matrices, *J. Opt. Soc. Am.* Vol. 73, No. 7, July 1983.

[55] Percy Jucheng Fang and Lucy Guinong J. Fang, *Zhou Enlai —— A Profile*, Foreign Language Press, 1986, Beijing.

[56] Lian Tongshu, Three-Axis Image-Stabilizing Reflectiong Prism Assembly, *Yunguang Jishu*, 1989, No. 3 − 4, Kunming.

[57] Hilda G. Kingslake (née Conrady), *The First Fifty Years, The Institute of Optics 1929 − 1979*, College of Engineering and Applied Science, University of Rochester, 1979, Rochester, New York.

[58] Ole Lokberg, Electronic Speckle Pattern Interferometry, *Phys. Technol.*, Vol. 11, 1980, Great Britain.

[59] R. S. Ilin, G. I. Fedotov, L. A. Fedin *Laboratorial Optical Instruments*, Machine Building Publishing House, 1966, Moscow.

[60] M. M. Rusinov, *Adjustment of Optical Instruments*, Nedra, 1969, Moscow.

[61] Mai Weilin, Errors in Optical Parallelism of Reflecting Prisms, *Letters of Standardization*, Fifth Ministry of Machine Building, 1965, No. 2, Beijing.

[62] Fritz Deumlich, *Surveying Instruments*, Walter de Gruyter, 1982, Berlin.

[63] Mu Guoguang and Zhan Yuanling, *Optics*, People's Education Press, 1978, Beijing.

[64] Wang Zhijiang, *Theory of Optical Design*, Science Press, 1965, Beijing.

[65] A. I. Tydorovskee, *Theory of Optical Instruments* , Vol. 1, Publishing House of Academy of Sciences, 1948, Moscow .

[66] A. E. Conrady, *Applied Optics and Optical Design, Part One*, Oxford University Press, 1929, London.

[67] Francis, T. S. Yu and Iam-Choon Khoo, *Principles of Optical Engineering* , John Wiley & Sons, 1990, New York .

[68] J. W. Goodman, *Introduction to Fourier Optics*, McGraw-Hill, 1968, New York .

[69] Wang Daheng, Progresses of Optical Science and Technology in China, Acta Optica Sinica, Vol. 1, No. 1, 1981, Shanghai.

[70] Chien Weizang, ed., *Proceedings of the International Conference on Nonlinear Mechanics*, Science Press, 1985, Beijing.

[71] Zhou Peiyuan, *Theoretical Mechanics*, People's Education Press, 1952, Beijing.

Subject Index

448